Renal Physiology

E. Koushanpour   W. Kriz

Renal Physiology
Principles, Structure, and Function

Second Edition

With 190 Illustrations and 3 Color Plates

Springer-Verlag
New York Berlin Heidelberg
London Paris Tokyo

ESMAIL KOUSHANPOUR, PH.D.
Associate Professor of Physiology and Anesthesia, Northwestern University Medical School, Chicago, Illinois, USA

WILHELM KRIZ, M.D.
Professor of Anatomy, University of Heidelberg, 6900 Heidelberg, Federal Republic of Germany

On the front cover: FIGURE 4.3. Organization of the intrarenal vasculature of the human kidney, p. xi.

Library of Congress Cataloging in Publication Data
Koushanpour, Esmail.
  Renal physiology.
  Includes bibliographies and index.
  1. Kidneys. I. Kriz, Wilhelm, 1936–
II. Title. [DNLM: 1. Body Fluids—physiology.
2. Kidney—physiology. WJ 300 K867r]
QP249.K68 1986   612′.463   86-15431

The first edition was published by W.B. Saunders Company © 1976.

© 1986 by Springer-Verlag New York Inc.
All rights reserved. No part of this book may be translated or reproduced in any form without written permission from Springer-Verlag, 175 Fifth Avenue, New York, New York 10010, U.S.A.
The use of general descriptive names, trade names, trademarks, etc. in this publication, even if the former are not especially identified, is not to be taken as a sign that such names, as understood by the Trade Marks and Merchandise Marks Act, may accordingly be used freely by anyone.
While the advice and information of this book is believed to be true and accurate at the date of going to press, neither the authors nor the editors nor the publisher can accept any legal responsibility for any errors or omissions that may be made. The publisher makes no warranty, express or implied, with respect to material contained herein.

Media conversion by Worldcomp, Leesburg, Virginia.
Printed and bound by Arcata Graphics/Halliday Lithograph, West Hanover, Massachusetts.
Printed in the United States of America.

9 8 7 6 5 4 3 2 1

ISBN 0-387-96304-9 Springer-Verlag New York Berlin Heidelberg
ISBN 3-540-96304-9 Springer-Verlag Berlin New York Heidelberg

Preface to the Second Edition

The first edition of this book was well received by medical students, graduate students, and clinicians interested in furthering their understanding of basic principles of renal physiology. Most of the reviews of the first edition and comments from the various instructors who used the book were very positive and complimentary with regard to the presentation of the physiological information and the use of the system analysis approach to describe renal function. These positive and encouraging comments over the past nine years, since the publication of the first edition, gave us the impetus to consider the preparation of a second edition. The explosive expansion of our knowledge of renal function, particularly the correlation of function with structure of the kidney, necessitated the complete reorganization of the book and substantial revision of its content. With this in mind, we undertook a comprehensive revision of the book integrating the structure of the mammalian kidney with renal function.

The major planning for revising this book was carried out in 1983–84, while Esmail Koushanpour was on an eight-month sabbatical as a Senior Fulbright Professor, doing collaborative research on the effect of some diuretics on the structure and function of the kidney with Professor Wilhelm Kriz, Anatomy Department, University of Heidelberg, Federal Republic of Germany.

As was mentioned in the preface to the first edition of this textbook, insofar as possible, an attempt was made to integrate the anatomical information with the functional description of the kidney. This approach represents a marked departure from similar textbooks, in which the anatomical information is usually treated separately. In this second edition, the anatomical sections have been greatly enhanced and updated. The two of us have made further collaborative efforts to present a better understanding of the function of the kidney in conjunction with the most recent anatomical findings.

The second edition consists of 13 Chapters and 3 Appendices. As in the first edition, the anatomical description of the kidney is incorporated into the various chapters dealing with kidney functions. Most of the anatomical information was written by Wilhelm Kriz. The physiological information was written by Esmail Koushanpour, except for Chapter 12 which was jointly written. Chapters 1 through 3 were partly revised for clarity and to include new information about capillary permeability and a more detailed treatment of the application of the Darrow-Yannet diagram. Chapters 4, 5, and 6 are completely new. Chapters 7 and 8 were partly revised to include new information. Chapter 9 was completely revised to reflect the current understanding of tubular processing of filtrate along the nephron. Chapters 10, 11, and 13 were also partly revised to include new information. Appendices A and B are the same as in the first edition. Appendix C is new. It provides a summary of nomenclature used to describe the structure of the kidney.

We wish to thank many people who helped in the revision of this book. In particular, we wish to thank Ms. Ingrid Ertel (Heidelberg) and the staff of Kascot Media, Inc. (Chicago) for the photographic work. We wish to thank Mr. Rolf Nonnenmacher (Heidelberg) and the staff of Kascot Media, Inc. (Chicago) for much of the original color and black and white art work. We wish to thank Mrs. Jenny Lee Koushanpour (Chicago) and Mrs. Helene Dehoust (Heidelberg) for typing the manuscript, and Mrs. Jenny Lee Koushanpour for proofreading the manuscript. We wish to express our appreciation to all the authors and publishers who kindly permitted the reproduction of borrowed illustrations. Finally, we wish to thank the staff of Springer-Verlag for their assistance in the production of this book.

ESMAIL KOUSHANPOUR
Chicago

WILHELM KRIZ
Heidelberg

Preface to the First Edition*

Physiologists and clinicians are keenly aware of the truth in the century-old statement by Claude Bernard that constancy of internal milieu is prerequisite to a normal life. The mechanisms by which this constancy of internal environment or "homeostasis" is maintained involve the dynamic interplay of several organ systems, of which the kidney is the most prominent. However, few textbooks on renal function have even attempted to, much less succeeded in, explaining this dynamic interaction to medical and allied health students.

This interdependence is well demonstrated by patients with primary or secondary renal disease, who often exhibit a wide variety of clinical symptoms, seemingly unrelated to the failure of the kidney. Included in these symptoms are: hypertension, of both the arterial and the portal vein variety; fluid retention in dependent extremities, often accompanied by acute renal failure; congestive heart failure; and liver cirrhosis; as well as acid-base disturbances associated with abnormal metabolism, such as diabetes mellitus and gastrointestinal disorders. Although not readily apparent, a careful analysis of the patient's physical and laboratory findings, as well as of the physiology of renal function, would reveal that most, if not all, of these unrelated symptoms can be traced to some disturbance in normal renal regulatory functions. Therefore, to facilitate an intelligent approach to the diagnosis of the underlying cause and the management of the clinical symptoms, it is necessary to acquire a thorough understanding of the renal function in relation to its dynamic interaction with other major organ systems involved in homeostatic regulation.

This book has developed from a 20-lecture course in renal physiology given for the past 12 years by the author to medical students at Northwestern University Medical School. It is an attempt to present an integrated, quantitative analysis of renal function and its role in body fluid homeostasis. The book uses the systems analysis and synthesis approach, which represents a significant departure from the traditional and conventional presentation of the subject. The understanding that such an approach provides is not descriptive, but mechanistic; it imposes mathematical rigor on conceptual processes. It facilitates the search for key factors, alternate possibilities, and missing links that guide experimentation in fruitful directions. At each stage of progress, it summarizes in unambiguous form the current state of understanding so that deficiencies are well exposed to prod further efforts. This approach does not replace experimental ingenuity nor depth of knowledge of the subject, but facilitates and stimulates both. Therefore, application of systems analysis to the study of renal function developed in the present book represents a new and novel approach to the description of this complex physiological system.

This book consists of 13 chapters and 2 appendices. The first chapter presents an overview of the renal-body fluid regulating system. It not only introduces the reader to the author's approach to the subject, but also brings into focus the unique role the kidney plays in the regulation of body fluid homeostasis. Chapters 2 through 11 are devoted to a rigorous and mechanistic description of body fluids and renal function. Where appropriate, sufficient background materials are included in each chapter so as to minimize the need for review. For example, Chapter 7 gives a detailed analysis of the biochemical and quantitative concepts necessary to understand the mechanisms of renal transport and the concen-

*Amended

tration and dilution of urine discussed in Chapters 8, 9, and 11. To better understand the role of the kidney in the regulation of acid-base balance, an extensive discussion of buffers and associated concepts as well as respiratory regulation of acid-base are included in Chapter 10. In this way, the materials in each chapter not only introduce and develop systematically some aspects of renal function, but they also provide the necessary background for the materials presented in the succeeding chapters. Furthermore, unlike other books on the subject, in which the anatomy of the kidney is treated separately, we have integrated the anatomical information with the discussion of kidney function. Also, at the end of each chapter, numerous problems are included, which are designed to further the students' understanding of the materials covered in the text.

Chapters 12 and 13 are devoted to a detailed and integrated analysis and synthesis of the renal-body fluid regulating system, in the light of what has been presented before, and from the standpoints of both normal and pathophysiological disturbances. Included are a mechanistic description of renal function in disease and the extent of its involvement in conditions such as acute glomerulonephritis, pyelonephritis, nephrotic syndrome, hypertension, liver cirrhosis, and congestive heart failure. It is hoped that these clinical examples will provide a clear demonstration to the reader of the utility and relevance of materials presented earlier and contribute to his understanding of the diverse processes which underlie a disease state.

Appendix A is an attempt to introduce the student to the principles of systems analysis and synthesis and its potential application to physiological systems. Appendix B presents a mathematical background for the principle of dilution used in this book. Finally, at the end of the book we have provided answers to some of the problems given at the end of each chapter, designed to increase the understanding of the student of the principles presented in the text.

As written, this book should fulfill the needs of all types of students, including those with little or no mathematical background. At first glance, the quantitative and rigorous approach to the subject may be considered somewhat beyond the need of the medical students. Our experience at Northwestern University Medical School has proved otherwise. The materials presented and the systems analysis approach were received enthusiastically not only by the medical students, but also by the physical therapy and medical technology students. Of course, for the latter group we minimized the extent of mathematical notations, but we made no compromise in the flow and functional diagram approach.

Finally, although the author's primary goal has been to write a book which satisfies the needs of several groups of students, it could be of special interest to researchers in renal physiology as well as medical practioners. For the latter audience, it should provide a fresh approach to a complex field hitherto not within easy grasp.

It would be impossible to acknowledge and adequately thank everyone who has helped make this book possible. The author is indebted to Professor John S. Gray, who not only introduced him to systems analysis and its application to physiological systems, but also helped with the development of some aspects of the book, especially the acid-base chapter. I wish to express my sincere appreciation to many former medical students, who made valuable contributions to the development of this book by their enthusiastic and critical feedback as well as their continuous encouragement. I can only say that they made the effort very much worthwhile. I wish to specially thank Miss Jenny L. Forman, who, as a devoted secretary, both encouraged me in the writing of the book and diligently undertook the typing of the manuscript, during all phases of its development. She also meticulously and patiently typed the final manuscript and assisted in proofreading. Special thanks are extended to Mr. Donald Z. Shutters, who skillfully rendered all the original illustrations. I also wish to express my appreciation to all the authors and publishers who kindly permitted the reproduction of the borrowed illustrations. I would like to thank the National Institutes of Health for their generous support of my research, mentioned in the book.

ESMAIL KOUSHANPOUR
Chicago

Contents

Preface to the Second Edition................. v
Preface to the First Edition vii
Color Plates.. xi

1 Introduction to the Renal-Body Fluid Regulating System 1
2 Body Fluids: Normal Volumes and Compositions 8
3 Body Fluids: Turnover Rates and Dynamics of Fluid Shifts........................... 21
4 An Overview of the Structural and Functional Organization of the Kidney..... 41
5 Formation of Glomerular Ultrafiltrate 53
6 Regulation of Renal Blood Flow and Glomerular Filtration Rate 73
7 Renal Clearance: Measurements of Glomerular Filtration Rate and Renal Blood Flow 96
8 Structural and Biophysical Basis of Tubular Transport 112
9 Tubular Processing of Glomerular Ultrafiltrate: Mechanisms of Electrolyte and Water Transport 132
10 Tubular Reabsorption and Secretion: Classification Based on Overall Clearance Measurements 214
11 Regulation of Acid-Base Balance 240
12 Mechanism of Concentration and Dilution of Urine....................................270
13 Renal Regulation of Extracellular Volume and Osmolality............................310

Appendices

A Introduction to Quantitative Description of Biological Control Systems 353
B Mathematical Basis of Dilution Principle ...367
C Anatomical Nomenclatures..................370

Answers to Problems 372
Index 375

FIGURE 4.2. Structural organization of the nephrons and collecting ducts of the human kidney. There are three nephron types depicted: short loop nephron *(left)*, cortical nephron *(middle)*, and long loop nephron *(right)*. The renal parenchyma is divided into the cortex (C), outer stripe (OS), inner stripe (IS), and inner medulla (IM). The cortex consists of two parts: the cortical labyrinth (containing the vessels, the glomeruli, and the convoluted tubules), and the medullary rays (containing the straight tubules). Note that the outer stripe does not include the medullary rays of the cortex. Two interlobular arteries are drawn on either site within the cortical labyrinth; together with their afferent and efferent arterioles they are shown in red. The renal corpuscles and the proximal tubules (convoluted and straight parts) are shown in brown. The thin limbs are shown in pale brown. The thick ascending limb (straight part of the distal tubule including the macula densa, which is shown as a bulging portion of the tubule) and the distal convoluted tubule are shown in yellow. The connecting tubules are shown in light green. Note that the connecting tubule of the long loop nephron forms an arcade. Finally, the collecting duct system is shown in dark green.

FIGURE 4.3. Organization of the intrarenal vasculature of the human kidney. The renal parenchyma is divided into the cortex (C), the outer stripe (OS), the inner stripe (IS), and the inner medulla (IM). At the corticomedullary border, short segments of arcuate artery (shown in red) and vein (shown in blue) are depicted. The interlobular artery arises from the arcuate artery and ascends within the cortical labyrinth. The interlobular artery gives rise to three afferent arterioles which supply, respectively, a superficial glomerulus, a midcortical glomerulus, and a juxtamedullary glomerulus. The efferent arterioles of the superficial and midcortical glomeruli supply both capillary plexuses of the cortex, namely, the round-meshed capillaries of the cortical labyrinth and the long-meshed capillaries of the medullary ray. The cortex is drained by interlobular veins, some of them begin as stellate veins at the surface of the kidney, as depicted in this figure. The efferent arterioles of the juxtamedullary glomeruli descend into the outer stripe and divide into the descending vasa recta, which descend within the vascular bundles. At successive levels within the medulla, descending vasa recta leave the bundles to supply the adjacent capillary plexus. In the inner stripe the capillary plexus is dense round-meshed, whereas in the inner medulla the capillary plexus is long-meshed. The drainage of the medulla is achieved by ascending vasa recta. Those originating in the inner medulla together with the descending vasa recta form the vascular bundles. Most of those originating from the inner stripe ascend independently from the vascular bundles. In the outer stripe, the ascending vasa recta form a dense pattern of vessels that supply the tubules. True capillaries are sparse within the outer stripe.

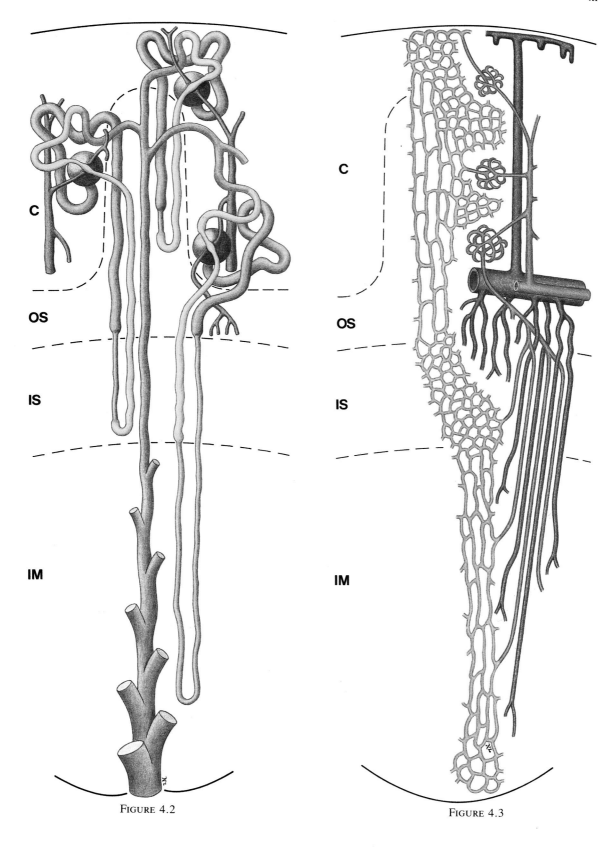

Figure 4.2

Figure 4.3

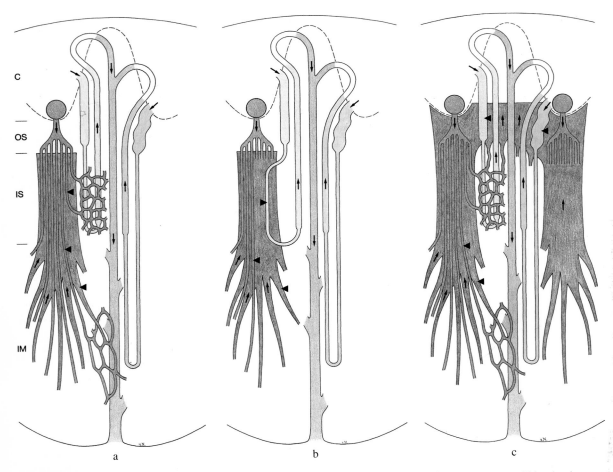

Figure 12.6. Possible connections established by the medullary circulation. In each of the three panels a short loop *(yellow)*, a long loop *(green)*, and a collecting duct *(brownish)* are shown. Descending vasa recta (DVR) (derived from efferent arterioles of juxtamedullary glomeruli) and capillaries are shown in *red;* ascending vasa recta (AVR), drawn en bloc, are shown in *blue*. C = cortex, OS = outer stripe, IS = inner stripe, and IM = inner medulla. The *left panel* (a) shows the simple type of medulla, where vascular bundles represent a countercurrent arrangement between DVR and AVR. Countercurrent exchange of some substance (e.g., urea) from ascending vasa recta to descending vasa recta (middle *arrowhead*) will trap the substance in the inner medulla. As AVR from IM are also arranged along DVR destined for IS, there is a possibility that some substance (e.g., urea, originating from IM) will also be transferred to the IS capillary plexus (upper *arrowhead*). Entry into tubules of IS (descending thin limbs of short loops!) will open a recycling possibility back to the inner medulla via the distal tubule/collecting duct route. Reentry into the ascending vasa recta within the inner medulla (lower *arrowhead*) will restart the process. The *middle panel* (b) shows the complex type of medulla, where in addition to DVR, descending thin limbs of short loops descend within bundles in a countercurrent arrangement with AVR from IM. In addition to countercurrent trapping between AVR and DVR (middle *arrowhead*), a recycling route via short loops of Henle is seen. Countercurrent exchange from AVR to descending limbs of short loops (upper *arrowhead*) will return some substance to IM via the normal nephron and the collecting duct route. The *right panel* (c) shows the relationships between AVR and tubules in OS (valid for both simple and complex type of medulla). The total venous blood from the medulla traverses the outer stripe in wide tortuous channels contacting tubules of OS like capillaries. Whether this arrangement serves as an ultimate trap for medullary solutes and/or is part of a recycling route for substances coming up from the medulla and secreted into the proximal tubules is unknown. (Modified from Kriz.[47])

1

Introduction to the Renal-Body Fluid Regulating System

All cells of the body are bathed by a fluid called the interstitial fluid, which provides the internal environment of the cells. Both volume and composition of the interstitial fluid must remain within narrow limits, or malfunctions result. Abnormal volumes of vascular and interstitial fluids impair cardiovascular function, and abnormal composition of interstitial fluid impairs cell function. The relevant concentrations include those of electrolytes (e.g., sodium, potassium, bicarbonate, chloride, phosphate, sulfate, and hydrogen ions), metabolic waste products, and even water (osmotic effects).

Numerous disturbing factors tend to upset both the volumes and the compositions of these body fluids.[1] These factors include water ingestion, deprivation, or loss; electrolyte ingestion, deprivation, or loss; fortuitous fluxes of acid or alkali; and the metabolic production of waste products, or the administration of toxic substances.

Clearly, there must be active regulation to maintain the vital constancy of the internal environment in the face of such disturbing factors. The system, in fact, has two compensating components subject to regulatory control. One is the gastrointestinal (GI) system, which can appropriately adjust intakes (thirst, appetite, etc.). The other, which we focus on in this book, is the kidney, which can appropriately adjust outputs. In the renal-body fluid regulating system, the kidney plays much the same compensating role as the bone marrow does for the hemoglobin (Hb) regulator, or the lung for the blood gas regulator.

A Flow Diagram of the Renal-Body Fluid Regulating System

We can acquire an initial orientation by examining a flow diagram of this system in the steady state of normality, as shown in Figure 1.1, which identifies the major fluid compartments of the body and the principal channels of influx and efflux. Briefly, the *flow diagram* specifies the pathways of material flow into and out of a subsystem, depicted by a *box*, where material transformation may take place. The material flow into and out of each subsystem is shown by an *arrow* entering or leaving the box, on the same or opposite sides. In contrast, the *functional diagram* depicts the cause and effect, the input-output relationships for one or many subsystems and the factors influencing their functional relationships. Thus, such a diagram serves as a basis for a quantitative and eventually a mechanistic description of the system. For a detailed treatment of both the flow diagram and the functional diagram, which are liberally used whenever appropriate in this book, the interested reader is referred to the materials in Appendix A.

The plasma compartment consists of a volume of about 3.2 L in a healthy 70-kg man (See Table 2.2, Chapter 2 for volume distribution of body fluids). Since the plasma circulates throughout the body, it provides the medium for transporting water and solutes from influx to efflux channels and for exchanging water and solutes with the largely uncirculated interstitial fluid compartment.

The interstitial compartment, of about 8.4 L, is shown to have a two-way exchange with the plasma. This occurs in the tissue exchanger, i.e., the systemic capillaries, where partially deproteinized plasma escapes into the interstitium from the arterial end of the capillary and is then reabsorbed into the blood in the venous end of the capillary. The mechanism of this two-way exchange is filtration. The exchange fluxes amount to about 4.5 L/min, so that 3.2 L of plasma are turned over every 0.7 min, and the 8.4 L of interstitial fluid every 1.9 min. In the steady state the escape and reabsorption rates are equal, but

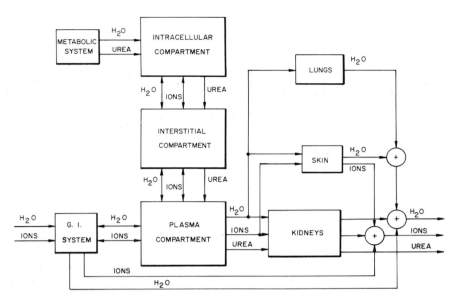

FIGURE 1.1. Flow diagram of the normal renal-body fluid regulating system. (From Koushanpour.[5])

in a transient state they may be temporarily unequal, thus yielding a net shift of fluid from one compartment to the other.

The large intracellular fluid compartment of about 23.1 L is also shown to have a two-way exchange with the interstitial compartment. In the steady state these two-way fluxes for water are enormous, but they are much smaller for ions whose penetration of the cell membrane is subject to severe constraints. Water is exchanged by passive diffusion and the ions by diffusion and/or active transport. In transient states the two-way fluxes of water may be temporarily unequal, thus yielding a net shift of water between the two compartments. The mechanism for such shifts is osmosis. Since the cells also constitute a source for metabolic water and waste products, such as urea, the metabolic system makes an extra one-way flux for water and a one-way flux for urea, which also moves by simple diffusion.

We thus see that the three major fluid compartments, which represent body stores of water and ions, are all dynamic turnover pools. Only the circulating plasma is subject to extracorporeal influxes and effluxes, but the interstitial and intracellular compartments can respond quickly, though more or less passively, to changes in the volume and composition of the plasma.

The main channel for influxes of water and solutes into body fluids is ingestion via the GI tract. On the average, these influxes amount to about 2.5 L of water and 7.0 g of sodium chloride, NaCl, per day. To these must be added the 0.3 L of metabolic water (from oxidation of nutrients) and 30 g of metabolic urea (from deamination of amino acids) per day. In the steady state these influxes are matched by equal effluxes of 2.8 L of water and 7.0 g of NaCl and 30 g of urea per day. Of the four channels for these effluxes, the kidneys are by far the most important, for they eliminate 1.5 L of water, 6.2 g of NaCl, and essentially all the urea. There is also some loss of water through the respiratory tract and small losses of water and ions via feces and skin. The water loss by these extrarenal routes amounts to about 1.3 L in 24 hours.

Clearly, any temporary inequality between total influxes and total effluxes will alter the volume and/or composition of the critical body fluid compartments. In the present context, all influx and efflux channels, except the renal, constitute possible disturbance forcings, to which the kidneys respond by making compensatory adjustments of their own controlled effluxes.

Basic Biophysical Chemistry

Before proceeding further, we must review some basic biophysical chemistry.

The concentrations of substances in body fluids

are expressed in several ways[3], each with special applications:

1. Volume % (vol%), e.g., milliliters (mL) of substance per 100 mL of fluid. This is often used for blood gases, and for the water content of body fluids. For example, plasma contains 94 vol% of water, but red cells only 72 vol%.

2. Weight%, e.g., grams of solute per 100 mL (g%) of fluid. This mixed weight/volume unit is still widely used. Plasma proteins average 7 g/100 mL of plasma and hemoglobin 35 g/100 mL of red cells. These proteins largely account for the different water contents of cells and plasma.

3. Molar concentration, e.g., millimoles per liter (mM/L) of fluid. Physiological saline solution contains 0.9 g% NaCl, which is [(0.9 x 10 g/L)/(58.5 g/mole)] × 1000 mM/mole = 154 mM/L. Note that molecular weight of NaCl is 58.5.

4. Equivalent concentration, e.g., milliequivalents per liter (mEq/L) of fluid. The equivalent weight (or combining weight) is defined as the atomic, radical, or molecular weight divided by valence:

1 mole of urea	=	1 equivalent
1 mole of NaCl	=	1 equivalent
1 mole of $CaCl_2$	=	2 equivalents
1 mole of Na_2SO_4	=	2 equivalents

In the case of electrolytes, the sum of negative charges must equal the sum of positive charges. Since ion valence corresponds to ion charges, the sum of anion equivalents will equal the sum of cation equivalents. For this reason, the ions of body fluids are best expressed in mEq/L.

5. Osmolar concentration, e.g., milliosmoles per liter (mOsm/L) of fluid. Osmolality is defined as the number of moles multiplied by the number of dissociating ions:

1 mole urea	=	1 osmole
1 mole Na^+Cl^-	=	2 osmoles
1 mole $Ca^{2+}Cl^-Cl^-$	=	3 osmoles
1 mole $Na^+Na^+SO_4^=$	=	3 osmoles

The *colligative properties* of a solution (freezing point depression, boiling point elevation, potential osmotic pressure, etc.) are functions of osmolar concentrations. The normal osmolar concentration of plasma is about 290 ± 10 mOsm/L. Physiological saline (0.9 g%, 154 mM/L, 154 mEq/L) has an osmolality of 308 mOsm/L and therefore is not iso-osmolar with plasma. The above were all expressed as "bulk" concentrations, i.e., quantities per unit volume of fluid bulk. Sometimes it is more meaningful to use "water" concentrations, i.e., quantities per unit volume of only the water portion of the fluid. The conversion is simply:

$$\frac{\text{"Water"}}{\text{concentration}} = \frac{\text{"Bulk" concentration}}{\text{Volumetric fraction of } H_2O \text{ in the fluid}} \quad (1.1)$$

Thus, the lower the water content of a fluid the more the "water" concentration exceeds the "bulk" concentration. We shall make these distinctions in the following way:

"Bulk" Concentration	"Water" Concentration
g%	g% in H_2O
mM/L (molar)	mM/L H_2O (molal)
mEq/L	mEq/L H_2O
mOsm/L (osmolar)	mOsm/L H_2O (osmolal)

Table 1.1 lists the major plasma electrolytes and their concentrations expressed in different units.

Osmosis is the flow of water across a membrane from a solution on one side to a solution on the other side, the latter containing a higher osmolality of solutes to which the membrane is impermeable. The water moves from the higher to the lower concentration of water just like a diffusion process. But since the water is the solvent, and not just a dissolved solute, the water flows as a convection process, analogous to filtration.

The dependency of osmosis on solute concentration is related to the change in the chemical potential of water caused by the addition of solute. The chemical potential or molar free energy for pure water (μ) is defined as the ratio of the change in total free energy of water (ΔF_{H_2O}) to a change in the number of moles of water (Δn_{H_2O}), at constant ambient temperature and pressure.[4] Expressed mathematically,

$$\mu = \frac{\Delta F_{H_2O}}{\Delta n_{H_2O}} \quad (1.2)$$

It so happens that the chemical potential of water in a solution is lower than that of pure water. Therefore, when a solution is separated from pure water by a membrane (permeable only to water), a *chemical potential* difference between the two sides develops. This difference in chemical potential can be abolished by at least three processes: (1) The free distribution of solutes on both sides of the membrane. This has the effect of equalizing the chemical po-

TABLE 1.1 Conversion of plasma electrolyte concentrations to mEq/L, or mg%.

Electrolytes	Calculated as	Atomic weight	Valence	Equivalent weight	Conversion factors: (mEq/L from mg%: divide; mg% from mEq/L: multiply)	Plasma concentration (normal ranges)	
						mg/100 mL	mEq/L
Cations							
Sodium	Sodium	23	1	23	2.3	310–335	136–145
Potassium	Potassium	39	1	39	3.9	14–21.5	3.5–5.5
Calcium	Total calcium	40	2	20	2.0	9–11	4.5–5.5
Magnesium	Magnesium	24	2	12	1.2	1.8–3.6	1.5–3.0
Anions							
Bicarbonate	CO_2 content			22.26	2.2	53–75 (av. 62) vol%	24–33 (av. 28)
Bicarbonate	CO_2 combining power			22.26	2.2	53–78 (av. 65) vol%	24–35 (av. 30)
Chloride	Chloride	35.5	1	35.5	3.5	350–375	98–106
Chloride	Sodium chloride	58.5	1	58.5	5.8	570–620	98–106
Phosphate	Phosphorus	31.0	1.8	17.2	1.7	2–4.5	1.2–3.0
Sulfate	Sulfur	32.0	2	16.0	1.6	0.5–2.5	0.3–1.5
Protein	Protein				0.41	6–8 grams	14.6–19.4

The phosphate is calculated as phosphorus with a valence of 1.8. The reason for this is that at normal pH of the extracellular water, 20% of the phosphate ions are in a form with one sodium equivalent (NaH_2PO_4), and 80% are in a form with two sodium equivalents (Na_2HPO_4). The total valence is therefore $(0.2 \times 1) + (0.8 \times 2) = 1.8$.

Source: Goldberger.[2]

tential of water on both sides. However, this is not possible if the membrane is permeable only to water. (2) The diffusion of water through the membrane, thereby equalizing the chemical potentials on both sides. (3) The application of mechanical pressure to the solution to increase its chemical potential to the level equal to that for pure water. The mechanical pressure thus applied is called the *osmotic pressure* and is a measure of the difference between the chemical potential of the solution and that of pure water.

The osmotic pressure of a solution as defined above is a measure of the lowering of the chemical potential of pure water by the addition of solute. Since the lowering of the chemical potential depends only on the number of solute particles added, the osmotic pressure depends on the number of particles in that solution and not on their size or weight.

If the solution on one side is pure water and that on the other side is one osmolal strength of a completely impermeable solute, the osmotic pressure that develops is 22.4 atmospheres (atm), or (760 mm Hg/atm $\times$ 22.4 atm) = 17,024 mm Hg. Normal saline of 308 mOsm/L H_2O thus has a *potential* osmotic pressure of (0.308 Osm $\times$ 17,024 mm Hg/Osm) = 5,244 mm Hg. We say potential, because the osmotic pressure that can actually develop depends not only on osmolality, but also on the permeability characteristics of the membrane system used.

The membranes of body cells have permeability constraints, such that NaCl cannot pass through the membrane, although H_2O can easily pass. Hence, cells with an intracellular concentration of impermeable solutes of 154 mOsm/L H_2O are said to be in osmotic equilibrium with the surrounding interstitial fluid having a concentration of impermeable solutes of 154 mOsm/L H_2O, so that there is no osmotic flow, or shift of water, across the cell membrane. Such a solution of impermeable solute having an osmolal concentration of 154 mOsm/L H_2O is said to be an *isotonic* solution because it did not allow changes in *cell volume*. However, a solution of 154 mOsm/L H_2O of a *permeable* solute, such as urea, is *iso-osmolar*, but not *isotonic*, because urea enters the cell thereby inducing osmotic flow of water and hence a change in the cell volume. In short, the term "iso-osmotic" implies that the solution in question has the same osmotic pressure as that of the extracellular fluid, whereas the term "isotonic" implies that the solution in question does not cause a change in cell volume. Thus, an iso-osmotic solution may or may not be isotonic, depending on the permeability properties of the membrane in question.

Whenever the intracellular fluid is exposed to hypertonic (or hypotonic) interstitial fluid, water will flow out of (or into) the cells until the osmolality becomes equal again on both sides. In short, any inequality of impermeable osmolal concentrations can be rectified only by the osmotic shift of water into or out of the cells. Or, stated another way: *All changes in intracellular fluid volume* (except growth, of course) *are the result of changes in the osmolality of the interstitial fluid.*

The normal efflux channels already identified vary in the osmolality of their fluids. For example, the pulmonary efflux consists of water vapor, with zero osmolality. Skin efflux, even in heavy sweating, is hypotonic. Effluxes from the GI tract (vomiting and diarrhea, for example) are generally isotonic. Since the kidneys must be able to compensate for both hypotonic and hypertonic fluxes, it is not surprising to find that urine osmolality may be adjusted, as needed, from one-sixth isotonicity to five times isotonicity.

Typical Forcings and Responses of the Renal-Body Fluid Regulating System

The disturbance forcings that produce water and electrolyte imbalances include fortuitous gain of fluids via the influx channels, fortuitous loss of fluids through the efflux channels, and combinations of these. In practice, most of these forcings occur intermittently and are usually short-lived. Hence, they are properly called *pulse* forcings (see Appendix A for classification of forcings and their characteristics), rather than *step* forcings, and are followed by recovery. The compensatory response of the kidneys, therefore, is to accelerate the recovery process, thereby speeding up the restoration of volumes and compositions of body fluids toward normal.

Since fortuitous fluid gained or lost may contain different proportions of water and electrolytes, the above forcings are subclassified as *isotonic, hypotonic,* or *hypertonic* forcings. Thus, on the influx side, we may have fortuitous gain of isotonic, hypotonic, or hypertonic fluids, and on the efflux side, we may have fortuitous loss of isotonic or hypotonic fluids. The fortuitous loss of hypertonic fluid occurs only in patients with the syndrome of inappropriate secretion of antidiuretic hormone (SIADH), which is discussed in Chapter 3.

Table 1.2 summarizes the responses of the renal-body fluid regulating system to the forcings just described. For each forcing, the deviation from normal is indicated by (+) for increase, (−) for decrease, and (0) for no change. The direction of water shift between the extracellular (interstitial plus plasma compartments) and intracellular compartments is indicated by a horizontal arrow (→ or ←).

Influxes

Isotonic

Oral intake or parenteral infusion of a large volume of isotonic saline increases the plasma volume, causing secondary transfer of fluid into the interstitium. The net result is a uniform expansion of the extracellular fluid volume. Since the ingested fluid is isotonic, there is no change in osmolality of the interstitial fluid and hence no net osmotic shift of water into or out of the cells. These characteristic changes in the body fluid compartments produced by fortuitous isotonic fluid influx are termed *isotonic hydration*. Unless otherwise specified, both "tonicity" and "hydration" refer to the extracellular fluid compartment.

The kidneys respond to this extracellular volume expansion by increasing their excretion of both salt and water, producing an increase in urine volume (*diuresis*). This controlled diuresis rapidly returns the extracellular volume back to normal.

Hypotonic

Ingestion of a large volume of plain water increases the plasma volume and dilutes plasma osmolality. Fluid then shifts from the plasma into the interstitium. This increases the extracellular volume and decreases its osmolality (*hypotonic hydration*). The reduced interstitial osmolality causes osmotic shift of water into the cells, causing them to swell and diluting their osmolality. This is called *water intoxication* of the cells. The kidneys respond by increasing the excretion of a dilute urine (reduced osmolality), thereby returning the intracellular and extracellular volumes and osmolalities back to normal.

Hypertonic

Oral or parenteral intake of large amounts of hypertonic fluid increases plasma volume and osmolality. This causes osmotic shift of water into the plasma from the interstitium and diffusion of salt in the opposite direction. The net result is an increase in the volume and osmolality of the extracellular fluid (*hypertonic hydration*). This induces osmotic shift of water out of the cells, thus reducing their volumes but increasing their osmolality. The kidneys respond

TABLE 1.2. Changes in volume and osmolality from normal in fluid compartments and controlled renal effluxes in response to typical forcings.

	Effects on fluid compartments						Renal effluxes	
	Extracellular			Intracellular				
Forcings	(Vol)	(Osmol)	Water shift	(Vol)	(Osmol)	Terminology	(Vol)	(Osmol)
Influxes								
Isotonic	+	0	0	0	0	Isotonic hydration	+	0
Hypotonic	+	−	→	+	−	Hypotonic hydration	+	−
Hypertonic	+	+	←	−	+	Hypertonic hydration	+	+
Effluxes								
Isotonic	−	0	0	0	0	Isotonic dehydration	−	+
Hypotonic	−	+	←	−	+	Hypertonic dehydration	−	+

Source: Koushanpour.[5]

by excreting a concentrated urine, thereby restoring the normal state.

Effluxes

Since the kidneys can only moderate the effects of abnormal effluxes, correction requires adjustment of influxes.

Isotonic

Abnormal loss of water and electrolytes in isotonic concentration leads to reduced extracellular volume, but no change in intracellular volume or tonicity *(isotonic dehydration)*. This may occur with hemorrhage, plasma loss through burned skin, and GI fluid losses (vomiting and diarrhea, for example). In all these conditions the kidneys respond by conserving both salt and water.

Hypotonic

Heavy loss of hypotonic sweat leads to a decrease in extracellular volume and an increase in its osmolality. This will induce osmotic shift of water from cells into the interstitium. The net result is cellular dehydration accompanied by extracellular *hypertonic dehydration*. The kidneys respond by conserving water and excreting salt.

The foregoing flow diagram analysis reveals *three* important operational features of the renal-body fluid regulating system: (1) the *renal system* plays a central role in maintaining the constancy of the internal environment, a direct consequence of stabilizing the volume and composition of the circulating blood; (2) since the *blood* is an integral part of the extracellular fluid compartment, its regulation is indispensable to the ultimate regulation of the volume and composition of body fluids; and (3) the function of the renal regulator is partly modified by other organ systems.

With this general background serving as the framework, let us now proceed with a systematic analysis and synthesis of the renal-body fluid regulating system, beginning with a consideration of the body fluid component.

Problems

1.1. Calculate the osmolar concentration (mOsm/L) and the osmotic pressure (mm Hg) exerted by:
   a. 0.9 g% saline solution (mol wt = 58.5)
   b. 1.8 g% urea solution (mol wt = 60)

1.2. Calculate the number of grams of glucose (mol wt = 180) required to make a glucose solution iso-osmotic with the solutions (a) and (b) above.

1.3. Calculate the mM/L, mEq/L, and mOsm/L in a liter solution of each of the following solutes:
   a. 180 g of urea
   b. 175.5 g of sodium chloride, NaCl
   c. 90 g of glucose

1.4. Calculate the osmolality of a 2.5 L solution containing 7.0 g of salt. Note that this solution represents the daily intake of water and salt via the GI system.

1.5. Calculate the osmolality of a 1.5 L solution containing 7.0 g of salt and 30 g of urea. Note that this solution represents the daily excretion of water, salt, and urea by the kidney.

References

1. Gamble JL: *Chemical Anatomy, Physiology and Pathology of Extracellular Fluid. A Lecture Syllabus*, ed 6. Cambridge, Harvard University Press, 1967.
2. Goldberger E: *A Primer of Water, Electrolytes and Acid-Base Syndrome*, ed 5. Philadelphia, Lea & Febiger, 1975.
3. Benson SW: *Chemical Calculations*. New York, John Wiley & Sons, Inc, 1954.
4. Hammett LP: *Introduction to the Study of Physical Chemistry*. New York, McGraw-Hill Book Co Inc, 1952.
5. Koushanpour E: *Renal Physiology: Principles and Functions*, ed 1. Philadelphia, WB Saunders Co, 1976.

2

Body Fluids: Normal Volumes and Compositions

We have just learned that the kidneys play a major role in stabilizing the volume and composition of body fluids. To proceed further, we must acquire a basic understanding of fluid and electrolyte balance, a subject that is of great clinical importance.

Postoperative patients; patients with severe vomiting, diarrhea, or excessive sweating; and patients with edema, shock, diabetic coma, or adrenocortical insufficiency all present problems in fluid and electrolyte balance. This chapter provides essential information concerning the methods of measurement and the normal distribution of volumes and compositions of body fluids as a necessary background for understanding the dynamics of their exchange under normal and abnormal conditions.

Methods of Measurement

The *volumes* of various body fluid compartments are measured by applying the *dilution principle*.[1] In theory, using this method, we can measure the volume of any fluid compartment if we have a test substance that upon injection into the compartment will penetrate it homogeneously without being excreted in urine. In practice, we inject a known quantity of a test substance into the compartment and allow it to penetrate uniformly; we then measure its concentration in a sample drawn from the compartment. The penetrated volume is then calculated from the rearranged definition of concentration:

$$\text{Penetrated volume} = \frac{\text{Known quantity of injected substance}}{\text{Measured concentration of substance}} \quad (2.1)$$

In applying this method to measure the volume of various body fluid compartments in intact subjects, the test substance that is usually injected into the plasma may pentrate into one or more other compartments, as well as be excreted by the kidneys. Consequently, the plasma concentration of the test substance decreases continuously, making it difficult to apply the dilution principle in the standard form. To overcome this difficulty, two test procedures have been devised, depending on the known renal excretion rate of the test substance. Regardless of which procedure is used, the test substance employed should (a) not be toxic, (b) distribute uniformly within the compartment of interest, (c) not be metabolized during the test period, and (d) not alter the volume of the compartment being measured.

The *single dose injection method* is used for substances with slow renal excretion rate. In this procedure, when a known quantity of a test substance is injected into the plasma, two factors determine its subsequent plasma concentration: renal excretion and possible penetration into other compartment(s). When a substance that penetrates into other compartment(s) has achieved uniform distribution, the rate of decrease in its plasma concentration will, thereafter, be constant and equal to its renal excretion rate.

In practice, to find the volume of a body fluid compartment by this method, a known quantity of a test substance is injected intravenously and its plasma concentration is then determined at several successive time intervals. The logarithm of the plasma concentration is then plotted against time, as illustrated in Figure 2.1.

To find the volume of the compartment(s) into which the test substance has penetrated, we must determine the plasma concentration that would have

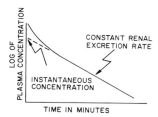

FIGURE 2.1. Time course of plasma concentration of a test substance following a single dose injection. (From Koushanpour.[10])

been obtained had the test substance penetrated uniformly and instantaneously throughout the compartment(s) without being excreted by the kidneys. This *instantaneous* concentration is obtained by extrapolating the linear portion of the curve, which reflects constant renal excretion rate, back to zero. Dividing the quantity of test substance initially injected by this concentration yields the volume of the compartment(s) penetrated by the test substance.

A detailed discussion of the mathematical basis of the dilution principle and of its applications is given in Appendix B.

The *constant-infusion equilibrium method* is used for substances with rapid renal excretion rate. In this procedure we first inject a large dose of the test substance, called the *priming dose*, to increase its plasma concentration for ease of measurement. Then, more test substance is slowly infused at a rate matching its renal excretion rate. As a result of this maneuver, changes in plasma concentration of the test substance will be due only to its penetration into other body fluid compartment(s). Once the substance has uniformly penetrated throughout the compartment(s), its plasma concentration should remain constant, at which time the infusion is stopped. Then the urine is collected until all the test substance that was in the body at the time the infusion was ended has been excreted. This amount of test substance contained in the volume of collected urine represents the quantity of the test substance that was retained in the body and yielded the plasma concentration at the time the infusion was stopped. Dividing this quantity by the plasma concentration yields the volume of the compartment(s) into which the test substance has penetrated uniformly.

The *compositions* of various body fluid compartments have also been measured by the dilution principle.[1] The concentrations of various substances present in the accessible body fluid compartments, such as plasma, are measured directly by chemical analysis. However, the concentrations of solutes in the nonaccessible compartments, such as interstitial space, are measured indirectly by using the dilution principle. This method involves measuring the quantity of *exchangeable* solute in the body and the corresponding volume that contains this solute. Then, by partitioning both quantity and volume between the accessible and nonaccessible compartments, the solute concentration in the nonaccessible compartment(s) may be determined.

The *exchangeable* quantity of any solute, such as sodium, is measured by introducing a tracer amount of labeled (radioactive) sodium, Na^* (measured in counts per minute, cpm), and then determining its *specific activity*, defined as Na^{**}/Na (cpm/mL/mEq), in the plasma after an adequate equilibration period.[4,19] Na^{**} is the quantity of labeled sodium (cpm/ml) in a sample of plasma drawn for analysis and Na is the total number of milliequivalents of sodium (labeled plus unlabeled) per milliliter of that sample as determined by chemical analysis. Then, using the dilution principle, the total number of milliequivalents of exchangeable sodium in the body is given by

$$\text{Exchangeable Na (mEq)} = \frac{\text{Amount of labeled Na injected (cpm)}}{\text{Specific activity of Na in plasma (cpm/mEq)}} = \frac{Na^*}{Na^{**}/Na} \quad (2.2)$$

Volumes of Body Fluid Compartments

Total Body Water

The total body water in intact man has been measured using three test substances: *antipyrine* and its derivatives, *deuterium oxide* (D_2O), and *tritiated water* (^3H_2O). Because of the ease of measurement and the rapid rate of distribution, antipyrine has become the substance of choice.

Total body water, as measured by antipyrine, constitutes about 70% of body weight at birth and falls to about 60% of body weight within the first two years of life.[6] This is due to a large extracellular volume that decreases gradually with growth as a result of three factors: (1) an increase in the number of cells; (2) an increase in the size of those tissues with greater intracellular water, such as muscle; and (3) an increase in the amount of body fat.

In adults, total body water, measured by antipyrine, averages about 42.0 L (60% of body weight) in men and 35.0 L (50% of body weight) in women. Similar values have been obtained with D_2O and ^3H_2O as test substances.

The distribution, by tissues, of total body water is given in Table 2.1. Note that most of the body water is distributed in muscle (32% of body weight), skin (13%), and blood (7%), with very little in skeleton (3.5%) and adipose tissues (0.01%).

The total body water varies with sex. This variation is due principally to the amount of body fat, which is normally about 15% of the body weight. Since fatty tissue contains less water per unit weight, an obese individual has relatively less water than a lean person. In women, after puberty, the total body water per unit of body weight is less compared with that for men. This is due to a greater quantity of fat in women and is related to the blood levels of female sex hormones.

The normal variation in total body water from person to person is due primarily to variation in *body fat*. In both sexes the ratio of total body water to body weight varies inversely with the amount of fatty tissues. However, the percentage of water in lean body mass is essentially constant at 73%. (The *lean body mass* is defined as 15% bone, 10% fat, and 75% tissue.) This constancy is the basis of the following empirical formula for determining the percentage of excess fat[12]:

$$\%\text{Excess fat} = 100 - \frac{\% \text{ water}}{0.732} \times 100 \qquad (2.3)$$

Another method for determining the quantity of stored fat is to measure the total body *specific gravity* (sp gr). The specific gravity of a substance is defined as the weight of the substance in air divided by the difference between the weight in air and the weight in water. The specific gravity of human fat is 0.918. This value is much lower than the specific gravity of bone (1.56) or other tissues (1.06). Since in very lean individuals the upper limit of the specific gravity is 1.10, a specific gravity of less than 1.10 can be attributed to an increase in the body fat content. Using the values of 0.918 and 1.10 as the lower and upper limits of specific gravity, Rathbun and Pace[14] derived the following empirical formula for determining the percentage of excess fat from the total body specific gravity in man:

$$\% \text{ Excess fat} = 100 \left(\frac{5.548}{\text{sp gr}} - 5.044 \right) \qquad (2.4)$$

TABLE 2.1. Distribution of water and kinetics of water movement in various tissues of a 70-kg man.

Tissues	Percent body weight	Percent water	Water in tissues as % body weight	Water (L)	Time for D_2O equilibration (min)
Muscle	41.7	75.6	31.53	22.10	38
Skin	18.0	72.0	12.96	9.07	120–180
Blood	8.0	83.0	6.64	4.65	RBC 1/60
Skeleton	15.9	22.0	3.50	2.45	120–180
Brain	2.0	74.8	1.50	1.05	
Liver	2.3	68.3	1.57	1.03	10–20
Intestine	1.8	74.5	1.34	0.94	Gastric juice 20–30
Adipose tissues	± 10.0	10.0	0.01	0.70	
Lungs	0.7	79.0	0.55	0.39	
Heart	0.5	79.2	0.40	0.28	
Kidneys	0.4	82.7	0.33	0.25	
Spleen	0.2	75.8	0.15	0.10	
Total body	100.0	62.0	60.48	43.40	180

Column 3 of the table was obtained by multiplying column 1 by column 2 and dividing by 100. Column 4 of the table was obtained by multiplying column 3 by 70 kg and dividing by 100. Modified from Ruch and Patton[15]. Water values were taken from Skelton.[18] D_2O equilibration values were taken from Edelman.[3]

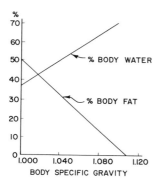

FIGURE 2.2. Relation of percentage of total body water and excess fat to body specific gravity. (From Koushanpour.[10])

Equations 2.3 and 2.4 describe two different methods of determining the percentage of excess body fat. Hence, setting them equal to each other yields another equation relating the percentage of the total body water to the specific gravity:

$$\% \text{ Water} = 100 \left(4.424 - \frac{4.061}{\text{sp gr}}\right) \quad (2.5)$$

This equation can be used to predict from the whole body specific gravity both the fat and the water contents with a reasonable degree of accuracy.

Using Equations 2.3 and 2.5, we can obtain a graphic relation of the percentages of the total body water and the excess fat to the body specific gravity. This is illustrated in Figure 2.2. It is evident that the variation in body fat is the chief determinant of the total body specific gravity.

It should be noted that both Equations 2.3 and 2.5 describe a rectangular hyperbola displaced along the principal Y-axis. However, for the range of specific gravity values plotted, the linear plots in Figure 2.2 represent a very small segment of the hyperbola.

Total body water is subdivided into two major compartments separated by the cell membrane: the intracellular and the extracellular fluid compartments. The extracellular compartment is divided into two subcompartments separated by the capillary membrane: (a) *intravascular fluid* or blood plasma and (b) *interstitial fluid*, including lymph, fluids in the dense connective tissues, cartilage and bone, and cavity fluids. The last, also called *transcellular fluid*, comprises the digestive, cerebrospinal, intraocular, pleural, peritoneal, and synovial fluids.

The volumes of these fluid compartments can be measured in the intact subject by virtue of the fact that antipyrine and heavy water will diffuse freely throughout the total body water, thiocyanate throughout the extracellular fluid only, and certain large molecular dyes (T-1824) throughout the blood plasma only. By injecting known quantities of these substances intravenously, the volumes of the separate compartments can be calculated approximately from their resulting concentration in the blood plasma by the dilution principle.

Table 2.2 summarizes the distribution of body water in a 70-kg healthy young man. The entries corresponding to the capitalized headings are the normal values obtained by application of the dilution principle. For convenience and purposes of comparison, the values are compiled under three categories: entries in column 1 give the volume of total body water and its distribution in each compartment; entries in column 2 give the volume of water in each compartment as a percentage of body weight; and entries in column 3 give the volume of water in each compartment as a percentage of total body water.

Extracellular Volume

The extracellular volume is very difficult to measure owing to the unavailability of a test substance that does not penetrate the cells. There is no known substance that will remain extracellular and distribute rapidly and uniformly throughout the plasma, interstitial, dense connective tissue, cartilage, bone, and transcellular fluids. Furthermore, the differing rates of equilibration of a test substance in these widely different tissues have made it difficult to interpret the results of dilution tests and have led to confusion of the anatomical space with the "physiological" quantities.

Because of these difficulties, it has become customary to measure the volume distribution of a specific substance and to refer to it as, for example, the *inulin space*, if the test substance happens to be inulin.

Two general types of test substances may be used to measure the volume distribution of the extracellular compartment[20]: (1) saccharides, such as inulin, sucrose, and mannitol, and (2) ions, such as thiosulfate, sulfate, thiocyanate, chloride, bromide, and sodium. Because of different rates of penetration of these test substances into the various components of the extracellular space, two phases can be distinguished. One is a *fast-equilibrating phase*, with a

TABLE 2.2. Body water distribution in a healthy 70-kg young man.

Compartments	Total volume (L)	% of body weight	% of total body water
Total body volume (TBV) (by antipyrine method)	42.0	60.0	100.0
Total extracellular volume (ECV)			
Fast ECV (by rapid equilibration of inulin or radioactive sulfate)	11.6	16.5	27.5
Slow ECV (by slow equilibration of thiocyanate)	18.9	27.0	45.0
Plasma volume (Vp) (T-1824 method)	3.2	4.5	7.5
Interstitial volume (V_I) (Fast ECV minus Vp)	8.4	12.0	20.0
Dense connective tissue and cartilage volume	3.2	4.5	7.5
Inaccessible bone volume	3.2	4.5	7.5
Transcellular fluid volume	1.0	1.5	2.5
Total intracellular volume (ICV) (TBV minus slow ECV)	23.1	33.0	55.0

Values taken from Edelman and Leibman.[5]

half-time of 20 minutes or less, representing the penetration into plasma and interstitial spaces. The other is a *slow-equilibrating phase*, with a half-time of five to nine hours, representing the penetration into the remaining extracellular compartments.

The *extracellular volume* ranges from about 12 L, determined from the fast-equilibrating phase, using inulin or radioactive sulfate as tracer, to 19 L, determined from the slow-equilibrating phase, using thiocyanate as a tracer. The former value includes only plasma and interstitial volumes, whereas the latter value includes these volumes plus the volumes of fluids in the dense connective tissue and cartilage. In the present context, we shall use the volume distribution of *thiocyanate* as a measure of the total extracellular volume (slow ECV, Table 2.2), which constitutes about 27% of the body weight.

Plasma Volume

Plasma volume may be measured using two methods[17]: (1) plasma albumin or globulin labeling by dye (T-1824) and radioactive iodine (^{131}I), and (2) red cell labeling using isotopes of phosphorus (^{32}P) and chromium (^{51}Cr). In both methods, using the dilution principle, the volume distribution of the labeled protein or cells is determined. In the case of red cell labeling the plasma volume is determined after correcting for the hematocrit. The plasma volume estimated by T-1824 labeled protein, in a healthy young man, averages about 3.2 L, representing about 7.5% of the total body water.

Interstitial Volume

Interstitial volume, including lymph, can be determined from the fast-equilibrating phase with inulin as tracer after correcting for the plasma volume. The estimated water content in this compartment averages about 8.4 L, which is equal to 12% of the body weight.

Dense Connective Tissue and Cartilage Volumes

Dense connective tissue and cartilage fluids differ from interstitial fluids in that they equilibrate very slowly with test substances used to measure the extracellular volume. The water content of cartilage and dense connective tissues has been estimated to be 3.2 L, or about 4.5% of body weight.

Transcellular Fluid Volume

Transcellular fluid is not a simple transudate; its composition differs from that of a simple ultrafiltrate (protein-free fraction) of plasma. Of the various components of the transcellular fluids mentioned earlier, the intraluminal gastrointestinal water constitutes the largest fraction; it is about 7.5 mL/kg of the body weight. The volume of cerebrospinal fluid is about 2.8 mL and that of bile is about 2.1 mL/kg of body weight. The total volume of the transcellular fluid compartment is estimated to be 1.0 L, or about 1.5% of body weight.

Intracellular Volume

The volume of intracellular water can be measured only as the difference between the total body volume and the extracellular volume. Hence, any error in the measurement of these quantities will introduce substantial error in the final estimate of the total intracellular volume. The intracellular volume is estimated to be about 23.1 L or 55% of the total body water.

Compositions of Body Fluid Compartments

Distribution of water within the various body fluid compartments ultimately depends on the quantities of ions in these compartments. Hence, any disturbance in body fluid distribution is always accompanied by displacement of electrolytes. Therefore, to understand the causes of fluid and electrolyte imbalance and to plan corrective therapeutic measures, it is imperative to have a basic understanding of the normal distribution of electrolytes among various compartments.

Like water, anions and cations are distributed between the extracellular and intracellular fluid compartments. The principal cations of the extracellular fluid, in the order of amount present, are sodium, potassium, calcium, and magnesium; the chief anions are chloride, bicarbonate, and phosphate. The major intracellular cations are potassium, sodium, magnesium, and calcium; the chief anions are phosphate, chloride, sulfate, and bicarbonate.

Table 2.3 summarizes the relative distribution of the electrolyte concentrations in plasma, interstitial, and intracellular fluids. It is apparent that the sum of the concentrations of the cations is equal to the sum of the concentrations of the anions in each body fluid compartment, making the solution electrically neutral.

The entries for water concentration of different solutes were calculated using Equation 1.1 by assuming that the plasma is 93% water by weight.

A close examination of Table 2.3 reveals that the intracellular fluid consists essentially of a solution of potassium and phosphate ions, while the extracellular fluid consists essentially of a solution of sodium and chloride ions. In addition, the intravascular and intracellular fluids contain proteins, whose concentration influences the distribution of ions, and hence water, between the various body fluid compartments. To illustrate this effect of protein, the intravascular ions are expressed in both "bulk" and "water" concentrations, while the interstitial and intracellular ions are expressed in "water" concentration. Note that regardless of units used to express the ionic concentration, the sum of the concentrations of the cations is equal to the sum of the concentrations of the anions in each body fluid compartment, making the solution electrically neutral.

Further study of this table reveals several important features of electrolyte concentration and distribution between the various body fluid compartments.

The extracellular cations have higher "water" concentrations in the plasma (IVF) than in the interstitial fluid (ISF). This difference is due to the plasma proteins, which are negatively charged (anions) at normal blood pH and which normally do not cross the capillary membrane. As a result, the impermeable protein anions bind some positively charged ions (sodium, potassium, etc.), yielding a higher "water" concentration of these ions in the plasma. Consequently, there will be unequal distribution of diffusible cations across the capillary wall, such that the *product of the diffusible cations and anions will be equal on both sides* (to maintain electrochemical neutrality). This is called the *Gibbs-Donnan rule*, and the unequal distribution of cations due to impermeable proteins is called the *Gibbs-Donnan effect*.

The essence of the *Gibbs-Donnan principle* may best be explained by the example illustrated in Figure 2.3. Initially, a solution of sodium proteinate, Na^+Pr^-, is separated from a saline solution, Na^+Cl^-, by a semipermeable membrane that is permeable to Na^+ and Cl^-, but impermeable to protein, Pr^-. To simplify the presentation, we assume that the vol-

TABLE 2.3. Electrolyte content of body fluid compartments.

Electrolyte	Intravascular (IVF) mEq/L of plasma[*]	Intravascular (IVF) mEq/L of water[†]	Interstitial (ISF) mEq/L of water	Intracellular (ICF) mEq/L of water
Cations				
Sodium	142	153.0	147.0	10
Potassium	5	5.4	4.0	140
Calcium	5	5.4	2.5	5
Magnesium	3	3.2	2.0	27
Total cations	155	167.0	155.5	182
Anions				
Bicarbonate	27	29.0	30.0	10
Chloride	103	111.0	114.0	25
Phosphate	2	2.2	2.0	80
Sulfate	1	1.1	1.0	20
Organic acids	6	6.5	7.5	—
Proteinate	16	17.2	1.0	47
Total anions	155	167.0	155.5	182

[*]A plasma water content of 93% was used in the calculations.
[†]Gibbs-Donnan factors used were 0.95 for monovalent anions and 1.05 for monovalent cations.
Values taken from Deane and Smith[2], Edelman and Leibman[5], Gamble[7], Goldberger[8], and Manery.[11]

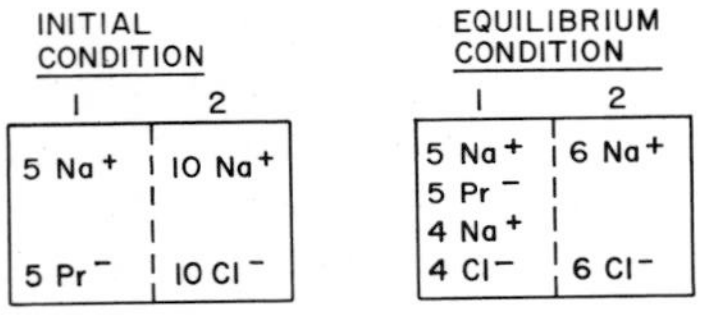

FIGURE 2.3. A theoretical scheme showing the effect of protein on the distribution of diffusible anions and cations. Modified and reproduced with permission from Pitts, R.F.: *Physiology of the Kidney and Body Fluids*, 3rd edition. Copyright © 1974 by Year Book Medical Publishers, Inc., Chicago.

umes of the two compartments will remain constant as the diffusible ions distribute between the two sides in accordance with their concentration gradients. Hence, any induced osmotic changes in the volumes will be ignored.

Since the initial concentration gradient for Cl^- is greater than that for Na^+, the rate of net movement of Cl^- from side 2 to side 1 will momentarily exceed that of Na^+. Because the membrane is impermeable to Pr^-, the transfer of Cl^- will lead to a momentary increase in the negative ions on side 1. This will generate an electrostatic force on side 1 that will induce a net transfer of Na^+ from side 2 to side 1, thereby making the rate of diffusion of both ions equal. Since Cl^- moves along its concentration gradient, but against the electrical potential gradient, the *work* required to move one equivalent of Cl^- is given by

$$\text{Work} = R \cdot T \cdot \log \frac{[Cl^-]_1}{[Cl^-]_2} + F \cdot E \quad (2.6)$$

where R is the gas constant, T is the absolute temperature, F is the Faraday constant, E is the electrical potential difference between the two sides, and $[Cl^-]_1$ and $[Cl^-]_2$ are the chloride concentrations on sides 1 and 2, respectively. Likewise, since Na^+ moves against its concentration gradient, but along the electrical potential gradient, the *work* required to move one equivalent of Na^+ is given by

$$\text{Work} = R \cdot T \cdot \log \frac{[Na^+]_1}{[Na^+]_2} - F \cdot E \quad (2.7)$$

At equilibrium, no work is done by the system, and therefore the sum of the two work equations must be zero. Hence, setting Equations 2.6 and 2.7 equal, and cancelling terms, we get

$$[Na^+]_2 \times [Cl^-]_2 = [Na^+]_1 \times [Cl^-]_1 \quad (2.8)$$

This equation defines the Gibbs-Donnan rule, which states that at equilibrium, the *products of the diffusible ions on both sides of the membrane are equal*.

To maintain electroneutrality in each compartment, the sum of anion concentrations (ions with net negative charges, such as Cl^- and Pr^-) must be equal to the sum of cation concentrations (ions with net positive charges, such as Na^+). This may be expressed as

$$[Na^+]_1 = [Cl^-]_1 + [Pr^-]_1 \quad (2.9)$$

$$[Na^+]_2 = [Cl^-]_2 \quad (2.10)$$

Substituting for $[Na^+]_2$ in Equation 2.8 its equivalent from Equation 2.10, we obtain

$$[Cl^-]_2^2 = [Na^+]_1 \times [Cl^-]_1 \quad (2.11)$$

From Equation 2.9 it is evident that some of the sodium ions on side 1 are associated with chloride and some with protein anions. Hence, $[Na^+]_1$ in Equation 2.11 is clearly greater than $[Cl^-]_1$. Equation 2.11 states that the product of two unequal quantities is equal to a square of a quantity. Therefore, if we represent the left side of Equation 2.11 by a square and the right side by a rectangle, it would be easy to show that the sum of the sides of a rectangle is greater than the sum of the sides of a square of equal area. Hence, in terms of the sum of the concentrations, Equation 2.8 may be written as

$$[Na^+]_1 + [Cl^-]_1 > [Na^+]_2 + [Cl^-]_2 \quad (2.12)$$

This inequality is due to the presence of protein anions in the plasma. Thus, the presence of protein results in an *increase* in the concentration of diffusible cations and a *decrease* in the concentration of diffusible anions, on the side containing the protein (side 1), compared with the side lacking it. Furthermore, since the sum of the diffusible ions is greater on the side containing the protein (Equation 2.12), the *osmolality* of this side will also be greater, compared with the side lacking protein. The total osmotic pressure on the side containing protein is called the *oncotic pressure*. It is the sum of the protein osmotic pressure and the osmotic pressure of the obligated (bound) cations.

The electrolyte content of the intracellular fluid markedly differs from that of the extracellular fluid both in the prevalent ionic species and in the total quantity of ions. First, in the intracellular fluid potassium and magnesium have replaced sodium and calcium as the cations, and phosphate and sulfate ions have replaced chloride as anions. Also, there is a reduction in the cellular bicarbonate and a marked increase in the cellular proteinate concentrations. Second, the total "water" concentration of the intracellular constituents is much greater compared with that of intravascular or interstitial compartments. This difference is due to impermeable proteins (Gibbs-Donnan effect), which maintain an osmotic equilibrium between a virtually protein-free solution (interstitial fluid) and the intracellular fluid separated by the cell membrane, which is impermeable to proteins but permeable to water and all other solutes.

Despite a high concentration of sodium in the ISF, sodium concentration in ICF is low, whereas the reverse condition holds for potassium ions. Similarly, despite high concentration of chloride and bicarbonate ions in ISF, the concentration of these ions is low in the ICF. Despite leakage of sodium into and potassium out of the cells, the sodium-potassium concentration gradients across the cell membrane are maintained both by active transport and as a consequence of asymmetrical permeability of the cell membrane to potassium and sodium ions and the differing cellular mechanisms involved for their transport. These cellular transport mechanisms have been classified according to whether the net movement of the ion is along or against an electrochemical potential gradient, or whether the transport requires expenditure of cellular energy, or both.[16] Accordingly, three major transport mechanisms have been identified: (1) simple diffusion, (2) convection, and (3) carrier-mediated. The last is further subdivided into (a) facilitated diffusion, (b) active transport, and (c) cotransport. The following is a brief account of these cellular transport processes. A more detailed description is given in Chapter 8.

Simple Diffusion. A number of substances, such as urea, creatinine, and other organic solutes of small molecular weight, are transported across the cell membrane by simple diffusion. For a given substance, the rate of transport by diffusion is directly proportional to the *concentration gradient* of the substance across the membrane. No expenditure of cellular energy is required for this type of transport.

Convection (or solute drag). Osmotic flow of water may *entrain* some solutes and carry them through a porous membrane. This type of solute transport is called convection. The smaller the molecular weight, the greater would be the ease of its transport by convection. Like simple diffusion, no expenditure of cellular energy is required for this type of transport.

Carrier-mediated. Ions, especially sodium and potassium, and some important metabolites, such as glucose and amino acids, are transported across the cell membrane by carrier-mediated mechanisms. The major characteristic of this type of transport is a temporary binding of the transported solutes to the carrier molecules or binding sites within the membrane.

Facilitated diffusion involves solute transport resulting from combination of the solute with a finite number of membrane sites called *carriers*. The rate of transport of the solute is a function of the solute concentration gradient and the availability of the carriers specific for that solute. An important feature of this type of transport is that as the solute concentration gradient increases, the available binding sites will be *saturated*, so that the rate of transport approaches an asymptote. Furthermore, some expenditure of cellular energy is required for this type of transport. The relationship between the rate of solute transport ($\dot{T}_x$), where a dot over the T denotes rate of change with respect to time (see also Appendix A), and solute concentration ($[x]$) is defined by a right-rectangular hyperbola of the form

$$\dot{T}_x = \frac{\dot{T}_{max} \cdot [x]}{[x] + K_m} \qquad (2.13)$$

where $[x]$ is the solute concentration, $\dot{T}_{max}$ is the maximum rate of transport (the asymptote when $\dot{T}_x$ is plotted against $[x]$), and K_m is the substrate concentration yielding half the maximum rate of transport.

Active transport is defined as the movement of solutes *against* the electrochemical gradient, exhibiting saturation at high solute concentration and requiring expenditure of cellular energy (adenosine triphosphate, ATP, hydrolysis). The sodium-potassium adenosine triphosphatase (ATPase) (the so-called *"sodium pump"*) is the membrane-bound enzyme that is responsible for the exchange of sodium for potassium across the basolateral cell membrane. The activation of the sodium pump results in transfer of more sodium than potassium per cycle, thereby hyperpolarizing the cell membrane beyond E_K, the K equilibrium potential. Because of this unequal sodium-potassium exchange, the sodium pump is said to be *electrogenic*. Like facilitated diffusion, the rate of transport asymptotically approaches a maximum as the solute concentration increases.

Cotransport is defined as the uphill (active) transport of one substance (e.g., glucose, amino acid, or chloride ion) coupled to the downhill transport of another substance, such as sodium, across the apical cell membrane. The electrochemical gradient for sodium across the apical membrane is the force that causes the inward movement of both sodium (downhill) and glucose (uphill). The sodium gradient is maintained by the Na-K-ATPase enzyme in the basolateral membrane, which lowers the sodium concentration inside the cell by active extrusion of sodium out of the cell across the basolateral membrane. Since the exchange of sodium for potassium at the basolateral membrane is the "primary active transport," the cotransport of a substance, such as glucose, with sodium at the apical membrane is called the "secondary active transport," because it depends on the active transport of sodium. This sodium-glucose cotransport system at the apical membrane is called a *"symport."* Another symport system is the cotransport of chloride with sodium and potassium in the thick ascending limb of Henle's loop,[9] which plays an important role in the concentration and dilution of urine (see Chapter 12). Another type of cotransport is the *"antiport"* system in which two ions of the same charge are exchanged across the cell membrane. The exchange of sodium for potassium across the basolateral membrane, mediated by the Na-K-ATPase enzyme, is an example of such an antiport transport mechanism.

Since changes in plasma concentration of sodium, potassium, chloride, and bicarbonate ions and proteins are often used as clinical guides to understanding the causes of fluid and electrolyte imbalance as well as to evaluating the success of therapy, a basic understanding of their normal distribution and factors affecting them is in order.

Sodium

Nearly all of the exchangeable sodium resides in the extracellular compartment. The exchangeable sodium pool is much higher in infant, compared with that in the adult, a finding compatible with the high total body water observed in the infant.

The combined plasma and interstitial fluids contain about 60% of the total exchangeable sodium. The latter is about 70% of the total body sodium, which consists of about 58 mEq/kg of body weight (Table 2.4). Since these fluids account for 70% of the sodium lost during acute sodium depletion, their functional significance cannot be underestimated.

In short, the relative size of the sodium pool of these compartments is subject to considerable variations with disease. For example, in patients with liver, heart, or kidney disease the edema that occurs constitutes a 20% to 100% increase in the total exchangeable sodium pool. Therefore, these diseases impose a considerable displacement and translocation of sodium among various body fluid compartments and reflect serious disturbances in the mechanisms involved in maintaining normal distribution.

Potassium

The bulk of body potassium is distributed within the intracellular compartment (Table 2.4), especially muscle cells. Nearly 90% of the total body potassium is exchangeable. In contrast to sodium, this exchangeable potassium pool remains constant up to adulthood. However, its distribution is subject to considerable variations with disease. In chronic illness, the potassium loss may be as high as 25% of the total body potassium. These losses occur at the expense of intracellular stores, while the extracellular potassium concentration is maintained relatively constant. However, a disturbance in extracellular pH could bring about an acute distortion of extracellular potassium content. This response is the basis of the dissociated metabolic acidosis and is considered in Chapter 11.

Chloride

Normally, the chloride content of the extracellular fluid constitutes about 40% of the total exchangeable chloride (Table 2.4). Its distribution and metabolism are influenced by the same factors that stabilize the body content of sodium.

In acid-base disturbance, plasma chloride may vary independently of plasma volume. Thus, hyperchloremia often is seen in metabolic acidosis, and hypochloremia often is present in patients with either metabolic alkalosis or respiratory acidosis. Renal regulation of body chloride is considered in Chapter 11.

Bicarbonate

Bicarbonate content of the body depends on two factors: (1) the metabolic cellular production of carbon dioxide (MR_{CO_2}), and (2) the relative excess of fixed cations, such as sodium, potassium, calcium, and magnesium over fixed anions, such as chloride, sulfate, phosphate, and proteinates. Therefore, the size of the bicarbonate pool is determined by the difference between the sum of the fixed cations and the sum of the fixed anions. Any cation excess, imposed by the formation of alkali within the body, is balanced by the hydration of carbon dioxide and subsequent formation of bicarbonate ions. Conversely, any anion excess imposed by the liberation of acid in the body is balanced by the conversion of the latter to carbon dioxide and water. The carbon dioxide thus formed is eliminated by the lungs. Therefore, the bicarbonate pool is labile and its size is regulated by the respiratory and renal systems.

TABLE 2.4. Comparative distribution of body sodium, potassium, and chloride among fluid compartments in a healthy young man.

Compartments	mEq per kg of Body Weight			% of Total Body		
	Na	K	Cl	Na	K	Cl
Plasma	6.5	0.2	4.5	11.2	0.4	13.6
Interstitial-lymph	16.8	0.5	12.3	29.0	1.0	37.3
Dense connective tissue and cartilage	6.8	0.2	5.6	11.7	0.4	17.0
Exchangeable bone	8.0	—	—	13.8	—	—
Total bone	25.0	4.1	5.0	43.1	7.6	15.2
Transcellular	1.5	1.0	1.5	2.6	1.0	4.5
Total exchangeable extracellular	39.6	—	—	68.3	—	—
Total extracellular	56.6	5.5	28.9	97.6	10.4	87.6
Total body	58.0	53.8	33.0	100.0	100.0	100.0
Total intracellular	1.4	48.3	4.1	2.4	89.6	12.4

Values taken from Edelman and Leibman.[5]

Proteins

The important plasma proteins from the standpoint of body fluid regulation are albumin and globulin. *Albumin* is the smaller molecule and is present in greater concentration. It is manufactured exclusively by the liver. Albumin plays an important role in the maintenance of the plasma volume. It influences the distribution of the diffusible anions of plasma and binds certain cations such as calcium. A low plasma albumin concentration reflects a low plasma protein concentration. Failure of hepatic function and/or excessive loss of albumin in urine are common explanations for a decreased plasma albumin concentration.

Globulin fraction is subdivided into three components:

1. α_1- and α_2-fractions include the iron-binding globulins, serum esterases, and angiotensinogen.
2. β-globulin fraction contains lipoproteins, including cholesterol, phosphatide, fatty acid, and vitamin A.
3. γ-globulin fraction includes immune substances, such as are formed in response to infectious hepatitis and measles, and many other antibodies that develop as one grows from infancy to maturity. Included in this fraction are the various factors concerned in the coagulation of blood, including prothrombin, antihemophilic globulin, and fibrinogen.

The origins of some of the plasma protein fractions are obscure. However, immune globulins are thought to be derived in part from lymphoid tissues and the reticuloendothelial system. Fibrinogen and prothrombin are largely derived from the liver, although some fibrinogen may also be formed in bone marrow. Certain other globulins are likewise hepatic in origin.

The normal plasma protein concentration is about 7 g/100 mL of plasma. Of the total plasma proteins, albumin fraction amounts to 4.4 g and globulin fraction is 2.6 g, giving an albumin/globulin (A/G) ratio of 1.67. The critical plasma protein concentration for normal body fluid distribution is about 5.5 g%. In nephrotic syndrome, in which there is excessive albuminuria (> 3.5 g/day) the A/G ratio may fall to 0.84 and even lower. At such low plasma protein concentration there would be a marked distortion of water and electrolyte distribution in the body fluid compartments. Details of the cardiovascular and renal compensatory mechanisms involved are discussed in Chapter 13.

Finally, it should be pointed out that the distribution of anions and cations within the body fluid compartments exhibits both daily fluctuation and variation with age. Among the cations, the plasma sodium concentration varies slightly with age. Extracellular potassium levels are distinctly high in the infant, with gradual decline in concentration with age. Calcium levels are comparable in different age groups, but rise slightly with age. In the case of anions, both chloride and bicarbonate levels vary with age, with a definite sex difference occurring in healthy young adults; the chloride concentration is lower in men and the bicarbonate concentration lower in women. However, there is no apparent sex difference in infants with respect to the concentration of these two ions. The plasma concentration of inorganic phosphate varies with age, being higher just after birth and declining steadily with age. The concentrations of various proteins are low in infants and gradually rise with age.

Because of day-to-day fluctuations of the plasma electrolytes, even during fasting, the stability of the electrolyte concentration in the body fluids is only relative, so that there is an oscillation of the plasma electrolytes under normal conditions.

Problems

2.1. 100 mL of deuterium oxide (D_2O) in isotonic saline was injected intravenously into a normal man weighing 90-kg. After an equilibration period of two hours the concentration of D_2O in plasma water was 0.2 mL%. The urinary, respiratory, and cutaneous losses of D_2O were averaged to be 0.4% of the administered dose. Calculate (a) the total body water as a percentage of the body weight, (b) the percentage of excess fat, (c) the amount of excess fat in kilograms, and (d) the total body specific gravity.

2.2. 10 mg of thiocyanate per kilogram of body weight was injected into a 70-kg subject. Thirty minutes later a sample of plasma was drawn and urine was collected. The concentration of thiocyanate in plasma was 4.6 mg%, and the total amount of thiocyanate excreted was 30 mg. Calculate the thiocyanate space as a percentage of body weight.

2.3. A 70-kg patient received an intravenous dose of radioactive sodium having an activity of 42×10^5 counts/minute (cpm). After allowing adequate time for equilibration, a sample of plasma drawn had an activity of 2×10^2 cpm/mL of plasma and a total plasma sodium content (labeled plus unlabeled) of 0.145 mEq/mL. If the urinary loss of radioactive sodium amounts to 2×10^5 cpm, calculate the exchangeable sodium per kilogram of body weight.

2.4. To determine the extracellular volume of an 80-kg patient, 7.0 g of inulin was injected intravenously, and the plasma concentration was subsequently measured at successive time intervals. The values obtained are reproduced in the following table. Calculate (a) the volume of the extracellular space measured by inulin and (b) the rate of renal excretion of inulin after it is uniformly distributed in the compartment(s).

Time blood sample was taken (min)	Plasma inulin concentration (mg%)
12	81
18	70
38	45
48	40
65	32
100	24
123	21
175	17
242	13
310	10
370	8
400	7
490	5

2.5. A test substance not metabolized in the body is infused at a constant rate of 90 mg/min. After four hours, the plasma concentration attained a steady-state value of 85 mg%. Calculate the volume distribution of the test substance.

2.6. List *four* essential properties of any substance that would make it suitable as an agent for measuring the volume of a body fluid compartment.
   a. _____
   b. _____
   c. _____
   d. _____

2.7. In an 80-kg patient the total body water as measured by deuterium oxide was 60% of the body weight, the inulin space was 20% of the body weight, and the total solute content was 22,800 mOsm. Calculate (a) the amount of excess fat in this patient and (b) the osmolality of the extracellular fluid.

2.8. The most prevalent diffusible cation and anion of the extracellular fluid are (a) _____ and (b) _____. The most prevalent diffusible cation and anion of the intracellular fluid are (c) _____ and (d) _____. An important plasma protein required for maintenance of normal plasma volume is (e) _____. The large total body water in an infant prior to aged 2 is due to (f) _____. The percentage of the total body water in the infant decreases with growth. This is believed to be due to three processes: (g) _____, (h) _____, and (i) _____.

References

1. Deane N: Methods of study of body water compartments, in Corcoran AC (ed): *Methods in Medical Research*. Chicago, Year Book Medical Publishers, Inc, 1952, vol 5.
2. Deane N, Smith HW: The distribution of sodium and potassium in man. *J Clin Invest* 1952; 31:197–199.
3. Edelman IS: Exchange of water between blood and tissues. Characteristics of deuterium oxide equilibration in body water. *Am J Physiol* 1952; 171:279–296.
4. Edelman IS, James AH, Brooks L, et al: Body sodium and potassium. IV. The normal exchangeable sodium, its measurement and magnitude. *Metabolism* 1954; 3:530–538.
5. Edelman IS, Leibman J: Anatomy of body water and electrolytes. *Am J Med* 1959; 27:256–277.
6. Frus-Hansen B: Changes in body water compartments during growth. *Acta Paediatr* 1957; 46 (suppl. 110):1–68.
7. Gamble JL: *Chemical Anatomy, Physiology and Pathology of Extracellular Fluid. A Lecture Syllabus*, ed 6. Cambridge, Mass, Harvard University Press, 1967.
8. Goldberger E: *A Primer of Water, Electrolytes and Acid-Base Syndrome*. Philadelphia, Lea & Febiger, 1975.
9. Greger R, Schlafler E, Lang F: Evidence for electroneutral sodium chloride cotransport in the cortical thick ascending limb of Henle's loop of rabbit kidney. *Pfluegers Arch* 1983; 396:308–314.
10. Koushanpour E: *Renal Physiology: Principles and Functions*, ed 1. Philadelphia, WB Saunders Co, 1976.
11. Manery JF: Water and electrolyte metabolism. *Physiol Rev* 1954; 34:334–417.
12. Pace N, Rathbun EN: Studies on body composition. III. The body water and chemically combined nitrogen content in relation to fat content. *J Biol Chem* 1945; 158: 685–691.
13. Pitts RF: *Physiology of the Kidney and Body Fluids*, ed 3. Chicago, Year Book Medical Publishers, Inc, 1974.
14. Rathbun EN, Pace N: Studies on body composition. I. The determination of total body fat by means of the body specific gravity. *J Biol Chem* 1945; 158:667–676.
15. Ruch TC, Patton HD: *Physiology and Biophysics*, ed 20. Philadelphia, WB Saunders Co, 1974.
16. Schultz S: Sodium-coupled solute transport by small intestine: A status report. *Am J Physiol* 1977; 233:E249–E254.

17. Schultz AL, Hammarsten JF, Heller BI, et al: A critical comparison of the T-1824 dye and iodinated albumin methods for plasma volume measurement. *J Clin Invest* 1953; 32:107–112.
18. Skelton H: The storage of water by various tissues of the body. *Arch Intern Med* 1927; 40:140–52.
19. Solomon AK: Equations for tracer experiments. *J Clin Invest* 1949; 28:1297–1307.
20. Walser M, Seldin DW, Grollman A: An evaluation of radiosulfate for the determination of the volume of extracellular fluid in man and dogs. *J Clin Invest* 1953; 32:299–311.

3

Body Fluids: Turnover Rates and Dynamics of Fluid Shifts

In the previous chapter, we learned about the normal volumes and compositions of body fluids and their distribution between various compartments. With this background, we are now in a position to consider the dynamics of fluid balance and the factors controlling their turnover, distribution, and shift between body fluid compartments.

External Fluid Exchange

Maintenance of the normal balance of volume and composition of body fluids depends on the regulation of the influx and efflux quantities. Recalling the flow diagram of the renal-body fluid regulating system in Chapter 1, we see that the exchange of fluid between the body and the environment, or the external fluid exchange, occurs through several channels. On the influx side, the gastrointestinal and metabolic systems are the only avenues of normal fluid intake. On the efflux side, fluids leave the body via the gastrointestinal system, lungs, skin, and kidneys. Let us now examine the functional characteristics of each of these influx and efflux routes.

Gastrointestinal System

The gastrointestinal (GI) system plays a dual role in fluid and electrolyte homeostasis. It is the most important source of fluid influx as well as providing a route for fluid efflux. Oral intake is the primary source of water and electrolytes; daily ingestion normally is 2.5 L of water and 7 g of salt. Including the metabolic water, the daily fluid turnover for a healthy young man is approximately 2.8 L. In addition, over 8 L of fluids are secreted into and reabsorbed from the GI system daily. This is more than twice the total plasma volume and is equal to more than half the total volume of extracellular fluid. These secretions are all isotonic with the blood plasma but vary somewhat in composition. Also, about 1,000 mEq of sodium is secreted daily into the GI tract. This represents about one third of the total exchangeable sodium and about six times the daily dietary intake of sodium.

Minor daily fluctuations in body weight are largely due to slight variations in the hydration of the body, particularly the extracellular fluid volume. In a 24-hour period, in addition to intake, a considerable amount of fluid is exchanged between the GI tract and the extracellular compartment. The fluid turnover in the GI tract amounts to 20% to 25% of the total body water. Hence, alteration in normal fluid intake and abnormal losses of GI secretions could precipitate profound changes in water and electrolyte balance. For instance, excessive loss of GI secretions, as in diarrhea or vomiting, rapidly decreases the volume of extracellular water, with concomitant loss of sodium, potassium, and associated anions. This results in a fall of the extracellular crystalloid osmotic pressure, a decrease in the plasma volume, and eventual disruption of cellular function. For these reasons, abnormal loss of fluids from the GI system leads to rapid alteration in fluid balance and must be replaced promptly.

Metabolic System

Oxidation of food provides a secondary but important source of water. The ordinary mixed diet will yield approximately 300 mL of metabolic water in 24 hours. Of the various constituents of the diet, oxidation of

100 g each of protein, fat, and starch yields 41 g, 107 g, and 55 g of water, respectively, whereas oxidation of 100 g of alcohol yields 117 g of water.

In addition to water, the metabolic system produces urea, a nontoxic product of protein metabolism, and other toxic wastes, which are eliminated by the kidneys.

Skin and Lungs

Skin plays a highly important role in temperature regulation, a function that, on occasion, may lead to considerable loss of fluids. Hence, the fluid efflux from the skin poses as a possible disturbance forcing in the renal-body fluid regulating system.

A nonperspiring young man in basal state loses approximately 30 g of water per hour by insensible perspiration. This, together with the water lost from the lungs, constitutes the *insensible loss* of water. A healthy nonperspiring adult in a resting state will have an insensible moisture loss of 800 to 1,400 mL/24 h, depending on body surface area and metabolic rate. Feverish states will increase the insensible loss of moisture according to the extent to which metabolic rate is augmented. The greater the surface area in relation to mass, the greater will be the relative turnover of fluid in relation to the total fluid content of the body. For example, the daily turnover of fluid by a baby will represent a far higher proportion of its total body water than for an adult. The small reserve in comparison to the large daily turnover of the infant explains why dehydration may develop so rapidly at this time in life.

Sensible sweat is a distinctly *hypotonic* solution, with a salt concentration ranging from 10 to 70 mEq/L of H_2O. Thus, moderate sweating tends to deplete body water in excess of salt depletion. However, with continuous activity, the loss of sodium and chloride may reach proportions detrimental to normal body fluid homeostasis. Persons doing hard labor in a hot environment may lose 10 to 12 L of hypotonic fluid per 24 hours. The excessive salt lost during hard labor may lead to a condition called *heat stroke*. If the lost fluid is replaced with salt-free water, another condition, known as *water intoxication*, may develop.

Kidney

Normally, the kidneys excrete about 1,500 mL of *hypertonic* urine in 24 hours. This represents about 60% of the normal daily fluid turnover. In addition, the kidneys excrete 30 g of urea per day, which is produced from deamination of amino acids. The obligatory minimal daily urine volume, for normal kidneys that must excrete an average load of solutes, is approximately 500 mL. Adding this volume to about 1 L of daily insensible water loss yields a minimal daily obligatory water loss of 1.5 L. Thus, to maintain normal fluid balance a minimum daily fluid intake of 1.5 L is necessary.

The ability of the kidney to adjust its excretion of water and solutes makes this organ the most important of all the avenues of fluid efflux. Unlike the GI system, skin, and lungs, the primary function of the kidneys is to conserve water and electrolytes, thereby stabilizing the volume and composition of body fluids. For instance, in diarrhea and vomiting, the GI system fails to reabsorb the secreted fluids and to retain the ingested fluids. During dehydration of any degree, the skin and lungs continue to lose water by vaporization. In fact, in metabolic acidosis the increased rate and depth of breathing (respiratory compensation, see Chapter 11) result in considerable water loss by vaporization. In contrast, in the face of sodium depletion or lack of intake, the kidneys virtually cease excreting this ion. Similarly, in water deprivation, the kidneys excrete the urinary solutes in the smallest possible volume of water. In short, by adjusting its excretion rate of water and electrolytes, the kidney actively stabilizes the body stores of these substances and indirectly stabilizes body fluid homeostasis.

One way to assess this ability of the kidneys to adjust their excretion rate is to measure urine *osmolality*. The most direct method of measuring osmolality is by an osmometer, an instrument that measures the freezing point depression of the sample and relates it to osmolality. However, because such an instrument is not widely available, the ease of measuring specific gravity has made it a commonly used substitute clinical test.

Specific gravity of urine relates the weight of equal volumes of the urine in question and distilled water under standard conditions. Hence, it reflects the *weight* of the solute in the urine and is not a true measure of the *number* of solute particles present (osmolality), as would be obtained by measuring the freezing point depression or vapor pressure elevation. Nevertheless, the specific gravity correlates with the osmolality and so it is a useful clinical tool. It indicates the quantity of water relative to solute removed, thus providing a means of testing the concentrating and diluting ability of the kidney.

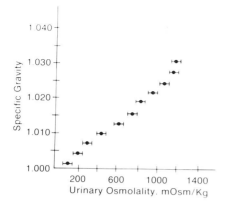

FIGURE 3.1. Relationship between urinary specific gravity and osmolality. (From Chapman, et al.[3])

Figure 3.1 shows the relationship between the specific gravity and urine osmolality. Note that the relationship is not linear and that at a given osmolality, there is considerable variation in the measured specific gravity. However, to obtain a useful correlation between these two quantities, the mean values read from this graph were fitted by least-squares method to the following equation:

Urinary specific gravity = 1.000 +
[25 × 10^{-6} × (Urinary osmolality, mOsm/L)] (3.1)

Thus, if the urinary specific gravity is known, we can estimate the total urinary solute output (osmolality) from this equation. For example, suppose that the average urinary specific gravity in a 24-hour urine sample is 1.020 and that the 24-hour urine output is 1,200 mL. Then, the total urinary solute output would be 800 mOsm/L × 1.2 L/24 hour = 960 mOsm/24 hour.

Internal Fluid Exchange

The continuous flow and turnover of water and solutes between the circulating blood and the metabolizing cells is called the *internal fluid exchange*. The circulation of this internal fluid takes place across two separate membranes, each with its own specialized exchange processes. First, there is an enormous bidirectional flow of water and solutes, with each cardiac output, across the capillary endothelium and its basement membrane, between the plasma and interstitial compartments. This is called the *capillary fluid exchange*. The capillary filtrate thus formed is the source of the interstitial circulation. Second, there is an enormous bidirectional flow of water and solutes across the cell membrane, between the interstitial and intracellular compartments. This is called the *cellular fluid exchange*. Although both of these exchange processes operate concomitantly to serve the needs of various cells and tissues, the mechanisms and forces involved in each process are different and have different origins. Let us now consider these two exchange processes and their mechanism of operation.

Capillary Fluid Exchange

In the steady state, defined as unchanging fluid balance between the plasma and interstitial compartments, water and electrolytes move continuously between the two compartments, in both directions and at equal rates.[17] This total fluid exchange flux has two components: (a) an enormous bidirectional diffusive flux of water and electrolytes, and (b) a small bidirectional convective (filtration) flow of fluid.

The diffusive flux of fluid is estimated at 120 L/min. This means that the 3.2 L of plasma water is turned over once every 1.6 seconds by this process alone. The *turnover time* for the 8.4 L of interstitial water is about 4.1 seconds. It is this enormous diffusive flux that allows the plasma and interstitial fluids to reach diffusion equilibrium in the capillary exchanger (capillaries and venules)—a condition so necessary for the rapid exchange of respiratory gases.

This bidirectional diffusive flow depends primarily on the physical properties of the capillary wall (capillary endothelium and its basement membrane) and the size and the charge of the solute molecules. The lipid-soluble substances, such as O_2 and CO_2, can diffuse freely across the entire surface of the capillary membrane. However, only 0.2% of the capillary surface is available for diffusion of water and lipid-insoluble molecules, such as sodium, chloride, urea, and glucose. Because of the high turnover rate of both the lipid-soluble and lipid-insoluble molecules, a *sieving process* has been suggested as a mechanism of exchange. The limiting factors in the sieving process are thought to be the size of the molecules, the size of the capillary pores, and the rate of water filtration. Under normal physiological conditions, it appears that both diffusion and the sieving process account for the high rate of exchange.

In contrast, the bidirectional filtration flow of water and solutes across the capillary is small, only about 16 mL/min. This means that the plasma water is

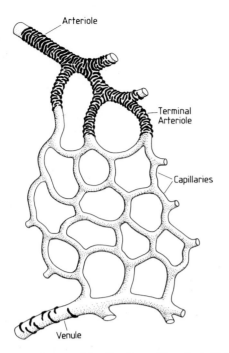

FIGURE 3.2. Structural details of a capillary bed. (Adapted from Wiedemann, et al.[23])

turned over once every 200 minutes and the interstitial water once every 525 minutes by this process alone. It is the circulation of this capillary transudate that provides the essential nutrients for the maintenance of the normal cellular metabolism. Although both diffusive and convective fluxes occur concomitantly, the physiological significance of the filtration flow is of special interest, because it is primarily concerned with *volume shift* between the intravascular and extravascular fluid compartments. Whenever the bidirectional components of this flow are unequal, a transient is initiated, which will shift fluid flow from one compartment to the other. This transient may lead to a new steady state in which the two fluid volumes have changed, but are no longer changing. Such shifts of fluid are important in stabilizing the more critical blood volume, at the expense of the interstitial volume, as occurs in hypovolemia and hypervolemia. If this fluid shift mechanism is disturbed, the interstitial volume may become abnormally large, a condition called *edema*. For these reasons, we shall now examine in more detail the mechanisms of volume shift by filtration in the steady state of normality.

As illustrated in Figure 3.2, the vessels that establish the microcirculation consist of the arteriole, the terminal arteriole, the capillaries (including the postcapillary venules), and the venules. Terminal arterioles are defined as the ultimate branches of arterioles that terminate in a capillary network without connections to any other type of vessel.[23] The terminal arterioles are the site of the final muscular investment which is often called the precapillary sphincter (precapillary sphincter area). Beyond this point the vessel continues as a pure endothelial tube, as a capillary.

Blood flow through capillaries is not uniform, but is dependent on rhythmic contraction of the preceding terminal arterioles (and possibly arterioles). The contraction of smooth muscles in these vessels produces a periodic opening and closing of the arteriolar-capillary communication (called *vasomotion*), resulting in an intermittent flow of blood through the capillary bed.[24] At rest, the average blood flow velocity in capillaries is about 1 mm/s, allowing sufficient time for equilibration of water and solute concentrations across the capillary membrane.

At the venous site, capillaries converge to form the venules where smooth-muscle cells gradually reappear. Sometimes the first joinings of capillaries, which are slightly larger in diameter than the parent capillaries but are still devoid of any encircling smooth-muscle cells, are called postcapillary venules; like the capillaries they serve as exchange vessels. Regarding the kidney, even large venules and small veins do not have any muscular sheet and—in addition to their collecting function—they serve as capillaries. This specificity of the kidney agrees with the general observation that—unlike the boundary between the arterioles and capillaries which is marked by the disappearance of the muscle tissue—the transition from capillary to venule is always gradual and cannot be defined as well.

As a consequence of this anatomical arrangement of the microvasculature, the turnover of fluid across the capillary wall, in the steady state, is determined by (a) the cardiovascular hemodynamics, which determine the arteriovenous hydrostatic and oncotic pressure differences, (b) the temporal or periodic opening and closing of the precapillary sphincters, which modify the volume of blood flowing through the capillary exchanger, (c) the permeability properties of the capillary wall, and (c) the lymphatic system, which plays a significant role in returning plasma proteins from interstitium back to the circulating plasma. Let us now consider how each of these factors influence the fluid shift by filtration across the capillary.

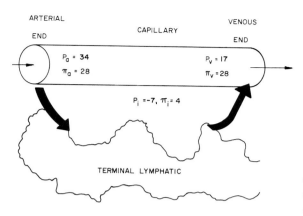

FIGURE 3.3. Schematic diagram of a capillary bed. (Redrawn and slightly modified from Deane.[6])

Hemodynamic Factors

Mechanism of Tissue Fluid Formation. The net shift of fluid by filtration across the capillary and its steady state distribution between the intravascular and extravascular compartments as first described by Starling[22] depend, in part, on the algebraic sum of four pressures: (1) the capillary hydrostatic pressure, P_c; (2) the interstitial hydrostatic pressure, P_i; (3) the capillary protein oncotic pressure, π_c; and (4) the interstitial protein oncotic pressure, π_i. The protein oncotic pressure is the sum of the protein osmotic pressure and the osmotic pressure of the bound cations (Gibbs-Donnan effect). This is why the protein oncotic pressure increases disproportionately with plasma protein concentration.

To better understand how each of these hemodynamic factors influences the capillary fluid exchange, consider the simplified diagram of a capillary bed shown in Figure 3.3. Although, in reality the values of the indicated pressures vary somewhat, for ease of presentation, we have only chosen their midrange values.

As the blood flows through the capillary, outward filtration of fluid causes a decrease in the capillary hydrostatic pressure from a value of 34 mm Hg at the arterial end (P_a) to a value of 17 mm Hg at the venous end (P_v). The outward filtration is further enhanced by a relatively constant interstitial protein oncotic pressure (π_i) of 4 mm Hg. Thus, in the steady state there is a net driving force of 21 mm Hg (34 − 17 + 4), which favors outward filtration of fluid across the capillary. Concomitantly, as the outward filtration proceeds, the capillary protein oncotic pressure (π_c) progressively rises from a value of 28 mm Hg at the arterial end (P_a) to a maximum value near the midcapillary region. As the blood flows toward the venous side, the elevated π_c causes an inward filtration of fluid, thereby restoring π_c to its arterial value by the time the blood reaches the venous end. Thus, in the steady state the capillary protein oncotic pressure provides a net reabsorptive force of 28 mm Hg, favoring inward filtration of fluid back to the blood. The magnitude of this oncotic pressure, primarily due to plasma albumin and globulin, is small compared with about 5,100 mm Hg pressure developed by the crystalloids of plasma. However, in contrast to crystalloids, because proteins normally do not cross the capillary wall in large amounts, the physiological significance of the plasma protein oncotic pressure is out of proportion to its small value.

In the steady state the algebraic sum of the forces favoring outward and inward filtration yields a value of −7, which is exactly equal to the interstitial hydrostatic pressure (P_i) of −7 mm Hg required to balance the other forces. The physiological significance of this *negative* interstitial hydrostatic pressure is discussed later in this chapter.

Thus, we see that as the blood enters the aterial end of an exchanging capillary, the algebraic sum of all the hydrostatic pressures (34 + 7 + 4 − 28) yields an *effective outward filtration pressure* (ΔP) of 17 mm Hg, which forces a protein-poor fluid out of the capillary into the interstitium. As this *outward filtration* continues, the capillary hydrostatic pressure falls while its protein oncotic pressure rises. By the time the blood reaches the venous end of this capillary, the effective filtration pressure has *reversed* its direction, so as to force the interstitial fluid back into the capillary. This hydrostatic pressure reversal accounts for the return of 90% of the filtered fluid back into the blood. The remaining 10% is returned to the circulation via the lymphatic system, whose function is discussed below. Thus, if 16 mL/min is filtered out in the arterial half of the capillaries, about 14.5 mL/min is reabsorbed in the venous half; the remaining 1.5 mL/min is returned to the blood via the lymphatic system.

What factors give rise to this effective filtration pressure, and which influence its direction reversal along the capillary? The effective filtration pressure is the algebraic sum of several pressures acting in opposite directions, which are either hydrostatic or colloid osmotic (oncotic) in origin. The main hydrostatic pressure is the capillary blood pressure.

What are the components of this effective pressure, and why does its direction (algebraic sign)

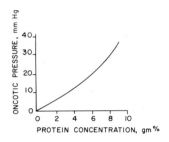

FIGURE 3.4. Relationship of plasma protein oncotic pressure to plasma protein concentration. Data were fitted, via least-squares forced through the origin, to the following equation:

$$\pi_p = 2.1[Pr]_p + 0.16[Pr]_p^2 + 0.009[Pr]_p^3$$

(Redrawn from Landis and Pappenheimer.[14])

reverse along the capillary? The pressure component that acts in the direction of *outward filtration* is *hydrostatic* in origin; it is a *net* hydrostatic pressure defined as the excess of the intracapillary pressure (34 mm Hg) over the smaller (negative) interstitial fluid pressure (− 7 mm Hg). This net pressure (41 mm Hg) decreases as the blood pressure decreases along the length of the capillary. The pressure component that acts in the direction of *inward filtration*, or absorption, is *oncotic* (colloid osmotic) in origin; it, too, is a net pressure defined as the excess of plasma oncotic pressure (28 mm Hg) over the smaller interstitial fluid oncotic pressure (4 mm Hg). This net pressure (24 mm Hg) increases along the capillary, as a result of outward filtration of protein-poor filtrate.

We now have two oppositely directed net pressures whose algebraic sum produces the effective pressure. The net *hydrostatic, filtration* pressure decreases along the capillary, while the net *oncotic, absorptive* pressure increases along the capillary. Toward the arterial end the excess of filtration over absorption pressures induces outward filtration; toward the venous end the excess of absorption over filtration pressure induces inward filtration, or absorption. In the steady state the filtration rate and the absorption and lymphatic return rates are equal, so that no shift of fluid volume between plasma and interstitium occurs.

An important factor contributing to this fluid balance is the protein oncotic pressure of the plasma proteins. As mentioned above and illustrated in Figure 3.4, the plasma protein oncotic pressure (π_p) increases disproportionately with plasma protein concentration ($[Pr]_p$). The equation fitted to the data

in Figure 3.4 is valid for rat as well as human plasma.[1] The physiological significance of this nonlinear relationship is that at low plasma protein concentration, such as exists in the interstitial fluids, the oncotic pressure increases linearly with protein concentration. However, at normal plasma protein concentration the oncotic pressure rises markedly with small increases in protein concentration.

Hence, because of this nonlinear relationship, loss of protein-free fluid from the plasma results in a greater rise in oncotic pressure and thus a greater inward filtration force than would be expected. Conversely, the gain of fluid would have a greater effect in diluting the plasma and in reducing the oncotic pressure than would be expected.

From the foregoing analysis, we see that the net shift of fluid from plasma into the interstitium, designated as $\dot{Q}_{p \rightarrow i}$, is largely determined by the magnitude of the net filtration forces, ΔP (the algebraic sum of the hydrostatic pressure and protein oncotic pressures). This relationship may be expressed as,

$$\dot{Q}_{p \rightarrow i} = k_f (P_c - P_i - \pi_c + \pi_i) = k_f \cdot \Delta P \quad (3.2)$$

where k_f is the capillary filtration coefficient and the other symbols have already been defined. From this relationship, it is evident that any factor that causes an increase in the capillary hydrostatic pressure or a decrease in the plasma protein concentration, and hence in π_c, favors increased outward filtration (from plasma to interstitium). Likewise, any factor that causes a decrease in P_c or an increase in π_c tends to reduce outward filtration, and hence increases inward filtration (from interstitium into plasma).

Pressure-Flow Relationship in a Capillary. To appreciate further the influence of changes in pressures and resistances of the arteriole and venule on the capillary pressure, and hence capillary filtration, consider the idealized capillary model shown in Figure 3.5. According to this scheme, the magnitude of the capillary hydrostatic pressure (P_c) assuming a normal plasma and tissue protein oncotic pressure and tissue hydrostatic pressure, is determined by the arterial (P_a) and venous (P_v) pressures and their respective resistances to blood flow. This dependence may be expressed quantitatively if we first write defining equations for precapillary and postcapillary resistances (R_a and R_v, respectively), and then combine these equations to obtain a defining equation for P_c.

The precapillary resistance, analogous to Ohm's law in electricity, may be defined as the ratio of the

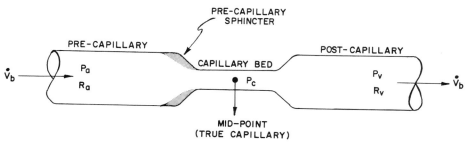

FIGURE 3.5. An idealized scheme of the precapillary, capillary, and postcapillary vascular tree. (From Koushanpour.[13])

hydrostatic pressure drop across the precapillary vessel to the blood flow, $\dot{V}_b$:

$$R_a = \frac{P_a - P_c}{\dot{V}_b} \qquad (3.3)$$

A similar equation may be written for the postcapillary resistance:

$$R_v = \frac{P_c - P_v}{\dot{V}_b} \qquad (3.4)$$

Combining Equations 3.3 and 3.4, eliminating the $\dot{V}_b$ term and solving for P_c, we get

$$P_c = \frac{\dfrac{R_v}{R_a} P_a + P_v}{1 + \dfrac{R_v}{R_a}} \qquad (3.5)$$

From this equation it is evident that for given arterial and venous pressures, the mean capillary hydrostatic pressure depends on the ratio of the postcapillary to precapillary resistances to blood flow. Rearranging Equations 3.3 and 3.4, in terms of P_c, yields

$$P_c = P_a - R_a \cdot \dot{V}_b \qquad (3.6)$$
$$P_c = P_v + R_v \cdot \dot{V}_b \qquad (3.7)$$

Equations 3.6 and 3.7 have the form of a linear equation in which P_c is the dependent variable, $\dot{V}_b$ is the independent variable, R_a and R_v are the slope, and P_a and P_v are the ordinate-intercept that become equal to P_c when $\dot{V}_b$ is zero.

Using the theoretical Equations 3.6 and 3.7, Pappenheimer and Soto-Rivera estimated the capillary hydrostatic pressure in the isolated perfused hindlimbs of cats and dogs by *isogravimetric* method.[16] The experimental procedure consisted of stepwise elevation of the venous pressure to compensate for the stepwise decrement in the arterial pressure, caused by outward fluid filtration, until the limb weight remained constant. When this was achieved, there were no further pressure drops across the precapillary, capillary, and postcapillary vasculatures, and therefore no changes in flow. Hence, the arterial and venous pressures became equal to the capillary hydrostatic pressure. This pressure was called the *isogravimetric* capillary pressure, P_{iso}. Since during the isogravimetric state there was presumably no net shift of fluid across the capillary membrane, P_{iso} was equal to the sum of all pressures opposing outward fluid filtration. Hence, Equation 3.2 may be written as

$$\dot{Q}_{p \to i} = k_f(P_c - P_{iso}) \qquad (3.8)$$

where $P_{iso} = (P_i + \pi_c - \pi_i)$ and k_f has already been defined.

Experimentally, Pappenheimer and Soto-Rivera altered the blood flow into the perfused hindlimb by varying the arterial pressure followed by compensatory adjustment of the venous pressure until the limb weight became constant.[16] The value of P_c was then estimated from P_{iso} by plotting changes in the arterial or venous pressures against the changes in blood flow. They found that a change in venous pressure of 0.5 mm Hg caused a detectable effect on fluid movement across the capillaries, while a similar fluid shift could be seen only after a 2 to 4 mm Hg change in the arterial pressure. Furthermore, the capillary pressure was found to be more sensitive to a change in venous pressure than to a change in arterial pressure. Also, the postcapillary resistance (R_v) was independent of the changes in the blood flow, while the precapillary resistance (R_a) increased monotonically as the blood flow decreased. In addition, the isogravimetric method allowed measurement of the capillary filtration coefficient (k_f).

Figure 3.6 summarizes the effects of the hemodynamic factors on the net fluid shift across the cap-

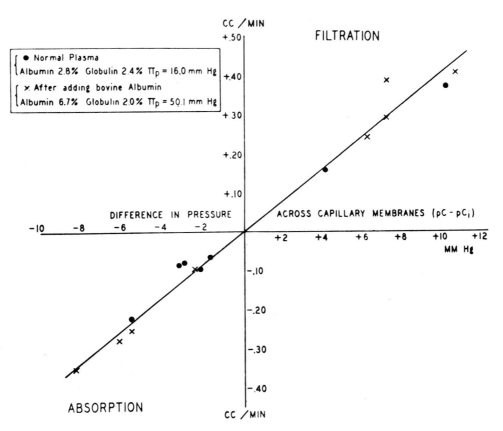

FIGURE 3.6. Relation of net fluid shift in perfused hindlimb of cat to the difference between mean capillary hydrostatic pressure, pC (= $P_c - P_i$), and the sum of all pressures opposing filtration, pC$_i$ (= P_{iso}). (From Pappenheimer and Soto-Rivera.[16])

illary membrane. It depicts the net fluid movement, i.e., filtration or absorption, as a function of the hydrostatic pressure gradient across the capillary membrane. We see that both fluid filtration and absorption are directly proportional to the difference between the calculated mean capillary blood pressure and the isogravimetric capillary pressure. The filtration coefficient, obtained from the slope of the plotted regression line, was 0.014 mL/100-g tissue/mm Hg change in capillary blood pressure. The constancy of this filtration coefficient at low and high capillary pressures indicated that under these conditions, the capillary porosity and surface area were not influenced by blood pressure. These findings have been confirmed by applying a similar method to the hindlimbs of rats.[19]

Significance of Negative Interstitial Pressure. The finding that the isogravimetric capillary pressure equals the sum of all pressures opposing filtration was based on the assumption that the interstitial hydrostatic pressure remained unchanged and was unaffected by the experimental forcings. This, however, was found not to be the case by Guyton who was able to measure the interstitial pressure directly in similar experiments.[9, 10] Guyton found that contrary to expectations, the interstitial pressure was not normally positive, but rather ranged between -6 to -7 mm Hg and rose to positive values only in edematous conditions. He also observed that the interstitial pressure was markedly influenced by intravenous fluid infusion as well as changes in arterial and venous pressures.

To ascertain the physiological significance of the normally observed negative (subatmospheric) interstitial pressure, Guyton determined the pressure-volume relationship of the interstitial space as follows. He perfused the isolated hindlimbs of dogs with 10% dextran (a plasma volume expander), while simultaneously measuring the changes in the interstitial

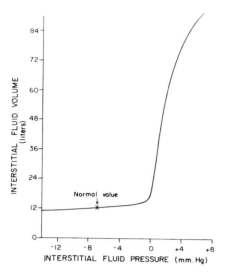

FIGURE 3.7. Pressure-volume curve of interstitial spaces. The curve is extrapolated to humans from data obtained in dogs. (From Guyton.[11])

pressure and hindlimb weight. The interstitial pressure was measured via a needle catheter inserted into previously implanted perforated plastic spheres. Then, assuming that changes in limb weight reflect changes in the interstitial volume, Guyton constructed pressure-volume curves for the interstitial space. Figure 3.7 depicts such a curve, which is extrapolated to humans from data obtained in dogs. The slope of such a curve, defined as the ratio of the incremental increase in interstitial volume (ΔV) to incremental increase in interstitial pressure (ΔP), depicts the interstitial *compliance* ($\Delta V/\Delta P$). We can see that the interstitial compliance is quite low at normal negative interstitial pressures, but it increases markedly as the interstitial pressure rises to positive values.

The pressure-volume curve, thus obtained, reveals two distinct operating regions for the interstitial space: (1) a low-compliance region representing the normal pressure-volume changes and (2) a high-compliance region reflecting the adaptive behavior of the interstitial space to abnormal pressure-volume changes.

According to Guyton, the normally negative interstitial pressure serves two important physiological functions: (1) It creates a partial vacuum environment that serves to hold the different body tissues together, thereby complementing connective tissue function. When interstitial pressure increases to positive values, fluid accumulates in the interstitial spaces with subsequent edema formation. (2) It also serves to minimize the tissue space across which the nutrients and metabolic waste must traverse. Since nutrient-metabolic waste exchange between blood and cells occurs primarily by diffusion, the compactness of tissue spaces will enormously enhance this exchange.

Temporal and Spatial Factors

The periodic opening and closing of the precapillary sphincters produces phasic changes in capillary hydrostatic pressure, allowing the plasma oncotic pressure to predominate intermittently. Hence, an increase in the period during which sphincters are closed tends to decrease capillary blood pressure relative to capillary oncotic pressure. The resulting hydrostatic pressure gradient favors inward filtration of fluid from the interstitial space to the intravascular compartment. Conversely, an increase in the period during which sphincters are open favors outward filtration of fluid into the interstitium.

At least three conditions are known to increase the period during which the sphincters are closed: (1) hemorrhage, (2) sympathetic stimulation and (3) epinephrine injection. The net effect of each of these stimuli is to increase fluid absorption into the capillary, thereby producing *hemodilution*, as manifested by a reduced hematocrit.

At least three conditions are known to increase the period during which the sphincters are open: (1) a rise in body temperature above normal, (2) direct trauma, such as surgery, and (3) prolonged anoxia. The net effect of each of these stimuli is to increase outward filtration of fluid into the interstitium, thereby producing *hemoconcentration*, as manifested by an elevation of hematocrit.

In short, any factor that alters the arteriolar pressure or interferes with free venous outflow affects the process of fluid exchange. Thus, arteriolar dilation or venous obstruction favors filtration from the intravascular to the interstitial compartment. The force that opposes this outward filtration, when protein oncotic pressure is constant, is the interstitial hydrostatic pressure. This pressure depends on the lymphatic flow, the amount of interstitial fluid, the volume of the interstitial space, and the permeability of the capillary membrane. The abnormal operation of any of these factors favors the accumulation of edema fluid.

Permeability Properties of the Capillary Wall

The capillary wall is established by the capillary endothelium resting on a basement membrane. The luminal membrane of the endothelium is generally covered by a cell coat (glycocalix) that is negatively charged. On the basis of the ultrastructure of the endothelium, three main types of capillaries have been described[21]: *continuous, fenestrated,* and *discontinuous* (sinusoid). The fenestrated capillaries are subdivided into those in which the fenestrae are covered by a diaphragm (such as the peritubular capillaries of the kidney) and those in which the fenestrae are not covered by a diaphragm and are in effect open (the best known example is the glomerular capillary).

In the following discussion of the capillary permeability, we will exclude considerations of discontinuous capillaries (sinusoid) because their endothelium is quite leaky and offers no impediment for transcapillary movement of solutes. In addition, we will exclude discussion of glomerular capillaries because their endothelium is part of the glomerular filter, a subject that is fully discussed in Chapter 5.

The transcapillary movement of solutes across the *continuous* and *fenestrated* (covered with diaphragm) capillaries is in many respects similar to the transport across an epithelium. Thus, like the epithelial transport, solute and water transport across a capillary endothelium involves both the transcellular and paracellular pathways (for further details see Chapter 8). In endothelia, however, characteristic transcellular routes are provided by plasmalemmal vesicles, vesicle channels, and fenestrae. These routes exclude any transport across cell membranes and may therefore be classified as "extramembranous" transcellular transport routes. Vesicular transport is characteristic of endothelia and is generally of only minor importance in epithelia.

As depicted in Figure 3.8, transendothelial transport of water and solutes involves four major pathways: the transcellular transport route, subdivided into a transmembranous and two extramembranous pathways, and the paracellular transport route.

The *"transmembranous transport"* route (Figure 3.8a), which is the hallmark of transepithelial transport, involves transport across the luminal membrane, through the cytoplasm, and finally across the basolateral membrane. As in epithelia, the two plasma membranes establish the main barriers along this transport route. It may be assumed that as in epithelia, the endothelial plasma membranes are permeable to water, small nonpolar solutes, and lipid-soluble molecules. Little is known about the contribution of active transmembranous transport steps to the overall transendothelial transport. The unusually high permeability to water, found in capillary endothelia, appears to be due to the extreme attenuation of the capillary wall, including the formation of fenestrae.

The second type of transcellular transport of solutes and water (Figure 3.8b) involves *shuttling of membrane-bound vesicles (plasmalemmal vesicles)* through the cytoplasm from the luminal to the basolateral site or in the opposite direction. This type of vesicular transport is also called *"transcytosis."* The plasmalemmal vesicles are obviously not formed by invagination of the luminal or basolateral plasma membrane,[21] but it is suggested that they represent a separate population of membranous organelles that shuttle and temporarily fuse with the two opposite cell membranes. Vesicles may also fuse with each other and form patent transendothelial channels, through which plasma and interstitial fluid can communicate directly. Like the fenestrae (see below), the openings of the plasmalemmal vesicles are mostly closed by a thin nonmembranous (i.e., no lipid bilayer) diaphragm. In contrast to the fenestral diaphragms, the vesicle openings are not negatively charged and therefore allow the uptake and transcytosis of polyanionic and polycationic macromolecules, such as plasma proteins.

The third type of transcellular transport involves movement of solutes and water through *fenestrae* (Figure 3.8c). It is obvious that this type of transport can occur only across the fenestrated capillaries. The only structural barriers in this type of transport are the fenestral diaphragms. They lack a lipid layer, are hydrophillic, and have a high density of negative charge. Fenestrae have often been regarded as fixed vesicular channels that are only bridged by one single diaphragm. The fenestrae are obviously part of the hydrophillic transendothelial transport system represented by vesicles, channels, fenestrae, and diaphragms. There are, however, functional differences. Vesicles, channels, and their uncharged diaphragms appear to be instrumental in transporting proteins (polyanions). Fenestrae and their negatively charged diaphragms are associated with a high permeability to water and small solutes. Thus, differences between the continuous and the fenestrated type of capillaries concern important functional parameters. Capillaries, established by a continuous endothelium, owing to the high quantity of plasmalemmal vesicles, are highly permeable to macromolecules such as albumins. This is in contrast to the fenestrated type of capillaries, which have a limited capacity for vesicular transport. On the other hand, the fenestrated cap-

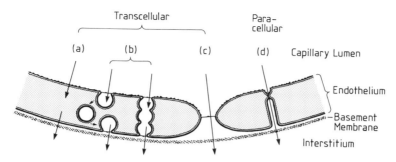

FIGURE 3.8. Schematic representation of the capillary endothelium showing four major pathways for water and solute transport through the capillary wall: (a) transmembranous pathway; (b) vesicular transport route including the formation of patent channels; (c) transfenestral pathway; and (d) paracellular pathway through tight junction and lateral intercellular spaces. The wavy line on the luminal surface depicts the negative charge covering the luminal membrane of the endothelium. Although not shown in the figure, the fenestral diaphragms are strongly negatively charged toward the luminal site. In contrast, the thin diaphragms covering the openings of vesicles are devoid of negative charges. (Modified from Simionescu.[21])

illaries are highly hydrophillic and therefore are well suited for the transport of water, electrolytes, and all other small solutes.

The *paracellular transport* route (Figure 3.8d) is another pathway for transport of water and solutes across the endothelium. As in epithelia, the major barrier along this route is represented by the tight junction. Tight junctions of capillary endothelia may be found to be differentlly elaborated in different capillary beds, but generally they are of the "leaky type."[20] It appears that in the continuous capillaries the presence of a paracellular route may be more important than in the fenestrated capillaries. Detailed knowledge about the contribution of the paracellular pathway to water and solute transport across endothelia is so far lacking.

In summary, capillary endothelia, unlike other simple epithelia, have an unusually high permeability for water and solutes, including macromolecules, such as proteins and lipoproteins. Capillary endothelia permeability coefficients for most substances and hydraulic conductivity are many times higher than those of simple epithelia.[18]

Lymphatic System

The lymphatic system provides a normal route for the return of the plasma proteins that have been filtered out of the capillary back into the blood. The amount of proteins leaking out of the capillary in 24 hours is estimated to be equal to that in the circulating blood itself. The normal passage of plasma proteins (both albumin and globulin) through the capillary wall is very important for cellular metabolism and for defense against infection. Impairment of lymphatic drainage leads to excessive fluid accumulation in the interstitial space, with high protein content. A rise in interstitial protein concentration may also result from damage to capillary endothelium from burns or hypoxia.

The normal leakage of a small but potentially significant amount of protein into the interstitial space yields a filtrate protein concentration of about 0.2 to 0.4 g%. This is somewhat less than the interstitial protein concentration of about 2 g% and the lymph protein concentration of about 4 g%. This interstitial protein concentration yields an oncotic pressure of about 4 to 5 mm Hg. Depending on the protein concentration, the interstitial oncotic pressure ranges from a low value of 0.1 mm Hg in the limb interstitium to a high value of 5 mm Hg in the interstitial fluids of intestine and liver.[15]

The quantity of proteins in the interstitial space is determined by the extent of the lymphatic flow, normally about 100 mL/h at rest. The progressive accumulation of protein in the interstitial space increases the oncotic pressure, causing an expansion of the interstitial volume at the expense of a reduced fluid absorption by the capillary. The expansion of interstitial volume increases the interstitial hydrostatic pressure, forcing the interstitial fluid into the lymphatic channels, carrying with it the excess accumulated plasma proteins. The overall effect is the restoration of the capillary hemodynamics to normal. In short, an increase in the tissue fluid proteins increases the rate of lymph flow, and this, by washing the proteins out of the tissue spaces, automatically returns the interstitial protein concentration back to the normal low level.

The lymphatic drainage is increased by factors that elevate either the interstitial hydrostatic pressure (P_i) or the lymphatic flow rate. The interstitial pressure may be increased by one or a combination of the following: (1) an increase in the capillary hydrostatic pressure (P_c), (2) a decrease in the plasma protein oncotic pressure (π_c), and (3) an increase in the permeability of the capillary wall. The lymphatic flow rate may be increased by the action of lymphatogogues—substances that increase lymphatic flow rate. There are two general classes of lymphatogogues: (a) any extracellular expanding agent such as isotonic saline or glucose solution, and (b) histamine and similar substances that bind to the H_2 receptors in the venular endothelium, thereby causing an increase in the capillary permeability by opening the endothelial junctions in venules.[21] The histamine response, which is a major component of the inflammatory response, can be abolished by administering heparin. Thus, lymphatogogues increase the lymphatic flow rate secondary to their effects on the interstitial hydrostatic pressure and capillary permeability.

Cerebrospinal Fluid

The cerebrospinal fluid, an important component of the transcellular fluid, has at least two main functions: (1) it provides a medium of exchange for nutrients and metabolic wastes between brain and blood, and (2) it regulates the volume of the cranium by manipulating its own volume in response to changes in brain blood volume.

The total volume of the cerebrospinal fluid is about 100 to 150 mL in man. Its composition resembles that of protein-free plasma, but has a somewhat lower concentration of the important plasma crystalloids. The cerebrospinal fluid is formed primarily by *secretion* from the blood at the choroid plexuses, at a rate of 20 mL/h or about 500 mL/d. About four fifth of this secreted fluid is absorbed via the cerebral arachnoid villi, which are fingerlike projections, and most of the rest via the spinal villi. The mechanisms of absorption are *filtration* and *osmosis*.

Because of its osmotic effect, saline infusion markedly influences the rate of secretion and absorption of the cerebrospinal fluid. For instance, intravenous infusion of *isotonic* saline causes a temporary rise of the cerebrospinal fluid pressure, which, along with the dilution of plasma proteins, promotes cerebrospinal fluid formation. Similarly, infusion of *hypotonic* saline raises the cerebrospinal fluid pressure, but decreases both the plasma oncotic (dilution of proteins) and crystalloid osmotic (hypotonic salt) pressures. The net effect is the osmotic flow of water from plasma into the cerebrospinal fluid and brain cells. Thus, there would be a prolonged rise in the cerebrospinal fluid pressure, a rise in the intracranial pressure, and a swelling of the brain. In contrast, infusion of *hypertonic* saline causes a marked fall in the cerebrospinal fluid pressure, a fall in the intracranial pressure, and a shrinking of the brain. These effects are the direct consequence of the elevated plasma crystalloid osmotic pressure following hypertonic saline infusion.

Clinically, hypertonic solutions of mannitol or urea are usually infused to lower the abnormally elevated intracranial pressure caused, for instance, by cerebral tumor. Such treatment helps to restore consciousness, relieve headache, and reduce the swelling of the optic disc.

Abnormal elevation of the intracranial pressure may also occur as a result of excessive accumulation of the cerebrospinal fluid due to obstruction of the outflow. Such a pathological fluid accumulation is called *hydrocephalus*. The cause of obstruction may be intravascular or extravascular, and it may be fatal if it is not diagnosed and treated promptly.

Edema State

Edema is defined as the excessive accumulation of fluid in the interstitial space. The immediate cause of edema is an abnormal balance in the hydrostatic and oncotic pressures across the capillary wall. Such an imbalance may result from (1) an abnormal increase in the capillary hydrostatic pressure as a result of venous obstruction, (2) an abnormal decrease in the plasma protein concentration, and hence oncotic pressure, caused by a decrease in intake or synthesis and/or increased renal loss, and (3) an abnormal accumulation of proteins in the interstitial space owing to increased capillary permeability to proteins and/or obstruction of lymphatic return. As a result of any of these factors or their combinations, the rate of outward filtration of plasma would exceed the rate of their absorption from the interstitium.

Edema may be formed in both extracellular and intracellular compartments. *Extracellular edema* is the usual way in which extra fluid accumulates in the body. This accumulation represents retention of both water and sodium chloride, as an isotonic solution. Both the interstitial and intravascular compartments share in this increase. The distribution of the extra fluid between the intravascular and extravascular compartments will depend, in part, on the normal functioning of the cardiovascular and renal systems as well as the hemodynamic factors operating across the capillary wall. Hence, altered cardiac, renal, hepatic, or endocrine function may be of primary importance in the etiology of edema state.[12]

Retention of fluid without corresponding retention of extracellular electrolytes (NaCl) will lead to an increase in both intracellular and extracellular fluids. *Intracellular edema* is poorly tolerated. Loss of adrenal function or an excess of antidiuretic hormone (ADH) predisposes one to intracellular edema. Infusion of electrolyte-poor fluids into anuric or electrolyte-depleted patients may also cause intracellular edema.

Clinically, formation of edema is usually accompanied by an abnormal elevation of the net hydrostatic pressure (ΔP) owing to (1) reduced protein synthesis and hence reduced π_c, as in liver cirrhosis and (2) reduced plasma protein concentration and hence reduced π_c, caused by excessive renal protein excretion, as in nephrotic syndrome, or by plasma volume expansion, as in cardiac and renal failures. In all these cases, edema is self-limiting. With the establishment of a new steady state, water and electrolyte exchange may be normal even though the extracellular fluid volume is increased.

Cellular Fluid Exchange

The continuous, rapid bidirectional flow of water between the interstitial and intracellular compartments depends on the transcellular *osmolar concentration* gradient and *cellular metabolism*. However, for a given metabolic rate, the net rate of water flow from the interstitium into the cells, designated as $\dot{W}_{i \rightarrow c}$, is largely determined by the difference between the cellular ($[Os]_c$) and interstitial ($[Os]_i$) osmolalities:

$$\dot{W}_{i \rightarrow c} = k_c([Os]_c - [Os]_i) \quad (3.9)$$

where k_c is the cell membrane permeability coefficient. Thus, an increase in the osmolality of the intracellular fluid above that of the interstitial fluid causes a net shift of water into the cell. Conversely, an increase in the interstitial osmolality results in a net transfer of water out of the cell. These osmotic shifts of water between the intracellular and extracellular fluid compartments are relative, and an equilibrium may be established in disease states in which both compartments become hypertonic or hypotonic compared with normal.

Cellular metabolism could upset the osmotic equilibrium between the intracellular and extracellular fluid compartments. Intracellular water is mobilized when cellular constituents are metabolized. As the amount of cellular proteins and electrolytes decreases, so does the capacity of the cell to hold water. It has been estimated that for every milliequivalent of potassium mobilized, 6 g of intracellular water are liberated, whereas mobilization of 1 g of protein liberates 3 g of intracellular water. Glycogen, like protein, holds three times its weight of water, whereas fat holds one tenth its weight of water. These values refer to the "preformed" water and must not be confused with the water derived from oxidation of food.

Disturbances in Fluid and Electrolyte Balance

As mentioned in Chapter 1, the various disturbance forcings affecting the renal-body fluid regulating system may be classified as (1) those that alter the normal *influx* and hence impose a *load* on the system and (2) those that affect the normal *efflux* and therefore result in abnormal *loss* from the system. Additionally, any of these disturbances may involve changes in either *volume, composition, distribution*, or a combination of these, leading to states of *hydration* or *dehydration* of different tonicities in the various body fluid compartments. The underlying causes include either (a) the failure to maintain a normal ratio between influx and efflux or (b) the failure of the influx-efflux organs, or both of these.[8]

Clinically, disturbances in body fluid distribution may result from one or a combination of several conditions:

1. *Cardiac failure.* The manifestations of altered fluid distribution may vary, depending on what part of the heart has failed. Fluid tends to leave the intravascular compartment or to be unequally distributed between lungs, viscera, and extremities.

2. *Altered permeability of the capillary wall.* This may result from surgical shock or burns. Plasma volume decreases due to loss of plasma proteins, and

fluid tends to accumulate in the interstitial space.

3. *Reduced plasma protein oncotic pressure.* Marked reduction in plasma albumin concentration may result from either reduced protein intake or decreased synthesis, as in liver cirrhosis, or increased urinary excretion, as in nephrotic syndrome, or a combination of these. In all cases, the reduced plasma oncotic pressure favors outward filtration of fluid into the interstitium, thereby producing extracellular edema.

4. *Interference with venous return or lymphatic drainage.* Obstruction of venous and/or lymphatic return of fluid back to the cardiovascular system causes accumulation of fluid in the interstitial space at the expense of the intravascular volume.

To understand the underlying causes of the *disturbances* in the quantity and distribution of body fluid that occur clinically, the following facts must be kept in mind:

1. *Water can freely penetrate the cell membrane in either direction.* Hence, distilled water admitted to the body is distributed throughout all body fluid compartments.

2. *In general, electrolytes (sodium, potassium, chloride, etc.) cannot penetrate normal cell membranes in either direction.* (There are several notable exceptions to this general rule, which may be disregarded for present purposes.) Hence, isotonic saline injected intravenously remains entirely extracellular.

3. *Normally, all fluid compartments are isotonic with each other.* Water can freely diffuse between compartments, thus preventing long-lasting anisotonicity.

4. *Osmotic adjustments can be accomplished only by a shift of water.* This follows from the fact that only water is freely diffusible. Hence, if hypertonic saline is injected intravenously, water leaves the cells to dilute the extracellular fluid to isotonicity.

5. *The quantity of sodium in the body largely determines the volume of extracellular fluid.* Water will be added to or subtracted from the extracellular fluid until its sodium concentration is rendered isotonic with the cells.

6. *The protein oncotic pressure of the plasma largely determines the relative volumes of the intravascular and interstitial fluids.* If the plasma proteins are decreased, fluid will escape into the interstitial spaces, producing edema.

Pathological states may arise from deficiencies or excesses in *total* body water. For example, dehydration easily results from restriction of fluid intake. Excess fluid given intravenously may cause cardiac embarassment, as well as peripheral, pulmonary, and cerebral edema.

Of equal importance are pathological states that arise from improper distribution of fluid between the various compartments. For example, a rapid shift of plasma into the interstitial spaces leads to secondary shock; if the shift occurs so gradually that plasma volume can be maintained by the water and salt intake, clinical edema results.

It is not so widely appreciated that maldistribution of fluid and dehydration can result from a primary electrolyte depletion as well as from a primary water depletion and that such conditions cannot be corrected by measures that fail to restore proper electrolyte balance. In general, water is never lost without electrolytes, nor electrolytes without water, although the relative proportions lost may vary, depending on the condition responsible. Hence, in treating disturbances of fluid balance, one should bear in mind several factors: (1) water requirements, (2) electrolyte requirements, (3) acid-base balance requirements, and (4) blood protein oncotic pressure requirements.

Failure to consider all these factors may at times be disastrous. For example, the administration of large amounts of water by mouth to a patient who is both demineralized and dehydrated as a result of excessive sweating will dilute the extracellular fluid, cause excessive cellular hydration, and precipitate nausea, vomiting, muscular cramps, and cerebral edema. Also, intravenous administration of saline solution to a patient in secondary shock will not restore the plasma volume because the saline, lacking protein oncotic pressure, cannot be retained intravascularly and merely adds to the edema.

The widespread use of intravenous fluid in clinical practice necessitates an understanding of the possible dangers involved in this form of therapy. Such dangers may arise from several sources, for example, the use of incompatible blood, contamination with pyrogens, and the like. The rate at which intravenous fluid can safely be given will vary with the state of hydration of the patient, the efficiency of his myocardium, and the nature of the fluid.

When large quantities of fluid are rapidly introduced into the venous system, several compensatory mechanisms work to prevent an excessive rise in circulating blood volume and venous pressure. These include:

1. Storage in the blood reservoirs of the skin,

spleen, liver, and splanchnic area. This fluid is retained in the intravascular compartment, but the increased capacity of the vascular system produced by dilation of these reservoir areas prevents an excessive rise in venous pressure.

2. Storage in the interstitial and/or intracellular compartment. Isotonic saline solution is limited to the extracellular compartment. Hypotonic solutions may increase intracellular water. Excessive accumulation of interstitial fluid constitutes edema, and this may occur in the skin, lungs, brain, or abdominal viscera. Hypertonic solutions pull water into the intravascular compartment and produce a greater rise in blood volume and venous pressure than do isotonic solutions given at the same rate.

3. Passage into the interstitial lumen, pleural cavity, or peritoneal cavity.

4. Excretion by the kidneys.

If fluid is given so rapidly that the operation of these compensatory mechanisms cannot prevent the venous pressure from rising to a critical level of about 25 to 30 cm H_2O, death will occur from acute cardiac decompensation as the heart dilates beyond its physiological limit. If the myocardium were weak to begin with, a lesser rise in venous pressure would precipitate decompensation. This indicates caution in administering intravenous fluids to patients with damaged hearts.

If the fluid is not given rapidly enough to exceed the capacity of the compensatory mechanisms, acute cardiac decompensation may be avoided, but death may occur from pulmonary or cerebral edema if enough fluid is given.

Obviously, a patient whose body fluids are depleted by acute dehydration, shock, or hemorrhage will tolerate much more rapid rates of fluid infusion than one in whom the water content and distribution is essentially normal to begin with.

We conclude this chapter by examining in some detail the response of the renal-body fluid regulating system to some common forcings resulting from oral or parenteral fluid administration or loss. Depending on whether excessive fluid enters or leaves the body, the overall response is classified as hydration or dehydration, respectively.

Hydration

Hydration, or fluid and salt retention, results from excessive influx of water and sodium chloride. Once in the blood, these substances diffuse uniformly between the vascular and interstitial compartments, but no salt enters the intracellular compartment. The final effects depend on the volume, the amount of salt, and the route and rate of administration.

Isotonic Hydration

Rapid oral intake or intravenous infusion of 2 L of isotonic saline results in about 10% increase in plasma volume, a decrease in plasma protein oncotic pressure, owing to dilution of protein concentration, and an increase in both arterial and venous pressures. The resulting increase in the capillary hydrostatic pressure and decrease in protein oncotic pressure lead to an increase in the effective filtration pressure (ΔP) and eventual shift of fluid from plasma to interstitial compartment. However, because the administered fluid is isotonic, there will be no change in crystalloid (nonprotein solutes) osmotic pressure, and hence no change in the intracellular volume. The expansion of the extracellular volume leads to increased urinary excretion of salt and water (diuresis), which eliminates about half the added load in about four hours. The remaining fluid is excreted gradually during the next few days.

Infusion of a large volume of isotonic saline at a rate that exceeds urinary excretion results in fluid retention or edema. Fluid retention may also result from either depressed renal excretion of salt and water (renal failure) or excessive expansion of vascular and interstitial volumes (congestive heart failure and hepatic cirrhosis). For a more detailed account of renal function in these disease states the reader should consult a textbook dealing with renal disease.

Hypotonic Hydration

An oral intake of 2 L of water, especially on an empty stomach, is rapidly absorbed into the blood. This results in about 10% increase in plasma volume, but only 3% fall in the crystalloid osmotic pressure, and a smaller decrease in protein oncotic pressure. The increase in plasma volume elevates the capillary hydrostatic pressure, which, along with the reduced crystalloid osmotic and protein oncotic pressures, leads to a shift of water from plasma into the interstitial compartment.

Concomitantly, because the interstitial crystalloid osmotic pressure is higher relative to that of plasma, electrolytes, (namely, sodium chloride) will diffuse from the interstitial compartment to the plasma. However, in terms of the overall effect, the rate of movement of water from plasma to interstitium is much faster and more important than the diffusion of electrolytes in the opposite direction.

The net effect of water intake is a decrease in the osmolality of the extracellular fluid, thus disturbing the normal volume and osmolality balance between the intracellular and extracellular compartments. Owing to the fall in the crystalloid osmotic pressure of the extracellular fluid, water enters the cells. Therefore, the final effect of the ingested water is an increase in the total body water and a decrease in the crystalloid osmotic pressure.

To preserve normal water and electrolyte balance the kidneys begin to eliminate the excess ingested water approximately 30 minutes after its intake. The water diuresis reaches its peak in about one hour, then declines and is virtually over in about three hours.

Hypertonic Hydration

Oral intake of large amounts of salt or intravenous infusion of a hypertonic saline solution leads to an increase in the plasma crystalloid osmotic pressure. The rise in plasma osmolality causes water to flow from the interstitium into the plasma, thereby initially increasing plasma volume. Concomitantly, the increase in plasma salt concentration causes sodium chloride to diffuse into the interstitium. The net result is a uniform increase in the crystalloid osmotic pressure of the extracellular compartment without a change in the volume. The increase in the extracellular fluid osmolality causes water to flow out of the cells, which eventually decreases the intracellular volume and increases the extracellular volume.

The renal response to this salt retention is an increase in sodium chloride excretion spread over several hours. This slow rate of renal excretion is due to inability of the kidneys to eliminate urine having a salt concentration in excess of 450 to 500 mEq/L (roughly three times normal). A survivor of a shipwreck, in a raft at sea, faces a similar danger of salt retention should he choose to drink the sea water; depending on the salt concentration, drinking of the sea water to quench thirst leads eventually to cellular dehydration and death.

Dehydration

Dehydration, or fluid and salt depletion, involves both extracellular and intracellular fluids in the majority of instances. Like hydration, the final effects of fluid and salt depletion depend on the volume of fluid, the amount of salt, and the route and rate of loss.

Isotonic Dehydration

Loss of water and electrolytes in isotonic concentration leads to extracellular isotonic dehydration. Examples of isotonic dehydration with reduced extracellular volume include hemorrhage, plasma loss through burned skin, starvation, and gastrointestinal fluid loss, including GI bleeding. Since in all of these conditions insensible and fecal losses of water and electrolytes occur, the kidneys are the only organs that can moderate the effects of fluid loss by conserving salt and water. In fact, in all these cases, renal excretion of salt and water is reduced markedly. However, because of the limited ability of the kidneys and the necessity of excreting a minimum urine volume (about 400 to 500 mL/d), an attempt must be made to replace the fluids and electrolytes lost.

Hypotonic Dehydration

Hypotonic dehydration results from efflux of hypertonic fluid. This condition occurs only in patients with the syndrome of inappropriate secretion of antidiuretic hormone (SIADH). This hormone increases the rate of osmotic water reabsorption from the collecting duct of the nephron. The mechanism of cellular action of antidiuretic hormone (ADH) is detailed in Chapter 9.

This syndrome is characterized by excessive retention of water, which persists despite a concomitant reduction in the osmolality of the extracellular fluid. Clinical findings include (1) hyponatremia along with plasma hypo-osmolality; (2) continued renal excretion of sodium; (3) a markedly below normal urine osmolality, but less than maximally dilute; and (4) normal renal and adrenal functions. For a further discussion of the pathophysiology of this syndrome and its treatment, the reader is referred to the excellent paper of Bartter and Schwartz[2] and a recent clinical study by Decaux and co-workers.[7]

Hypertonic Dehydration

Hypertonic dehydration may be produced by efflux of hypotonic fluid, such as in severe sweating. This results in an increased plasma crystalloid osmotic pressure, which causes water to move by osmosis from cells to interstitium to plasma. If the dehydration is too severe, the induced osmotic flow of water

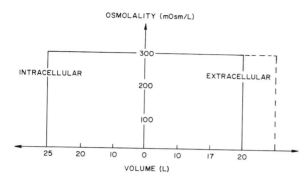

FIGURE 3.9. Schematic representation of changes from normal in the volume and osmolality of the extra- and intracellular fluid compartments after intravenous infusion of isotonic saline. (From Koushanpour.[13])

into the plasma may not be sufficient to prevent the onset of circulatory failure.

Renal response to severe sweating is a marked decrease in water excretion, thereby producing a very concentrated urine. To aid the kidneys in their efforts, ingestion or intravenous infusion of hypotonic saline or oral intake of water along with salt tablets is helpful.

Deviations from normal in volume and osmolality of the intracellular and extracellular compartments, the direction of fluid shift, and the controlled renal excretion in response to the above forcings (with the exception of hypertonic dehydration) are given elsewhere (Table 1.2).

Representation of Fluid Shift: Darrow-Yannet Diagram

A convenient method of depicting the changes in volume and osmolality of the extra- and intracellular fluid compartments in response to the above forcings is the use of the classic Darrow-Yannet diagram.[4, 5] The procedure consists of representing each of the fluid compartments by a rectangle whose width and height indicate, respectively, the volume and osmolality of that compartment. Also, to illustrate the comparative effects of a forcing on a given compartment and the possible interaction between compartments, the rectangles representing the extra- and intracellular compartments are placed side by side.

Figure 3.9 shows such a diagram, depicting changes from normal in the extra- and intracellular fluid compartments following intravenous infusion of a large volume of isotonic saline solution. In this diagram solid lines represent the normal pattern of fluid distribution, whereas the dashed lines indicate changes from normal in the fluid distribution in response to the forcing. Note also that the area of the rectangle, for each compartment, represents the total milliosmoles of solutes in that compartment:

$$\text{Area (mOsm)} = \text{Volume (L)} \times \text{Osmolality (mOsm/L)} \quad (3.10)$$

This expression is useful when determining the direction of fluid shift as well as the final steady-state values of the volume and osmolality of the compartment.

To better appreciate the application of the Darrow-Yannet diagram, let us consider the effect of intravenous infusion of 2 L of hypotonic solution (osmolality = 150 mOsm/L) on the osmolalities and the distribution of the volumes of the extra- and intracellular compartments. For this purpose, we assume that the initial plasma osmolality is 300 mOsm/L and that the volumes of the extra- and intracellular compartments are 15 L and 25 L, respectively.

For the purpose of this and similar calculations, it is important to remember that the solute (whether injected into or removed from the body) always enters or leaves the extracellular fluid compartment. Thus, the only effect that presence or absence of solute in the extracellular fluid compartment has is its effect on the osmotic flow of water between the extra- and intracellular fluid compartments across the cell membrane.

To proceed with the solution of the above problem, we first determine the total volume and number of osmoles in the body fluids:

$$\text{Total fluid volume (L)} = \text{Extracellular volume} + \text{Intracellular volume}$$
$$= 15 \text{ L} + 25 \text{ L} = 40 \text{ L} \quad (3.11)$$

$$\text{Total osmoles (mOsm)} = \text{Extracellular osmoles} + \text{Intracellular osmoles} \quad (3.12)$$

From the rearrangement of the concentration equation (Equation 2.1), we can calculate the number of osmoles in the extra- and intracellular fluid compartments as follows:

$$\text{Extracellular osmoles} = \text{Extracellular volume} \times \text{Extracelluar osmolality}$$
$$= 15 \text{ L} \times 300 \text{ mOsm/L} = 4{,}500 \text{ mOsm} \quad (3.13)$$

$$\text{Intracellular osmoles} = \text{Intracellular volume} \times \text{Intracellular osmolality}$$

$$= 25 \text{ L} \times 300 \text{ mOsm/L} = 7{,}500 \text{ mOsm} \quad (3.14)$$

Therefore, the total number of osmoles in the body is equal to

$$\text{Total body osmoles} = 4{,}500 \text{ mOsm} + 7{,}500 \text{ mOsm} = 12{,}000 \text{ mOsm} \quad (3.15)$$

Intravenous infusion of 2 L of hypotonic solution having an osmolality of 150 mOsm/L would increase the volume of total body water by 2 L and add 300 mOsm (2 L × 150 mOsm/L) to the number of osmoles in the extracellular fluid compartment. Thus, after the completion of infusion, the volume of total body water and the number of osmoles in the extracellular fluid would be:

$$\text{Total fluid volume} = 40 \text{ L} + 2 \text{ L} = 42 \text{ L} \quad (3.16)$$

$$\text{Extracellular osmoles} = 4{,}500 \text{ mOsm} + 300 \text{ mOsm} = 4{,}800 \text{ mOsm} \quad (3.17)$$

Since the injected solute remains in the extracellular fluid compartment and only water flows by osmosis across the cell membrane, the final distribution of volume and osmolality between the extra- and intracellular fluid compartments can be determined by first calculating the final steady-state osmolality (after osmotic flow of water has occurred) and then determining the distribution of volume between the extra- and intracellular fluid compartments from the ratio of the number of osmoles in each compartment and the final osmolality as follows:

$$\text{Final osmolality} = \frac{\text{Total osmoles}}{\text{Total fluid volume}} \quad (3.18)$$

$$= \frac{4{,}500 \text{ mOsm} + 7{,}500 \text{ mOsm} + 300 \text{ mOsm}}{42 \text{ L}}$$

$$= \frac{12{,}300 \text{ mOsm}}{42 \text{ L}} = 293 \text{ mOsm/L} \quad (3.19)$$

Since after steady state, the final osmolality of 293 mOsm/L will be the osmolality for both the extra- and intracellular fluid compartments, then the final volume in each compartment would be equal to the ratio of the total number of osmoles in each compartment and the new osmolality:

$$\text{Final extracellular volume} = \frac{\text{Total extracellular osmoles}}{\text{Final osmolality}}$$

$$= \frac{4{,}800 \text{ mOsm}}{293 \text{ mOsm/L}} = 16.4 \text{ L} \quad (3.20)$$

$$\text{Final intracellular volume} = \frac{\text{Total intracellular osmoles}}{\text{Final osmolality}}$$

$$= \frac{7{,}500 \text{ mOsm}}{293 \text{ mOsm/L}} = 25.6 \text{ L} \quad (3.21)$$

Thus, we see that intravenous infusion of 2 L of hypotonic solution (osmolality = 150 mOsm/L) results in 1.4 L (16.4 L − 15 L) expansion of the extracellular fluid volume and only 0.6 L (25.6 L − 25 L) expansion of the intracellular fluid volume. Note that *the ratio of volume distribution (25.6 L/16.4 L = 1.56) is equal to the ratio of the number of osmoles between the two compartments (7,500 mOsm/4,800 mOsm)*.

To gain more experience in using such a diagram, solve problems 3.3 and 3.4 following this chapter by showing with a Darrow-Yannet diagram the change in volumes and osmolalities of the extra- and intracellular fluid compartments after each forcing.

Summary

The materials presented thus far have revealed four important operational features of the renal-body fluid regulating system: (1) The kidneys play a central role in maintaining the constancy of the internal environment, a direct consequence of stabilizing the volume and composition of the extracellular fluid. (2) Since the blood volume is an integral part of the extracellular compartment, its regulation is indispensable to homeostasis. (3) On the input side we could exercise some degree of control over the rate of fluid intake, both orally and parenterally. (4) However, on the output side, the kidneys are the only system that can exert some control over the rate of fluid loss; the extrarenal losses, both gastrointestinal and insensible, would presumably occur with little or no control.

In clinical practice you will be confronted with a more challenging and difficult problem, notwithstanding its enormous reward. You must deduce the causes from the effects and diagnose disease from the symptoms. The keenness of such a deductive effort will depend, in large part, on your knowledge of the various factors affecting the distribution of body fluids and the renal and extrarenal mechanisms involved in maintaining the normal water and electrolyte balance.

Let us now proceed to detail these mechanisms, beginning with an overview of the structural and functional organization of the kidney.

Problems

3.1. The total solute excreted in a 24-hour urine sample is 1,200 mOsm. Calculate the total 24-hour excreted urine volume that has to contain this amount of solute if the average urine specific gravity were (a) 1.015, (b) 1.020, and (c) 1.035.

3.2. A patient was given 620 g of human serum albumin over a period of 16 days. Analysis of urinary nitrogen excreted revealed that 468 g of exogenous albumin was metabolized. Other laboratory data showed that the plasma volume rose from 3.86 L to 4.29 L; plasma albumin concentration rose from 3.86 to 5.11 g%; total plasma protein rose from 6.54 to 7.63 g%; total circulating plasma albumin rose from 149 to 219 g; and hematocrit fell from 47.5% to 41.7%. Calculate the amount of unmetabolized exogenous albumin in (a) the plasma and (b) the interstitial fluids.

3.3. Consider a 70-kg man whose total body water is 60% of his body weight. If the ratio of the intracellular volume (V_C) to extracellular volume (V_E) is 1.47, and the initial osmolar concentration ([Os]) is 300 mOsm/L, calculate the final osmolar concentration and volume of each of the two compartments after the following forcings, assuming no renal excretion:
   a. Infusion of 300 mL of a 15 g% saline solution.
   b. Infusion of 2 L of a 0.9 g% saline solution.
   c. Infusion of 4 L of a 5 g% glucose solution.

3.4. Calculate the final volume and osmolality (equivalent to sodium chloride) of the extracellular and intracellular compartments after the following forcings:
   a. Severe sweating resulting in the loss of 3 L of water having a sodium chloride concentration of 75 mEq/L.
   b. Oral ingestion of 3 L of hypertonic saline having a concentration of 450 mEq/L.
   c. Oral intake of 2 L of plain water (zero osmolality).

In these calculations assume the following initial normal values:
Intracellular volume (V_C) = 30 L
Extracellular volume (V_E) = 15 L
Plasma osmolality $[Os]_p$ = 300 mOsm/L
Plasma sodium concentration $[Na]_p$ = 150 mEq/L

3.5. The accompanying table presents deviations from normal findings for five patients.
   a. In the first column, enter your choice of an appropriate forcing combination (isotonic, hypotonic, or hypertonic, and efflux or influx; for example, isotonic influx) that accounts for the findings.
   b. In the last column, enter the name of one condition (or disease) that is characterized by the findings.

Definition of symbols used:
V_p = Plasma volume
$[Na]_p$ = Plasma sodium concentration
$[Pr]_p$ = Plasma protein concentration
Hct = Hematocrit
$[Na]_u$ = Urinary sodium concentration

Patient	Forcings	V_p	$[Na]_p$	$[Pr]_p$	Hct	$[Na]_u$	Condition or disease
1	_____	+	0	−	−	+	_____
2	_____	−	+	+	+	+	_____
3	_____	+	+	−	−	+	_____
4	_____	−	0	+	+	−	_____
5	_____	+	−	−	−	−	_____

3.6. Briefly explain the role of proteins in the differential distribution of anions and cations between plasma and interstitial fluid compartments.

3.7. Briefly discuss the current concepts of tissue fluid formation and list the forces involved and their modes of operation.

3.8. State concisely why the effective plasma protein oncotic pressure increases disproportionately as the plasma protein concentration increases.

3.9. Define edema and briefly discuss the various conditions that precipitate edema formation.

3.10. Define osmosis, osmolality, and osmotic pressure.

References

1. Allison MM, Lipham EM, Gottschalk CW: Hydrostatic pressure in the rat kidney. *Am J Physiol* 1972; 223:975–983.
2. Bartter FC, Schwartz WB: The syndrome of inappropriate secretion of antidiuretic hormone. *Am J Med* 1967; 42:790–806.
3. Chapman WH, Bulger RE, Cutler RE, et al : *The*

Urinary System, An Integrated Approach. Philadelphia, WB Saunders Co, 1973.
4. Darrow DC, Hellerstein S: Interpretation of certain changes in body water and electrolytes. *Physiol Rev* 1958; 38:114–137.
5. Darrow DC, Yannet H: The change in the distribution of body water accompanying increase and decrease in extracellular electrolytes. *J Clin Invest* 1935; 14:266–275.
6. Deane N: *Kidney and Electrolytes.* Englewood Cliffs, NJ, Prentice-Hall, Inc, 1966.
7. Decaux G, Waterlot Y, Genette F, et al: Inappropriate secretion of antidiuretic hormone treated with frusemide. *Br Med J* 1982; 285:89–90.
8. Elkinton JR, Danowski T: *The Body Fluids: Basic Physiology and Practical Therapeutics.* Baltimore, The Williams & Wilkins Co, 1955.
9. Guyton AC: A concept of negative interstitial pressure based on pressures in implanted perforated capsules. *Circ Res* 1963; 12:399–414.
10. Guyton AC: Interstitial fluid pressure. II. Pressure-volume curves of interstitial spaces. *Circ Res* 1965; 16:452–460.
11. Guyton AC: *Textbook of Medical Physiology*, ed 6. Philadelphia, WB Saunders Co, 1981.
12. Guyton AC, Coleman TG, Granger HJ: Circulation: overall regulation. *Ann Rev Physiol* 1972; 34:13–46.
13. Koushanpour E: *Renal Physiology: Principles and Functions*, ed 1. Philadelphia, WB Saunders Co, 1976.
14. Landis EM, Pappenheimer JR: Exchange of substances through the capillary walls, in Hamilton WF, Dow P (eds): *Handbook of Physiology. vol 2, sec 2, Circulation.* American Physiological Society, 1963, pp 961–1034.
15. Mayerson HS: Physiologic importance of lymph, in Hamilton WF, Dow P (eds): *Handbook of Physiology. vol 2, sec 2, Circulation.* Washington DC, American Physiological Society, 1963, pp 1035–1073.
16. Pappenheimer HR, Soto-Rivera A: Effective osmotic pressure of the plasma proteins and other quantities associated with the capillary circulation in the hind limbs of cats and dogs. *Am J Physiol* 1948; 152:471–491.
17. Renkin EM: Capillary blood flow and transcapillary exchange. *Physiologist* 1966; 9:361–366.
18. Renkin EM, Curry FE: Transport of water and solutes across capillary endothelium, in Giebisch G, Tosteson DC, Ussing HH (eds): *Transport Organs*, (Membr. Transport Biol. Ser.).Berlin, Springer-Verlag, 1978, vol 4, pp 1–45.
19. Renkin EM, Zaun BD: Effects of adrenal hormones on capillary permeability in perfused rat tissues. *Am J Physiol* 1955; 180:498–502.
20. Schneeberger EE, Lynch RD: Tight junctions: Their structure, composition, and function. *Circ Res* 1984; 55:723–733.
21. Simionescu N: Cellular aspects of transcapillary exchange. *Physiol Rev* 1983; 63:1536–1579.
22. Starling EH: On the absorption of fluids from the connective tissue spaces. *J Physiol* 1896; 19:312–326.
23. Wiedemann MP, Tuma RF, Mayrovitz HN: *An Introduction to Microcirculation. Biophysics and Bioengineering series, vol 2.* New York, Academic Press, 1981, p 12 ff.
24. Zweifach BW, Intaglietta M: Mechanics of fluid movement across single capillaries in the rabbit. *Microvasc Res* 1968; 1:83–101.

4

An Overview of the Structural and Functional Organization of the Kidney

Our understanding of the kidney's function and of its role in the regulation of the volume and composition of body fluid is related to an understanding of the structure of the kidney. In this chapter, we provide a detailed description of the structural organization of the whole kidney, the nephrons and their layout within the kidney, the blood supply of the kidney, the renal interstitium, lymphatics, and innervation, including the architectural organization of the cortex. We conclude the chapter by briefly considering an overview of the functional organization of the kidney. The materials presented provide a framework for subsequent analysis and synthesis of the renal function described in the ensuing chapters.

Kidney Structure

Subdivision of Renal Parenchyma

A cross section of the human kidney reveals a complex structural pattern. To better understand this complex pattern, it is necessary to examine the structure of more simple mammalian kidneys, such as mouse, rat, rabbit, and also small primates. A longitudinal section through such an unipapillary kidney (Figure 4.1a) reveals two distinct parts: the renal cortex and the renal medulla. The renal medulla is shaped like a pyramid whose base is surrounded by the cortex. The apex of the pyramid corresponds to the renal papilla, which projects into the renal pelvis. The renal pelvis, together with the large branches of arteries and veins, lymphatics and nerves, is situated in the renal sinus, which opens into the renal hilus.

Unipapillary kidneys obviously have a limited possibility to grow. If more renal tissue is necessary in larger species the unit of an unipapillary kidney is multiplied. The multiple units may stay fully separated from each other, as in the whale, or they may grow together, establishing a compound multipapillary kidney as in human (Figure 4.1b). During growth, primarily the cortical tissue fuses but the medullary pyramids may also fuse to some extent. That is the reason why the human kidney does not have a constant number of individual papillae, but one that varies between 4 and 14. The most frequent type of human kidney has 7, 8, or 9 individual papillae.[18] Note that in such a kidney, cortical tissue is also found in the midst of the kidney facing the renal sinus. Table 4.1 summarizes the subdivisions of renal parenchyma.

Nephrons and Collecting Ducts

The renal parenchyma is composed of nephrons, collecting ducts, blood vessels, lymphatics, and nerves. Generally, the nephron is said to be the structural and functional unit of the kidney. However, the collecting ducts, despite their names, are very active tubules that are responsible for the final adjustment of renal excretion of water and electrolytes. Thus, the name "collecting duct" underestimates the functional importance of these tubules. The term "renal tubule" includes both the tubular parts of the nephron and the collecting ducts.

A nephron begins with a renal corpuscle (which contains the glomerulus) followed by a complicated and twisted tubule that finally empties into a collecting duct. The renal corpuscles are situated in the cortex. They are spherical corpuscles with a diameter of about 200 μm in man. Functionally, a renal cor-

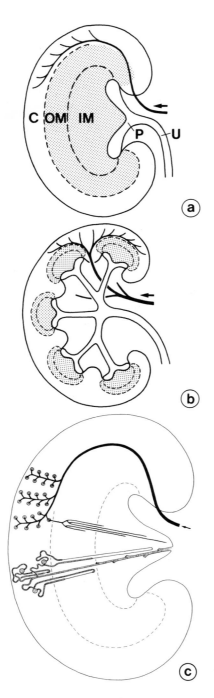

TABLE 4.1. Subdivision of renal parenchyma.

I. Renal cortex
   A. Cortical labyrinth (including the cortex corticis)
   B. Medullary rays of the cortex
II. Renal medulla
   A. Outer medulla
      1. Outer stripe
      2. Inner stripe
   B. Inner medulla (including the papilla)

puscle is a filtration device, where a constant fraction of the plasma flowing to the kidney is filtered and delivered to the tubule for processing.

A renal corpuscle is supplied by an afferent arteriole that, at the vascular pole of the corpuscle, splits into a tuft of specialized capillaries called glomerulus or glomerular tuft. This tuft is drained by an efferent arteriole that leaves the renal corpuscle also at the vascular pole.

The tubule of a nephron may be described as a thin tube with a complicated structure. According to its configuration we can distinguish a highly convoluted proximal part situated around the corresponding renal corpuscle, followed by a hairpin-shaped loop (Henle's loop) that passes down into the renal medulla and returns into the cortex, and a distal convoluted tubule, which again is situated in the neighborhood of its glomerulus. The distal convoluted tubule passes over into the connecting tubule, which connects the nephron to a cortical collecting duct.

The collecting ducts begin in the renal cortex and descend into the medulla. Within the inner medulla, they fuse successively (a total of eight times) before emptying as papillary ducts at the tip of a papilla into the renal pelvis. In the rat, each kidney contains 30,000 to 35,000 nephrons, whereas a human kidney contains roughly one million nephrons. In man, an average of 11 nephrons (six in rat) drain into a cortical

FIGURE 4.1. Different types of mammalian kidneys: (a) Unipapillary kidney (rat) showing the three major zones: cortex (C), outer medulla (OM), and inner medulla (IM), including pelvis (P) and ureter (U). (b) A compound multipapillary kidney (human). The medulla is shaded and the dark line, as in (a), depicts the course of the arterial vessels. (c) The arrangement of arterial vessels (black) and of the nephron (shaded) within an unipapillary kidney (corresponding to a lobe in a multipapillary kidney) is shown. An arcuate artery, running at the corticomedullary border, gives rise to three interlobular arteries, which ascend in the renal cortex. Their branches are afferent arterioles, which run to the renal corpuscles. The efferent arteriole of a juxtamedullary renal corpuscle splits into the descending vasa recta supplying the renal medulla. Note that there is no direct arterial branch to supply the renal medulla. Cortical efferent arterioles are not shown. In addition, three nephrons: a cortical, a short looped, and a long looped nephron together with their corresponding collecting duct are illustrated.

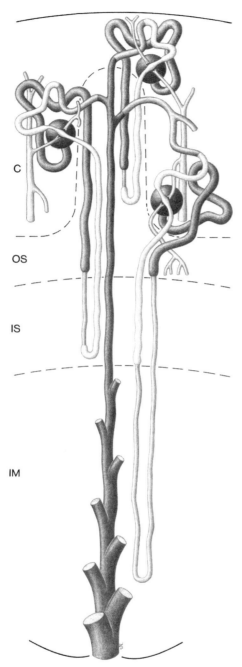

TABLE 4.2. Subdivision of the nephron and the collecting duct system.

I. Nephron
  A. Renal corpuscle
    1. Glomerulus (a term frequently used to refer to the entire renal corpuscle)
    2. Bowman's capsule
  B. Tubule
    1. Proximal tubule
      (a) Convoluted part
      (b) Straight Part (pars recta) or descending thick limb of Henle's loop
    2. Intermediate tubule
      (a) Descending part or thin descending limb of Henle's loop
      (b) Ascending part or thin ascending limb of Henle's loop
    3. Distal Tubule
      (a) Straight part or thick ascending limb of Henle's loop (which contains the Macula Densa within its most terminal portion)
      (b) Convoluted part
II. Collecting duct system
  A. Connecting tubule (the arcades belong to this tubular part in most species)
  B. Collecting duct
    1. Cortical collecting duct
    2. Outer medullary collecting duct
    3. Inner medullary collecting duct

collecting duct. Successive fusions of the collecting ducts in man results in roughly 350 papillary ducts (20 in rat), which empty into the pelvis. The renal tubule, including the collecting ducts, consists of several segments that are outlined by different epithelia. This segmentation together with the location of the various nephron segments within the kidney is depicted in Figure 4.2 and outlined in Table 4.2.

Further subdivision of the renal cortex and medulla

FIGURE 4.2. Structural organization of the nephrons and collecting ducts of the human kidney. There are three nephron types depicted: short loop nephron *(left)*, cortical nephron *(middle)*, and long loop nephron *(right)*. The renal parenchyma is divided into the cortex (C), outer stripe (OS), inner stripe (IS), and inner medulla (IM). The cortex consists of two parts: the cortical labyrinth (containing the vessels, the glomeruli, and the convoluted tubules), and the medullary rays (containing the straight tubules). Note that the outer stripe does not include the medullary rays of the cortex. Two interlobular arteries are drawn on either site within the cortical labyrinth; together with their afferent and efferent arterioles they are shown in pale gray. The renal corpuscles and the proximal tubules (convoluted and straight parts) are shown in dark gray. The thin limbs are shown in pale gray. The thick ascending limb (straight part of the distal tubule including the macula densa, which is shown as a bulging portion of the tubule) and the distal convoluted tubule are shown in an even paler gray. The connecting tubules are shown in slightly darker. Note that the connecting tubule of the long loop nephron forms an arcade. Finally, the collecting duct system is shown in dark gray. *For color art see front matter.*

is based on the location of the various tubular segments (Tables 4.1 and 4.2 and Figure 4.2). The cortex is subdivided into the cortical labyrinth and the medullary rays. The cortical labyrinth contains the renal corpuscles and the convoluted parts of the proximal and distal tubules. The straight parts of the proximal and distal tubules (pars recta of the proximal tubule and the cortical thick ascending limb of Henle's loop) together with the cortical collecting ducts establish the medullary rays. In the human kidney, a medullary ray is established by roughly 40 to 60 proximal as well as distal tubules together with 4 to 6 collecting ducts.

The medullary rays pass over into the outer stripe of the outer medulla without a visible border. The border between cortex and outer stripe (cortico-medullary border) is defined as a vaulted area on which the innermost parts of the cortical labyrinth are based. The outer stripe of the renal medulla, like the medullary rays of the cortex, contains the straight parts of the proximal and distal tubules and collecting ducts. The border between outer and inner stripes of the outer medulla is established by transition of straight proximal tubules into the thin descending limbs of Henle's loops. Thus, the inner stripe of the outer medulla contains thin descending limbs of Henle's loops, thick ascending limbs (straight parts of the distal tubule), and collecting ducts. Only the long loops descend into the inner medulla. Consequently, the inner medulla contains the thin descending and thin ascending limbs of long loops and collecting ducts which continuously fuse toward the tip of the papilla. Histologically, the papilla and thus also the tip of the papilla are part of the inner medulla.

There are several types of nephrons, those with long loops, which turn back in the inner medulla and those with short loops, which turn back in the outer medulla. In the human kidney there is a third type of nephron, which already turns back within the medullary rays of the cortex and which are known as nephrons with "cortical loops." These nephrons are rarely found in other species. The nephrons with cortical loops alter slightly the composition of the medullary rays of the cortex. Since the cortical loops have a small thin tubule, the medullary rays of the cortex in the human kidney also contain some thin tubules. The amount of cortical loops is not known with certainty, but they are a regular component of a human kidney.

The long loops belong to those nephrons whose renal corpuscles are situated in the innermost cortex near the medulla (juxtamedullary nephrons). The shortest loops of Henle belong to the most superficially situated renal corpuscles. This distribution may be explained by the development of the kidney, since the innermost renal corpuscles develop first and acquire the longest loops. The most superficially situated renal corpuscles are the latest to develop and their loops will grow only down to the outer medulla or, as in the human kidney, they will never reach the medulla and already have turned back within the medullary rays. Thus, it has been said that nephron growth in the human kidney terminates earlier compared with other species. The length of the loops correlates to some extent with the size of the renal corpuscle: *the longest loops have the largest renal corpuscles.*

There are differences in the manner by which the nephrons reach their respective collecting ducts. Juxtamedullary and, depending on species, also midcortical nephrons do not drain individually into a collecting duct but their connecting tubules fuse to form a fused connecting tubule, which is called an arcade. Arcades ascend within the cortical labyrinth before emptying into a cortical collecting duct within the superficial cortex. The human kidney has only a few arcades. Most nephrons in the human kidney drain individually into a collecting duct by an individual connecting tubule.

Blood Supply of the Kidney

In general, each kidney receives its blood supply from a single renal artery that arises directly from the abdominal aorta. Depending on species, before entering the renal hilus, within the hilar tunnel or within the renal sinus, the renal artery undergoes several divisions until finally the interlobar ateries are established (Figure 4.1). The interlobar arteries enter the renal tissue at the border between the cortex and the medulla. In a multipapillary kidney this occurs between the cortical tissue of a renal column and the medulla. They then follow an arch-like course at the cortico-medullary border where they are called arcuate arteries. Despite their name and unlike their accompanying veins, they never form true arterial arches (by which arteries anastomose) but they become endarteries. The arcuate arteries give rise to the interlobular arteries which ascend radially within the renal cortex where they generally terminate. In some species (man and dog) a few of them reach the renal capsule and these may join with other capsular arteries. No arteries penetrate into the renal medulla.

The intrarenal veins accompany the arteries. Most

important for the venous drainage are the arcuate veins that form real anastomosing arches at the corticomedullary border. They accept the veins from the cortex as well as the draining vessels of the medulla. In some species, including man, there are two types of cortical interlobular veins. One type starts at the renal surface as a stellate vein, which collects the blood from the outer cortex. Such a stellate vein then continues as an interlobular vein traversing the entire cortex to empty into an arcuate vein. The majority of interlobular veins belong to the second type, which begin within the cortex (without ever touching the surface) and, like the other types, generally accompany an interlobular artery and drain into an arcuate vein (Figure 4.3).

Interlobular arteries (and to some extent also arcuate arteries) give rise to the afferent arterioles which supply the glomeruli (Figures 4.3 and 4.4). An interlobular artery finally fully splits into afferent arterioles. In kidneys of healthy young individuals no other branches exist. In older individuals some aglomerular branches are found which directly supply the peritubular capillaries. They are interpreted as the persistent component of degenerated nephrons.[16] Those branches, however, do not appear to be of much functional relevance. The blood supply of the peritubular capillaries is derived from postglomerular vessels.

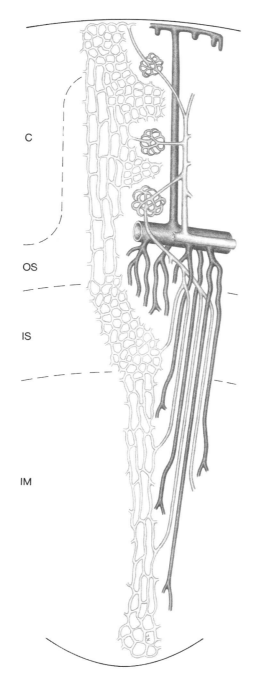

FIGURE 4.3. Organization of the intrarenal vasculature of the human kidney. The renal parenchyma is divided into the cortex (C), the outer stripe (OS), the inner stripe (IS), and the inner medulla (IM). At the corticomedullary border, short segments of arcuate artery (shown in pale gray) and vein (shown in dark gray) are depicted. The interlobular artery arises from the arcuate artery and ascends within the cortical labyrinth. The interlobular artery gives rise to three afferent arterioles which supply, respectively, a superficial glomerulus, a midcortical glomerulus, and a juxtamedullary glomerulus. The efferent arterioles of the superficial and midcortical glomeruli supply both capillary plexuses of the cortex, namely, the round-meshed capillaries of the cortical labyrinth and the long-meshed capillaries of the medullary ray. The cortex is drained by interlobular veins, some of them begin as stellate veins at the surface of the kidney, as depicted in this figure. The efferent arterioles of the juxtamedullary glomeruli descend into the outer stripe and divide into the descending vasa recta, which descend within the vascular bundles. At successive levels within the medulla, descending vasa recta leave the bundles to supply the adjacent capillary plexus. In the inner stripe the capillary plexus is dense round-meshed, whereas in the inner medulla the capillary plexus is long-meshed. The drainage of the medulla is achieved by ascending vasa recta. Those originating in the inner medulla together with the descending vasa recta form the vascular bundles. Most of those originating from the inner stripe ascend independently from the vascular bundles. In the outer stripe, the ascending vasa recta form a dense pattern of vessels that supply the tubules. True capillaries are sparse within the outer stripe. *For color art see front matter.*

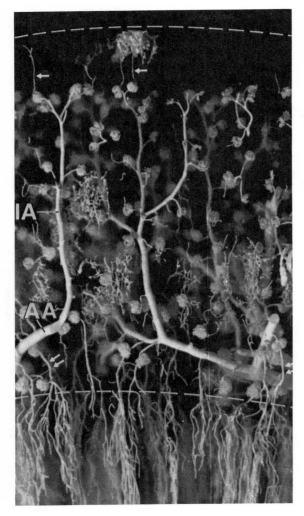

FIGURE 4.4. A longitudinal section through a rabbit kidney whose arterial vessels have been filled with silicon rubber. The broken lines show the surface *(top)* and the corticomedullary border *(bottom)*. The section shows several interlobular arteries (IA) that arise from an arcuate artery (AA). Each interlobular artery gives rise to many afferent arterioles supplying the glomeruli. The efferent arterioles of superficial glomeruli (one *arrow*) ascend unbranched to the surface before splitting into capillaries. The efferent arterioles of the juxtamedullary nephron (two *arrows*) descend into the outer stripe and divide into the descending vasa recta. (From Kaissling and Kriz.[10])

The glomeruli of the renal corpuscles are drained by efferent arterioles. According to the location of the renal corpuscles, there are several types of efferent arterioles.[5] Basically, two types are distinguished: cortical and juxtamedullary (Figures 4.3 and 4.5). The cortical efferent arterioles arise from superficial and midcortical glomeruli, and they supply the peritubular capillaries of the entire cortex. The capillary pattern of the cortical labyrinth and the medullary ray of the cortex are different, adapting to the course of the tubules in each region. The convoluted tubules in the cortical labyrinth are perfused by dense round-meshed capillary plexus, whereas the straight tubules in the medullary ray of the cortex are perfused with a less dense long-meshed capillary plexus. Both capillary plexuses drain into interlobular or arcuate veins.

The efferent arterioles of juxtamedullary glomeruli are the supplying vessels of the medullary circulation.[10, 17, 24] They descend into the renal medulla to divide, within the outer stripe, into the descending vasa recta, which descend to successive levels of the medulla to supply the dense round-meshed capillary plexus of the inner stripe and the less dense long-meshed capillary plexus of the inner medulla. Ascending vasa recta drain the medulla, and together with descending vasa recta they establish the vascular bundles of the renal medulla. In these bundles descending and ascending vasa recta are arranged in a checkerboard pattern establishing a perfect countercurrent exchange system (Figures 4.5 and 4.7).

Not all ascending vasa recta ascend within the bundles. The capillary plexus from most of the inner stripe is drained by the ascending vasa recta, which do not join the bundles but ascend directly to the outer stripe. The outer stripe is a very specific region with respect to its blood supply.[12] True capillaries, originating from small side branches of juxtamedullary efferent arterioles, are sparse. The blood supply to the tubules of this region is mainly provided by the ascending vasa recta, which traverse the outer stripe as individual vessels among the tubules (Figure 4.6). They have an identical wall structure, and they contact the tubules like the capillaries proper. Finally, the ascending vasa recta empty into arcuate veins or into the basal parts of interlobular veins.

For a functional evaluation of the renal circulation it is necessary to have some information about the wall structure of the intrarenal vessels. Structurally, the intrarenal arteries and the proximal portions of the afferent arterioles are similar to arteries and arterioles of the same size elsewhere in the body. The specific structural alterations of the terminal portions of the afferent arterioles, containing the renin producing granular cells, is considered together with the juxtaglomerular apparatus (JGA) in Chapter 6. The structural organization of the glomerular capillaries

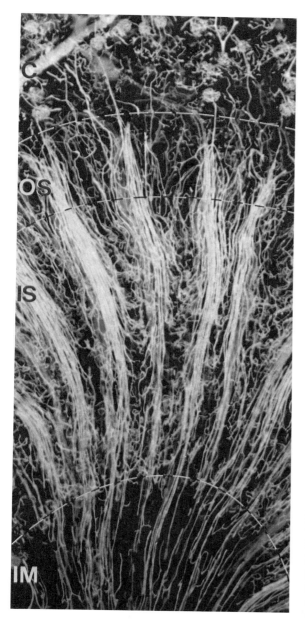

FIGURE 4.5. A longitudinal section through a rat kidney showing the development of the vascular bundles, as revealed by arterial injection of silicon rubber. Efferent arterioles of the juxtamedullary glomeruli situated in the deep cortex (C) divide into the descending vasa recta which form the vascular bundles of the medulla. OS = outer stripe, IS = inner stripe, and IM = inner medulla.

FIGURE 4.6. A longitudinal section through rabbit kidney whose venous vessels are filled with silicon rubber. The broken lines indicate the surface (*top*), the corticomedullary border (*middle*), and the border between the outer and inner stripes (*bottom*). The section shows several interlobular veins (IV) which drain the cortex and empty into an arcuate vein (AV). Note the dense pattern of ascending vasa recta in the outer stripe of the medulla, which finally empty into either an interlobular or an arcuate vein. C = cortex, OS = outer stripe, and IS = inner stripe.

is described along with the glomerulus in Chapter 5.

The efferent arterioles of the cortical and juxtamedullary nephrons are developed differently.[7,8] Juxtamedullary efferent arterioles are larger in diameter than the cortical efferent arterioles, and they are equipped with more layers of smooth muscle cells than the cortical ones (two to three layers in the

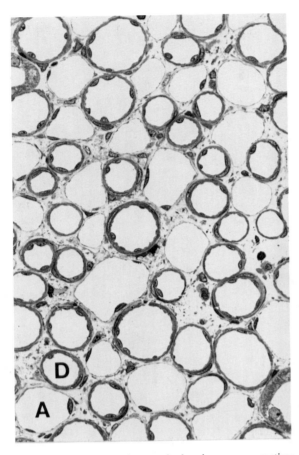

FIGURE 4.7. A light micrograph showing a cross-section through a vascular bundle of the human kidney. The descending (D) and ascending (A) vasa recta are arranged in a checkerboard pattern.

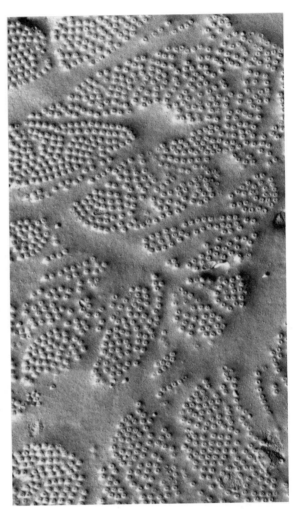

FIGURE 4.8. An electron micrograph of a freeze-fracture replica showing the dense distribution of fenestrae within the capillary endothelium. The spider-shaped structures correspond to the thicker cytoplasmic strands separating the fenestrated area.

former, generally not more than one layer in the latter arterioles). The heaviest coat of smooth muscle cells is generally found at the very beginning of the efferent arterioles at the vascular pole of a glomerulus. The descending vasa recta are still arterioles. The smooth-muscle cells, however, are gradually replaced by pericytes. Although pericytes may be regarded as contractile cells, they lack innervation. The ultrastructure of the peritubular capillaries in the kidney is identical in the cortex and the medulla, with the exception of glomerular capillaries. The capillaries of the kidney are of the fenestrated type. The capillary wall consists of an extremely flat endothelium surrounded by a thin basement membrane. In contrast to glomerular capillaries, the fenestrations of peritubular capillaries are bridged by a thin diaphragm, which is a rather complex structure. The fenestrations are arranged in the very thin non-nuclear regions of the endothelial cells. The areas with fenestrations comprise roughly 50% of the capillary circumference (Figure 4.8).

As already mentioned, the ascending vasa recta are a capillary type of vessel. They are formed by the same fenestrated capillary endothelium as the capillaries proper. This holds true for all the interlobular veins and, in small species, also for the arcuate veins. These wide-lumen venous vessels, called veins, judged from their wall structure, are capillaries. In addition to their function as draining vessels, they are believed to play a reabsorptive role in the renal circulation.[9, 13]

Interstitium and Lymphatics

The renal interstitium consists of the interstitial cells (mainly fibrocytes) and the interstitial spaces, which contain various amounts of collagenous fibrils. In the kidney the interstitium is differently developed in various regions.[19] The interstitium of the renal medulla exhibits some specific features that are described in Chapter 12.

Lymphatics do not occur within the medullary interstitium.[14] Within the renal cortex the peritubular interstitium must be distinguished from the periarterial interstitial spaces; only the latter contain lymphatics.

The *peritubular interstitium* is sparsely developed. Tubules and peritubular capillaries are closely packed together with little space left for an interstitium. At many places the basement membranes of tubules and those of adjacent capillaries may fuse or, at least, may come very close together, so that an interstitial space is narrowed to the interstitial space which is occupied by the basement membranes. Nevertheless, and this has to be stressed, all transports between tubules and capillaries have to pass through an interstitial compartment.

In contrast to the sparse peritubular interstitium, the *periarterial interstitium* is well developed. This is a circular layer of loose connective tissue around the intrarenal arteries (Figure 4.9). It is continuous on all sides with the peritubular interstitium. It attenuates toward the end of the arterial system and terminates at the vascular pole of a glomerulus.

Lymphatics within the kidney are exclusively found within this periarterial connective tissue. Lymphatic capillaries begin within this tissue at the afferent arterioles or, more frequently, somewhere along the interlobular arteries. The lymphatics then leave the kidney accompanying the arteries toward the hilus. One may suggest that this interstitial tissue together with its lymphatics are a functional entity by which a hilus-directed drainage system of the kidney is established.[10]

Renal Nerves

Nerves accompany the intrarenal arteries embedded in the periarterial connective tissue. For the major part they are postganglionic sympathetic efferents. Epinephrine and dopamine have been shown to be the transmitters[4,6]; several peptides (neurotensin; vasoactive intestinal polypeptides, VIP) have been identified as cotransmitters.[3,22] A recent study suggests the presence of chemosensitive and mechan-

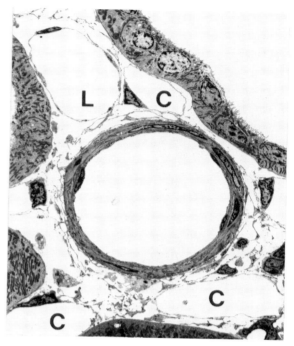

FIGURE 4.9. An electron micrograph of a cross-section through an interlobular artery of the rat kidney showing the periarterial connective tissue including a lymph capillary (L) and some blood capillaries (C).

osensitive elements among the afferent renal nerves, which are presumed to coordinate the activity of the two kidneys through neurally mediated renorenal reflexes.[20,21,23]

What are the targets of the efferent nerves? There is agreement that the smooth-muscle cells of the arteries, of the afferent and efferent arterioles, and of the descending vasa recta are innervated. Moreover, the smooth-muscle cell derived renin-producing cells of the afferent arterioles are reached by terminal axons; renin secretion can be initiated by a sympathetic stimulus (see JGA). There is disagreement, however, about whether the renal tubules are innervated. Applying morphological criteria, the overwhelming majority of tubular portions have no direct relationships to nerve terminals. Terminal nerves are only found adjacent to those tubules that are situated around arteries and arterioles.[1,2] Nerves and nerve terminals generally do not leave the periarterial spaces to penetrate between the tubules.

The tubules of the medullary rays of the cortex never come into contact with nerve terminals. This holds true also for most of the tubules within the renal medulla. Tubular catecholamine receptors have

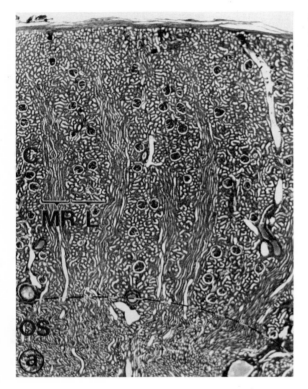

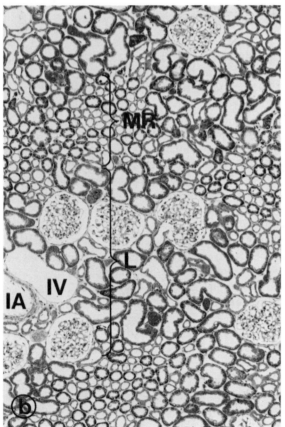

◁ FIGURE 4.10. A longitudinal section (*panel a*) and a cross section (*panel b*) through the human kidney showing both components of the cortex, namely, the cortical labyrinth (L) and the medullary ray (MR). The dashed line represents the corticomedullary border. In the cross section (*panel b*), note the interlobular artery (IA) and vein (IV). C = cortex, and OS = outer stripe.

so far only been demonstrated at connecting tubules (arcades), which frequently are situated alongside the interlobular arteries. Thus, from a morphological point of view a direct neural influence on proximal tubules, as suggested by several investigators, must be doubted, and an innervation of Henle's loop appears to be impossible.[10]

Architectural Organization of the Cortex

So far the various components of the renal parenchyma have been described, but nothing has been said about how they are topographically related to each other. In the kidney the knowledge of the topographical relationship of tubules and vessels is of utmost importance as many specific functions result from the cooperative possibilities provided by the morphology. In this chapter the topographical relationships within the cortex is described; the medulla is dealt with in Chapter 12.

The architectural pattern of the cortex[11, 15], can easily be understood when looking at a cross section through the midst of the cortex (Figures 4.10 and 12.1a). Two portions of the cortical tissue, the *cortical labyrinth* and the *medullary rays*, may be clearly distinguished. The medullary rays, which are fingerlike structures pushing into the renal cortex from the cortico-medullary border, are found to be regularly distributed within the cortical labyrinth. The medullary rays are individual structures, separated from one another by labyrinth tissue. In contrast, the cortical labyrinth is continuous throughout the cortex.

The interlobular arteries and veins run within the cortical labyrinth and together establish a vascular axis. The arteries are surrounded by their loose connective tissue, which contains the lymphatics and nerves. The vein comes very close to the surrounding tubules. Around each axis the corresponding renal corpuscles are situated and around each corpuscle are the corresponding convoluted tubules. Barriers separating the population of renal corpuscles and convoluted tubules belonging to another axis are not visible.

The medullary rays contain straight proximal and

distal tubules (i.e., descending and ascending thick limbs of superficial and midcortical nephrons) and cortical collecting ducts. The straight tubules of an individual medullary ray generally belong to nephrons of several vascular axes. The collecting ducts of a given medullary ray drain the nephrons whose straight proximal and distal tubules are found in the same medullary ray.

The number of straight tubules contained in one medullary ray increases because deeper nephrons enter toward the corticomedullary border. In the human kidney medullary rays near the corticomedullary border may contain 40 to 60 loops and 5 to 6 collecting ducts. These loops all belong to superficial and midcortical nephrons. The loops of juxtamedullary nephrons directly penetrate into the medulla.

A regular pattern of the proximal and distal convoluted tubules within the cortical labyrinth is not apparent. If arcades are present (in the human kidney there are only a few) they consistently ascend within the cortical labyrinth alongside the interlobular arteries. The topographical relationships within the juxtaglomerular apparatus is presented in Chapter 6.

Within the medullary rays the straight tubules of superficial nephrons generally are located more centrally than those of deeper nephrons. The collecting ducts are situated between both groups.

The tubules within the cortical labyrinth are perfused by the dense capillary plexus of this region. The plexus surrounding the straight tubules within the medullary ray is less dense. Both plexuses receive blood directly from primary branches of efferent arterioles. With respect to the venous drainage, however, there is a difference between the two plexuses. Since cortical veins are all situated within the cortical labyrinth, the blood from the medullary ray plexus has to pass through the plexus of the cortical labyrinth to reach an interlobular vein. Therefore, the blood that has perfused the straight tubules of the medullary rays mixes afterwards with blood that perfuses the convoluted tubules of the labyrinth.[10]

Kidney Function

The primary function of the kidney is to stabilize the volume and composition of the extracellular fluid compartment, and indirectly those of the intracellular fluid compartment, by forming urine. Specifically, the kidneys:

1. Conserve water and osmoles normally present in the body.
2. Conserve the major electrolytes of body fluids, especially sodium, potassium, bicarbonate, chloride, and hydrogen ions.
3. Eliminate excess water, electrolytes (especially hydrogen ions), and osmoles resulting from fortuitous influxes.
4. Eliminate metabolic waste products of which the body is the source and toxic materials that may gain access to the body.

The functional unit of the kidney is the *nephron*, which consists of two major components: (1) a *renal corpuscle* or *ultrafilter*, consisting of the glomerular capillaries and Bowman's capsule, and (2) a *renal tubule* or *tubular exchanger*, consisting of a sequence of a proximal tubule, Henle's loop, distal and connecting tubules, and collecting ducts.

The overall mechanism of regulation of volume and composition of body fluids at the level of a single nephron consists of the operation of three interrelated processes:

1. First, the renal ultrafilter forms a large volume (about 180 L/24 h) of essentially protein-free *filtrate*, having the same osmolar concentration of all the crystalloids as that in the aqueous phase of plasma. The filtrate, thus formed, contains all the plasma solutes regardless of whether they are to be conserved or eliminated.

2. The renal tubules then *selectively* process this filtrate. Approximately 99% of the filtered water (178.5 L/24 h) is conserved, allowing only 1.5 L/24 h to be excreted in urine. The filtered crystalloids are selectively conserved by *two* tubular exchange processes: (a) *reabsorbing* from tubules into the blood those substances that the body needs, and (b) adding to the filtrate, by *secreting* from blood into the tubules, those substances that must be eliminated from the body, but were present in small concentrations in the filtrate.

3. The renal tubules also selectively conserve the total quantity of filtered water and osmoles, thereby adjusting *urine osmolality* to the body needs.

Each one of these processes is examined in detail in the ensuing chapters.

Problems

4.1. Describe the subdivisions of the renal parenchyma and indicate the topographic locations of different segments of the three nephron types within the human kidney.

4.2. Describe the subdivisions of the cortex and the nephron structures found in each subdivision.

4.3. What are the distinguishing features of the inner stripe as compared with those of the outer stripe and inner medulla?

4.4. What are the main tubular segments of a nephron?

4.5. What are the nephron segments that establish (a) a short loop of Henle, and (b) a long loop of Henle?

4.6. To which nephron segment does the macula densa belong?

4.7. Describe the blood supply of the renal medulla.

4.8. Describe the differences between the efferent arterioles of a midcortical glomerulus and a juxtamedullary glomerulus.

4.9. Describe the structural organization of a vascular bundle.

4.10. What kind of vessels are (a) the descending vasa recta, and (b) the ascending vasa recta?

4.11. What type of capillary comprises the peritubular capillaries?

4.12. What is the difference between the endothelium of peritubular and glomerular capillaries?

4.13. Describe the anatomical location and course of lymphatics within the renal cortex.

4.14. Describe the lymphatics of the renal medulla.

4.15. Describe the distribution of efferent nerves within the kidney.

References

1. Barajas L: Innervation of the renal cortex. *Fed Proc* 1978; 37:1192–1201.
2. Barajas L, Powers K, Wang P: Innervation of the renal cortical tubules: a quantitative study. *Am J Physiol* 1984; 247:F50–F60.
3. Barajas L, Sokolski KN, Lechago J: Vasoactive intestinal polypeptide-immunoreactive nerves in the kidney. *Neurosci Lett* 1983; 43:263–269.
4. Barajas L, Wang P: Localization of tritiated norepinephrine in the renal arteriolar nerves. *Anat Rec* 1979; 195:525–534.
5. Beeuwkes R: Efferent vascular patterns and early vascular-tubular relations in the dog kidney. *Am J Physiol* 1971; 221:1361–1374.
6. Bell C: Regulation of renal function by endogenous dopamine. in Imbs JL, Schwartz L (eds): *Peripheral Dopamine Receptors*. New York, Pergamon Press, 1979; pp 381–390.
7. Dieterich H J: Die Structur der Blutgefasse in der Ratten-niere. Norm u *Patho Anat* 1978; 35:1–127. Stuttgart; Thieme-Verlag.
8. Edwards JG: Efferent arterioles of glomeruli in the juxtamedullary zone of the human kidney. *Anat Rec* 1956; 125:521–529.
9. Jones WR, O'Morchoe CC: Ultrastructural evidence for a reabsorptive role by intrarenal veins. *Anat Rec* 1983; 207:253–262.
10. Kaissling B, Kriz W: Structural analysis of the rabbit kidney. *Adv Anat Embryol Cell Biol* 1979; 56:1–123.
11. Kriz W: Der architektonische und funktionelle Aufbau der Rattenniere. *Z Zellforsch* 1967; 82:495–535.
12. Kriz W: Structural organization of the renal medulla: comparative and functional aspects. *Am J Physiol* 1981; 241:R3–R16.
13. Kriz W: Structural organization of renal medullary circulation. *Nephron* 1983; 31:290–295.
14. Kriz W, Dieterich HJ: Das Lymphgefasssystem der Niere bei einegen Saugetieren. Licht-und elektronenmikroskopische Untersuchungen. *Z Anat Entwickl Gesch* 1970; 131:111–147.
15. Mollendorff Wv: Der Exkretionsapparat. in: *Handbuch der mikroskopischen Anatomie des Menschen*, vol 7. Berlin, Springer Verlag, 1930, pp 1–327.
16. Moffat DB: *The mammalian Kidney*. London, Cambridge University Press, 1975.
17. Moffat DB, Fourman J: The vascular pattern of the rat kidney. *J Anat* 1963; 97:543–553.
18. Oliver J: *Nephrons and Kidneys*. New York-Evanston-London, Harper & Row, Hoeber Medical Division, 1968.
19. Pinter GG, Gaertner K: Peritubular capillary, interstitium and lymph of the renal cortex. *Rev Physiol Biochem Pharmacol* 1984; 99:183–202.
20. Recordati GM, Moss NG, Waselkov L: Renal chemoreceptors in the rat. *Circ Res* 1978; 43:534–543.
21. Recordati G, Genovesi S, Cerati D: Renorenal reflexes in the rat elicited upon stimulation of renal chemoreceptors. *J Auton Nerv Syst* 1982; 6:127–142.
22. Reinecke M: Neurotensin. Immunohistochemical localization in central and peripheral nervous system and in endocrine cells and its functional role as neurotransmitter and hormone. *Progress in Histochemistry and Cytochemistry*, vol 16, No 1. Stuttgart-New York, Fischer, 1985.
23. Rogenes P: Single-unit and multiunit analyses of renorenal reflexes elicited by stimulation of renal chemoreceptors in the rat. *J Auton Nerv Syst* 1982; 6:143–156.
24. Rollhauser H, Kriz W, Heinke W: Das Gefasssystem der Rattenniere. *Z Zellforsch* 1964; 64:381–403.

5

Formation of Glomerular Ultrafiltrate

The first step in urine formation begins with a passive process of ultrafiltration at the glomerulus. The term "ultrafiltration" refers to the passage of protein-free fluid from the glomerular capillaries into Bowman's space. We begin this chapter with a description of the structure of the glomerulus followed by a detailed analysis of the passive forces governing the process of glomerular ultrafiltration, including the functional characteristics of the filtration barrier and the composition of the ultrafiltrate.

The Structure of the Glomerulus (Renal Corpuscle)

A renal corpuscle consists of a tuft of specialized capillaries supplied by the afferent and drained by the efferent arterioles (Figure 5.1). During development the glomerular tuft is pushed into the blind end of the tubule anlage, which later forms the Bowman's capsule of a renal corpuscle. Therefore, the glomerular capillaries are covered by epithelial cells (podocytes) representing the visceral layer of Bowman's capsule. At the vascular pole the visceral layer is reflected to become the parietal layer of Bowman's capsule (a simple squamous epithelium), which at the urinary pole of the renal corpuscle transforms into the tubular epithelium. The space between both layers of Bowman's capsule is called Bowman's space, which at the urinary pole passes over into the tubule lumen. Filtrate is delivered into Bowman's space (Figure 5.2).

The term "glomerulus" was originally used to designate the glomerular tuft with its cover of Bowman's visceral epithelium. At present, the term "glomerulus" is frequently used for the entire renal corpuscle.

Renal corpuscles are spherical in shape. In the rat they have a diameter of approximately 120 μm, whereas in man their diameter is roughly 200 μm.[41] Juxtamedullary corpuscles are generally somewhat larger than midcortical and superficial ones.[3, 33]

The glomerular tuft derives from the afferent arteriole, which divides into several primary capillary branches (five in rat), each of which establishes a lobule.[1, 37] Within these lobules, the capillaries—frequently branching and anastomosing—run toward the urinary pole and then turn back to join at the vascular pole of the efferent arteriole. In the center of the glomerular tuft, anastomoses between the lobules also occur. Consequently, a strict lobulation only exists in the peripheral portions of the glomerular tuft.

The glomerular tuft is composed of several components (Figures 5.3–5.5): (1) the mesangium consisting of the mesangial cells and the mesangial matrix located within the axes of the individual lobules; (2) the endothelium lining the lumen of the capillaries; (3) the epithelial layer (Bowman's visceral layer or podocytes), which covers the capillaries from the outside, and (4) the glomerular basement membrane (GBM), which separates the epithelium from the endothelium or, at other sites, from the mesangium.

The mesangial cells together with the mesangial matrix form the *mesangium* of the glomerulus. The mesangium is generally thought to be the stabilizing core tissue establishing the axes of the individual lobules of the glomerular tuft. It is continuous with the extraglomerular mesangium (see Chapter 6, JGA), which may be interpreted as an anchor outside the glomerulus.

The *mesangial cells* are heavily branched. They contain contractile filaments and are connected to each other as well as to the extraglomerular mes-

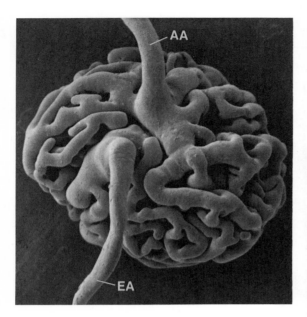

FIGURE 5.1. A scanning electron micrograph showing a vascular cast of the rat glomerulus with an afferent arteriole (AA) and an efferent arteriole (EA). (From Murakami.[31])

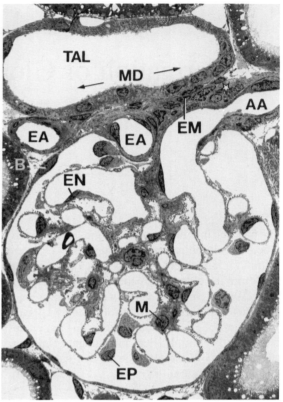

FIGURE 5.2. An electron micrograph of a section through a rat renal corpuscle showing an afferent arteriole (AA) splitting into glomerular capillaries and an efferent arteriole (EA) leaving at the vascular pole. Note the four cell types, namely, the endothelial cell (EN), the epithelial cell (EP), the mesangial cell (M), and the parietal layer cells of Bowman's capsule (B). The upper part of the figure shows the components of the juxtaglomerular apparatus (JGA). Between the afferent and efferent arterioles are the extraglomerular mesangial cells (EM), which are covered by the macula densa (MD) of the thick ascending limb (TAL). A granular cell within the media of the afferent arteriole is marked by an *asterisk* (*).

angial cells by gap junctions.[39] They are believed to be derived from smooth-muscle cells. In culture, mesangial cells have been shown to be able to contract.[26] By their contractility they are presumed to contribute to the regulation of glomerular blood flow.[30] Their pinocytic and phagocytic abilities[20] appear to be important for "cleansing" the filtration barrier of macromolecules which have been trapped within the barrier (see below).

The mesangial cells are embedded into the *mesangial matrix*, which is continuous with a similar matrix surrounding the extraglomerular mesangial cells. The chemical composition of this matrix is different from the GBM; fibronectin is concentrated in the mesangial matrix.[30] The functional relevance of this matrix is unknown.

The *endothelium* of the glomerular capillaries is established by large cells that, except for the nuclear regions, are extremely attenuated. These thin films of cytoplasm, which form the greater part of the capillary wall, are densely perforated by large round pores (50–100 nm in diameter), generally called fenestrae, which, in contrast to the fenestrae of the fenestrated peritubular capillaries, are not bridged by diaphragms. In the adult kidney the glomerular fenestrae are open. At their luminal side the endothelial cells are covered by a negatively charged cell coat.[23] Basally, the endothelium rests on the GBM. Toward the mesangium, where the GBM deviates from the endothelium, the endothelial cells (generally their nuclear regions) directly face the mesangial matrix.

The *podocytes* (or *glomerular epithelial cells*) collectively form the visceral layer of Bowman's capsule, which covers the glomerular capillaries from the capsular lumen side. These are uniquely differentiated cells. In the adult, mitotic divisions have never been found to occur. They have large cell

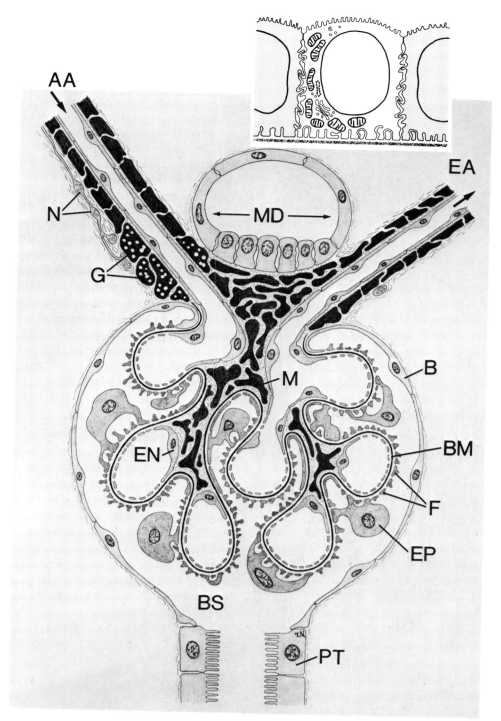

FIGURE 5.3. An overview of a renal corpuscle and JGA. At the vascular pole an afferent arteriole (AA) enters and an efferent arteriole (EA) leaves the glomerulus. At the urinary pole Bowman's space (BS) becomes the tubular lumen of the proximal tubule. The epithelial cells comprising Bowman's capsule (B) enclose Bowman's space. Smooth muscle cells proper of the arterioles and all cells derived from smooth muscle are shown in black, including the granular cells (G). The afferent arteriole is innervated by sympathetic nerve terminals (N). The extraglomerular mesangial cells are located at the angle between AA and EA, and continue into the mesangial cells (M) of the glomerular tuft. The glomerular capillaries are outlined by fenestrated endothelial cells (EN) and covered from the outside by the epithelial cells (EP) with foot processes (F). The glomerular basement membrane (BM) is continuous throughout the glomerulus. At the vascular pole, the thick ascending limb touches with the macula densa (MD) the extraglomerular mesangium. The *inset* shown at the top depicts the ultrastructural organization of the macula densa epithelium.

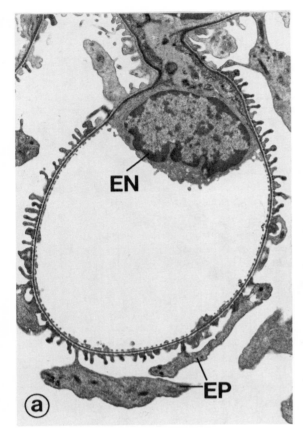

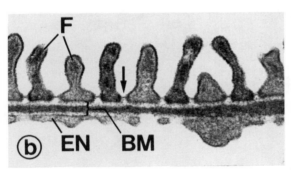

FIGURE 5.4. An electron micrograph of the glomerular capillary (a) and the filtration barrier (b) which is part of the capillary wall. Note the various cell types: endothelium (EN) and epithelium (EP) with foot processes (F). The glomerular basement membrane (BM) consists of three distinct layers, namely, the lamina rara interna (facing the endothelium), the lamina densa, and the lamina rara externa (facing the epithelium). Note the slit membrane (*arrow*) between the foot processes.

bodies, which bulge into the capsular lumen, and many cytoplasmic processes (primary processes), which fix the cells to the capillaries.

The cytoplasm of the podocyte cell body is filled with abundant organelles, indicating metabolically very active cells. It contains a very prominent Golgi apparatus, many profiles of rough and smooth endoplasmic reticulum, and mitochondria. The many lysosomal elements correlate with the podocyte's phagocytic ability. The peculiarity of these cells is established by their processes. Large primary processes embrace the capillaries. At the outer surface of the capillaries they ramify into slender finger-like processes that interdigitate, as clasped fingers, with those originating from the next podocyte, thereby covering the entire outer surface of a capillary (Figure 5.5). The finger-like processes are called foot processes because of their appearance in cross section (Figure 5.4). Also, the name podocytes, i.e. "foot cells," relates to this appearance. The foot processes are anchored in the GBM. They do not touch each other but are separated by narrow spaces of a fairly constant width (20–30 nm) which mostly form a meandering channel, the so-called filtration slit. The floor of the filtration slit is made up of the slit diaphragm, which rests on the GBM. The slit diaphragm is an extracellular structure (therefore it is sometimes regarded as part of the BM) with a regular substructure.[36] Like a zipper, it consists of two rows of subunits that extend from both sides of the slit to its center, where they are linked by a central bar. In between the subunits on both sides are rectangular "open" spaces, 4x14 nm in area, which is approximately the size of an albumin molecule. All membranes of the podocytes that face the urinary space (including the filtration slits) carry a thick negatively charged cell coat that also covers the slit membrane.[23]

The *glomerular basement membrane* is a specific structure that largely contributes to the specific permeability characteristics of the filtration barrier.[21,43] It cannot simply be regarded as the basement membrane of the capillaries, as it is absent at those sites of the capillaries that face the mesangium. Throughout the glomerular tuft it follows the podocytes; it consistently separates podocytes either from the capillary endothelium or from the mesangium.

According to the appearance in conventionally stained transmission electron micrographs (TEM), the BM consists of three layers: the electronlucent *lamina rara interna* (facing the capillary endothelium), the electrondense *lamina densa*, and the electronlucent *lamina rara externa*. In humans, the

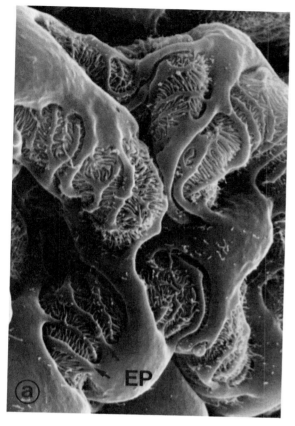

thickness of the entire GBM amounts to roughly 200 nm.

The entire GBM may be viewed as a gelatinous structure consisting of a cross-linked mesh of polymer chains with hydrated interstices. It contains collagenous (type VI collagen) and noncollagenous (sialoglycoproteins such as laminin and fibronectin) proteins; a third component is proteoglycans of the heparin sulfate type. The nonpolar collagenous components are concentrated in the lamina densa. Both lamina rarae contain more of the polar noncollagenous components (heparin sulfate) and are therefore strongly anionic. Laminin and fibronectin are suggested to play a role in the attachment of the endothelium and the epithelium to the GBM.[21]

In summary, the glomerular filtration barrier consists of (1) the endothelium with large open pores, (2) the basement membrane (GBM), and (3) the slits between the foot processes of the podocytes, which are closed by the slit diaphragms. In addition, the negatively charged cell coats on both sides of the filter are essential for the barrier function. Also, both lamina rarae of the GBM are strongly anionic. The knowledge of this composite structure is essential to an understanding of its functional characteristics as well as of the various sites in which immune deposits can accumulate in glomerulonephritis.

Compared with the barrier established in usual capillaries (see Chapter 3), the glomerular barrier has a unique combination of functional characteristics. *It combines an extremely high permeability for water and ions with a very low permeability for plasma proteins.* The high water and ion permeability appears to be based on the fact that the barrier includes no continuous cellular layer: filtration occurs via extracellular routes. The major hydraulic barrier apparently is established by the slit diaphragm,[38] which comprises only about 3% of the entire GBM area.

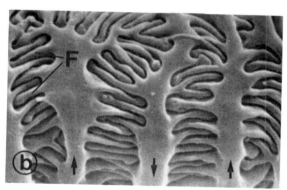

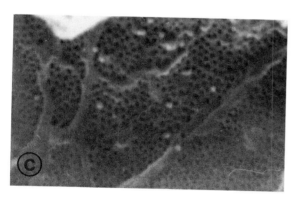

FIGURE 5.5. A scanning electron micrograph of a rat glomerular capillary. *Panel a* is an overview from the Bowman's space side showing the epithelial cell (EP) giving rise to primary processes (*arrows*) embracing several capillaries. *Panel b* is an enlargement showing the splitting pattern of the primary processes (*arrows*) into foot processes (F). Between the foot processes the filtration slits are visible. *Panel c* shows the surface aspect of the endothelium viewed from the capillary lumen. The fenestrations appear as real pores and are not closed by a diaphragm.

The barrier function for macromolecules is selective with respect to size (including configuration) and to charge.[35] The many negative charges accumulated within both cell coats of the barrier and within the GBM itself repel molecules that carry a negative charges, such as plasma proteins, which are polyanionic. This electro-negative shield seems to be of great importance since it minimizes a clogging of the filter. Under experimental conditions, a blocking or removal of the anionic sites of the filter membrane results in proteinuria.[9]

The structural equivalent for size selectivity is the dense network of the GBM. Neutral (uncharged) compounds up to an "effective radius" of 18 angstroms freely pass through the filter. Larger compounds are more and more restricted up to molecules with an effective radius of 40 angstroms, at which size the absolute restriction begins.[16] The term "effective radius" is an empirical value, measured in artificial membranes, which compares the configuration of molecules and attributes a "radius" also to nonspherical molecules. Plasma albumin has an effective radius of 36 angstroms; without the repulsion due to the negative charge, plasma albumin would pass through the filter in considerable amounts.

The final trap for macromolecules that have escaped restriction in more proximal layers is apparently established by the slit membrane.[23] Molecules that have been trapped by the slit membrane are phagocytized by the podocytes preventing a clogging of the peripheral layers of the filter. Substances that have been trapped at more proximal sites (in the subendothelial space or within the GBM) are believed to be taken up by the mesangial cells.

Determinants of Glomerular Ultrafiltration

Historical Background

In 1843 Carl Ludwig proposed that urine is formed by a passive process of ultrafiltration at the glomerulus.[28] However, it was not until 1924 when Wearn and Richards,[44] using micropuncture and microchemical methods, confirmed that the fluid in Bowman's space is an ultrafiltrate of plasma.

Briefly, the micropuncture techniques involved puncturing surface glomeruli with glass micropipettes having an outer diameter of 6 to 10 μm. The glomeruli were identified under stereomicroscope with illumination provided by passing a high-intensity light through a rod of fused quartz.

Three important discoveries made in the 1960s led to a better understanding of the forces and flows governing the process of glomerular ultrafiltration. In 1964 Wiederhelm and co-workers[45] reported the development of a micropipette servo-null transducer system that was well suited to measure the hydraulic pressures in the microvasculature of the kidney surface. This technical development, coupled with the subsequent discovery of a mutant strain of Wistar rat (the so-called Munich-Wistar rat) in Thurau's laboratory[40] having surface glomeruli readily accessible to micropuncture, made it possible to directly measure the glomerular hydraulic pressure in mammals. Finally, in 1969 Brenner and colleagues[8] reported the development of an ultramicromethod that allowed estimation of protein concentration in the blood of the efferent arteriole, a quantity necessary to assess the increase in protein concentration in the efferent arteriole and hence the decrease in the net driving force for ultrafiltration along the glomerular capillary network. Application of these techniques in the Munich-Wistar rat has led to a better understanding of the dynamics of glomerular ultrafiltration in this species and, by extrapolation, in man.

Hemodynamics of Glomerular Ultrafiltration

Normal human kidneys receive a *blood flow* of about 1,200 mL/min, which, assuming a hematocrit (Hct) of 0.45, corresponds to a *plasma flow* of about 660 mL/min. At the same time, the glomerular ultrafilter forms 125 mL/min of glomerular filtrate. The *filtration fraction*, defined as the ratio of the glomerular filtration rate (GFR) to renal plasma flow (RPF), is thus $125/660 = 0.19$. This means that 19% of the entering plasma volume is removed as filtrate. Note that the filtration fraction refers to the bulk volume of fluids, not to their solutes, and only to the filtration stage.

The filtrate, thus formed, is essentially protein-free and containing all the crystalloids in the same osmolar concentration as that in the aqueous phase of plasma. The glomerular ultrafiltration mechanism is governed by the same passive Starling forces that determine the translocation of fluid across other capillaries in the body (see Chapter 3), namely, the imbalance between transcapillary hydraulic and colloid osmotic pressures. Hence, for ultrafiltration to occur, sufficient energy, supplied by the cardiovas-

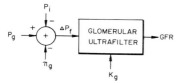

FIGURE 5.6. The hemodynamic forces involved in glomerular ultrafiltration in the whole kidney. (From Koushanpour.[25])

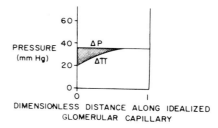

FIGURE 5.7. Hydraulic and colloid osmotic pressures in the afferent and efferent arterioles of a glomerulus and their axial profiles along an idealized glomerular capillary in the rat. $P_{GC} = P_g$, $P_T = P_i$, $\pi_{GC} = \pi_g$, and $P_{UF} = P_f$. (From Brenner.[7])

cular system, is required to develop (1) an osmotic force (π) to separate proteins from the plasma, and (2) an hydraulic force (P) to overcome the frictional resistance of the ultrafilter to filtrate flow. Since the ultrafiltrate separates primarily proteins, the magnitude of the required osmotic force is small and equal to the plasma protein oncotic pressure in the glomerular capillaries, π_g. The magnitude of the hydraulic force, on the other hand, is large and equal to the difference between the glomerular capillary blood pressure, P_g, and the pressure in Bowman's space and connected tubules, P_i. Thus, for a given glomerular filtration coefficient, K_g, the single-nephron glomerular filtration rate (SNGFR) bears a direct and positive relationship to the algebraic sum of these forces, called the net driving or ultrafiltration pressure, $\overline{P_f}$:

$$\begin{aligned}\text{SNGFR} &= K_g\,[(P_g - P_i) - (\pi_g - \pi_i)] \\ &= K_g\,[\overline{\Delta P} - \overline{\Delta \pi}] \\ &= K_g \cdot \overline{P_f}\end{aligned} \quad (5.1)$$

where K_g has a dimension of mL/min per mm Hg. Since SNGFR is the integrated mean, for the entire glomerulus, of the local filtration rate at different points along the glomerular capillary, the transcapillary hydraulic (ΔP) and protein oncotic ($\Delta \pi$) pressure differences and, therefore, the net driving pressure ($\overline{P_f}$) in Equation 5.1 are averaged over the length of the glomerular capillary network. Also, because protein concentration in the glomerular ultrafiltrate is extremely small, the contribution of π_i is negligible, and $\overline{\Delta \pi}$ is practically equal to $\overline{\pi_g}$. Figure 5.6 depicts diagramatically the determinants of glomerular ultrafiltration for the whole kidney GFR.

From both Equation 5.1 and Figure 5.6, we see that the rate of formation of glomerular filtrate depends on two factors: (1) the *structural* characteristics of the ultrafilter and the *permeability* of its filtering membranes, symbolized by K_g, and (2) the *imbalance of hemodynamic forces* in the blood supplying the nephron, symbolized by ΔP_f. Let us now examine the magnitude of each of the determinants of SNGFR as revealed by direct measurements.

Brenner and co-workers[11] were the first to publish the determinants of the glomerular ultrafiltration given in Equation 5.1 in Munich-Wistar rats under conditions of hydropenia (reduced body fluid volume) and euvolemia (normal body fluid fluid volume). They found that P_g averaged about 45 mm Hg, some 40% of the mean aortic pressure. More significantly, they found that the protein oncotic pressure in the efferent arterioles did not differ from the net transcapillary hydrostatic pressure difference, indicating that *filtration equilibrium* was achieved at some point along the glomerular capillaries. Similar values have been found in the squirrel monkey,[29] a small primate having surface glomeruli. From these measurements it was possible to define the net driving force for ultrafiltration at the afferent and efferent ends of the glomerular capillary network. Figure 5.7 (*upper panel*) summarizes the results of these studies.

As shown in Figure 5.7 (*upper panel*), the value of P_g given as 45 mm Hg is measured in the afferent arteriole end of the glomerulus and is assumed for the efferent arteriole end. This assumption is based on the observation that the axial hydraulic pressure drop along the glomerular capillary network is very small, of the order of 2 to 3 mm Hg. The value of P_i, the hydraulic pressure in the Bowman's space, was found to be nearly the same as that in sites further downstream in the proximal tubule, where it is measured routinely, and to average about 10 mm Hg. Thus, the difference between $\overline{P}_g$ and $\overline{P}_i$ yielded a net mean transcapillary hydraulic pressure, $\overline{\Delta P}$, of about 35 mm Hg. Measurements of plasma protein concentrations in the afferent and efferent arterioles revealed that owing to formation of protein-free filtrate, plasma protein concentration increased as the blood flowed along the length of the glomerular capillary network. Thus, as depicted in Figure 5.7, the colloid osmotic pressure increased from a minimum value of about 20 mm Hg at the afferent end of the capillary network to a value of about 35 mm Hg by the time the blood entered the efferent arteriole. Consequently, the value of ΔP_f decreased from a maximum value of about 15 at the afferent end of the glomerulus to zero at the efferent end. In other words, by the time the blood reached the efferent end of the glomerular capillary network, the net transcapillary colloid osmotic pressure, $\overline{\Delta \pi}$, increased to a value that, on the average, exactly equaled $\overline{\Delta P}$, the net transcapillary hydraulic pressure. This equality of $\overline{\Delta P}$ and $\overline{\Delta \pi}$ is referred to as the *filtration pressure equilibrium*.

In a subsequent study, Brenner and co-workers[12] showed that changes in the contour of the $\Delta \pi$ curve (Fig. 5.7, *lower panel*) are due to changes in the initial rate of glomerular plasma flow ($\dot{Q}_A$). Thus, an increase in $\dot{Q}_A$ will tend to displace the point at which filtration pressure equilibrium is reached further toward the efferent end of the glomerular capillary. The resulting increase in P_f (the shaded area between the ΔP and $\Delta \pi$ curves) will cause an increase in SNGFR. Conversely, a decrease in $\dot{Q}_A$ will bring about the opposite results; it will displace the filtration pressure equilibrium toward the afferent end of the glomerular capillary network, thereby reducing P_f and SNGFR. These observations have led to the conclusion that, at least in Munich-Wistar rat, the *glomerular filtration rate is strongly dependent on the glomerular plasma flow rate*. Therefore, expansion of the plasma volume will increase the glomerular plasma flow ($\dot{Q}_A$), thereby increasing ΔP, which provides a larger driving pressure for filtration all along the glomerular capillaries. The resulting higher rate of glomerular filtration slows the rate of rise in $\Delta \pi$, thereby decreasing $\Delta \pi$ all along the glomerular capillaries. In short, any factor that increases extracellular fluid volume, such as intravenous fluid administration, will increase the glomerular plasma flow, which will result in a decrease in the rate of rise of glomerular colloid osmotic pressure, thereby increasing SNGFR. Conversely, any factor that decreases the extracellular fluid volume, such as hemorrhage, will have the opposite effects on $\dot{Q}_A$, $\Delta \pi$, and SNGFR. In short, the most important feature of these findings is that changes in SNGFR appear to be largely due to changes in the rate of rise in $\Delta \pi$ and not to large or important changes in ΔP. Indeed, it appears that ΔP is quite effectively autoregulated (see Chapter 6).

Despite the experimentally observed profound impact of changes in $\dot{Q}_A$ on SNGFR, the glomerular plasma flow does not explicitly appear as a term in Equation 5.1. If SNGFR does, indeed, depend on $\dot{Q}_A$, then one or more terms on the right side of Equation 5.1 must vary with the glomerular plasma flow. To explore these possibilities, Deen and co-workers[17] formulated a mathematical model of glomerular ultrafiltration by combining the conservation of mass and the Starling principle expressed in Equation 5.1. The resulting model yielded a differential equation that described the rate of change of protein concentration with distance along an ideaized glomerular capillary network. The numerical solution of this equation was used to compute $\Delta \pi$ profiles for given values of protein concentration in the afferent arteriole, $[Pr]_A$, K_g, $\overline{\Delta P}$, and $\dot{Q}_A$. The results are summarized in the lower panel of Figure 5.7, which depicts graphically the relationship between ΔP and $\Delta \pi$ as a function of the fractional distance along an idealized glomerular capillary, with 0 being the afferent end and 1 being the efferent end. Note that as stated earlier, ΔP remains essentially constant along the length of the capillary, whereas $\Delta \pi$ increases, owing to removal of a protein-free filtrate by filtration, from a minimum value of about 20 mm Hg to a maximum value of about 35 mm Hg. The vertical distance between the ΔP and $\Delta \pi$ lines is the net ultrafiltration pressure, P_f, at any point along the glomerular capillary. Accordingly, the gradual decrease in P_f from a maximum value of 15 mm Hg to 0 is due to progressive rise in $\Delta \pi$ with distance along the length of the glomerular capillary. The nonlinearity in the rate of increase in $\Delta \pi$ is attributed to two factors. First, since the rate of ultrafiltration at any point along the capillary is proportional to the net driving force (P_f), then one would expect that the ultrafiltrate would be formed more rapidly at the afferent end of the glomerular capillary. This would, in turn, result in a more rapid increase in protein concentration, $[Pr]$, in the beginning of the glomerular capillary network. Second, it is well known that the change in $\Delta \pi$ is a nonlinear function of $[Pr]$ (see Chapter 3): $\Delta \pi = a_1 [Pr] + a_2 [Pr]^2$. Both of these

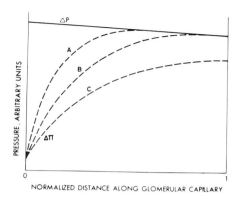

FIGURE 5.8. Effect of changes in transcapillary colloid osmotic pressure ($\Delta\pi$) depicted by curves A, B, and C on filtration pressure equilibrium. (From Brenner.[7])

reasons lead to the nonlinear increase in $\Delta\pi$ along the length of the glomerular capillary network, as depicted in the lower panel of Figure 5.7. It should be noted that evaluation of the colloid osmotic pressure in these experiments was based on measurement of protein concentration in samples obtained from systemic blood to represent the afferent arteriole, and samples of blood from the efferent arteriole. Therefore, the profile for $\Delta\pi$ depicted in Figure 5.7 represents one of an infinite number of profiles consistent with the measurements of colloid osmotic pressure in the afferent and efferent arterioles. Hence, the exact point along the glomerular capillary network where $P_f = 0$ (*filtration pressure equilibrium*) can not be determined from these measurements. Nevertheless, extension of Equation 5.1 to the glomerular capillary network for a single nephron will allow us to explore some important characteristics of the dynamics of glomerular ultrafiltration as revealed by the results depicted in Figure 5.7. Thus, the rate of SNGFR may be expressed as:

$$\text{SNGFR} = K_g [\overline{\Delta P} - \overline{\Delta\pi}] = K_g \cdot \overline{P_f} \\ = k \cdot S \cdot \overline{P_f} \quad (5.2)$$

Note that K_g is equal to the product of the effective hydraulic permeability of the capillary wall (k) and its surface area (S) available for filtration. The quantity enclosed in brackets is equal to the shaded area shown in the lower panel of Figure 5.7.

In a subsequent study, Deen and co-workers[18] increased the initial glomerular plasma flow rate, $\dot{Q}_A$, by iso-oncotic fluid infusion to levels high enough (approximately 200 nL/min) to obtain filtration pressure disequilibrium (e.g., Figure 5.8, curve C), a circumstance that allowed calculation of the ultrafiltration coefficient, K_g. They found a value of 0.08 nL/(s · mm Hg) for K_g which remained essentially unchanged within a twofold increase in $\dot{Q}_A$, suggesting that K_g is independent of $\dot{Q}_A$. This was found to be true also under conditions of primary glomerular injury where K_g is less than normal. Taking this value of K_g together with the measured value of SNGFR, they found that P_f averaged about 4 to 6 mm Hg in the normal hydropenic rat and increased to about 8 to 12 mm Hg after marked expansion of the plasma volume. The rise in P_f during plasma expansion accounted for the large increase in SNGFR observed in such a condition. Figure 5.8 shows how variations in the contour of the $\Delta\pi$ curve produced by changes in glomerular plasma flow rate ($\dot{Q}_A$) affect the profile of P_f along the length of the glomerular capillary network.

Since $K_g = k \cdot S$ (Equation 5.2), substituting the above value for K_g they calculated the value of k from the available estimate for S. Thus, taking a value of 0.08 nL/(s · mm Hg) for K_g and a value of 0.0019 cm² for S, they obtained a value of 42.1 µL/(s · cm² · mm Hg) for k, the effective hydraulic permeability for the glomerular capillary network. This calculated value for k is about one to two orders of magnitude greater than that estimated for other capillaries. Brenner and co-workers[7] suggested that this large value for k may be the reason that glomerular ultrafiltration proceeds at a very rapid rate despite a low net driving force.

Factors Affecting the Determinants of SNGFR

According to Equation 5.2 and as depicted in Figure 5.6, the magnitude of SNGFR is determined by two factors, namely, K_g and P_f. The magnitude of P_f in turn is determined by the algebraic sum of four variables, namely, ΔP, $\Delta\pi_A$, $\dot{Q}_A$, and K_g, where $\Delta\pi_A$ is the protein osmotic pressure of the blood in the afferent arteriole. To explore the effects of various factors on SNGFR, it is best to express SNGFR as a function of glomerular plasma flow ($\dot{Q}_A$) and single-nephron filtration fraction (SNFF), the fraction of the initial glomerular plasma flow that is filtered:

$$\text{SNFF} = \text{SNGFR}/\dot{Q}_A \quad (5.3)$$

In practice, SNFF is determined from the measurements of protein concentration in the afferent ($[\text{Pr}]_A$) and efferent ($[\text{Pr}]_E$) arterioles:

$$\text{SNFF} = 1 - [\text{Pr}]_A/[\text{Pr}]_E \qquad (5.4)$$

Let us now explore the effects of changes in $\dot{Q}_A$, K_g, $\overline{\Delta P}$, and $[\text{Pr}]_A$ on SNFF (Figure 5.9) and on SNGFR (Figure 5.10), respectively. The relationships depicted in these figures are based on the predictions obtained from the previously described mathematical model of glomerular ultrafiltration developed by Deen and co-workers.[17] To generate these theoretical curves, they used the following data representing hydropenic state: $\dot{Q}_A = 75$ nL/min, $K_g = 0.08$ nL/(s · mm Hg), $\overline{\Delta P} = 35$ mm Hg, and $\pi_A = 19$ mm Hg ($[\text{Pr}]_A = 5.7$ g/100 mL).

Effects of Changes in Initial Glomerular Plasma Flow Rate

Figure 5.9 (panel A) depicts the theoretical dependence of SNFF on $\dot{Q}_A$. As shown, SNFF remains constant at a value of 0.33 until $\dot{Q}_A$ exceeds a value of 100 nL/min. Thereafter, an increase in $\dot{Q}_A$ results in a progressive fall in SNFF. A similar qualitative relationship has been found in the rat. However, because of the difficulty of measuring $\overline{\Delta P}$ and K_g in dog and man, it is not possible to verify the existence of such a relationship in these species at present.

The effects of changes in $\dot{Q}_A$ on SNGFR are shown in panel A of Figure 5.10. The dashed line represents the relationship between SNGFR and $\dot{Q}_A$ at constant SNFF = 0.33 (SNGFR = 0.33 · $\dot{Q}_A$), indicating that as long as SNFF remains constant, SNGFR changes linearly with changes in $\dot{Q}_A$. However, as $\dot{Q}_A$ exceeds 100 nL/min resulting in filtration pressure disequilibrium, SNGFR increases less in proportion to increases in $\dot{Q}_A$, a finding that has been verified in experiments in the rat. The linear relationship between SNGFR and $\dot{Q}_A$ at filtration pressure equilibrium ($\dot{Q}_A < 100$ nL/min), indicates that the SNGFR is highly dependent on glomerular plasma flow rate. This dependency of SNGFR on $\dot{Q}_A$ holds true also at filtration pressure disequilibrium where $\dot{Q}_A > 100$ nL/min, but to a lesser extent. These theoretical predictions have also been verified in experiments with volume-expanded rats.

Effects of Changes in Ultrafiltration Coefficient

The theoretical dependence of SNFF on K_g is shown in *panel B* of Figure 5.9. As can be seen, a low value of K_g leads to filtration pressure disequilibrium, whereas at values of K_g exceeding 0.08 nL/(s · mm Hg) SNFF remains relatively constant. Qualitatively, a similar relationship between SNGFR and K_g is

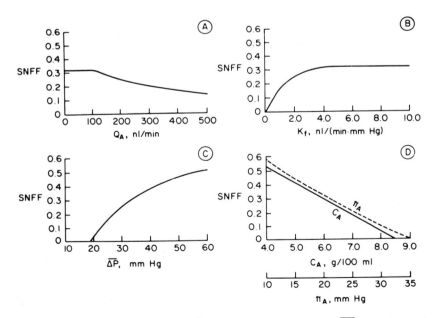

FIGURE 5.9. Theoretical dependence of SNFF on (A) $\dot{Q}_A$, (B) K_f (same as K_g), (C) $\overline{\Delta P}$, and (D) π_A (or $[\text{Pr}]_A$). (From Brenner.[7])

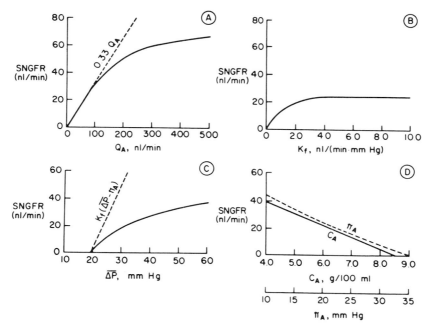

FIGURE 5.10. Theoretical dependence of SNGFR on (A) $\dot{Q}_A$, (B) K_f (same as K_g), (C) $\overline{\Delta P}$, and (D) π_A (or $[Pr]_A$). (From Brenner.[7])

shown in panel B of Figure 5.10. These two relationships indicate that once K_g achieves a value large enough to produce filtration pressure equilibrium, further increases in K_g will have no appreciable effect on SNFF and SNGFR. Thus, except in situations where the capillary wall is damaged, such as in glomerulonephritis, small changes in K_g will have no significant impact on SNGFR.

Effects of Changes in Mean Glomerular Transcapillary Hydraulic Pressure Gradient

Theoretical dependence of SNFF and SNGFR on $\overline{\Delta P}$ are shown in panel D of Figures 5.9 and 5.10, respectively. Since π_A is about 19 mm Hg in the rat, ultrafiltration occurs only when $\overline{\Delta P}$ exceeds this value. Note that both SNFF and SNGFR increase in a nonlinear fashion as $\overline{\Delta P}$ increases. This is due to the fact that as $\overline{\Delta P}$ increases, the resulting increase in ultrafiltration produces a smaller increase in $\overline{\Delta \pi}$. Since experimental maneuvers (such as partial occlusion of carotid arteries or constriction of the aorta above the renal arteries in Munich-Wistar rat) that were designed to alter $\overline{\Delta P}$ failed to change $\overline{\Delta P}$ by more than 5 mm Hg, these results indicate that $\overline{\Delta P}$ is effectively autoregulated (see Chapter 6).

Effects of Changes in Afferent Arteriole Plasma Protein Concentration

Theoretical dependence of SNFF and SNGFR on π_A (or $[Pr]_A$), shown in panel D of Figures 5.9 and 5.10, is in many respects similar to the dependence of SNFF and SNGFR on changes in $\overline{\Delta P}$ alone. Thus, a reduction in $\Delta \pi_A$ is nearly equivalent to an increase in $\overline{\Delta P}$, thereby increasing $\overline{P}_f$. Although changes in π_A are shown to vary with changes in $[Pr]_A$, a recent study[7] showed that this relationship is not always valid. In fact, a reduction in $[Pr]_A$ from about 6.5 to 3 g/100 mL resulted in concomitant reduction in K_g. Therefore, the predicted increase in $\overline{\Delta P}_f$ subsequent to a reduction in π_A is blunted by this reduction in K_g.

In a recent review Tucker and Blantz[42] made a comprehensive analysis of the available data on glomerular ultrafiltration dynamics and assessed the relative contributions of various determinants of the glomerular filtration rate. Their analysis revealed that (1) the computed glomerular permeability coefficient, K_g, was much higher than previously proposed, (2) the ultrafiltration process was characterized by a tendency toward filtration pressure equilibrium in which changes in the glomerular protein osmotic pressure play a dominant role, and (3) the glomerular filtration rate was highly plasma flow-dependent, a phenomenon associated with the filtration pressure equilibrium, and that renal plasma flow is a much more significant regulator of the SNGFR than the transcapillary hydraulic pressure gradient, ΔP. Furthermore, in animals with filtration pressure equilibrium, about 60% of the nephron filtration was determined by changes in renal plasma flow (RPF), whereas changes in systemic plasma protein oncotic pressure (π_A) and transcapillary hydraulic pressure gradient (ΔP) accounted for 27% and 13% of changes in SNGFR, respectively. In contrast, in animals with filtration pressure disequilibrium, RPF accounted for 33%, π_A for 21%, ΔP for 25%, and glomerular filtration coefficient, K_g, for 21%, respectively.

However, the universality of some of these conclusions have recently been questioned by Arendshorst and Gottschalk.[2] Comparing the available data on glomerular ultrafiltration dynamics obtained from different species, including man, they made the following conclusions: (1) The glomerular filtration coefficient, K_g, is not a constant parameter and is responsive to physiological conditions of the animal. (2) Despite wide variations in K_g between species, SNGFR remains relatively constant. This is primarily due to a strong inverse relationship between K_g and $\overline{P}_f$. (3) The glomerular filtration dynamics in most rat strains, including some colonies of Munich-Wistar rats, are characterized by a relatively low K_g and filtration pressure disequilibrium. (4) The available data in dog indicate that the glomerular dynamics are characterized by filtration pressure disequilibrium in this species.

Although the issue of the relative contribution of the various determinants of nephron filtration is not completely resolved, most investigators agree that changes in RPF, π_A, and K_g are the major regulators of the glomerular filtration rate. The presence or absence of filtration pressure equilibrium appears to be secondary to these variables and the physiological state of the animal.

Effects of Hormones and Vasoactive Agents on Determinants of SNGFR

A number of studies, using clearance techniques (see Chapter 7) to measure whole kidney renal plasma flow (RPF), have shown that renal vasodilation with prostaglandin E_1 (PGE$_1$), acetylcholine (ACh), or bradykinin (BK) produces an increase in RBF. However, whole kidney GFR and SNGFR in both dog and man failed to change significantly despite the increase in RBF.[24] This latter observation has been taken as evidence against the existence of filtration pressure equilibrium in these species. Direct measurements of determinants of SNGFR in Munich-Wistar rats[4] showed that administration of PGE$_1$, ACh, and BK resulted in an increase in $\dot{Q}_A$ without any change in SNGFR or $\overline{\Delta P}$. This was accompanied by a significant reduction in K_g and a proportionately greater fall in the efferent arteriole resistance (R_E) than in the afferent arteriole resistance (R_A).

Norepinephrine (NE) and angiotensin II (ANG II) have been found to produce profound reduction in RBF with little change in GFR. In Munich-Wistar rat, Myers and co-workers[32] found that administration of pressor doses of NE and ANG II resulted in an increase in $\overline{\Delta P}$ and no change in SNGFR, despite a reduction in $\dot{Q}_A$. Filtration pressure equilibrium, as evidenced by equality of π_E and $\overline{\Delta P}$, was maintained before and during the infusion of both drugs. It should be noted that given filtration pressure equilibrium, SNGFR is determined by three parameters, namely, $\dot{Q}_A$, $\overline{\Delta P}$, and π_A. Therefore, they postulated that since π_A did not change, despite the increase in $\overline{\Delta P}$, SNGFR was unchanged with both drugs because of the offsetting decrease in $\dot{Q}_A$.

The apparent paradoxical effects of vasoactive substances on the whole kidney RPF and determinants of SNGFR have been clarified by more recent experiments. A recent review of these studies[19] indicates that the epithelial, the endothelial, and the mesangial elements of the glomerular capillary play an *active* role in the process of ultrafiltration. This concept is based on the finding that the glomerular capillary is both target organ and site for synthesis of vasoactive hormones, which profoundly influence the filtration process. It has been found that the glomerulus contains receptors for a number of hormones, namely, angiotensin II, insulin, antidiuretic hormone (ADH), parathyroid hormone (PTH), prostaglandins (E series, I_2, thromboxanes), histamine, serotonin, norepinephrine, and dopamine. In addition, the glomerulus is capable of de novo synthesis of cyclic AMP (cAMP), ANG II, and prostaglandins. Thus, systemic administration of these vasoactive substances can produce both global hemodynamic effects and local effects on the glomerular ultrafiltration. Available evidence suggest that the local glomerular

Glomerular Permselectivity: Factors Governing Filtration of Macromolecules

Size Selectivity of Glomerular Capillary Wall

Direct measurements of the composition of the glomerular filtrate have been carried out only in amphibians (frog and *Necturus*),[44] the Munich-Wistar rat[13], and the squirrel monkey,[22] species that have surface glomeruli where fluid in Bowman's space can be sampled by micropuncture techniques. Results of such measurements in the rat clearly indicate that inulin (a fructo-polysaccharide) and substances smaller than inulin appear in the glomerular ultrafiltrate in essentially the same concentration as that in plasma water. However, substances with large molecular weights, such as albumin, are filtered to a lesser extent; their glomerular clearance (Chapter 7) amounts to less than 1% of that of inulin.

Our current concept of the mechanism of permselectivity of the glomerular capillaries is based on the comparison of the clearance of test macromolecules, which are only filtered and are neither reabsorbed nor secreted by renal tubules, with that of inulin. The ratio of clearances, referred to as the *fractional clearance* (clearance of test substance/clearance of inulin) thus obtained is equivalent to the ratio of the concentration of the test macromolecule in Bowman's space, $[S]_f$, to that in plasma water, $[S]_p$. This ratio provides a convenient measure of glomerular permselectivity, varying from a value of 0 for impermeable solutes to a value of 1 for solutes that are filtered without restriction. Table 5.1 compares the effect of molecular size of different substances on the filtrate-to-plasma concentration ratio $[S]_f/[S]_p$, a quanity equivalent to clearance ratio for that substance at the same urine flow rate (see Chapter 7). Note that the larger the substance, the smaller is the ratio of its filtrate-to-plasma concentration.

Theoretical analysis of the effects of variations of the individual determinants of SNGFR on fractional solute clearance[13–15] has revealed that fractional clearance of macromolecules decreases with increasing plasma flow rate, which increases SNGFR. Since transport of solutes across the glomerular capillary is coupled to water transport, an increase in plasma flow rate and hence SNGFR will have disproportionate effects on solute and water transport. Thus,

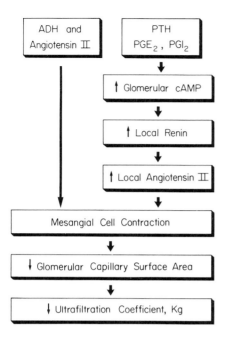

FIGURE 5.11. Theoretical pathways involved in reducing ultrafiltration coefficient, K_g, by angiotensin II (ANG II), antidiuretic hormone (ADH), parathyroid hormone (PTH), and prostaglandins E_2 and I_2 (PGE_2, PGI_2). (Modified from Dworkin.[19])

effects of these substances are mediated via stimulation of the mesangial cell contraction, thereby reducing the surface area for ultrafiltration and hence the glomerular filtration coefficient, K_g.

Dworkin and associates[19] have proposed two pathways for the local control of SNGFR by hormones and vasoactive substances. As shown in Figure 5.11, the first pathway involves direct stimulation of mesangial cell contraction. Since these cells are known to have specific receptors for ANG II and ADH, local concentration of these substances would lead to mesangial cell contraction. The second pathway involves renin release and local synthesis of ANG II by cAMP stimulation. Thus, any substance that stimulates glomerular cAMP production could lead to mesangial cell contraction and a fall in K_g. In this manner the glomerular mesangial cells act to regulate the glomerular surface area and thereby control the ultrafiltration coefficient and SNGFR. The physiological consequence of this local regulation of SNGFR and its impact on renal regulation of volume and composition of body fluids are described in Chapter 6.

TABLE 5.1. Relationships between molecular weight, molecular dimensions, and filtrate-to-plasma concentration ratio.

Substance	Molecular weight (g)	Radius from diffusion coefficient, Kg (angstroms)	$\dfrac{[S]_f}{[S]_p}$
Water	18	1.0	1.0
Urea	60	1.6	1.0
Glucose	180	3.6	1.0
Sucrose	342	4.4	1.0
Inulin	5,500	14.8	0.98
Myoglobin	17,000	19.5	0.75
Egg albumin	43,000	28.5	0.22
Hemoglobin	68,000	32.5	0.03
Serum albumin	69,000	35.5	<0.01

Source: Pitts.[34]

solute filtration increases less as compared with the increase in filtration of water. These theoretical predictions, which have been confirmed experimentally, indicate that fractional solute clearance varies inversely with GFR.

Charge-Selectivity of the Glomerular Capillary Wall

For a neutral substance the molecular size, as estimated from the fractional clearance, becomes the only determinant of its filtration across the glomerular capillary wall. However, when comparing the fractional clearance of a neutral versus a polyanionic molecule having the same molecular size, the fractional clearance is considerably smaller for a charged molecule. The effect of electrostatic properties of the glomerular capillary wall on the filtration of charged molecules is illustrated for dextran sulfate, an anionic polymer of dextran, in Figure 5.12, which compares the fractional clearances of neutral dextran (D) and charged dextran sulfate (DS) as a function of the effective molecular radius. A value of one on the ordinate corresponds to a dextran clearance equal to that of inulin. As can be seen, neutral dextran clearance is measurably restricted after the effective radius exceeds 20 angstroms. Fractional clearance decreases progressively with increasing effective radius, approaching zero with neutral dextran radii greater than about 42 angstroms. In contrast, over the entire range of molecular sizes studied, the fractional clearance of dextran sulfate is considerably less than that of neutral dextran. These results suggest that glomerular filter selectively restricts the transcapillary movement of polyanions.

The identities of the responsible substances with negative charge covering the glomerular endothelia, epithelia, and their foot processes have been known for some time. As mentioned earlier in this chapter, studies with polyanionic tracers[21, 23, 27] have shown that substances bearing negative charges are found on the surface of the endothelium, the lamina rara interna, the lamina rara externa, and the surface coat of the podocytes. The presence of negative charge on these surfaces restrict the filtration of polyanionic macromolecules, such as albumin. For neutral substances, whose glomerular filtration is determined entirely by their molecular size and, possibly, molecular shape, the available ultrastructural studies point to the GBM as the main barrier for filtration.[10]

Effect of Disease on Permselectivity of Glomerular Capillary Wall

Of the various factors affecting fractional clearance the effect of molecular charge appears to have the most clinical relevance. Studies in Munich-Wistar rats with experimental *glomerulonephritis*[5] have shown that the fractional clearance of dextran sulfate was greater over the entire range of molecular size tested than that in the normal rat. These findings, shown in Figure 5.13, suggest that the loss of negative charge from the glomerular capillary wall may be a primary factor in determining the degree of proteinuria after experimental or spontaneous glomerular injury. Similar mechanisms may be responsible for *postexercise proteinuria* in humans.

Composition of the Glomerular Ultrafiltrate

The ultrafiltrate, as described above, is virtually a protein-free solution. It consists of about 94% water

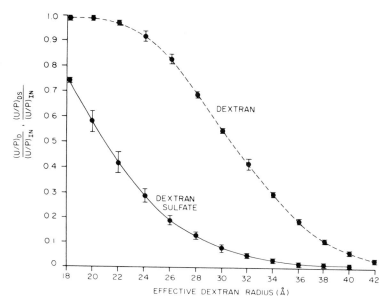

FIGURE 5.12. Fractional clearance of dextran sulfate (DS) and neutral dextran (D) plotted as a function of effective molecular radius. (U/P) denotes urine-to-plasma concentration ratio for a given substance. Values are mean ± SE. (From Chang.[13–15])

by volume and 6% solutes. It contains two types of solutes: electrolytes and nonelectrolytes. Chemical analyses of the glomerular filtrate fluid, collected by micropuncturing the early portions of the proximal convoluted tubules, have revealed that it is *iso-osmotic* and *isotonic* with the similarly collected protein-free plasma fluid from the glomerulus.[46] Table 5.2 compares the concentration of important electrolytes in the ultrafiltrate with that of plasma. Note that there is a slight but significant difference in the concentration of cations and anions in the plasma and ultrafiltrate. This difference, as explained in Chapter 2, is due to the retention of protein anions on the plasma side. The ratio of anion or cation concentration of filtrate to that of plasma is called the Gibbs-Donnan ratio. Since proteins are negatively charged at normal hydrogen ion concentration, the Gibbs-Donnan ratio for cations is less than unity and that for anions is greater than unity. Typically, the Gibbs-Donnan ratios for the most prevalent cations and anions of the plasma and filtrate, namely, sodium and chloride, are 0.95 and 1.05, respectively.

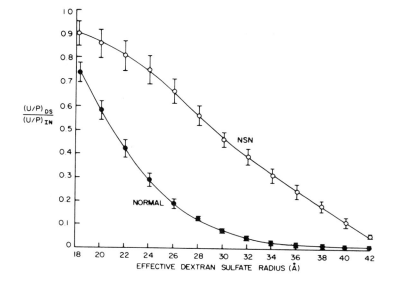

FIGURE 5.13. Comparison of fractional dextran sulfate (DS) clearances for rats with experimental glomerulonephritis (nephrotoxic serum nephritis, NSN) and normal rats as a function of effective DS radius. Values are mean ± SE. Reproduced from *The Journal of Clinical Investigation*, 1976, vol. 57, pp. 1287–1294 by copyright permission of The American Society for Clinical Investigation.

TABLE 5.2. A Comparison of electrolyte concentration in plasma and ultrafiltrate.

Electrolytes	Plasma concentration (mEq/L)	Ultrafiltrate concentration (mEq/L)	Gibbs-Donnan ratio
Cations			
Sodium	142	135.5	0.95
Potassium	4	3.8	0.95
Calcium	5	2.5	0.50
Magnesium	3	1.5	0.50
	154	142.8	
Anions			
Chloride	103	108.0	1.05
Bicarbonate	27	28.4	1.05
Phosphate	2	2.0	1.00
Sulfate	1	1.0	1.00
Organic Acids	5	5.0	1.00
Proteins	16	0.0	0.00
	154	144.4	

Source: Koushanpour.[25]

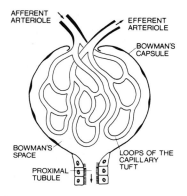

FIGURE 5.14. Structural details of a renal corpuscle.

Hemodynamics of Blood Flow in Glomerular Capillaries

As depicted in Figure 5.14, the afferent arteriole is the only source of blood supply to the glomerular capillaries. The structural organization of the glomerular capillaries, as described earlier in this chapter, allows for rapid blood flow through the glomerulus and provides maximum capillary surface area for filtration. After passing through the capillaries, the fraction of blood that has escaped filtration leaves through the efferent arteriole. Accordingly, in the steady state, the volume of the single-nephron glomerular filtration rate (SNGFR) is equal to the difference between the volume of blood flowing into ($\dot{V}_{in}$) and out of ($\dot{V}_{out}$) the glomerular capillaries:

$$SNGFR = \dot{V}_{in} - \dot{V}_{out} \quad (5.5)$$

The magnitude of these inflow and outflow volumes depends primarily on the *blood pressure* in the afferent (P_A) and efferent (P_E) arterioles and their respective *resistances* R_A and R_E. The arteriolar resistances are determined by the degree of constriction of smooth muscles in the pre- and postglomerular arterioles.

Assuming that the locus of glomerular capillary resistance, R_g, is midway between R_A and R_E, the magnitude of $\dot{V}_{in}$ is then determined by the pressure gradient ($P_A - P_g$) across the two resistances R_A and R_g in series:

$$\dot{V}_{in} = (P_A - P_g)/(R_A + R_g) \quad (5.6)$$

where P_g is the blood pressure in the glomerular capillaries, and the denominator is the algebraic sum of the two resistances in series.

Similarly, an expression for $\dot{V}_{out}$ in terms of the appropriate pressures and resistances would be

$$\dot{V}_{out} = (P_g - P_E)/(R_g + R_E) \quad (5.7)$$

Since potential changes in R_g are infinitesimal as compared with those in R_A and R_E, Equations 5.6 and 5.7 may be further simplified by eliminating the R_g terms.

Substituting the simplified Equations 5.6 and 5.7 into Equation 5.5, and consolidating the fractions, we get

$$SNGFR = [(P_A - P_g)/R_A] - [(P_g - P_E)/R_E] \quad (5.8)$$

Since P_g decreases from an initial value equal to P_A to a final value equal to P_E, then assuming a linear decrease in P_g along the length of the glomerular capillary, its value at midway in the glomerulus would be $P_g = (P_A - P_E)/2$. Substituting this in Equation 5.8, we get

$$\text{SNGFR} = (1/2) \cdot \{[(P_A - P_E)/R_A] - [(P_A - P_E)/R_E]\} \quad (5.9)$$

Equation 5.9 reveals that insofar as the hemodynamics of the glomerular circulation are concerned, the magnitude of SNGFR is determined by blood pressures and resistances of the afferent and efferent arterioles. Thus, a step increase in P_A, assuming the other parameters remain unchanged, increases SNGFR, whereas a step increase in R_A decreases SNGFR. In contrast, a step increase in R_E increases SNGFR. It should be remembered that these alterations in SNGFR are mediated by induced changes in P_g.

Normally changes in p_A and P_E are related to changes in systemic arterial (P_{AS}) and venous (P_{VS}) blood pressures. Hence, for a given intrarenal resistance, a step increase in P_{AS} causes a proportional increase in P_A and eventually in SNGFR. Conversely, a step increase in P_{VS} induces a proportional increase in P_E, thereby reducing SNGFR.

Changes in plasma protein concentration, and hence plasma protein oncotic pressure (π_p), markedly influence SNGFR. This effect is mediated by changes in π_g. As ultrafiltration proceeds, π_g increases nonlinearly (Figure 5.7) from an initial value equal to the protein oncotic pressure in the afferent arteriole (π_A) to a final value equal to that in the efferent arteriole (π_E). Since π_A is nearly identical to the plasma protein oncotic pressure in the systemic blood (π_p), changes in π_g may be expressed in terms of π_p:

$$\pi_g = [\text{RPF}/(\text{RPF-GFR})]/\pi_p \quad (5.10)$$

Equation 5.10 is an example of the application of the law of conservation of mass to plasma proteins. It states that for a given renal plasma flow (RPF) and π_p, the incremental increase in π_g is proportional to GFR. Thus, changes in GFR, resulting from pathological alterations in glomerular hemodynamics or ultrafilter permeability, have a profound influence on protein homeostasis as well as on the normal functioning of the renal tubules.

Micropuncture studies in the Munich-Wistar rat[11] showed that the systemic arterial blood pressure falls an average of 51% by the time the blood reaches the glomerulus. A second pressure drop occurs between

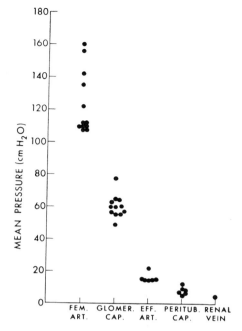

FIGURE 5.15. Hydrostatic pressure gradients along renal cortical microvasculature. Reproduced from *The Journal of Clinical Investigation*, 1971, Vol. 50, pp. 1776–1780 by copyright permission of The Rockefeller University Press.

the glomerulus and the efferent arteriole. Thus, as shown in Figure 5.15, and consistent with Equation 5.9, the major decrease in the intrarenal vascular pressure occurs on either side of the glomerular capillary bed.

The foregoing quantitative analysis reveals that the formation of the glomerular filtrate depends on the structural and permeability characteristics of the ultrafilter membrane (K_g) and the hemodynamic factors that determine the effective filtration pressure ($\overline{P_f}$). Thus, any factor that alters either of these variables will have a profound influence on GFR. The following are some examples of factors affecting $\overline{P_f}$ and hence GFR:

1. *Elevation of systemic arterial pressure*, following intravenous fluid infusion, leads to an increase in P_A, RPF, $\dot{Q}_A$, P_g, and hence $\overline{P_f}$, thereby increasing GFR.

2. In *hemorrhagic shock*, when the blood pressure falls below 50 mm Hg, *anuria* results. This is due to a marked reduction in P_g and hence GFR. If blood is reinfused, arterial blood pressure will rise, thereby increasing GFR and restarting urine formation.

3. Elevation of *ureteral pressure* or *intrapelvic*

pressure, due to obstruction or stone formation, leads to an increase in tubular interstitial pressure (P_i), thereby decreasing $\overline{P}_f$ and hence GFR. However, recent studies in Munich-Wistar rats have shown that the effect of ureteral obstruction on SNGFR is more complex than the simple rise in the tubular interstitial pressure.[6] Thus, in hydropenic rats, ureteral obstruction resulted in proportionate increase in both P_i and $\overline{P}_g$, with a concomitant fall in SNGFR secondary to a significant reduction in K_g. In contrast, in plasma-expanded rats, ureteral obstruction resulted in a decrease in SNGFR secondary to a fall in $\overline{P}_g$ caused by a rise in P_i. Thus, it appears that the mechanisms mediating the effect of ureteral obstruction on SNGFR are a function of the state of fluid balance of the animal.

4. *A rise in renal venous pressure* leads to retrograde distention of the venules and the peritubular capillary network. The resulting rise in the capillary blood pressure compresses adjacent tubules until pressure in them equals that in the capillaries. The net result is a decrease in $\overline{P}_f$ and hence GFR.

5. *A decrease in plasma protein concentration*, which might occur in protein deficiency, starvation, *cirrhosis of the liver* (decreased protein synthesis), or *nephrotic syndrome* (increased urinary loss of proteins) is expected to decrease π_g, thereby increasing $\overline{P}_f$ and hence GFR. However, this expected increase in GFR may not occur in early stages of these diseases. An explanation might be that the reduced plasma protein concentration in these conditions tends to have a much greater effect in the systemic circulation, where it reduces the effective circulating blood volume, owing to edema formation. Therefore, the effect of decrease in π_g on GFR will be somewhat blunted by a parallel reduction in P_g, secondary to reduced plasma volume.

Problems

5.1. In each of the following circumstances, if you expect glomerular filtration rate to increase, mark (+), if you expect it to decrease, mark (−), and if you expect no change to occur, mark (0). Assume that no other changes will occur.
   a. An increase in the blood pressure at the glomerulus. ____
   b. An increase in the protein osmotic (oncotic) pressure in the plasma of the systemic blood. ____
   c. A marked decrease in the concentration of proteins in the plasma of the systemic blood. ____
   d. Constriction of the afferent arteriole to the glomerulus. ____
   e. An intravenous injection of isotonic glucose solution sufficient to expand the extracellular fluid volume. ____
   f. An intravenous injection of epinephrine sufficient to raise the systemic arterial blood pressure. ____
   g. Drinking two cups of coffee. ____

5.2. If the plasma concentration of sodium is 144 mEq/L and that for chloride is 100 mEq/L,
   a. Calculate the ultrafiltrate concentration of sodium if its chloride concentration is 110 mEq/L.
   b. Calculate the Gibbs-Donnan ratio for both sodium and chloride.

5.3. Rank as higher or lower the difference between the capillary hydrostatic (P_c) and protein oncotic (π_c) pressures ($P_c - \pi_c$) in the lungs and kidneys relative to that in the gastrointestinal system. Briefly indicate the physiological benefits of your choice.

5.4. Complete the following table by indicating for the items listed what would be the steady-state deviation from normal (+ for increase; − for decrease; and 0 for no change), for each forcing.
   $\overline{P}_{AS}$ = mean systemic arterial pressure
   $\overline{P}_{VS}$ = mean systemic venous pressure
   GFR = glomerular filtration rate
   π_p = plasma protein oncotic pressure
   R_A = preglomerular capillary resistance
   R_E = postglomerular capillary resistance

Forcings		$\overline{P}_{AS}$	$\overline{P}_{VS}$	GFR	π_p
a.	Infusion of 2 L of isotonic saline solution.				
b.	Venous hemorrhage equal to 20% of blood volume.				
c.	Infusion of 2 L of plasma.				
d.	A step increase in R_A.				
e.	A step increase in R_E.				

5.5. Describe the structural components of the glomerular filtration barrier.

5.6. What are the differences between the glomerular basement membrane (GBM) and the peritubular basement membrane?

5.7. What are the four cell types found within a renal corpuscle?

5.8. Why are the glomerular epithelial cells also called podocytes?

5.9. What kind of membrane is the slit membrane?

5.10. Describe the course of the GBM with respect to the glomerular mesangium.

5.11. What is known about the function of mesangial cells?

5.12. The glomerular filtration barrier is known to be strongly negatively charged. Where are the negative charges located?

5.13. What are the three layers of the GBM?

5.14. What are the chemical components that establish the GBM?

5.15. What is the structural correlate for the size selective barrier of the glomerular filter?

5.16. How is clogging of the glomerular filter prevented?

5.17. Describe the hemodynamic forces governing the passive process of glomerular ultrafiltration.

5.18. What is meant by the term "filtration pressure equilibrium"? Explain its physiological significance, and the factors affecting it.

5.19. What are the mechanisms which normally restrict protein filtration at the glomerulus?

5.20. Describe the mechanism of proteinuria in glomerular disease, such as glomerulonephritis.

5.21. Explain the possible mechanisms involved in the humoral regulation of SNGFR.

References

1. Aikens B, Eenboom A, Bohle A: Untersuchungen zur struktur des Glomerulum. Rekonstruktion eines Rattenglomerulum an O,5 u dicken Serienschnitten. *Virchows Arch Pathol Anat* 1979; 381:283–293.
2. Arendshorst WJ, Gottschalk, WW: Glomerular ultrafiltration dynamics: Historical perspective. *Am J Physiol* 1985; 248:F163–F174.
3. Bankir L, Farman N: Heterogeneite des glomerulus chez le lapin. *Arch Anat Microsc* 1973; 62:281–291.
4. Baylis C, Deen WM, Myers BD, Brenner BM: Effect of some vasodilator drugs on transcapillary fluid exchange in the renal cortex. *Am J Physiol* 1976; 230:1148–1158.
5. Bennett CM, Glassock RJ, Chang RLS, et al: Permselectivity of the glomerular capillary wall. Studies of experimental glomerulonephritis in the rat using dextran sulfate. *J Clin Invest* 1976; 57:1287–1294.
6. Blantz RC, Konnen KS, Tucker BJ: Glomerular filtration response to elevated ureteral pressure in both the hydropenic and the plasma-expanded rats. *Circ Res* 1975; 37:819–829.
7. Brenner BM, Baylis C, Deen WM: Transport of molecules across renal glomerular capillaries. *Physiol Rev* 1976; 56:502–534.
8. Brenner BM, Falchuk KH, Keimowitz RI, Berliner RW: The relationship between peritubular capillary protein concentration and fluid reabsorption by the renal proximal tubule. *J Clin Invest* 1969; 48:1519–1531.
9. Brenner BM, Hostetter TH, Humes HD: Molecular basis of proteinuria of glomerular origin. *N Engl J Med* 1978; 298:826–833.
10. Brenner BM, Hostetter TH, Humes HD: Glomerular permselectivity: Barrier function based on discrimination of molecular size and charge. *Am J Physiol* 1978; 234:F455–F460.
11. Brenner BM, Troy JL, Daugharty TM: The dynamics of glomerular ultrafiltration in the rat. *J Clin Invest* 1971; 50:1776–1780.
12. Brenner BM, Troy JL, Daugharty TM, et al: Dynamics of glomerular ultrafiltration in the rat. II. Plasma-flow dependence of GFR. *Am J Physiol* 1972; 223:1184–1190.
13. Chang RLS, Robertson CR, Deen WM, Brenner BM: Permselectivity of the glomerular capillary wall to macromolecules. I. Theoretical considerations. *Biophys J* 1975; 15:861–886.
14. Chang RLS, Ueki IF, Troy JL, et al: Permselectivity of the glomerular capillary wall to macromolecules. II. Experimental observations in the rat. *Biophys J* 1975; 15:887–906.
15. Chang RLS, Deen WM, Robertson CR, Brenner BM: Permselectivity of the glomerular capillary wall: III. Restricted transport of polyanions. *Kidney Int* 1975; 8:212–218.
16. Deen WM, Bohrer MP, Brenner BM: Macromolecule transport across glomerular capillaries: Application of pore theory. *Kidney Int* 1979; 16:353–365.
17. Deen WM, Robertson CR, Brenner BM: A model of glomerular ultrafiltration in the rat. *Am J Physiol* 1972; 223:1178–1183.
18. Deen WM, Troy JL, Robertson CR, Brenner BM: Dynamics of glomerular ultrafiltration in the rat. IV. Determination of ultrafiltration coefficient. *J Clin Invest* 1973; 52:1500–1508.
19. Dworkin LD, Ichikawa I, Brenner BM: Hormonal modulation of glomerular function. *Am J Physiol* 1983; 244:F95–F104.
20. Elema JD, Hoyer JR, Vernier RL: The glomerular mesangium: Uptake and transport of intravenously injected colloidal carbon in rats. *Kidney Int* 1976; 9:395–406.
21. Farquhar, M. G.: The glomerular basement mem-

brane, in Hay ED (ed). *Cell Biology of the Extracellular Matrix*. New York, Plenum Press, 1982, pp 335–378.
22. Harris CA, Bauer PG, Chirito E, Dirks JH: Composition of mammalian glomerular ultrafiltrate. *Am J Physiol* 1974; 227:972–976.
23. Karnovsky MJ: The ultrastructure of glomerular filtration. *Ann Rev Med* 1979; 30:213–224.
24. Knox FG, Cuche JL, Ott CE, et al: Regulation of glomerular filtration and proximal tubule reabsorption. *Circ Res (suppl I)* 1975; 36–37:I107–I118.
25. Koushanpour E: *Renal Physiology: Principles and Functions*, ed 1. Philadelphia, WB Saunders Co, 1976.
26. Kreisberg JI: Contractile properties of the glomerular mesangium. *Fed Proc* 1983; 42:3053–3057.
27. Latta H, Johnston WH, Stanley TM: Sialoglycoproteins and filtration barriers in the glomerular capillary wall. *J Ultrastruct Res* 1975; 51:354–376.
28. Ludwig C: *Beitrage Zur Lehre vom Mechanismus der Harnsecretion*. Marburg, Elwert, 1843.
29. Maddox DA, Deen WM, Brenner BM: Dynamics of glomerular ultrafiltrtion. VI. Studies in the primate. *Kidney Int* 1974; 5:271–278.
30. Michael AF, Keane WF, Raij L, et el: The glomerular mesangium. *Kidney Int* 1980; 17:141–154.
31. Murakami T: Glomerular vessels of the rat kidney. A scanning electron microscope study of corrosion casts. *Arch Histol Jpn* 1972; 34:87.
32. Myers BD, Deen WM, Brenner BM: Effects of norepinephrine and angiotensin II on the determinants of glomerular ultrafiltration and proximal tubule fluid reabsorption in the rat. *Circ Res* 1975; 37:101–110.
33. Peter K: *Untersuchungen uber Bau und Entwicklung der Niere*. Jena, Gustav Fischer, 1909.
34. Pitts R F: *Physiology of the Kidney and Body Fluids*, ed 3. Chicago, Year Book Medical Publishers, Inc, 1974.
35. Rennke HG, Venkatachalam MA: Structural determinants of glomerular permselectivity. *Fed Proc* 1977; 36:2619–2626.
36. Rodewald R, Karnovsky MJ: Porous substructure of the glomerular slit diaphragm in the rat and mouse. *J Cell Biol* 1974; 60:423–433.
37. Shea SM: Glomerular hemodynamics and vascular structure: The pattern and dimensions of a single rat glomerular capillary network reconstructed from ultrathin sections. *Microvasc Res* 1979; 18:129–143.
38. Shea SM, Morrison AB: A stereological study of the glomerular filter in the rat. Morphometry of the slit diaphragm and basement membrane. *J Cell Biol* 1975; 67:436-443.
39. Taugner R, Schiller A, Kaissling B, Kriz W: Gap junctional coupling between the JGA and the glomerular tuft. *Cell Tissue Res* 1978; 186:279–285.
40. Thurau K: Renal hemodynamics. *Am J Med* 1964; 36:698–719.
41. Tisher CC: Anatomy of the kidney, in Brenner BM, Rector FC (eds): *The Kidney*, Philadelphia, WB Saunders Co, vol 1, 1976, pp 3–64.
42. Tucker BJ, Blantz RC: An analysis of the determinants of nephron filtration rate. *Am J Physiol* 1977; 232:F477–F483.
43. Venkatachalam MA, Rennke HG: The structural and molecular basis of glomerular filtration. *Circ Res* 1978; 43:337–347.
44. Wearn JT, Richards AN: Observations on the composition of glomerular urine, with particular reference to the problem of reabsorption in the renal tubule. *Am J Physiol* 1924; 71:209–227.
45. Wiederhelm CA, Woodbury JW, Kirk S, Rushmer RF: Pulsatile pressures in the microcirculation of frog's mesentery. *Am J Physiol* 1964; 207:173–176.
46. Windhager EE: *Micropuncture Technique and Nephron Function*. New York, Appleton-Century-Crofts, 1968.

6
Regulation of Renal Blood Flow and Glomerular Filtration Rate

In the preceding chapter we learned that the process of passive filtration at the glomerulus delivers a copious volume of virtually protein-free plasma to the proximal tubule for further processing. Since major alterations in the volume of plasma filtered at the glomerulus per minute, namely, the glomerular filtration rate (GFR), would have a profound effect on the subsequent tubular processing of this filtrate, there must be some intrinsic mechanisms whereby GFR is carefully controlled. In this chapter we briefly review some of the experiments that have led to the current concept of renal autoregulation and the intrinsic control of renal blood flow (RBF), including the possible role of the juxtaglomerular apparatus. In addition to these mechanisms, the roles of the intrarenal production of renin-angiotensin and prostaglandin in autoregulation and renal regulation of volume and composition of body fluids are also discussed. Finally, we conclude with a discussion of the extrinsic control of the renal circulation.

Autoregulation or Intrinsic Regulation of Renal Circulation

Autoregulation has been defined as the intrinsic property of an organ, such as the kidney, to maintain its blood flow nearly constant despite changes in arterial perfusion pressure.[54] This implies that regulation of renal blood flow can occur independently of external influences, such as neural activity and hormones. Furthermore, since renal oxygen extraction, relative to renal blood flow (Chapters 5 and 7) is small (10% to 15%), it is unlikely that renal metabolism per se is the underlying mechanism for autoregulation of renal blood flow. Because renal demand for the oxygen required for tubular transport processes (Chapters 9 and 10) is a function of renal blood flow, it is more likely that the goal of autoregulation is to regulate renal blood flow. Evidence amassed to date confirms this conclusion and indicates further that autoregulation of blood flow applies to both cortical and medullary regions of the kidney. However, a recent study[22] has shown that the autoregulation of the medullary blood flow occurs over a narrower range of arterial blood pressure, as compared with autoregulation for whole kidney blood flow (see below). This asymmetric effect of changes in arterial blood pressure on cortical versus medullary blood flow may in part account for pressure-diuresis and reduced urinary osmolality following extracellular fluid volume expansion. The effect of changes in medullary blood flow on the ability of the kidney to concentrate and dilute urine is detailed in Chapter 12.

Since changes in RBF have profound effect on GFR (Chapter 5), it is homeostatically beneficial to prevent large changes in GFR and hence salt and water excretion secondary to changes in renal perfusion pressure. Thus, it is likely that regulation of renal blood flow is a by-product of the autoregulation of GFR. This may be achieved by two major intrinsic mechanisms: The myogenic and the tubuloglomerular feedback (TGF) mechanisms. The myogenic mechanism involves the control of smooth-muscle contraction in the preglomerular renal vasculature in response to changes in renal perfusion pressure. The TGF mechanism involves the selective control of the resistance of the pre- and postglomerular arterioles in response to changes in distal tubule flow. The myogenic and TGF mechanisms, working in series, act as coarse- and fine-control mechanisms, respectively, to stabilize GFR and regulate the volume and composition of body fluids. In the following sections we briefly review the experimental evidence pertaining to how the intrinsic mechanisms are involved in regulation of renal circulation and glomerular filtration rate.

Myogenic Mechanism

In 1951 Shipley and Study demonstrated in dogs that the GFR increases linearly as the mean renal arterial pressure is increased from about 20 to 80 mm Hg.[88] However, when the renal arterial pressure was raised beyond 80 mm Hg, GFR remained relatively constant (Figure 6.1, *left panel*). Similarly, they showed that RBF increased linearly when renal arterial pressure was raised from about 10 to 80 mm Hg, but remained relatively constant as the pressure was increased from 80 to 180 mm Hg. However, in contrast to GFR, RBF increased again as the pressure was raised beyond 180 mm Hg. Also, as shown in this figure, it appears that RBF and GFR are not autoregulated in parallel and that the pressure ranges for the two variables are not the same. Furthermore, the yield pressure, or the minimum mean renal arterial pressure necessary before the blood begins to flow into the kidney, is somewhat less than that required for the formation of the glomerular filtrate.

This constency of GFR and RBF, when the mean renal arterial pressure is varied from 80 to 180 mm Hg (Figure 6.1, left panel), is called *renal autoregulation*. It has been observed in denervated and isolated kidneys, in nonfiltering, nontransporting kidney without fluid delivery to the macula densa,[77] during blockade of sympathetic ganglia, in the presence of angiotensin II antagonist,[1] and even in the absence of red blood cells. Recent studies, using a variety of experimental preparations, have confirmed that autoregulation of GFR and RBF is present, to a greater or lesser extent, in the kidneys of rat,[5, 23] dog,[29, 69–71] and man.[14]

An interesting corollary finding in the study of Shipley and Study[88] was that at renal arterial pressures between 0 and 60 mm Hg, there was no measurable urine output. However, after the pressure exceeded 60 mm Hg, there was a progressive increase in urine output, a phenomenon referred to as *pressure diuresis* (Figure 6.1, right panel). At present, the mechanism of this pressure diuresis, which has been postulated to play an important role in the regulation of the extracellular fluid volume,[42] remains unclear. The fact that urine output is progressively increased as the renal arterial pressure is increased from 80 to 180 mm Hg and beyond suggests that changes in renal perfusion pressure somehow affect renal tubular transport of salt and water independently of variations in GFR. Since previous studies have failed to establish clear-cut roles for pressure-induced changes in peritubular physical forces, intrarenal blood flow distribution, or renal hemodynamics in pressure-diuresis, this has led to the concept that some intrarenal humoral mechanisms may be responsible for the phenomenon of pressure diuresis.

Recent studies have implicated renal prostaglandin as the humoral agent mediating pressure natriuresis.

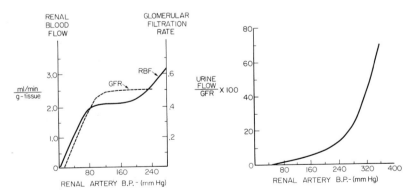

FIGURE 6.1. *Left panel*: Steady-state levels of GFR and RBF, both expressed as milliliter per minute per gram of kidney tissue, as a function of renal artery blood pressure (BP) in denervated, blood-perfused dog kidney. Note that the yield pressure for the RBF curve occurs at about 10 mm Hg, whereas that for the GFR curve occurs at about 20 mm Hg. *Right panel*: Changes in urine flow rate, expressed as a percentage of GFR, as a function of BP. Note the yield pressure for the urine flow curve occurs at about 60 mm Hg, a value considerably higher than that for both the GFR and RBF curves. (Redrawn from Shipley and Study.[88])

Thus, micropuncture studies have shown that infusion of prostaglandin inhibited sodium chloride reabsorption in the loops of Henle, distal tubule, and collecting duct.[76, 100] Studies in dogs further showed a positive correlation between sodium excretion and changes in renal perfusion pressure which was abolished by infusion of indomethacin, an inhibitor of prostaglandin synthetase, without affecting renal potasium transport.[20] Also, the pressure-induced changes in sodium excretion were found to parallel changes in prostaglandin (PGE_2) excretion. Taken together, these findings indicate the sodium-reabsorptive specificity of the prostaglandin-dependent pressure-natriuresis phenomenon. To further delineate the role of prostaglandin in pressure diuresis, Haas and coworkers implanted specially designed polyethylene matrix capsules in the dog kidney to measure renal interstitial pressure, and compared the effects of intrarenal infusion of prostaglandin with that of a newly developed prostaglandin analog.[44] They found that intrarenal infusion of PGE_2 increased RBF, renal interstitial pressure, and urinary sodium excretion. However, intrarenal infusion of the prostaglandin analog resulted in a similar increase in RBF without any effect on renal interstitial pressure or urinary sodium excretion. Intrarenal infusion of PGE_2, while keeping the renal interstitial pressure constant, produced an effect similar to that of the prostaglandin analog. They concluded that the increase in renal interstitial pressure is necessary for the prostaglandin-induced vasodilation and natriuresis. Despite this evidence, the nature of the stimulus that induces prostaglandin synthesis during changes in renal perfusion pressure remains to be identified.

What are the possible hemodynamic mechanisms and potential vascular sites that may account for the phenomenon of autoregulation of renal blood flow? To explore this question, we may relate the renal blood flow (RBF), the driving mean renal arterial pressure ($\bar{P}$), and the renal arterial resistance (R) by the equation RBF = $\bar{P}$/R. Thus, referring to Figure 6.1 (left panel), we note that at pressure values below 80 mm Hg, renal blood flow is a positive linear function of pressure, which according to the above equation requires the renal resistance to remain constant. At pressure values ranging from 80 to 180 mm Hg, the resistance must vary as a positive linear function of pressure for the renal blood flow to remain constant. Since over such a pressure range the GFR also remains unchanged, the increase in resistance with increasing pressure may be attributed primarily to an adaptive increase in the resistance of the *afferent arteriole*. This pressure-flow relationship is one aspect of autoregulation. The other aspect is that the filtration fraction (GFR/RPF) falls as the renal plasma flow (RPF) increases at pressure values beyond 180 mm Hg.

The intrinsic mechanism for the autoregulation of RBF, and hence GFR, as outlined above, is based on the concept, originally proposed by Bayliss,[9] that the vascular smooth-muscle contracts, thereby reducing its luminal radius, in response to the increased stretch produced by increased blood pressure. Extension of this so-called *"myogenic"* mechanism to the kidney in renal autoregulation is based on the premise that the renal resistance vessels are arranged as series-coupled independent myogenic units, each unit responding to changes in its own intraluminal pressure gradient.[53] Thus, a stepwise increase (from 80 to 180 mm Hg) or a stepwise decrease (from 180 to 80 mm Hg) in arterial blood pressure (and hence renal perfusion pressure) will induce the greatest pressure change in the interlobular arteries, thereby causing a lesser pressure change downstream in the afferent arterioles. In this manner, and in accordance with Laplace's law (T = P × r), small changes in the luminal radius (r) of the afferent arteriole subsequent to small changes in its transmural pressure (P) will result in a large change in its circumferential wall tension (T). Accordingly, in response to abrupt changes in arterial blood pressure, the myogenic mechanism serves to maintain a constant wall tension by reducing the vessel radius so that the *pressure radius product* is restored to normal.

A number of studies have provided direct as well as indirect support for the role of the myogenic mechanism in renal autoregulation. Kallskog and coworkers measured a pressure drop of 20 to 30 mm Hg between the renal artery and the occasional interlobular arteries that reach the surface of the rat kidney.[56] They found that this pressure drop was reduced as the arterial blood pressure was decreased, suggesting that the autoregulation occurs in the renal arterial system and is not confined to the arterioles. Recently, Gilmore and associates provided direct evidence in support of the concept that a wall-tension-regulating mechanism in the afferent arteriole accounts for the intrinsic mechanism of renal autoregulation.[33] They found that in kidney fragments transplanted to the hamster cheek pouch, changes in the diameter of the afferent arterioles varied inversely with changes in the transmural pressure. More recently, Young and Marsh found a biphasic pattern in the time-course of autoregulatory vascular resistance changes in response to a sudden change in arterial blood pressure.[102] Approximately half of the resistance change occurred within a "time constant" of 1 to 2 seconds and before any detectable change in the distal tubular flow, whereas the remaining change in resistance occurred within a time constant of 10 to 12 seconds, a time-course compatible with known responses of the tubuloglomerular feedback mechanism (see below).

What is the mechanism by which stretch of the smooth muscles in the afferent arteriole is transduced into vasoconstriction? It is well known that smooth-muscle contraction results from an increase in free cytoplasmic calcium. Since sarcoplasmic reticulum is rather sparse in smooth muscles, transport of calcium from the extracellular compartment may play a major role in the regulation of smooth-muscle contraction. Accordingly, calcium entry into the muscle is regulated by several proposed channels, one of which may be stretch-sensitive.[13] Fray[30] and Lush and Fray[60] recently developed a stretch receptor model to examine the implications of this concept as a possible explanation for the cellular processes involved in myogenic autoregulation of renal blood flow. According to this model, an increase in renal perfusion pressure increases the stretch of the afferent arteriole, which causes an increase in permeability of the granular cell membrane to calcium (opens the stretch-sensitive calcium channel), thereby depolarizing the granular cells and inhibiting renin release. Conversely, a decrease in renal perfusion pressure will produce the opposite effects: it reduces stretch, which decreases permeability to calcium (closing the calcium channel), thereby hyperpolarizing the granular cells, and stimulating renin release. In this manner the resistance in the afferent arteriole is controlled by an inverse relationship between renin release and renal perfusion pressure.

How are changes in renal perfusion pressure compensated for by the various determinants of SNGFR? This question has been addressed by extensive micropuncture studies in Munich-Wistar rats.[24, 74] The results of these studies are summarized in Figure 6.2.

The left-hand side of this figure summarizes the effects of reduction in mean arterial pressure $(\overline{AP})$ on the initial glomerular plasma flow rate $(\dot{Q}_A)$ and SNGFR (upper panel), net glomerular transcapillary hydraulic pressure $(\overline{\Delta P})$, afferent (π_A) and efferent (π_E) protein oncotic pressures (middle panel), and afferent or preglomerular (R_A) and efferent or postglomerular (R_E) resistances (lower panel). The right-hand side of this figure summarizes the effect of reduction in $\overline{AP}$ for the same variables during administration of papaverine, a potent inhibitor of smooth-muscle contraction.

As shown on the left-hand side of Figure 6.2, as $\overline{AP}$ is reduced from 115 mm Hg to 80 mm Hg (stippled area), there is a small fall in SNGFR (shown by the curve drawn through the filled circles), indicating attenuation of the autoregulatory ability of the kidney. As shown, over the entire range of arterial pressures studied, single-nephron filtration fraction (SNFF) remained relatively constant, indicating that the fall in SNGFR was accompanied by a proportional fall in $\dot{Q}_A$ (shown by the curve drawn through the open circles). Despite this effect, ΔP, as well as π_A and π_E, remained relatively unchanged (middle panel). As can be seen in the lower panel, this constancy of ΔP is achieved by a simultaneous decrease in R_A and increase in R_E. The results found during the administration of papaverine (a smooth-muscle relaxant), depicted on the right-hand side of Figure 6.2, indicate that these alterations in R_A and R_E were mediated by *adaptive* changes in the resistance of the afferent arteriole as reflected by changes in the arteriolar diameter. Administration of this drug paralyzes the smooth muscles lining the walls of the afferent and efferent arterioles, thereby abolishing the autoregulatory responses of R_A and R_E to reductions in $\overline{AP}$ and allowed ΔP to change in proportion to changes in $\overline{AP}$. These results clearly indicate that in Munich-Wistar rats, autoregulation of GFR is mainly the consequence of autoregulation of renal plasma flow, since GFR is strongly dependent on plasma flow (Chapter 5).

In summary, though the precise nature of the myogenic mechanism is uncertain, it may serve to *rapidly buffer* the effect of sudden changes in arterial blood pressure on GFR and hence the tubular flow rate, which would otherwise adversely affect the stable delivery of sodium chloride to the macula densa cells and the distal nephron. In contrast, the tubuloglomerular feedback mechanism (see below) serves to protect against changes in tubular reabsorption, which is a *slow* process.

Tubuloglomerular Feedback Mechanism

Recent studies have demonstrated the existence of an intrarenal, local feedback mechanism regulating GFR. This regulating system consists of a specialized structure within the nephron called the juxtaglomerular apparatus. Before we can proceed to describe

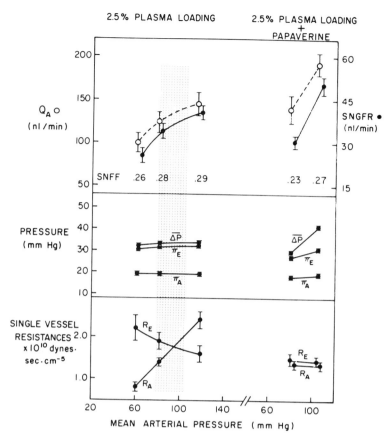

FIGURE 6.2. Summary of the effects of reducing mean arterial pressure ($\overline{AP}$) on measured determinants of glomerular ultrafiltration in mildly plasma-loaded Munich-Wistar rats before (*left*) and during (*right*) administration of papaverine. The stippled area on the left corresponds to the range of change in $\overline{AP}$ produced on the right. The mild plasma-loading used in these experiments is a maneuver that has been found to enhance the autoregulatory capability of the rat kidney. (From Deen.[24])

the function of this system and its contribution to renal autoregulation, a review of its morphology is in order.

Morphology of Juxtaglomerular Apparatus (JGA)

At the point where Henle's loop returns to its renal corpuscle, the end portion of the thick ascending limb passes through the angle of the afferent and efferent arterioles and contacts the vascular pole of the glomerulus. The juxtaglomerular apparatus comprises all structures that are present within this specific area of contact. These are (Figure 5.3): (1) the macula densa of the thick ascending limb; (2) the extraglomerular mesangial cells, which fill the angle between the afferent and efferent arterioles; and (3) the renin producing granular cells of the afferent arteriole. Because of their intimate connection to the JGA and probably due to their functional relationships, the mesangial cells proper as well as the ordinary smooth-muscle cells of the afferent and efferent arterioles are sometimes considered as part of the JGA.[95]

The *macula densa* is the tubular structure with which the thick ascending limb touches the vascular pole of its glomerulus (Figures 6.3 and 6.4). It is a plaque of 10 to 20 specialized epithelial cells replacing at this side the usual epithelium of the thick ascending limb. The term "macula densa", i.e., a "dense plaque," refers to the fact that the tall but slender cells have large nuclei, which are packed densely together.

The cells of the macula densa are polygonal in

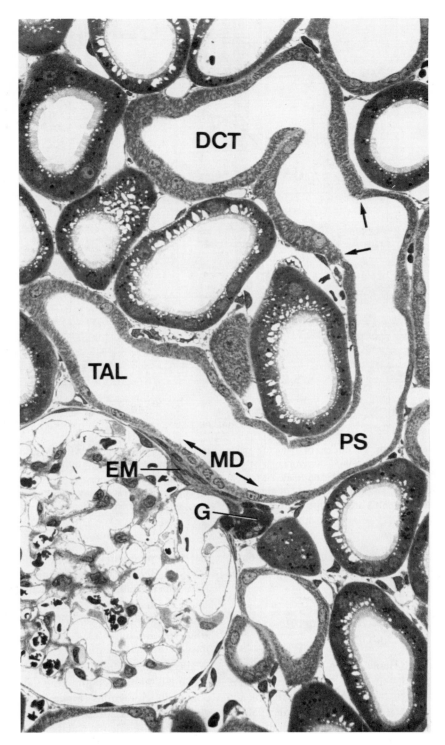

FIGURE 6.3. A light micrograph showing the relationship between the thick ascending limb (TAL) and the extraglomerular mesangium (EM) at the vascular pole of a glomerulus. At the point of contact, the epithelium of the TAL differentiates into macula densa (MD). After the MD, the TAL has a short postmacula segment (PS) which abruptly passes over (arrows) into the distal convoluted tubule (DCT). Note the clusters of granular cells (G) in the wall of the afferent arteriole.

cross sections (Figures 5.3 inset and 6.4) and do not interdigitate with each other by large processes, as do most other proximal and distal tubular cells (Chapter 9). In contrast, they are separated from each other by conspicuous intercellular spaces similar to those found between collecting duct cells (see below). These spaces vary in width according to function.[55] Lateral microplicae protrude into the lateral interspaces; some basal infoldings contribute to an overall moderate amplification of the basolateral cell membrane. Apically, the intercellular spaces are closed by deep tight junctions. The luminal membrane is increased in area by many short microvilli. The Tamm-Horsefall protein, which is believed to contribute to the water impermeability of the thick ascending limb, is absent from the macula densa cells.[48,87] The macula densa is, therefore, suggested to be a water permeable cell plaque within the otherwise water-impermeable thick ascending limb epithelium. Direct functional evidence, however, is missing. In summary, the macula densa is constructed like a transporting epithelium. There are no apparent structural features to suggest that the macula densa is a receptor.

The base of the macula densa covers the extraglomerular mesangium. This is the only consistent histotopographical relationship of the macula densa, even though it may touch variable portions of the afferent and efferent arterioles.[7,21,38] This arrangement suggests that the function of the JGA requires a total cover of the extraglomerular mesangium by the macula densa. To guarantee this cover, the macula densa overlaps and touches other structures. With respect to function, the macula densa and the extraglomerular mesangium appear to be intimately connected to each other.

The *extraglomerular mesangium* fills the space between both glomerular arterioles and the macula densa (Figure 5.3). The extraglomerular mesangial cells are piled up in several layers like a cone whose base is covered by the macula densa and whose apex passes over into the mesangium of the glomerular tuft.[38] Extraglomerular mesangial cells and the mesangial cells proper are the same type of cells. In addition, both are embedded into a conspicous extracellular matrix. Cells and matrix together establish the glomerular and the extraglomerular mesangium, respectively. The extraglomerular mesangium represents a solid cell complex that is not penetrated by vessels, neither by blood nor by lymphatic capillaries (Figure 6.4).

The extraglomerular mesangial cells are elongated

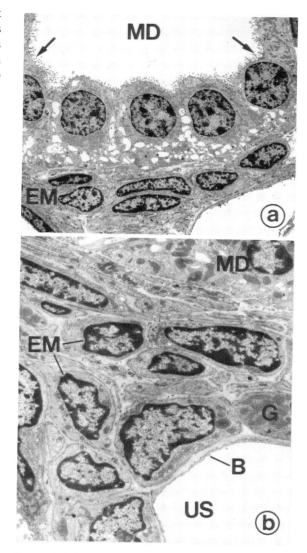

FIGURE 6.4. *Panel a*: An electron micrograph showing the structural details of the macula densa (MD) cells. Note the conspicuous intercellular spaces between the epithelial cells. The bases of the MD cells cover the extramesangial (EM) cells. *Panel b*: An electron micrograph showing further details of the EM cells. Note the extracellular spaces between the cells are filled with basement-membrane-like structures. B = Bowman's capsule; US = urinary space; G = granular cell.

and flattened cells with many processes and they contain contractile filaments. The cells and their processes are extensively coupled by gap junctions.[73] In addition, gap junction contacts are established with all other cells of the JGA (except the macula densa cells), i.e., to the mesangial cells of the glomerular tuft, to the granular cells, and to the ordinary smooth muscle cells of both glomerular arterioles. This fact is indicative that these cell types are all of smooth-muscle origin.

The extraglomerular mesangial cells are not smoothly adapted in shape to each other. The irregular spaces between them are filled by prominent extracellular matrix. It has recently been shown that this material is more sensitive to changes in total body fluid than to the common peritubular interstitium, shrinking in volume depletion and enlarging in volume expansion.[78]

The granular cells (often simply termed juxtaglomerular cells) are assembled in clusters within the terminal portion of the afferent arteriole replacing ordinary smooth-muscle cells (Figure 6.5). Their name refers to their many specific cytoplasmic granules. In usually stained sections for transmission electron microscopy (TEM), these granules are dark, membrane-bound, and irregular in size and shape. Immunocytochemical studies have clearly shown that they contain renin, which is obviously synthesized by these cells and stored in granular form.[94] Most probably, renin release occurs by exocytosis into the surrounding interstitium.

The *granular cells* clearly are modified smooth-muscle cells. In situations that require the synthesis of more renin than usual (e.g., volume depletion and stenosis of the renal artery) additional smooth-muscle cells located upstream in the wall of the afferent arteriole transform into granular cells. It has already been mentioned that the granular cells are extensively connected to adjacent smooth muscle cells and to extraglomerular mesangial cells by gap junctions. The granular cells are densely innervated by sympathetic nerve terminals.

The structural organization of the JGA strongly suggests a regulatory function. Goormaghtigh[37] was the first to propose that some component of the distal urine is sensed by the macula densa and that this information is used to adjust the tonus of the glomerular arterioles, thereby producing a change in glomerular blood flow and filtration rate. This hypothesis has been verified by many studies and is now known as the tubuloglomerular feedback mechanism.[79] Many details of this mechanism are still subject to debate (see below). The structural organization, however, allows some basic conclusions about the possible functional relevance of the individual components. (1) The macula densa is not a sensor but a transporting epithelium. (2) The macula densa is intimately related to the extraglomerular

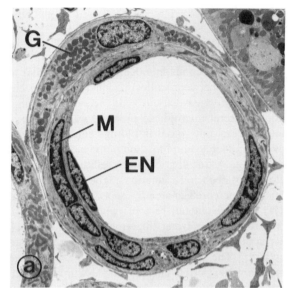

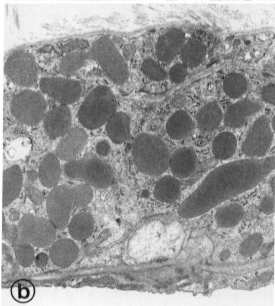

FIGURE 6.5. *Panel a*: An electron micrograph of a cross section through an afferent arteriole showing the endothelial (EN) cells, ordinary smooth muscle (M) cells, and granular (G) cells. *Panel b*: An electron micrograph showing the renin-storing granules of a granular cell.

mesangial cells: transports of the macula densa terminate or begin in the extraglomerular mesangium. (3) Since the extraglomerular mesangium has no direct blood supply and since its connections to the common peritubular interstitium are very sparse, the extracellular environment within the extraglomerular mesangium appears to be strongly dependent on the transport function of the macula densa. (4) The extraglomerular mesangial cells are coupled by gap junctions to all other smooth-muscle derived components of the JGA; these are the mesangial cells of the glomerular tuft, the granular cells, and the ordinary smooth-muscle cells of the afferent and efferent arterioles. All these cells are candidates as effectors within the tubuloglomerular feedback mechanism. (5) By exclusion, the most probable candidates for being the sensors within this regulatory circuit are the extraglomerular mesangial cells.

Theory of Local Regulation of GFR via JGA

Of the several modern hypotheses advanced to explain the role of the JGA in the feedback regulation of GFR,[43, 96] the one proposed by Thurau[96] has received the most attention and has been the subject of intensive investigation. Therefore, we use his proposal as the framework for the following discussion concerning the role of the JGA in the *local* feedback regulation of GFR.

On the basis of then known anatomical and physiological evidence concerning the structural organization of the JGA and its endocrine function, Thurau[96] formulated the following proposal, which has also been called the macula densa theory of renin activation and release, for the local feedback regulation of GFR in each nephron by the JGA.

A high concentration of sodium in the thick ascending limb fluid passing by the macula densa cells into the distal convoluted tubule, directly or presumably via the extramesangial cells, stimulates the granular cells to release small quantities of the enzyme renin into the afferent arteriole of the same nephron. This enzyme acts on its substrate angiotensinogen (found chiefly in the α_2-globulin fraction of plasma) to form the decapeptide angiotensin I. This substance is then converted into an octapeptide angiotensin II, a potent vasoconstrictor substance, by the action of an angiotensin I converting enzyme,[35, 68] which is assumed to also be present locally. The local increase in angiotensin II concentration in the afferent arteriole of the same nephron causes constriction of the afferent arteriole, thereby increasing its resistance to flow and decreasing glomerular capillary hydraulic pressure, and hence net filtration pressure (ΔP_f) and single-nephron GFR (SNGFR). In this manner, changes in SNGFR manifested by changes in sodium concentration and monitored by the macula densa cells, are kept within normal limits through this negative tubuloglomerular feedback mechanism. Furthermore, Thurau[96] proposed that this local mechanism for regulation of SNGFR is also responsible for the autoregulation of the GFR and RBF in the whole kidney as the renal perfusion pressure is increased from 80 to 180 mm Hg.

Before considering the merit and the experimental basis for this proposal, the following four premises of this theory should be kept in mind. First, there exists a feedback mechanism between the distal tubule and the SNGFR of the same nephron. Second, a change in sodium concentration in the distal tubule is the signal monitored by the macula densa cells. Third, the afferent arteriole is the effector structure and its tone (resistance) is controlled by the renin-angiotensin system. And fourth, the local activation of the renin-angiotensin system is the mediating signal to the effector structures. Let us now briefly review the studies that have examined the four assumptions of this hypothesis.

Experimental Basis for Local Regulation of SNGFR via JGA

1. Does a feedback mechanism exist between the distal tubule and SNGFR of the same nephron? The assumption that a feedback mechanism exists between the distal tubule and the SNGFR of the same nephron has been the subject of intensive investigation. Numerous studies have used both orthograde and retrograde microperfusion of Henle's loop to assess the presence or absence of SNGFR tubuloglomerular feedback. The orthograde perfusion technique involves microperfusion of Henle's loop from an end-proximal tubular segment with test solutions and subsequent measurement of SNGFR in an early proximal tubular segment. In contrast, the retrograde perfusion technique involves microperfusion of Henle's loop from an early distal tubular segment, in the opposite direction of normal tubular flow and subsequent measurement of SNGFR in an early proximal tubular segment. Using the orthograde microperfusion technique, a number of studies have provided supportive evidence for the existence of a feedback mechanism between the distal tubule and the SNGFR of the same nephron. Thus, Schnermann and co-workers[83, 84] showed that an increase in the rate of flow through Henle's loop from 8 to 25 nL/min resulted in a 10 nL/min fall in SNGFR. Further studies by Schnermann and colleagues[79] revealed that the feedback relationship between changes in SNGFR (ΔSNGFR) and changes in flow in Henle's loop

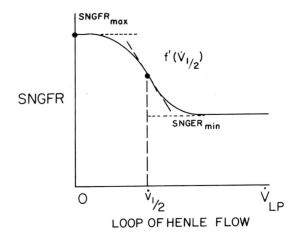

FIGURE 6.6. The sigmoidal relationship between changes in SNGFR and the flow rate in Henle's loop. The logistic equation describing the sigmoidal curve is given in the lower box. See text for explanation of symbols. (Redrawn from Schnermann and Briggs.[79])

($\Delta \dot{V}_{LP}$) is best described by a sigmoidal curve and defined by the following equation:

$$\Delta SNGFR = SNGFR_{max} - \frac{(SNGFR_{max} - SNGFR_{min})}{[1 + \exp k(A - \Delta \dot{V}_{LP})]} \quad (6.1)$$

where $SNGFR_{max}$ and $SNGFR_{min}$ are the measured maximum and minimum SNGFR, A is the loop flow at which the half-maximum response is seen ($\dot{V}_{1/2}$), $k = [4\, f'(\dot{V}_{1/2})]/(SNGFR_{max} - SNGFR_{min})$, and $f'(\dot{V}_{1/2})$ is the slope at the half-maximum point. This equation, which has frequently been called the logistic equation,[73] defines a sigmoidal relationship between two variables, $\Delta SNGFR$ and $\dot{V}_{LP}$, in terms of three parameters: the maximum change in SNGFR ($SNGFR_{max} - SNGFR_{min}$), the value of loop flow at which the half-maximum response is seen ($\dot{V}_{1/2}$), and the slope at the half-maximum point ($f'(\dot{V}_{1/2})$). Figure 6.6 depicts the salient features of such a sigmoidal relationship as defined by Equation 6.1. The experimental data, which were used to define the plot shown in Figure 6.6, revealed that changes in loop flow below 10 nL/min resulted in a small change in SNGFR; changes in loop flow from 14 to 24 nL/min resulted in a steep fall in SNGFR; and changes in loop flow above 30 nL/min resulted in no change in SNGFR. These results suggested that roughly 60% to 70% of the total SNGFR response occurs in a 10 nL/min flow range, with the normal operating point at the midpoint of this flow range.

Subsequent studies have shown that this negative feedback mechanism linking the distal sensing site to glomerular function has a variable sensitivity control (gain) that is affected by the status of the salt and water balance of the animal under study. It should be noted that most of the previous studies were carried out on surface glomeruli. Since the anatomical relationship between the macula densa and the glomerulus is similar for both deep and superficial nephrons, it was assumed that such a feedback mechanism exists for all nephrons throughout the kidney. The validity of this assumption was confirmed by recent studies that showed that the juxtamedullary nephrons have a tubuloglomerular feedback mechanism capable of regulating their filtration rate and that their feedback sensitivity is much greater as compared with the superficial nephrons.[67, 89] Thus, we may conclude that the tubuloglomerular feedback mechanism exerts its effect on the GFR of the whole kidney by regulating the SNGFR of both superficial and juxtamedullary nephrons.

2. What is the nature of the signal at the macula densa that initiates the feedback mechanism? Although sodium concentration in the thick ascending limb fluid near the macula densa appears to be an important factor in stimulating the macula densa cells, several studies have further clarified the nature of the signal perceived by the macula densa cells. Thus, when flow rate through Henle's loop is increased, early distal sodium concentration is found to increase.[65] However, subsequent studies showed that it is not the distal tubule sodium concentration that is the sensed variable, but rather it is the reabsorption of sodium or some other constituent of the distal tubule fluid that initiates the feedback mechanism. Using a retrograde perfusion technique, Schnermann and co-workers[82] found unequivocal feedback responses during perfusion of solutions containing the chloride salts of either sodium, potassium, rubidium, cesium, or ammonium, as well as bromide salts of either sodium or potassium. These results are consistent with the concept of sodium-dependent chloride cotransport recently proposed by Greger and co-workers.[40] Thus, during loop perfusion with sodium-free solution, enough Na enters the perfusate to saturate the cotransport system so that sodium concentration probably is never the rate-limiting step in the transport. This leaves chloride concentration as the critical variable that, under normal conditions, is likely to limit the magnitude of sodium chloride transport by the macula densa cells, thereby modulating the tubuloglomerular feedback response. From these studies, Schnermann and colleagues concluded that *chloride* rather than sodium was the critical constituent of the distal tubule fluid responsible for initiating the feedback.[15, 79, 82] These and other studies have led to the development of the concept of ion transport rather than the concentration per se as the initiating signal for the tubuloglomerular feedback mechanism. Thus, microperfusion studies with diuretic drugs that act in the ascending limb of Henle's loop, furosemide[18] and ethacrynic acid,[17] which apparently inhibit chloride transport also inhibited the feedback mechanism. In contrast, microperfusion studies with diuretic drugs which inhibit sodium chloride transport in either the proximal tubule (acetazolamide)[10] or the collecting duct (amiloride)[92] had no effect on the feedback response. The results of these studies, coupled with the ion-substitution studies of Schnermann and co-workers,[82] provide strong evidence that the reabsorption of sodium chloride by the macula densa cells is the signal initiating the feedback response.

3. Where is the locus of the effector structures? A number of studies have examined whether the adaptive changes in the resistance of the afferent arteriole is the locus of the negative feedback regulation of SNGFR. Thus, it has been shown that an increase in flow in Henle's loop is associated with a parallel decrease in both glomerular capillary pressure (P_g) and SNGFR. This finding has reduced the possible loci of resistance change in renal vasculature to either an increase in preglomerular resistance, or a decrease in postglomerular resistance, or a combination of these. Most anatomical studies favor the afferent arteriole as the effector site, since the afferent arteriole has a thicker layer of smooth-muscle cells and may be a more likely candidate for the action of vasoactive substances that might possibly mediate the response. One such substance might be calcium. Thus, Gutsche and co-workers[41] found that the vasoactive drug verapamil, which is believed to interfere with excitation-contraction coupling by blocking calcium entry into the smooth muscle cells, interfered with the feedback response. Direct evidence for changes in the afferent arteriolar resistance comes from histological studies of kidneys in which the feedback mechanism was stimulated or blocked in individual superficial nephrons.[80] After rapid freezing of the kidney, the glomeruli under study were traced, and the diameters of the afferent arterioles were examined. They found that the diameter of the afferent arteriole was decreased when the feedback mechanism was stimulated, prior to freezing, by perfusing the kidney with high flow rate. However, when furosemide was added to the perfusion solution, the narrowing of the diameter in response to high flow rate was abolished. This is consistent with earlier findings that furosemide blocks the feedback mechanism.

Although many studies (see Wright and Briggs[101] for details) have implicated the efferent arteriole and the extraglomerular mesangial cells as the possible effector sites, some recent studies have clearly confirmed the afferent arteriole as the primary locus of the effector mechanism for the negative feedback regulation of SNGFR.[16, 49] In these experiments glomerular pressure, flow, SNFF, and SNGFR were measured before and after increasing the flow through Henle's loop of the same nephron. The results showed that doubling the loop flow reduced SNFF by about 25% and SNGFR by about 20%. Quantitative estimates of possible resistance changes consistent with these data revealed that the feedback-mediated responses could best be accounted for by a 40% increase in the resistance of the afferent arteriole coupled with no change in the resistance of the efferent arteriole.

4. What is the nature of the mediating signal at the effector structures? The fourth assumption of the tubuloglomerular feedback mechanism is that changes in the afferent arteriolar resistance are controlled by local activation of the renin-angiotensin system (an increase in distal tubule-sodium chloride concentration stimulates local activation and release of renin). This assumption has been the subject of extensive studies. The compelling question is whether the feedback-mediated changes in SNGFR are mediated by a local change in renin activity and angiotensin II production capable of altering the afferent arteriolar resistance. A synthesis of available information (see below) neither supports nor excludes the renin-angiotensin system as the mediator of the feedback mechanism.

Numerous studies have established that the components of the renin-angiotensin system, namely, the renin, the renin substrate angiotensinogen (α_2-globulin), the converting enzyme for splitting of decapeptide angiotensin I into octapeptide angiotensin II, and angiotensinase are all present in the JGA. Further analysis of these data shows that a single JGA has angiotensin I activity equivalent to that found in 1 mL of plasma. Because the volume[98] of a single JGA is about 10^{-9} mL the local concentration of renin in the cells and surrounding interstitium could be very much higher than that present in the afferent arteriolar blood. Moreover, considering the renal plasma flow, the arteriovenous difference for plasma renin, and the number of glomeruli in a kidney, it has been estimated that the renin content of all glomeruli is about ten times the quantitiy of renin released into the renal vein in one day. Consequently, one would expect that any local intrarenal action of renin probably occurs through the interstitial side of the afferent arteriole rather than from the blood or luminal side.[64]

The most compelling evidence in support of renin-angiotensin as the mediator of the tubuloglomerular feedback is the observation that perfusion of the loops of Henle with isotonic saline resulted in a twofold to threefold increase in the renin content of the associated JGA.[34, 97] It is believed that this observed increase in renin is not due to de novo synthesis of renin, but rather to transformation of renin from an inactive form into an active enzyme.

That the local activation of the renin-angiotensin system may mediate the tubuloglomerular feedback response is supported by other lines of evidence. In many experimental circumstances the feedback response was found to vary directly with renal or JGA renin content. Thus, renin depletion by deoxycorticosterone acetate (DOCA) and salt administration or by contralateral renal artery stenosis resulted in a marked reduction in the feedback response. In contrast, feedback response was enhanced in kidneys with renal artery stenosis and in chronically adrenalectomized animals receiving dexamethasone (an anti-inflammatory adrenocortical steroid of the glucogenic type). These experimental maneuvers are associated with increasing renal renin content.

Several studies have examined the predictions of the tubuloglomerular feedback mechanism at both the whole-kidney level and single-nephron level. At the whole-kidney level, experimental maneuvers that have produced a reduction in GFR or tubular flow rate have resulted in feedback-mediated vasodilation of the afferent arterioles. Thus, an increase in intratubular pressure secondary to ureteral obstruction or renal vein constriction results in renal vasodilation, thereby maintaining GFR within normal limits despite elevated intratubular pressure in the proximal tubule.[70] Furthermore, an increase in plasma colloid osmotic pressure, a maneuver expected to decrease effective filtration pressure (ΔP_f) and hence GFR, also leads to renal vasodilation and increased GFR sufficient to offset the effect of increased colloid osmotic pressure.[70]

A number of studies have examined the role of JGA in the autoregulation of GFR and RBF of the whole kidney. As has been pointed out by Wright and Briggs,[101] difficulties arise in trying to compare the results from different types of studies. In some experiments systemic perturbations—such as changes in arterial, venous, or ureteral pressure, or renal nerve activity—were used to evoke changes in loop flows and activation of the renin-angiotensin system. In other studies the feedback response was determined while keeping the systemic variables constant. A careful review of these studies indicates that the changes in systemic or whole-kidney hemodynamics are more powerful stimuli for activation of the renin-angiotensin system and therefore may mask the local effect on the renin-angiotensin system attributable to changes in the composition of the distal tubule. With this in mind, Wright and Briggs[101] suggest the following explanations for reconciling these apparently conflicting observations. They suggest that if flow through Henle's loop is altered as a consequence of changes in systemic or whole-kidney hemodynamics (hereafter referred to as the *global stimuli* for renin activation and release), then renin activation and release may be expected to change in the opposite direction of the changes in loop flows. Thus, an increase in mean renal perfusion pressure, secondary to an increase in systemic arterial pressure within the autoregulatory range, would increase GFR and flow through Henle's loop but decrease renin activation and release. In contrast, if flow through Henle's loop is increased while keeping the global stimuli constant, the renin activation and release may be expected to increase. We should add that this expected difference may be due to the fact that the global stimuli act directly on the granular cells to alter renin activation and release, whereas *local stimuli* that alter loop flow, independent of global stimuli, act via the macula densa cells to alter release of renin from existing stable prorenin form. Such a concept suggests that renin activation and release by the global and local stimuli must serve different but related functions in the kidney in regulating the volume and composition of body fluid.

Physiological Role of Tubuloglomerular Feedback Mechanism

What is the physiological role of the macula densa feedback mechanism? Before attempting to answer this question, it should be recognized that the rate of delivery of solute and water into the distal convoluted tubules and collecting ducts (the so-called *"distal nephron"*) is determined by two interrelated mechanisms that stabilize the balance between the SNGFR and the rate of filtrate reabsorption in the proximal tubule and loops of Henle (the so-called *"proximal nephron"*). The first is the *tubuloglomerular feedback mechanism* (TGF), which lowers SNGFR in response to an increase in chloride concentration at the macula densa above its normal value caused by a decrease in water and sodium chloride reabsorption in the proximal nephron, or caused by a primary increase in SNGFR. The second is the adaptive increase in proximal water and sodium chloride reabsorption in response to a spontaneous increase in the filtered load, a phenomenon referred to as the *glomerulotubular balance* (GTB). In terms of nephron regulation, TGF acts as a *negative-feedback mechanism*, which adjusts SNGFR to compensate for inadequate filtrate reabsorption in the distal nephron, whereas the GTB acts as a *feedforward mechanism* to increase reabsorption in the proximal nephron in response to an increase in SNGFR. Both of these mechanisms are reset by maneuvers that reduce the extracellular volume. Thus, acute hemorrhage is expected to increase the strength of the TGF response.

To determine the effect of the filtration dynamics on SNGFR and the contribution of the TGF mechanism to the overall regulation of the filtrate delivery to the distal nephron, Moore and co-workers[62, 63] measured simultaneously the feedback response, filtration, and reabsorption in response to an external forcing. From these data, they developed an algebraic model of the TGF based on conservation of mass (Figure 6.7) to account for the interaction of the TGF and distal nephron reabsorption during perturbations. Since filtrate flow in the early distal nephron ($\dot{V}_d$) is known to be monotonically related to filtrate flow at the macula densa, it was assumed that changes in $\dot{V}_d$ reflect changes in the chloride concentration at the macula densa, the stabilized variable, and may be used as an indication of TGF activation. This model has provided an insight into the relative contributions of the intrinsic (TGF-independent) and TGF-dependent mechanisms to the autoregulation of SNGFR under normal conditions and in response to certain physiological perturbations.

As shown in Figure 6.7, the model consists of two major components: the intrinsic vascular controller, represented by the box labeled afferent arteriole, shown on the left, and the tubuloglomerular feedback (TGF), represented by the two boxes labeled proximal nephron (referring to the proximal tubule and the loops of Henle) and TGF, respectively, shown on the right. The model input is $\Delta SNGFR_{OL}$, the open-loop change in SNGFR that would occur if neither the TGF mechanism nor the TGF-independent mechanisms (the intrinsic vascular autoregulatory mechanisms) were active. The output of the vascular intrinsic controller mechanism is the $\Delta SNGFR_O$, which is the change in SNGFR that would be observed if the TGF-independent intrinsic vascular autoregulatory mechanism were active but the TGF mechanism were not. The *gain* (G) or *magnification* of the TGF-independent intrinsic vascular autoregulatory mechanism is defined as the ratio of the output to the input:

$$\text{Gain} = \text{Magnification} = \frac{\text{Output}}{\text{Input}} = \frac{\Delta SNGFR_O}{\Delta SNGFR_{OL}} \quad (6.2)$$

The gain was found to be equal to 0.46, indicating that the TGF-independent mechanism accounted for 46% of the autoregulation of SNGFR.

The input to the TGF mechanism (Figure 6.7, right panel) is Δ SNGFR, the actual change in filtration rate when both the TGF and the TGF-independent mechanisms are active. Its value is equal to the algebraic difference (shown by the summation symbol) between the output of the TGF-independent mechanism, $\Delta SNGFR_O$, and a fractional (represented by α) change in the early distal tubule flow ($\alpha \cdot \Delta \dot{V}_d$):

$$\Delta \text{ SNGFR} = \Delta SNGFR_0 - \alpha \cdot \Delta \dot{V}_d \quad (6.3)$$

α is the TGF coefficient; it expresses the magnitude of the TGF-dependent decrease in SNGFR per unit increase in $\dot{V}_d$. Stated another way, the coefficient α is an index of the *strength* of the TGF response. In this model, α is assumed to be independent of arterial blood pressure. In hydropenic rat, the value of α, at constant arterial blood pressure, was found to vary between 1.0 and 2.0, depending on the ex-

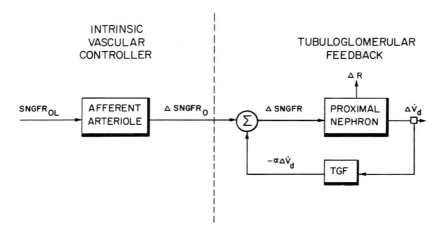

FIGURE 6.7. A control-system model depicting the functional features of the tubuloglomerular feedback mechanism and its contribution to regulation of SNGFR during experimental perturbations. Finally, the change in the early distal tubule flow ($\Delta \dot{V}_d$) is equal to the difference between Δ SNGFR and the change in filtrate volume reabsorption up to the macula densa (ΔR):

$$\Delta \dot{V}_d = \Delta SNGFR - \Delta R \qquad (6.4)$$

Substituting Equations 6.2 and 6.4 into Equation 6.3, and rearranging terms, yields:

$$\Delta SNGFR = \frac{[G \cdot \Delta SNGFR_{OL} + \alpha \cdot \Delta R]}{(1 + \alpha)} \qquad (6.5)$$

This equation predicts the autoregulation of SNGFR in response to a 30 mm Hg increase in arterial blood pressure. Thus, the tendency for a rise in arterial blood pressure to increase SNGFR is expressed by the open-loop response, $\Delta SNGFR_{OL}$, and the subsequent effect on reabsorption in the proximal nephron is expressed by ΔR. The stabilizing effects of the TGF and the TGF-independent intrinsic vascular autoregulatory mechanisms are expressed by the TGF coefficient α and the magnification or gain (G) factor, respectively.

On the basis of the results of simulation studies, using Equation 6.5, Moore[62] suggested that the net effect on SNGFR following an increase in arterial blood pressure depends on several factors: (1) the magnitude of the open-feedback-loop response, which in turn is influenced by the state of filtration dynamics (Chapter 5); (2) the magnitude of the signal sensed by the TGF at the macula densa, which is determined by the steady-state balance between $\Delta SNGFR$ and ΔR; (3) the strength of the TGF response; and (4) the efficiency of the intrinsic vascular autoregulatory mechanism. Moore proposed further that there are at least two mechanisms that may be involved in the SNGFR autoregulation.[62] One is the intrinsic mechanism that is entirely vascular, and the other mechanism involves the interaction between the tubular and vascular systems of the kidney. Furthermore, since the intrinsic mechanism operates in the absence of glomerular filtration and the TGF mechanism senses the tubular fluid composition at the macula densa, neither mechanism directly monitors the changes in SNGFR. Thus, it appears that the SNGFR is not the variable controlled during autoregulation, but rather regulation of SNGFR occurs as an important intermediate event secondary to the direct regulation of other variables.

Based on the simulation results obtained from the above model and in keeping with the experimental findings, Moore concluded (1) that an acute increase in filtered load resulted in a significant compensatory response tending to stabilize the filtrate delivery to the distal nephron, (2) that the TGF mechanism alone can provide significant regulation of the function in the intact nephron, and (3) that the TGF is reset to operate more effectively in response to a reduction in the extracellular volume, as in hemorrhage.[62] Furthermore, the model predicted and the data confirmed that the changes in tubular reabsorption strongly affect the TGF mechanism, thereby modulating the relationship between the distal nephron delivery and the SNGFR.

The interaction between the TGF mechanism and changes in proximal tubular reabsorption has several consequences. First, TGF and GTB act in parallel to stabilize the delivery of the filtrate to the distal nephron. Second, despite the operation of both TGF and GTB, the SNGFR is poorly regulated. This is because the nephron variable, which is stabilized by both mechanisms, is the distal delivery of the filtrate and not SNGFR. Third, the parallel resetting of TGF and GTB by changes in the extracellular fluid volume will exert a synergistic effect on the regulation of the distal nephron delivery and hence renal excretion of sodium. Thus, a reduction in extracellular volume following hemorrhage results in enhanced regulation of the distal nephron delivery, thereby reducing renal excretion of salt and water tending to restore the extracellular volume toward normal. In contrast, an expansion of the extracellular fluid volume following high salt diet and chronic DOCA administration inhibits both TGF and proximal GTB regulation of distal delivery, thereby increasing SNGFR and renal excretion of salt and water tending to restore the extracellular fluid volume toward normal.

A number of studies have provided compelling evidence that the tubuloglomerular feedback mechanism plays a major role in the autoregulation of renal blood flow and glomerular filtration rate. Thus, it has been shown that as the arterial blood pressure is varied from 80 to 180 mm Hg, renal blood flow, whole-kidney GFR, SNGFR, glomerular capillary pressure, proximal tubule pressure, and peritubular capillary pressure remain nearly constant.[101] In those studies that have examined the role of the JGA and the feedback mechanism in renal autoregulation, clear-cut and reproducible results were obtained only when the arterial pressure was held constant. This suggests that the mechanisms responsible for constancy of RBF, GFR, SNGFR, and glomerular pressures at constant arterial pressure may not be the same mechanisms responsible for maintaining constant RBF and GFR in response to changes in arterial blood pressure. Therefore, we may conclude that a single mechanism can not explain the phenomenon of autoregulation and that the feedback mechanism and other pressure-or flow-dependent mechanisms involving substances other than the renin-angiotensin system may serve as mediators. The list of such substances include renin-angiotensin, kallikrein-kinin, arachidonic acid-prostaglandin, and nucleoside-nucleotide systems. The precise role of each of these humoral systems as the possible mediator of renal autoregulation remains to be elucidated.

In summary, the results of many studies to date are consistent with the concept that renin activation and release from the granular cells of the JGA occurs in response to two separate but related mechanisms: (a) direct stimulation of the granular cells by systemic and whole-kidney perturbations (the so-called global stimuli), and (b) indirect stimulation of the granular cells, via the macula densa cells, by transient within single-nephron perturbations (the so-called local stimuli). The operation of these two mechanisms are somehow coordinated to preserve the constancy of the volume and composition of body fluids. Although the precise mechanism for renin release in response to changes in sodium chloride concentration in the tubular fluid at the macula densa is as yet unresolved, the involvement of the macula densa-renin-angiotensin system as a negative feedback control of SNGFR has not been excluded. Furthermore, owing to their strategic location, it has been suggested that the granular cells act as sensors for detecting changes in blood flow, and hence perfusion pressure in the afferent arteriole.[99] This aspect of JGA function has led to the formulation of the baroreceptor theory of renin activation and release. Therefore, these cells play an important role in the extra-renal (global) feedback mechanism regulating the volume and composition of the fluid entering and leaving the kidney and hence the entire body fluids. This role of JGA is considered in more detail in Chapter 13.

Roles of the Renin-Angiotensin and Prostaglandin Systems

The question of whether intrarenal angiotensin II plays a role in renal autoregulation has been addressed in a number of recent studies. Steiner and co-workers[91] found that during chronic sodium chloride depletion in the Munich-Wistar rat, a condition known to increase renal renin-angiotensin activity, the SNGFR was decreased secondary to reductions in both nephron plasma flow ($\dot{Q}_A$) and the glomerular permeability coefficient (K_g). The infusion of saralasin, an angiotensin II receptor antagonist, increased $\dot{Q}_A$ to control levels secondary to a decrease in the afferent arteriolar resistance, thereby restoring SNGFR to control levels. Saralasin infusion, however, did not restore K_g to control levels, but acute volume depletion did. These results demonstrate that in chronic sodium chloride depletion, intrarenal generation of angiotensin II plays an important role in the regulation of SNGFR by modulating both the glomerular resistance vessels (preglomerular and postglomerular vessels) and the glomerular permeability coefficient K_g (Figure 5.11, Chapter 5).

Recent quantitative studies, at the whole-kidney level, on the role of the renin-angiotensin system in the regulation of renal hemodynamics have supported the concept that when renal perfusion pressure falls below normal, the renal vasoconstrictor effect of angiotensin II is confined to the postglomerular vessels.[45] This effect is thought to play an important role in preventing reductions in GFR (Chapter 5) during certain physiological circumstances associated with reduction in mean systemic arterial blood pressure below normal (less than 90 mm Hg), such as chronic sodium deprivation or hemorrhage.

Kirchheim and co-workers recently investigated the pressure-dependent nature of intrarenal renin-angiotensin generation and its contribution to renal autoregulation and maintenance of systemic arterial blood pressure in conscious dogs.[29] They examined the effect of variation of renal artery pressure between 40 and 160 mm Hg on renal blood flow and renin release in conscious dogs receiving β-adrenergic blocking agent propranolol and normal sodium diet. The results, reproduced in Figure 6.8, demonstrate that as the renal perfusion pressure is reduced from 160 to about 90 mm Hg, there was no change in renal blood flow or renin release. However, as the renal artery pressure was progressively decreased below 90 mm Hg, there was a decrease in renal blood flow and a sharp increase in renin release. Quantitative examination of the slope of the pressure-renin release curve revealed that a fall in systemic arterial blood pressure for as little as 1.3 mm Hg below the "threshold" pressure of 90 mm Hg resulted in about 100% increase in renin release. These findings have recently been confirmed in conscious rats.[23] Taken together, these studies provide strong evidence in support of the concept that the pressure-dependent intrarenal renin-angiotensin generation serves to stabilize normal renal blood flow. Thus, it is suggested that the renin-angiotensin system is primarily mobilized to prevent the systemic arterial blood pressure from falling below a critical level. To maintain normal renal blood flow and GFR, it is postulated that angiotensin II causes a preferential constriction of the postglomerular vessels (efferent arterioles).

Prostaglandins are formed from oxygenation of arachidonic acid (a 20-carbon essential fatty acid). In the kidney, prostaglandins are beleved to act as chemical messengers that regulate cell function. Several studies[51] have suggested that prostaglandins play an important role in four major aspects of renal func-

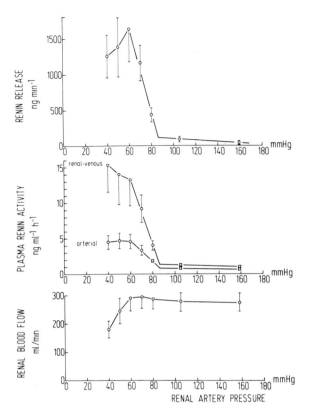

FIGURE 6.8. Effects of changes in renal artery pressure on the renal renin release (*upper panel*), the plasma renin activity (*middle panel*), and the renal blood flow (*lower Panel*). Data, expressed as mean ± SE, are from six conscious, chronically instrumented dogs. (From Finke.[29])

tion: (a) renal blood flow, (b) renin release, (c) sodium excretion, and (d) solute-free water clearance (see Chapter 12 for definition). Although these functions are interrelated, evidence indicates that the prostaglandin system can affect each function independently.

Intrarenal infusion of prostaglandins (e.g., PGE_2) causes renal vasodilation in both rat[52] and dog.[86] Microsphere studies have shown that, besides causing renal vadodilation, prostaglandins induce a redistribution of blood flow in the cortex, with a greater proportion flowing to the juxtamedullary nephrons.[32] Furthermore, it has been suggested that the high concentration of angiotensin II stimulates renal release of prostaglandins, which, via vasodilation, minimizes reduction in renal blood flow.[3] Despite these important effects of prostaglandins on renal circulation, the available evidence indicate that prostaglandins do not contribute to renal autoregulation and exert only a permissive role in the tubuloglomerular feedback mechanism.[81]

Intravenous infusion of prostaglandins in dogs causes an increse in sodium excretion, an effect attributed to a decrease in sodium reabsorption in the proximal tubules.[32] The effect of prostaglandins on sodium reabsorption was found to be independent of its effect on renal circulation. Thus, prostaglandins inhibited sodium reabsorption at a dose lower than that required to cause renal vasodilation.[93]

Prostagldins (PGE_2) has been found to inhibit antidiuretic hormone (ADH)-induced osmotic reabsorption of water in the collecting duct.[39] Although the precise mechanism mediating this effect is not known, it has been suggested that prostaglandins may inhbit ADH-induced cyclic adenosine monophosphate (AMP) formation in the collecting duct epithelium (Figure 9.38, Chapter 9). Intrarenal infusion of PGE_2 in dogs causes an increase in solute-free water clearance, suggesting a direct inhibitory effect of prostaglandins on ADH action. This effect was potentiated by angiotensin II and was blunted by indomethacin (a prostaglandin synthetase inhibitor) and saralasin (a competitive angiotensin II antagonist). These findings are consistent with the concept that angiotensin II stimulates PGE_2 formation, which in trun inhibits the action of ADH, thereby promoting enhanced solute-free water clearance.[31]

Summary

The accumulated experimental evidence to date clearly indicates that no single mechanism can completely explain the autoregulation of renal blood flow and GFR. The results of the various studies cited above provide compelling evidence that at least three mechanisms, working together, may explain the phenomenon of renal autoregulation depicted in Figure 6.1 (left panel). (1) The myogenic mechanism acts to buffer changes in RBF and GFR in response to abrupt changes in systemic arterial blood pressure. This mechanism appears to be most effective at mean systemic arterial blood pressures exceeding the normal threshold pressure of 90 mm Hg. (2) The tubuloglomerular feedback (TGF) mechanism acts to stabilize GFR at normal systemic arterial blood pressures ranging between 80 and 120 mm Hg. Under normal conditions and normal arterial blood pressure, the TGF mechanism accounts for about 50% of the autoregulatory compensation in response to changes in the blood pressure in the upper pressure interval (90 to 120 mm Hg) and about 30% in the lower pressure interval (90 to 80 mm Hg). The TGF mechanism serves to stabilize the distal tubule flow rate, a factor that may contribute to the preservation of the medullary longitudinal gradient and hence the regulation of the final osmolality of urine (Chapter 12). And (3) the renin-angiotensin mechanism is activated when the systemic arterial blood pressure falls below 90 mm Hg. This mechanism serves to protect the animal from life-threatening hypotension and contributes approximately 20% to the autoregulatory compensation at the lower end of the autoregulatory pressure range. The endogenous prostaglandins, although playing an important role in regulating renal circulation, sodium, and water reabsorption, do not contribute to the autoregulation of renal blood flow and glomerular filtration rate.

Of these mechanisms, the tubuloglomerular feedback system appears to be of greater physiological importance at the normal pressure range. The TGF is an adaptive mechanism serving to maintain a balance between the glomerular filtration rate and tubular reabsorption commensurate with the prevailing extracellular fluid volume of the animal.

Extrinsic Regulation of Renal Circulation

In addition to the three intrinsic mechanisms described above, the renal blood flow, and hence GFR, are influenced by extrinsic factors which may be classified as neural mechanisms, such as changes in renal sympathetic activity and humoral mechanisms, such as changes in the blood levels of vasoconstrictor and vasodilator substances. In this section, we shall briefly review the experimental evidence pertaining to how these extrinsic mechanisms modulate RBF and GFR.

Neural Mechanisms

Electrophysiological and ultrastructural studies have established that the kidney is innervated by mixed nerves containing both afferent and efferent fibers.[66] The efferent fibers, composed of postganglionic sympathetic fibers, enter the kidney together with the renal artery and extend along the arteries and terminate on the smooth muscle cells of the afferent and efferent arterioles.[8, 26] These nerves exert their effect by releasing norepinephrine onto the adrenergic receptors in the postsynaptic membranes.[50] There are also dopaminergic fibers innervating both afferent and efferent arterioles.[27] Dopamine administration causes renal vasodilation and natriuresis mediated by specific dopamine receptors.[36] Recent studies suggest that the renal afferents play an important role in coordinating the function of both kidneys via a centrally mediated renorenal reflexes.[46,47,75]

At rest, the spontaneous electrical activity of the renal nerves is about 1 to 2 Hz.[46,47] Moderate (2 to 5 Hz) to high (5 to 10 Hz) frequency stimulation of the renal sympathetic nerve leads to α-adrenergic receptor-mediated vasoconstriction, with subsequent reduction in both renal blood flow and GFR. In addition, the same stimulation regimen results in an increase in renin secretion, which is mediated partly by α- and partly by $α_1$-adrenergic receptors[59]. In general, renal blood flow will decrease in proportion to the frequency of stimulation, reaching its maximal effect at a frequency of 10 Hz.[28]

Recent studies[25] suggest that the renal vasoconstrictor effect begins at stimulation frequency of about 2 Hz, whereas in the frequency range of 0 to 2 Hz, the effects on renal hemodynamics are negligible. Stimulation at frequencies below 2 Hz resulted in a decrease in urinary sodium and water excretion, an effect attributed to a direct nervous stimulatory effect on tubular reabsorption of fluid.[11,90] However, as stated in Chapter 4, morphological studies do not support the existence of a direct innervation of renal tubules.[26,55] A recent micropuncture study in Munich-Wistar rats by Kon and Ichikawa demonstrated that the observed effects of low frequency renal nerve stimulation on tubular function are mediated by changes in the determinants of glomerular hemodynamics.[58] They found that high frequency (3 to 5 Hz) stimulation of renal nerve resulted in a decrease in SNGFR and glomerular plasma flow rate ($\dot{Q}_A$), whereas mean transcapillary hydraulic pressure difference ($\overline{\Delta P}$) remained unchanged. The decrease in $\dot{Q}_A$ was associated with a significant increase in both the afferent (R_A) and efferent (R_E) arteriolar resistance, which resulted in a fall in early peritubular capillary hydraulic pressure. There was also a marked decrease in the ultrafiltration coefficient (K_g). In contrast, low frequency (0.5 to 1.5 Hz) renal nerve stimulation resulted in a significant fall in the early peritubular capillary hydraulic pressure, but produced only small and variable effects on the other parameters. From the analysis of Starling forces acting across the glomerulus and the peritubular capillaries, they concluded that the effector loci for renal neural vasomotor control are the increases in the afferent and efferent arteriolar resistances, and the decreases in the glomerular ultrafiltration coefficient (K_g), and the peritubular capillary reabsorption coefficient.

Recent studies suggest that renal denervation may delay the development of hypertension in animal models in which enhanced sympathetic activity has been implicated.[57] Thus, it has been suggested that renal denervation may reduce systemic arterial blood pressure by interrupting the activity in the afferent renal nerves which modulate the sympathetic tone. These afferent neurons are found to project onto the hypothalamus, with the kidney serving as a sensory organ for modulating the activity of central adrenergic activity.[2] Thus, the activity of the renal nerves (afferent and efferents), which would be interrupted by renal denervation, may play an important role in the development of several models of hypertension.

In summary, the extent of contribution of renal nerves to the extrinsic regulation of renal circulation depends on the animal model and the experimental conditions used in their study. It is well established that stimulation of renal nerves increase the intrarenal resistance and activate the renin-angiotensin system, which contribute to the normal regulation of renal circulation. Although many previous studies have indicated that renal nerves may directly control sodium and water reabsorption by the proximal tubules, and even by the loops of Henle[25,66] the anatomical evidence for such a control has not yet been shown. At present, considering their basal activity, it appears that renal nerves play a more important role in influencing renal function in the pathological states, such as renal and essential hypertension, rather than in normal conditions.

Humoral Mechanisms

When considering the effects of humoral agents on renal circulation, we need to distinguish between the direct effect of these agents on the kidney (Figure 5.11, Chapter 5) versus their systemic effects. Thus, intravenous infusion of these agents may produce a renal effect different from that produced by intrarenal injection of the same agents. Thus, the increase in RBF observed with intravenous administration of a β-adrenergic agonist may result from an increase in cardiac output, a decrease in renal vascular resisitance, or both.[85] The renal response is further complicated by the fact that the kidney has the capacity to inactivate many substances, including catecholamines and angiotensin.[85] Therefore, to assess the role of humoral agents on renal hemodynamics, these agents must be injected directly into the renal circulation.

The humoral agents affecting renal circulation may be classified as those that produce vasoconstriction and those that cause vasodilation. We shall examine the effects of each separately.

Vasoconstrictor Agents

Catecholamines and the renin-angiotensin system represent the major humoral agents whose intrarenal infusion brings about renal vasoconstriction. Catecholamines consist of norepinephrine and epinephrine. Intrarenal infusion of small doses (3 to 4 μg/min) of either substance produces marked renal vasoconstriction without affecting systemic hemodynamics.[61] Infusion of norepinephrine increases filtration fraction (GFR/RPF), despite a decrease in RBF. This is attributed to the preferential effect of norepinephrine on the smooth muscles of the efferent arteriole. Referring to Figure 5.14 (Chapter 5), the increase in the efferent arteriolar resistance leads to a reduction in RBF accompanied by an increase in the glomerular capillary hydraulic pressure (P_g). The increase in P_g causes an increase in GFR, which, along with the decrease in RBF, results in an increase in filtration fraction. Recently, Anderson and co-workers[4] showed that a norepinephrine-induced increase in filtration fraction in dogs could be blocked with angiotensin I converting enzyme inhibitor. Since norepinephrine is known to stimulate renin release from the granular cells,[85] they concluded that the increase in the efferent arteriolar resistance is due to the vasoconstrictor action of locally generated angiotensin II. This conclusion has recently been confirmed in studies in rats.[6]

As reviewed earlier in this chapter, the renin-angiotensin system has been implicated in the regulation of renal circulation. The release of renin from the granular cells varies inversely with an increase in mean renal perfusion pressure. Since the substrate for renin, namely, angiotensinogen, and the angiotensin I converting enzyme are both present in the kidney, then an increase in renin release, secondary to a reduction in mean renal perfusion pressure, will lead to a proportional increase in renal angiotensin II level.

Although changes in mean renal perfusion pressure are the major stimulus for renin release, other factors, such as catecholamines and prostaglandins, also influence the release of renin from the granular cells. Catecholamines act on the β_1-adrenergic receptors of the granular cells to stimulate renin release.[61] Bing and Paulsen showed that inhibition of the angiotensin I converting enzyme results in an increase in plasma renin level, suggesting a possible negative feedback relationship between renin release and angiotensin II formation. This so-called short feedback loop is believed to be mediated by prostaglandins, which in turn are a potent stimulator of renin release.[19]

Vasodilator Agents

Several substances have been shown to cause renal vasodilation when injected into the renal artery. The following is a brief description of their effects. For a more detailed description of the pertinent literature, the reader is referred to an excellent review by Margolis and Stein.[61]

Intrarenal injection of isoproterenol in dogs causes vasodilation. Its effect is blocked by propranolol, a β-adrenergic receptor blocking agent. In contrast, injection of *dopamine* in low doses causes renal vasodilation, whereas high doses (40 μg/kg of body weight) cause renal vasoconstriction. The latter effect is blocked by phentolamine, an α-adrenergic receptor blocking agent. Renal denervation is known to reduce renal dopamine content, a finding that may explain denervation-induced renal vasodilation and natriureis.

Intrarenal infusion of acetylcholine at doses up to 25 μg/kg/min causes renal vasodilation. Despite the increase in RBF, filtration fraction falls secondary to a decrease in the resistance of the afferent and efferent arterioles. Atropine blocks acetylcholine-induced renal vasodilation, suggesting that activation of muscarinic receptors is responsible for the increase in RBF.

Summary

The available evidence to date indicate that the glomerular filtration rate and the various factors affecting it play a key role in the ultimate renal regulation of body fluids. To fully understand the changes in GFR and its effects on renal function, it is imperative to measure its value in normally functioning kidneys. Since measurement of GFR, based on the passive Starling forces operating across the glomerulus (Equation 5.1), poses numerous technical problems and is impractical in clinical practice, we shall resort to an empirical method of measuring GFR, based on the dilution principle. A description of the application of this principle in determining GFR, RPF, and tubular transport capacity is given in Chapter 7.

Problems

6.1. What are the three major components of the juxtaglomerular apparatus?

6.2. Describe the location of the macula densa.

6.3. What are the cytological differences between the macula densa cells and cells of the thick ascending limb?

6.4. What kinds of cells are the extraglomerular mesangial cells?

6.5. What kinds of structural communications exist between the extraglomerular mesangial cells and the mesangial cells proper as well as the granular cells?

6.6. Into which compartment is renin released from granular cells?

6.7. Are granular cells innervated?

6.8. Which changes occur in granular cells after long-term sodium depletion?

6.9. Define renal autoregulation and briefly describe the various mechanisms that have been proposed to explain this phenomenon.

6.10. Define the concept of pressure-diuresis and explain the probable underlying mechanisms.

6.11. What is the experimental evidence in support of the concept that adaptive changes in the resistance of the afferent arteriole are responsible for autoregulation of SNGFR?

6.12. What are the proposed functions of the renin-angiotensin system in renal autoregulation?

6.13. What is the experimental evidence that chloride transport by the macula densa cells is the signal that initiates the tubuloglomerular feedback mechanism?

References

1. Abe Y, Kishimoto T, Yamamoto K: Effect of angiotensin II antagonist infusion on autoregulation of renal blood flow. *Am J Physiol* 1976; 231:1267–1271.
2. Abboud FM: The sympathetic system in hypertension: State-of-the-art review. *Hypertension* 1982; 4 (suppl II):208–225.
3. Aiken JW, Vane JR: Intrarenal prostaglandin release attenuates the renal vasoconstrictor activity of angiotensin. *J Pharmacol Exp Ther* 1973; 184:678–687.
4. Anderson WP, Korner PI, Selig SE: Mechanisms involved in the renal response to intravenous and renal artery infusions of noradrenaline in conscious dogs. *J Physiol (Lond)* 1981; 321:21–30.
5. Arendshorst WJ, Finn WF, Gottschalk CW: Autoregulation of blood flow in the rat kidney. *Am J Physiol* 1975; 228:127–133.
6. Arundell LA, St Johns E: Effect of converting enzyme inhibition on the renal hemodynamic responses to noradrenaline infusion in rats. *Br J Pharmacol* 1982; 75:553–558.
7. Barajas L: The ultrastructure of the juxtaglomerular apparatus as disclosed by three-dimensional reconstructions from serial sections: The anatomical relationship between the tubular and vascular components. *J Ultrastruct Res* 1970; 33:116–147.
8. Barajas L, Powers K, Wang P: Innervation of the renal cortical tubules: A quantitative study. *Am J Physiol* 1984; 247:F50–F60.
9. Bayliss WM: On the local reactions of the arterial wall to changes of internal pressure. *J Physiol (Lond)* 1902; 28:220–231.
10. Beck LH, Senesky D, Goldberg M: Sodium-independent active potassium reabsorption in proximal tubule of the dog. *J Clin Invest* 1973; 52:2641–2645.
11. Bello-Reuss E, Trevino DL, Gottschalk GW: Effect of renal sympathetic nerve stimulation on proximal water and sodium reabsorption. *J Clin Invest* 1976; 57:1104–1107.
12. Bing J, Paulsen K: Time course of change in plasma renin after blockade of the renin system. *Acta Pathol Microbiol Scand (abstract)* 1975; 83:454.
13. Blaustein MP: Sodium ions, calcium ions, blood pressure regulation, and hypertension: A reassessment and a hypothesis. *Am J Physiol* 1977; 232:C165–C173.
14. Brenner BM, Baylis C, Deen WM:. Transport of molecules across renal glomerular capillaries. *Physiol Rev* 1976; 56:502–534.
15. Briggs J, Schubert G, Schnermann J: Evidence for an inverse relationship between macula densa NaCl concentration and filtration rate. *Pfluegers Arch* 1982; 392:372–378.
16. Briggs JP, Wright FS: Feedback control of glomerular filtration rate: site of the effector mechanism. *Am J Physiol* 1979; 236:F40–F47.
17. Burg MB, Green N: Effect of ethacrynic acid on the thick ascending limb of Henle's loop. *Kidney Int* 1973; 4:301–308.
18. Burg MB, Stoner L, Cardinal J, Green N: Furosemide effect on isolated perfused tubules. *Am J Physiol* 1973; 225:119–124.
19. Campbell WB, Jackson EK, Graham RM: Saralasin induced renin release: Its blockade by prostaglandin synthesis inhibitors in the conscious rat. *Hypertension* 1979; 1:637–642.
20. Carmines PK, Bell PD, Roman RJ, et al: Prosta-

glandins in the sodium excretory response to altered renal arterial pressure in dogs. *Am J Physiol* 1985; 248:F8–F14.
21. Christensen JA, Bohle A: The juxtaglomerular apparatus in the normal rat kidney. *Virchows Arch Pathol Anat* 1978; 379:143–150.
22. Cohen HJ, Marsh DJ, Kayser B: Autoregulation in vasa recta of the rat kidney. *Am J Physiol* 1983; 245:F32–F40.
23. Conrad KP, Brinck-Johnsen T, Gellai M, Valtin H: Renal autoregulation in chronically catheterized conscious rats. *Am J Physiol* 1984; 247:F229–F233.
24. Deen WM, Robertson CR, Brenner BM: Glomerular ultrafiltration. *Fed Proc* 1974; 33:14–20.
25. DiBona GF: The functions of the renal nerves. *Rev Physiol Biochem Pharmacol* 1982; 94:75–181.
26. Dietrich HJ: Electron microscopic studies of the innervation of the rat kidney. *Z Anat Entwicki Gesch* 1974; 145:169–186.
27. Dinerstein RJ, Jones RT, Goldberg LI: Evidence for dopamine-containing renal nerves. *Fed Proc* 1983; 42:3005–3008.
28. DiSalvo J, Fell C: Changes in renal blood flow during renal nerve stimulation. *Proc Soc Exp Biol Med* 1971; 136:150–153.
29. Finke R, Gross R, Hackenthal E, et al: Threshold pressure for the pressure-dependent renin release in the autoregulating kidney of conscious dogs. *Pfluegers Arch* 1983; 399:102–110.
30. Fray JCS: Stretch receptor model for renin release with evidence from perfused rat kidney. *Am J Physiol* 1976; 231:936–944.
31. Galvez OG, Roberts BW, Mishkind MH, et al: Studies of the mechanism of contralateral polyuria after renal artery stenosis. *J Clin Invest* 1977; 59:609–615.
32. Gerber JG, Nies AS, Friesinger GC, et al: The effects of PGI_2 on canine renal function and hemodynamics. *Prostaglandins* 1978; 16:519–528.
33. Gilmore JP, Cornish KG, Rogers SD, Joyner WL: Direct evidence for myogenic autoregulation of the renal microcirculation in the hamster. *Circ Res* 1980; 47:226–230.
34. Gillies A, Morgan T: Renin content of individual juxtaglomerular apparatuses and the effect of diet, changes in nephron flow rate and in vitro acidification on the renin content. *Pfluegers Arch* 1978; 375:105–110.
35. Gocke DJ, Gerter J, Sherwood LM, Laragh JH: Physiological and pathological variations of plasma angiotensin II in man: Correlation with renin activity and sodium balance. *Circ Res* 1969; 24 (suppl I):131–146.
36. Goldberg LI, Glock D, Kohli JD, Barnett A: Separation of peripheral dopamine receptors by a selective DA_1-antagonist. SCH 23390. *Hypertension* 1984; 6 (suppl I):125–130.
37. Goormaghtigh N: L'appareil neuro-myoarteriol juxtaglomerulaire du rein. *C R Soc Biol* 1937; 124:293–296.
38. Gorgas K: Structure and innervation of the juxtaglomerular apparatus of the rat. *Adv Anat Embryol Cell Biol* 1978; 54:5–84.
39. Grantham JJ, Orloff J: Effect of prostaglandin E_1 on the permeability response of the isolated collecting tubule to vasopressin, adenosine 3', 5'-monophosphate and theophylline. *J Clin Invest* 1968; 47:1154–1161.
40. Greger R, Schlafler E, Lang F: Evidence for electroneutral sodium chloride cotransport in the cortical thick ascending limb of Henle's loop of rabbit kidney. *Pfluegers Arch* 1983; 396:308–314.
41. Gutsche HU, Muller-Suur R, Schurek HJ: Ca^{++} antagonist prevents feedback-induced SNGFR decrease in rat kidney. *Kidney Int* 1975; 8:477.
42. Guyton AC, Coleman TG, Fourcade JC, Navar LG: Physiologic control of arterial pressure. *Bull NY Acad Med* 1969; 45:811–830.
43. Guyton AC, Langston JB, Navar LG: Theory for renal autoregulation by feedback of the juxtaglomerular apparatus. *Circ Res* 1964; 15 (suppl I):187–197.
44. Haas JA, Hammond TG, Granger JP, et al: Mechanism of natriuresis during intrarenal infusion of prostaglandin. *Am J Physiol* 1984; 247:F475–F479.
45. Hall JE, Coleman TG, Guyton AC, et al: Control of glomerular filtration rate by circulating angiotensin II. *Am J Physiol* 1981; 241:R190–R197.
46. Hermansson K, Ojteg G, Wolgast M: The cortical and medullary blood flow at different levels of renal nerve activity. *Acta Physiol Scand* 1984; 120:161–169.
47. Hermansson K, Ojteg G, Wolgast M: The reno-renal reflex; evaluation from renal blood flow measurements. *Acta Physiol Scand* 1984; 120:207–215.
48. Hoyer JR, Sisson SP, Vernier RL: Tamm-Horsfall glycoprotein: Ultrastructural immunoperoxidse localization in rat kidney. *Lab Invest* 1979; 41:168–173.
49. Ichikawa I: Direct analysis of the effector mechanism of the tubuloglomerular feedback system. *Am J Physiol* 1982; 243:F447–F455.
50. Insel PA, Snavely MD: Catecholamines and the kidney: Receptors and renal function. *Ann Rev Physiol* 1981; 43:625–636.
51. Jackson EK, Branch RA, Oates JA: Participation of prostaglandins in the control of renin release. in Oates JA (ed): *Prostaglandins and the Cardiovascular System*. New York, Raven Press, 1982, pp 255–276.
52. Jackson EK, Heidemann H, Branch RA, Gerkens JF: Low dose intrarenal infusions of PGE_2, PGI_2, and 6-keto-PGE_1 vasodilate the *in vivo* rat kidney. *Circ Res* 1982; 51:67–72.
53. Johnson PC: The myogenic response, in Bohr DF, Somlyo AP, Sparks HV Jr (eds): Handbook of Phys-

iology, Section II: *The Cardiovascular system, vol 2. Vascular smooth muscle*. Washington DC, American Physiological Society, 1980, pp 409–442.
54. Johnson PC: Review of previous studies and current theories of autoregulation. *Circ Res* 1964; 14/15 (suppl I):2–9.
55. Kaissling B, Kriz W: Variability of intercellular spaces between macula densa cells: A TEM study in rabbits and rats. *Kidney Int* 1982; 12:9–17.
56. Kallskog O, Lindbom LO, Ulfendahl HR, Wolgast M: Hydrostatic pressures within the vascular structures of the rat kidney. *Pfluegers Arch* 1976; 363:205–210.
57. Katholi RE: Renal nerves in the pathogenesis of hypertension in experimental animals and humans. *Am J Physiol* 1983; 245:F1–F14.
58. Kon V, Ichikawa I: Effector loci for renal nerve control of cortical microcirculation. *Am J Physiol* 1983; 245:F545–F553.
59. Kopp U, Aurell M, Sjolander M, Ablad B: The role of prostaglandins in the α-and β-adrenoceptor mediated renin release response to graded renal nerve stimulation. *Pfluegers Arch* 1981; 391:1–8.
60. Lush DJ, Fray JCS: Steady-state autoregulation of renal blood flow: A myogenic model. *Am J Physiol* 1984; 247:R89–R99.
61. Margolis BL, Stein JH: The renal circulation. in Priebe H-J (ed): *International Anesthesiology Clinics: The Kidney in Anesthesia*, Boston, Little Brown & Co, 1984, pp 35–63.
62. Moore LC: Tubuloglomerular feedback and SNGFR autoregulation in the rat. *Am J Physiol* 1984; 247:F267–F276.
63. Moore LC, Mason J: Perturbation analysis of tubuloglomerular feedback in hydropenic and hemorrhaged rats. *Am J Physiol* 1983; 245:F554–F563.
64. Morgan T: Factors controlling the release of renin. A micropuncture study in the rat. *Pfluegers Arch* 1977; 368:13–18.
65. Morgan T, Berliner RW: A study by continuous microperfusion of water and electrolyte movements in the loop of Henle and distal tubule of the rat. *Nephron* 1969; 6:388–405.
66. Moss NG: Renal function and renal afferent and efferent nerve activity. *Am J Physiol* 1982; 243:F425–F433.
67. Muller-Suur R, Ulfendahl HR, Persson AEG: Evidence for tubuloglomerular feedback in juxtamedullary nephrons of young rats. *Am J Physiol* 1983; 244:F425–F431.
68. Ng KKF, Vane JR: Conversion of angiotensin I to angiotensin II. *Nature* 1967; 216:762–766.
69. Navar LG, Bell PD, White RW, et al: Evaluation of the single nephron glomerular filtration coefficient in the dog kidney. *Kidney Int* 1977; 12:137–149.
70. Navar LG: Renal autoregulation: Perspectives from whole kidney and single nephron studies. *Am J Physiol* 1978; 234:F357–F370.
71. Navar LG, Ploth DW, Bell PD: Distal tubular feedback control of renal hemodynamics and autoregulation. *Ann Rev Physiol* 1980; 42:557–571.
72. Pricam C, Humbert F, Perrelet A, Orci L: Gap junctions in mesangial and lacis cells. *J Cell Biol* 1974; 63:349–354.
73. Riggs DS: *Control Theory and Physiological Feedback Mechanisms*. Baltimore, Williams & Wilkins Co, 1970.
74. Robertson CR, Deen WM, Troy JL, Brenner BM: Dynamics of glomerular ultrafiltration in the rat. III. Hemodynamics and autoregulation. *Am J Physiol* 1972; 223:1191–1200.
75. Rogenes PR: Single-unit and multiunit analyses of renorenal reflexes elicited by stimulation of renal chemoreceptors in the rat. *J Auton Nerv Sys* 1982; 6:143–156.
76. Roman RJ, Kauker ML: Renal effect of prostaglandin synthetase inhibition in rats: Micropuncture studies. *Am J Physiol* 1978; 235:F111–F118.
77. Sadowski J, Wocial B: Renin release and autoregulation of blood flow in a new model of non-filtering and non-transporting kidney. *J Physiol (Lond)* 1977; 266:219–233.
78. Schnabel E, Kriz W: Morphometric studies of the extraglomerular mesangial cell field in volume expanded and volume depleted rats. *Anat Embryol* 1984; 170:217–222.
79. Schnermann J, Briggs J: Function of the juxtaglomerular apparatus: Local control of glomerular hemodynamics. in Seldin DW, Giebisch G (eds): *The Kidney: Physiology and Pathophysiology*. New York, Raven Press, 1985, vol 1, pp 669–697.
80. Schnermann J, Briggs JP, Kriz W, Moore LC, Wright FS: Control of glomerular vascular resistance by the tubuloglomerular feedback mechanism. in Keaf A, Giebisch G, Bolis L, Gorini S (eds): *Renal Pathophysiology, Recent Advances*. New York, Raven Press, 1980, pp 165–182.
81. Schnermann J, Briggs J, Weber PC: Tubuloglomerular feedback, prostaglandins and angiotensin in the autoregulation of glomerular filtration rate. *Kidney Int* 1984; 25:53–64.
82. Schnermann J, Ploth DW, Hermle M: Activation of tubulo-glomrular feedback by chloride transport. *Pfluegers Arch* 1976; 362:229–240.
83. Schnermann J, Persson AEG, Agerup B: Tubuloglomerular feedback. Non-linear relationship between glomerular hydrostatic pressure and loop of Henle perfusion rate. *J Clin Invest* 1973; 52:862–869.
84. Schnermann J, Wright FS, Davis JM, et al: Regulation of superficial nephron filtration rate by tubuloglomerular feedback. *Pfluegers Arch* 1970; 318:147–175.

85. Schrier RW: Effects of the adrenergic nervous system and catecholamines on systemic and renal hemodynamics, sodium and water excretion and renin secretion. *Kidney Int* 1974; 6:291–306.
86. Seymour AA, Davis JO, Freeman RH, et al: Renin release from filtering and nonfiltering kidneys stimulated by PGI_2 and PGD_2. *Am J Physiol* 1979; 237:F285–F290.
87. Sikri KL, Foster CL, MacHugh N, Marshall RD: Localization of Tamm-Horsfall glycoprotein in the human kidney using immunofluorescene and immuno-electron microscopical techniques. *J Anat* 1981; 132:597–605.
88. Shipley RE, Study RS: Changes in renal blood flow, extraction of inulin, GFR, tissue pressure and urine flow with acute alterations of renal artery blood pressure. *Am J Physiol* 1951; 167:675–688.
89. Sjoquist M, Goransson A, Kallskog O, Ulfendahl HR: The influence of tubulo-glomerular feedback on the autoregulation of filtration rate in superficial and deep glomeruli. *Acta Physiol Scand* 1984; 122:235–242.
90. Slick GL, Aguilera AJ, Zambraski EJ, et al: Renal neuroadrenergic transmission. *Am J Physiol* 1975; 229:60–65.
91. Steiner RW, Tucker BJ, Blantz RC: Glomerular hemodynamics in rats with chronic sodium depletion: Effect of saralasin. *J Clin Invest* 1979; 64:503–512.
92. Stoner LC, Burg MB, Orloff J: Ion transport in cortical collecting tubule: effect of amiloride. *Am J Physiol* 1974; 227:453–459.
93. Tannenbaum J, Splawinski JA, Oates JA, Nies AS: Enhanced renal prostaglandin production in the dog. I. Effect on renal function. *Circ Res* 1975; 36:197–203.
94. Taugner C, Poulsen K, Hackenthal E, Taugner R: Immunocytochemical localization of renin in mouse kidney. *Histochemistry* 1979; 62:19–27.
95. Taugner R, Schiller A, Kaissling B, Kriz W: Gap junctional coupling between the JGA and the glomerular tuft. *Cell Tissue Res* 1978; 186:279–285.
96. Thurau K: Renal hemodynamics. *Am J Med* 1964; 36:698–719.
97. Thurau K, Dahlheim H, Gruner A, et al: Activation of renin in the single juxtaglomerular apparatus by sodium chloride in the tubular fluid at the macula densa. *Circ Res* 1972; 31 (suppl II):182–186.
98. Thurau K, Mason J: The intrarenal function of the juxtaglomerular apparatus, in Thurau K (ed): *MTP Int. Rev. Sci. Kidney and Urinary Tract Physiology.* Baltimore, University Park Press, 1974, Ser I, vol 6, pp 357–390.
99. Tobian L, Tomboulian A, Janecek J: Effect of high perfusion pressures on the granulation of juxtglomerular cells in an isolated kidney. *J Clin Invest* 1959; 38:605–610.
100. Work J, Baehler RW, Kotchen TA, et al: Effect of prostaglandin inhibition on sodium chloride reabsorption in the diluting segment of the conscious dog. *Kidney Int* 1980; 17:24–30.
101. Wright FS, Briggs JP: Feedback control of glomerular blood flow, pressure, and filtration rate. *Physiol Rev* 1979; 59:958–1006.
102. Young DK, Marsh DJ: Pulse wave propagation in rat renal tubules: Implications for GFR autoregulation. *Am J Physiol* 1981; 240:F446–F458.

7

Renal Clearance: Measurements of Glomerular Filtration Rate and Renal Blood Flow

In a limited sense the function of the kidney is to remove excess water and solutes as well as waste products of metabolism from the circulating plasma. To accomplish this, the renal corpuscle makes a copious ultrafiltrate of the renal plasma flow (RPF), which subsequently is modified both in volume and in composition by tubular reabsorption and secretion, yielding a small volume of hypertonic urine. Accordingly, the final volume and composition of the excreted urine depends largely on the volume of plasma filtered at the glomerulus (GFR) and on the fraction of renal plasma flow that escaped filtration (RPF-GFR) and was subsequently processed by the tubules.

It is of particular interest to know whether the urinary excretion of a given substance is due to ultrafiltration, secretion, reduced reabsorption, or a combination of these. Furthermore, it is of practical value to know in what segments of the nephron each of these tubular processes has occurred and to what extent each process determines the final amount of a given substance excreted in the urine. Such information not only provides a sound basis for understanding the normal functioning of the kidney, it also gives an insight into delineating the possible causes of abnormal kidney function, diagnosing the disease, and assessing the success of the therapy as well as the final prognosis.

Renal plasma flow and GFR as well as tubular functions are readily measured by applying the dilution principle. However, before describing the specifics of the method, we need to learn two important associated concepts.

Permeability Ratio

As described in Chapter 5, the composition of the glomerular filtrate is largely determined by the permeability of the ultrafilter. Small molecular size substances, such as electrolytes, glucose, and urea, which can freely pass through the ultrafilter membranes, will have the same concentrations in the ultrafiltrate and plasma. On the other hand, substances with sufficiently large molecular size, such as plasma proteins and lipids, will not freely pass through the ultrafilter. Therefore, their concentrations in the ultrafiltrate will be less than that in plasma. A convenient way to measure the functional permeability of the ultrafilter for a given substance (x) is to compare its concentration in the ultrafiltrate ($[x]_f$) with its "bulk" concentration in the arterial plasma ($[x]_p$). Such a comparison may be called the permeability ratio for that substance, designated by symbol K_x, and defined by Equation 7.1:

$$\text{Permeability Ratio } (K_x) = \frac{\text{Filtrate Concentration } ([x]_f)}{\text{Plasma Concentration } ([x]_p)} \tag{7.1}$$

Certain endogenous and exogenous substances used to study renal function are often bound to plasma proteins. Since only the free (unbound) portion can pass through the ultrafilter membranes, the filtrate concentration will be less than the plasma concentration. For these substances the permeability ratio is less than one.

It should be remembered that the use of plasma permeability ratio is meaningful only when the substance is confined to the plasma. Also, it should be noted that the whole blood permeability ratio, a sometimes useful quantity, will tend to vary with the hematocrit, depending on whether the substance penetrates red blood cells or not.

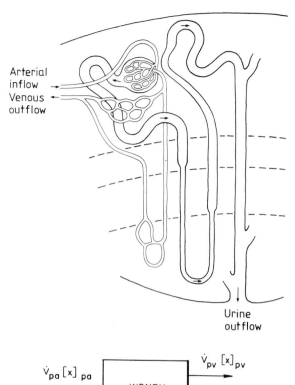

FIGURE 7.1. *Upper panel*: A representative nephron and its blood supply. *Lower panel*: Input-output mass flow through the kidney.

Extraction Ratio

The kidney removes varying fractions of the total quantity of a substance brought to it by the blood, as the arterial plasma makes one transit through the renal vasculature (Figure 7.1, upper panel). The ratio of the quantity of a substance excreted in urine to the total quantity brought to the kidney by the arterial plasma inflow is called the extraction ratio, symbolized E_x, and defined by Equation 7.2:

$$\begin{aligned}\text{Extraction Ratio } (E_x) &= \frac{\text{Quantity Excreted in Urine}}{\text{Total Quantity Brought to the Kidney}} \\ &= \frac{\dot{V}_u \cdot [x]_u}{\dot{V}_{pa} \cdot [x]_{pa}} \quad (7.2)\end{aligned}$$

where $\dot{V}_{pa}$ and $\dot{V}_u$ are the renal arterial plasma and urine volume flows in milliliters per minute, and $[x]_{pa}$ and $[x]_u$ are the renal arterial plasma and urinary concentrations of substance x in milligrams per milliliter, respectively.

For different substances E_x may have values ranging from 0 (not excreted at all, such as plasma proteins) to 1.0 (completely excreted). If E_x for a substance is zero, it means that the kidney excretes none, and the renal venous plasma outflow would have a higher concentration than the arterial plasma inflow (because of the concentrating effect of forming an x-free glomerular filtrate). On the other hand, if E_x for a substance approaches unity, it means that the kidney excretes all of it, leaving a zero concentration in the venous plasma outflow.

Experimental measurement of E_x, using Equation 7.2, is not always easy, for one of the terms in the denominator ($\dot{V}_{pa}$) may not be known, although the other three terms can be measured directly. But the equation may be written in another form, taking advantage of the fact that what appears in the urine must have disappeared from the blood, as illustrated in Figure 7.1 (lower panel), so that

$$\dot{V}_u \cdot [x]_u = \dot{V}_{pa} \cdot [x]_{pa} - \dot{V}_{pv} \cdot [x]_{pv} \quad (7.3)$$

where $\dot{V}_{pv}$ and $[x]_{pv}$ are the renal venous plasma outflow and concentration of substance x, respectively. Substituting Equation 7.3 for the numerator of Equation 7.2 yields:

$$E_x = \frac{\dot{V}_{pa} \cdot [x]_{pa} - \dot{V}_{pv} \cdot [x]_{pv}}{\dot{V}_{pa} \cdot [x]_{pa}} = 1 - \frac{\dot{V}_{pv} \cdot [x]_{pv}}{\dot{V}_{pa} \cdot [x]_{pa}} \quad (7.4)$$

Now, if we can find a substance whose concentration in the venous outflow is zero ($[x]_{pv} = 0$), Equation 7.4 reduces to

$$E_x = 1.0 \quad (7.5)$$

Such a substance is very useful, for it can be used to measure the arterial plasma inflow, $\dot{V}_{pa}$. But, to anticipate, it is known that $\dot{V}_{pa}$ is usually about 660 mL/min (Chapter 5), whereas $\dot{V}_u$ is normally 1 to 3 mL/min. Hence, $\dot{V}_{pv}$ is about 99.7% of $\dot{V}_{pa}$, allowing us to set $\dot{V}_{pa} = \dot{V}_{pv}$ with negligible error. Substitution of this equality in Equation 7.4 excludes all flow terms and reduces Equation 7.4 to

$$E_x = \frac{[x]_{pa} - [x]_{pv}}{[x]_{pa}} \quad (7.6)$$

This equation provides a practical method for measuring the extraction ratio of a substance with reasonable accuracy, especially if E_x approaches unity.

It should be remembered that the plasma extraction

ratio is meaningful only if the substance is confined to the plasma. Also, keep in mind that the whole blood extraction ratio is meaningful for a substance that is present in cells as well as plasma, but will tend to vary with the hematocrit.

Renal Clearance: Measurement of Plasma Volume Flow

Formation of urine ultimately involves the tubular reabsorption of water and solutes from the ultrafiltrate as it passes through the various nephron segments and secretion of selected solutes into tubules from the blood supplying these segments. Since the ultrafiltrate is virtually protein-free plasma, then, in the final analysis, it is the total plasma supplying the nephron that is the source of any substance excreted into the urine. Moreover, if a large quantity of a substance is excreted into the urine, it follows that a large fraction of the total plasma volume supplying the nephron and containing that substance must have been processed to yield the quantity excreted. Consequently, it is of interest to know the *volume of plasma that contains the same quantity of the substance as is excreted in urine in one minute*. The volume of plasma flowing into the kidney per minute and containing this substance is called the *plasma flow* for that substance. Note that its value may range from zero, for a substance not excreted in the urine, to a value equal to the actual renal plasma flow, for a substance that is completely removed from blood by the kidney.

In practice, to measure the plasma flow of any substance, we inject it intravenously continuously, so as to maintain a steady plasma level throughout the period. While the plasma level is steady, we collect urine over a specified number of minutes, and then determine the urinary concentration of the substance. The plasma concentration is also measured in a blood sample drawn in the middle of the urine collection period.

If $[x]_u$ is the urinary concentration of substance x in milligrams per milliliter and $\dot{V}_u$ is the urine flow rate in milliliters per minute, then $[x]_u \cdot \dot{V}_u$ is the urinary excretion rate of this substance in milligrams per minute. Now, if $[x]_p$ is the plasma concentration of x, the ratio of the urinary excretion rate to plasma concentration ($[x]_u \cdot \dot{V}_u/[x]_p$) yields the volume of plasma that contains the same quantity of substance x as is excreted in the urine in one minute. It is thus expressed in milliliters per minute. We shall designate this plasma flow by the symbol $\dot{V}_x$, where subscript x is a reminder that the plasma flow thus measured represents that fraction of the total renal plasma flow (RPF = $\dot{V}_p$) from which the substance has been cleared:

$$\dot{V}_x = \frac{[x]_u \cdot \dot{V}_u}{[x]_p} \qquad (7.7)$$

If the extraction ratio for a substance x as it passes through the kidney is 1, i.e., $[x]_{pv} = 0$, which according to Equation 7.4 implies that $[x]_{pa} \cdot \dot{V}_{pa} = [x]_u \cdot \dot{V}_u$, then its plasma flow ($\dot{V}_x$) is a measure of the *true* renal plasma flow ($\dot{V}_p$); if the extraction ratio is other than 1, it is only a *virtual* or apparent measure of renal plasma flow.

The plasma flow of a substance thus measured is also called *renal clearance* of that substance and is defined as the *virtual volume of the plasma from which a given substance is removed per minute*. Thus, renal clearance is an empiric measure of the abililty of the kidney to remove a substance from the blood plasma.

The clearance formula that is widely used is somewhat different in appearance from Equation 7.7. Therefore, to maintain continuity with the literature we shall write the above clearance equation in terms of the classical notations:

$$C_x = \frac{U_x \cdot V}{P_x} \qquad (7.8)$$

where U_x and P_x are the urinary and plasma concentrations of substance x in milligrams per milliliter, respectively, V is the urine flow rate in milliliters per minute, and C_x is the clearance of substance x in milliliters per minute.

Having thus described, in general terms, the meaning of renal plasma flow and the procedure for measuring it, we are now ready to apply this concept to determine the glomerular filtration rate, the renal plasma flow, and the renal blood flow, as well as tubular functions.

Measurement of Glomerular Filtration Rate

The plasma flow or clearance of a substance as such provides no information about the mechanisms by which the kidney removes that substance from the plasma. The substance may have been removed by ultrafiltration, tubular reabsorption, or secretion, or a combination of these. To determine the magnitude of each process, it is necessary to measure GFR and RPF.

For any blood solute (x) that passes through the glomerular ultrafilter, but is not created or destroyed by the kidney, nor secreted or reabsorbed by the renal tubule, a *filtrate-urine* mass balance must apply to the passage of filtrate through the tubule. This simply means that the rate of influx of x into the filtrate must equal its rate of efflux into the final urine. Since these fluxes may be expressed as the product of a volume flow rate ($\dot{V}$) and a concentration ([x]), we may write this filtrate-urine mass balance as follows:

$$\dot{V}_f \cdot [x]_f = \dot{V}_u \cdot [x]_u \qquad (7.9)$$

where the subscripts refer to filtrate and urine. Simple rearrangement yields a dilution principle formula for calculating filtrate volume flow rate:

$$\dot{V}_f = \dot{V}_u \cdot [x]_u / [x]_f \qquad (7.10)$$

which simply states that dividing the solute by its concentration yields the solvent. Note the similarity of this equation with Equation 7.7.

This formula is not yet suitable for practical use, because of difficulty in measuring $[x]_f$. But the latter can be determined indirectly by making use of the permeability ratio (K_x), defined as the ratio of filtrate concentration to arterial plasma "bulk" concentration:

$$K_x = [x]_f / [x]_{pa} \qquad (7.11)$$

Now, if x is not metabolized by peripheral tissues, its superficial venous concentration ($[x]_p$) will be the same in arterial plasma as in venous plasma obtained from any superficial (nonrenal) vein. Hence,

$$[x]_p = [x]_{pa} = [x]_{pv} \qquad (7.12)$$

Substituting Equations 7.11 and 7.12 into Equation 7.10 yields a practical formula for plasma flow rate ($\dot{V}_p$):

$$\dot{V}_p = \dot{V}_u \cdot [x]_u / K_x \cdot [x]_p \qquad (7.13)$$

It has been found that *inulin*, a fructo-polysaccharide (mol wt = 5,200), is a nontoxic substance that is not metabolized by tissues (hence $[In]_{pv} = [In]_{pa}$) and is neither created nor destroyed.[10] Also, it is neither secreted nor reabsorbed by the tubules (hence the filtrate-urine mass balance is valid) and has a known and constant K_x of 1.06 ("bulk"). When expressed in "water" concentration units, the K_x is unity, meaning that inulin has equal "water" concentration in plasma and filtrate. Hence, the practical

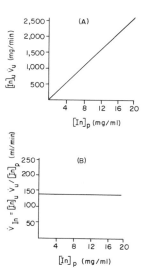

FIGURE 7.2. Relationships depicting the renal excretion rate (A) and clearance (B) of inulin as a function of plasma inulin concentration. (Modified and reproduced with permission from Pitts, R.F.: *Physiology of the Kidney and Body Fluids*, 3rd Edition. Copyright © 1974 by Year Book Medical Publishers, Inc., Chicago.)

formula for measuring the glomerular filtration rate is

$$\text{GFR} = \dot{V}_{In} = \frac{\dot{V}_u \cdot [In]_u}{1.06 [In]_p} \qquad (7.14)$$

where $[In]_u$ and $[In]_p$ are the inulin concentrations in the urine and plasma, respectively, and $\dot{V}_{In}$ is the fraction of renal plasma flow which contains the same amount of inulin as is excreted in urine in one minute.

For $\dot{V}_{In}$ to be a true measure of the glomerular filtration rate, in addition to the above requirements, its value must be *constant* and *independent* of inulin plasma concentration. This implies that the inulin excretion rate must be directly proportional to, and a linear function of, inulin plasma concentration. These relationships are summarized graphically in Figure 7.2. Since inulin fulfills all of these requirements, its clearance is a true measure of GFR.

In practice, to measure GFR from inulin clearance, we inject a large (priming) dose of inulin intravenously to increase its plasma concentration. Then, more inulin is slowly infused until its plasma concentration attains a steady state value. At this point, a timed sample of urine is collected (for measuring $\dot{V}_u$ and $[In]_u$), and midway through this period a venous blood sample is taken (for measuring $[In]_p$).

To illustrate the calculation, consider the following inulin clearance data obtained from a patient:

urine flow rate, $\dot{V}_u$ = 60 mL/h; urine concentration of inulin, $[In]_u$ = 100 mg/dL; and plasma inulin concentration, $[In]_p$ = 1 mg/dL. Before we can use Equation 7.14 to calculate the glomerular filtration rate in this patient, it is convenient to convert the units for these data to those expressed in this equation. Thus, we may express the volume of urine excreted per minute rather than per hour, and urine and plasma inulin concentrations in milligrams per milliliter rather than milligrams per deciliter. Carrying on these conversions yields: $\dot{V}_u$ = 60 mL/h = 60 mL/60 min = 1 mL/min; $[In]_u$ = 100 mg/dL = 100 mg/100 mL = 1 mg/mL; and $[In]_p$ = 1 mg/dL = 1 mg/100 mL = 0.01 mg/mL. Now, substituting these values into Equation 7.14, we get

$$\text{GFR} = \dot{V}_{In} = \frac{1 \text{ mL/min} \times 1 \text{ mg/mL}}{0.01 \text{ mg/mL} \times 1.06}$$
$$= 93.4 \text{ mL/min} \quad (7.15)$$

If, as is customary, we omit the K_{In} value of 1.06 from the above equation, we get a value of 100 mL/min for GFR in this patient.

In a healthy young man, inulin clearance yields an average value of 118 mL/min for GFR. You will find this normal value elsewhere cited as 125 mL/min (actually 125 mL/min/1.73 m² body surface area), because the K_{In} of 1.06 is usually omitted from Equation 7.14. The use of K_{In} would be more logical, however, since the fundamental principle of the method estimates filtrate concentration from the plasma ("bulk") concentration.

The glomerular filtration varies with body size and age. The GFR is low in newborn[24] owing to prominent visceral epithelial cell layers enfolding the glomerular capillary membrane. The GFR does not assume adult characteristics until one to three years after birth. Its value decreases in very old people, even in the absence of renal disease.[6]

It should be clear that the tubular system must reabsorb most of the 118 mL/min of glomerular filtrate, yielding only 1 to 3 mL/min of final urine. This large GFR is remarkably constant. Although it is subject to pathological variations, the kidneys do not adjust its rate as a means of compensating for disturbances affecting the renal-body fluid regulating system. Furthermore, as far as solutes are concerned, glomerular filtration is essentially nonselective (except for proteins and lipids) and nonadjustable. The controlled, compensatory processes of the kidney reside in the tubular system.

Partition of Total Renal Clearance

In general, the total renal excretion rate of a substance x ($[x]_u \cdot \dot{V}_u$) by the kidney is the algebraic sum of three processes: (1) the rate at which the substance is delivered to the tubules by glomerular filtration; (2) the rate at which it is reabsorbed by the tubules; and (3) the rate at which additional amount of the substance is added to the tubular fluid by secretion. The quantity of the substance filtered at the glomerulus is called the *filtered load*. It is equal to the product of the GFR, the arterial plasma concentration of the substance ($[x]_p$), and the permeability ratio for that substance (K_x). The net rate of excretion due to tubular transport per minute ($\dot{T}_x$) is determined by whether the substance is reabsorbed or secreted. Since secretion of a substance into the tubular lumen represents an *addition* to the amount already filtered, it may be designated by $+\dot{T}_x$. Conversely, because reabsorption of a substance represents a *reduction* in the amount already filtered, it may be designated by $-\dot{T}_x$. Using these notations, the total rate of excretion of a substance may now be expressed as

$$[x]_u \cdot \dot{V}_u = \text{GFR} \cdot [x]_p \cdot K_x \pm \dot{T}_x \quad (7.16)$$

Dividing the above equation by the plasma concentration ($[x]_p$), we obtain an expression that partitions the clearance for that substance ($\dot{V}_x$) into the fraction processed by filtration and that processed by tubular transport:

$$\dot{V}_x = \frac{[x]_u \cdot \dot{V}_u}{[x]_p} = \text{GFR} \cdot K_x \pm \frac{\dot{T}_x}{[x]_p} \quad (7.17)$$

Note that each term in this equation has a dimension of volume flow rate (milliliters per minute). Therefore, Equation 7.17 represents another definition for the total clearance of a substance in terms of the glomerular and tubular components:

Total Clearance = Glomerular ± Tubular
  Clearance   Clearance (7.18)

If the kidney removes a substance from the plasma by filtration only (such as inulin), the tubular clearance is zero and the excretion rate will be proportional to the plasma concentration.

If the clearance of a substance involves filtration and either tubular reabsorption or secretion, then the magnitude of its tubular clearance depends on the plasma concentration and the tubular transport capacity for reabsorption or secretion. The rate of tubular transport ($\dot{T}_x$) at a given plasma concentration depends on the kind of substance being transported. For the present discussion, the overall tubular transport—either secretion or reabsorption—may be classified according to whether the renal tubular cells have a *maximum tubular transport capacity ($\dot{T}_{mx}$)* for the particular substance or not. The maximum tubular transport capacity implies that as long as the plasma concentration is below the *threshold concentration* that saturates the transport process, all of the filtered quantity of the given substance is reabsorbed or a maximal fraction of the quantity of the substance entering peritubular capillaries (which escaped filtration) is secreted. The quantitative aspects of this classification of the renal transport mechanism is considered in Chapter 10. For the present purpose of understanding the influence of the tubular clearance on the overall clearance, we choose those substances, such as glucose (filtered and reabsorbed) and *p*-aminohippurate (PAH) (filtered and secreted), for which the renal tubular cells have a maximum transport capacity. Accordingly, Equation 7.17 becomes

$$\dot{V}_x = GFR \cdot K_x \pm \frac{\dot{T}_{mx}}{[x]_p} \quad (7.19)$$

The total clearance of a substance that is removed from the renal plasma by both filtration and tubular secretion decreases as the plasma concentration increases. This is because the right-hand side of Equation 7.19 represents the algebraic sum of a constant quantity (GFR · K_x) and a varying quantity ($\dot{T}_{mx}/[x]_p$). Since the value of $\dot{T}_{mx}$ (amount of the substance transported maximally per minute) is constant for a given substance, then the tubular clearance ($\dot{T}_{mx}/[x]_p$) decreases as $[x]_p$ increases. Therefore, at a high plasma concentration the total clearance of a substance that is either reabsorbed (glucose) or secreted (PAH) asymptotically approaches the GFR or the inulin clearance. In the case of glucose, which is filtered and subsequently reabsorbed maximally in the proximal tubule, the tubular clearance decreases as the plasma concentration increases. At a very high plasma concentration, the total clearance increases and asymptotically approaches the inulin clearance. In contrast, for a substance such as PAH, which is filtered and maximally secreted in the proximal tubule, at high plasma concentration the total clearance decreases and asymptotically approaches the inulin clearance. The relation of total clearance of PAH ($\dot{V}_{PAH}$) and glucose ($\dot{V}_G$) to that of inulin ($\dot{V}_{In}$) at high

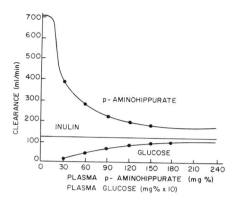

FIGURE 7.3. Comparison of clearances of *p*-aminohippurate (PAH), glucose, and inulin as a function of their plasma concentrations. (Modified and reproduced with permission from Pitts, R.F.: *Physiology of the Kidney and Body Fluids*, 3rd Edition. Copyright © 1974 by Year Book Medical Publishers, Inc., Chicago.)

plasma PAH and glucose concentrations in man are illustrated graphically in Figure 7.3.

The GFR is defined as the volume of plasma filtered per minute and not as the volume of plasma water. Since plasma is 93% to 94% water by weight and 6% to 7% solutes, including proteins, the volume of water filtered is approximately 93% to 94% of the GFR. The normal variations in urine flow are due to changes in water reabsorption and not to the GFR. Approximately 60% to 80% of the filtered water is reabsorbed in the proximal tubules. This is called obligatory water reabsorption. The remaining 20% to 40% is reabsorbed in variable quantities, mainly in the distal tubule and collecting duct. This is called nonobligatory or facultative water reabsorption and is regulated by the antidiuretic hormone (ADH). In the presence of normal plasma ADH concentration, most of this water is reabsorbed, reducing the urine flow rate to about 1 to 3 mL/min. The upper limit of facultative water reabsorption is determined by the osmotic activity of the nonreabsorbed solutes in the tubular urine.

Clinical Measurement of GFR

In most patients with suspected renal disease, for whom frequent assessment of GFR is required for both diagnosis and prognosis, it is clinically impractical to measure the inulin clearance. This technical difficulty in measuring renal function by inulin clearance in clinical situations, plus the ease and rapidity required to assess renal function in disease, has prompted a continuous search for suitable endogenous substances whose clearances approximate that of inulin. Of the various potential endogenous substances whose clearance might be of diagnostic value, the endogenous urea and creatinine clearances have received the most attention.

Although historically the endogenous urea clearance was the test of choice,[23] in recent years it has been largely replaced by the endogenous creatinine clearance.[11] Therefore, to better appreciate the potential value of the endogenous creatinine clearance in assessing renal malfunction, in this section we discuss the clearance of both these substances and delineate their advantages and shortcomings.

Endogenous Urea Clearance

Urea, the nitrogenous product of protein metabolism, is, apart from water, the chief constituent of urine. The renal transport of urea involves a rather complex intrarenal recycling (Chapter 12). Urea is freely filtered at the glomerulus and is reabsorbed from the tubular fluid by back-diffusion. The fraction of filtered urea reabsorbed increases from 35% in the hydrated states to 60% in the dehydrated states.[9] Therefore, depending on the state of hydration, urea clearance accounts for only 40% to 65% of the true GFR, as measured by inulin clearance.

The tubular reabsorption rate of urea depends largely on the urine flow rate. The greater the urine flow rate, the smaller is the urea reabsorption rate by back-diffusion and the greater is its renal clearance. At urine flow rate above 2 mL/min, urea reabsorption rate diminishes greatly and its renal clearance approaches that of inulin. Therefore, at very high urine flow rate, urea clearance gives an approximate measure of GFR.

In practice, urea clearance is determined by measuring urea excretion rate during one hour and then drawing a blood sample at the halfway point of urine collection. Thus, urea clearance gives the volume of plasma that contains the same quantity of urea as is excreted in urine in one hour (not one minute).

The results of the urea clearance are expressed as percent of normal. The endogenous urea clearance measured in man varies between 64 and 99 mL/min/1.73 m² body surface area, when urine flow rate is greater than 2 mL/min. A urea clearance of 74 mL/min is considered 100% of maximal urea clearance, yielding a normal range of 86% to 132%. A urea clearance of 70% to 80% of normal cannot always be considered low, particularly if the adult individual is small.

If it becomes necessary to perform the urea clearance in patients with a urine flow rate less than 2 mL/min, the so-called standard urea clearance (C_s) may be calculated. The standard urea clearance is an empirically derived formula[18] that states that the clearance of urea changes in proportion to the square root of the urine flow rate, when the latter is less than 2 mL/min:

$$\text{Standard Clearance } (C_s) = \frac{U}{P}\sqrt{V} \quad (7.20)$$

The standard urea clearance in normal man determined at a urine flow rate of 1 mL/min is 54 mL/min. It should be noted that the standard clearance defined by Equation 7.20 is no longer a "clearance" as defined earlier, but rather a mathematical attempt to correct for the variations in urea excretion rate when urine flow rate is less than 2 mL/min. Hence, the standard urea clearance cannot be compared with clelarances performed in the same individual at other urine flow rates.

Since the renal clearance of urea, at urine flow rates above 2 mL/min, depends primarily on the GFR, it will be reduced in those situations in which the GFR may be reduced in the absence of parenchymal renal disease; i.e., heart failure, severe dehydration, or shock. The endogenous urea clearance is low in the newborn, becomes normal by about 2 years of age, and again declines in the aged to approximately 50% to 60% of normal in the ninth decade of life.

Despite its earlier clinical use, the endogenous urea clearance is now largely replaced by the endogenous creatinine clearance, which is described below. This is because there are at least three major limitations in the routine clinical application of the endogenous urea clearance as a measure of GFR. First, since the rate of urea production is variable, its blood level changes in response to dietary protein intake, hepatic function, catabolic processes, and gastric hemorrhage. Second, renal excretion of urea depends on GFR and the rate of tubular reabsorption by back-diffusion. The reabsorption rate is found to be related directly to the urine flow rate. And third, at low urine flow rates (less than 2 mL/min) the clearance values obtained are inaccurate, even when the correction formula (Equation 7.20) is used. In contrast, creatinine is not reabsorbed, and its renal excretion rate is not related directly to urine flow rate.

Endogenous Creatinine Clearance

In a creatinine-free diet, muscle creatinine and phosphocreatinine are the only source of urinary creatinine. Since normally the muscle mass is relatively constant, the creatinine turnover is also relatively constant. This constancy is reflected in the stable plasma concentration and urinary excretion of creatinine. In addition, unlike urea, the endogenous creatinine clearance does not depend on urine flow rate or the dietary protein intake.

Creatinine, when given intravenously, is freely filtered at the glomerulus and is also secreted by the tubules. Hence, like PAH, creatinine clearance at normal plasma concentration (0.64 to 1.1 mg/dL) is substantially greater than inulin clearance. However, at high plasma concentration (in excess of 100 mg/dL), creatinine clearance asymptotically approaches inulin clearance. This is believed to be a consequence of the existence of maximal tubular secretion rate, $\dot{T}_m$, for creatinine. In man, $\dot{T}_m$ for creatinine is about 13 mg/dL GFR.

Endogenous creatinine clearances for normal subjects varying in age from 20 to 99 years are listed in Table 7.1.

Since 1940 endogenous creatinine clearance has become a widely accepted measure of GFR. It was shown then by Steinitz and Turkland[20] and subsequently by Brod and Sirata[4] that the renal excretion of endogenous creatinine was largely dependent on the glomerular filtration rate and that its renal clearance closely approximated that of inulin in both normal subjects and those with impaired renal function.

Although inulin clearance is generally accepted as the most accurate method of measuring GFR, the techniques required are complicated to perform and therefore useful only as a research tool. The endogenous creatinine clearance, on the other hand, is not subject to these practical limitations and therefore is well suited for general clinical use. The simplicity, and the advantages and shortcomings of the routine clinical application of the endogenous creatinine clearance, compared with inulin and urea clearance methods, have been well documented.[7, 22]

As pointed out by these authors, for practical clinical purposes endogenous creatinine clearance has been found to be equal to inulin clearance, both as a diagnostic test and as a prognostic guide. Furthermore, when performed at regular intervals, endogenous creatinine clearance was found to be a sensitive test over an unusually wide range of renal damage, especially in following the prognosis of patients with advanced renal insufficiency. The main

TABLE 7.1. Serum creatinine concentration, urinary creatinine excretion, and endogenous creatinine clearance in adult men and women at different age groups.

	Age (yr)	Serum creatinine (mg/dL ± SD)	Urinary creatinine (mg/kg/24 h ± SD)	Creatinine clearance (mL/1.73 m^2)
Men	20–29	0.99 ± 0.16	23.8 ± 2.3	110
	30–39	1.14 ± 0.22	21.9 ± 1.5	97
	40–49	1.10 ± 0.20	19.7 ± 3.2	88
	50–59	1.16 ± 0.17	19.3 ± 2.9	81
	60–69	1.15 ± 0.14	16.9 ± 2.9	72
	70–79	1.03 ± 0.22	14.2 ± 3.0	64
	80–89	1.06 ± 0.25	11.7 ± 4.0	47
	90–99	1.20 ± 0.16	9.4 ± 3.2	34
Women	20–29	0.89 ± 0.17	19.7 ± 3.9	95
	30–39	0.91 ± 0.17	20.4 ± 3.9	103
	40–49	1.00 ± 0.24	17.6 ± 3.9	81
	50–59	0.99 ± 0.26	14.9 ± 3.6	74
	60–69	0.97 ± 0.17	12.9 ± 2.6	63
	70–79	1.02 ± 0.23	11.8 ± 2.2	54
	80–89	1.05 ± 0.22	10.7 ± 2.5	46
	90–99	0.91 ± 0.12	8.4 ± 1.4	39

Source: Kampmann, et al.[11]

disadvantage is that the normally low plasma concentration of endogenous creatinine plus the presence of significant amounts of nonspecific chromogen in the plasma makes the measurement of the endogenous plasma creatinine concentration somewhat imprecise.[17]

Measurement of RPF

For any (noncreated, nondestroyed) blood solute (x) a *blood-urine* mass balance must apply to each single passage of the blood through the kidneys. This simply means that the arterial influx must equal the sum of venous and urinary effluxes (Figure 7.1, lower panel). This balance is valid regardless of the degree of glomerular filtration, of tubular reabsorption or secretion, and of whether the substance has access to red blood cells or is confined to the plasma.

Of more practical interest is the *plasma-urine* mass balance, which is valid *only* if the solute is confined to the plasma phase of blood. This mass balance may be written by rearranging Equation 7.3 as

$$\dot{V}_{pa} \cdot [x]_{pa} = \dot{V}_{pv} \cdot [x]_{pv} + \dot{V}_u \cdot [x]_u \quad (7.21)$$

In the case of mass balance for volume flow rates, it turns out that the plasma flow rates are enormous compared with the ultimate urine flow rate, so that we can equate the arterial and venous plasma flows to the RPF ($\dot{V}_p$) and write, with only a fraction of percent error,

$$\dot{V}_p = \dot{V}_{pa} = \dot{V}_{pv} \quad (7.22)$$

Substituting Equation 7.22 into Equation 7.21 yields a simplified mass balance expression:

$$\dot{V}_p \cdot [x]_{pa} = \dot{V}_p \cdot [x]_{pv} + \dot{V}_u \cdot [x]_u \quad (7.23)$$

Simple rearrangement now yields a Fick principle formula for calculating the renal plasma flow:

$$\dot{V}_p = \frac{\dot{V}_u \cdot [x]_u}{[x]_{pa} - [x]_{pv}} \quad (7.24)$$

This equation states that analogous to measuring the cardiac output from the ratio of the oxygen consumed to the arteriovenous oxygen concentration difference, the RPF may be determined from the ratio of the excretion rate of a substance (x) to its renal arteriovenous concentration difference. The arteriovenous concentration difference is a measure of the amount of substance (x) removed from each milliliter of plasma as it perfuses the kidney. Therefore, the ratio of the excretion rate to the arteriovenous concentration difference gives the total volume of plasma that contains the same quantity of substance x as is excreted in urine in one minute.

There are at least two difficulties in measuring RPF in man by applying the Fick principle. First, the renal venous outflow ($\dot{V}_{pv}$) is not equal to the renal arterial inflow ($\dot{V}_{pa}$), as required by the Fick principle. This inequality is due to a small but significant urine flow rate, which may be negligible if $\dot{V}_u$ is small. Second, it is impractical in clinical medicine to procure a renal venous blood sample in man. However, if the urine flow rate is small and the extraction ratio (E_x) for the test substance is very high (i.e., $[x]_{pv} \to 0$), the Fick equation (Equation 7.24) then reduces to the clearance equation for that substance.

In practice, the assumption that $\dot{V}_u$ is small (hence, $\dot{V}_{pa} = \dot{V}_{pv}$) has a less serious effect on the accuracy of calculation of RPF from Fick equation than the substitution of superficial venous blood sample for the renal venous blood sample. However, the difficult sampling of renal venous blood can be avoided by using the extraction ratio (E_x), defined earlier (Equation 7.2) as the fraction of the arterial influx that the kidneys excrete. This is simply the ratio of two terms in Equation 7.23:

$$E_x = \frac{\dot{V}_u \cdot [x]_u}{\dot{V}_p \cdot [x]_{pa}} \quad (7.25)$$

If we replace $\dot{V}_u \cdot [x]_u$ in this equation by the difference between the arterial influx and venous efflux (Equation 7.22), we obtain an expression, analogous to Equation 7.6, that excludes all flow terms:

$$E_x = \frac{[x]_{pa} - [x]_{pv}}{[x]_{pa}} = \frac{[x]_p - [x]_{pv}}{[x]_p} \quad (7.26)$$

The second form of Equation 7.26 replaces the superficial venous plasma for the arterial plasma sample, a valid substitution if substance x is not metabolized by nonrenal tissues. With this formula we can experimentally search for a substance having a known and constant E_x.

Substituting Equation 7.26 into Equation 7.24 yields a practical clearance formula for measuring RPF:

$$\dot{V}_p = \frac{\dot{V}_u \cdot [x]_u}{E_x \cdot [x]_p} = \frac{\dot{V}_x}{E_x} \quad (7.27)$$

where $\dot{V}_x$ is the renal clearance of substance x.

It has been found that PAH is a nontoxic substance that is not metabolized by nonrenal tissues. Furthermore, PAH is neither created nor destroyed by the kidney, and it is confined to plasma (making the plasma-urine mass balance valid). In addition, within certain plasma concentrations (1 to 10 mg/dL), extraction ratio of PAH is nearly constant between 0.70 to 0.90 in dogs and 0.85 to 0.95 in man,[18] revealing that it is both readily filtered and also secreted by renal tubules. Hence, the particular formula for measuring RPF from the clearance of PAH is

$$RPF = \frac{\dot{V}_u \cdot [PAH]_u}{0.9 \, [PAH]_p} = \frac{\dot{V}_{PAH}}{0.9} \quad (7.28)$$

where 0.9 is the average PAH extraction ratio in man. Since PAH is primarily secreted by the proximal tubules, its renal clearance ($\dot{V}_{PAH}$) without correcting for the extraction ratio yields an approximate measure of the cortical RPF.

The PAH clearance method (Equation 7.28) yields an average RPF for healthy young man of 660 mL/min. By also measuring the hematocrit (Hct), the RPF may be converted to renal blood flow (RBF):

$$RBF = \frac{RPF}{1 - Hct} \quad (7.29)$$

The normal hematocrit of 0.45 thus yields a normal RBF of 1,200 mL/min, which, it can be seen, is 20% of the total resting cardiac output of 6.0 L/min.

Filtration Fraction

The fraction of RPF that is filtered through the glomeruli is called the filtration fraction. This is estimated from the ratio of inulin clearance to PAH clearance.

$$\text{Filtration Fraction} = \frac{\text{GFR, mL/min}}{\text{Renal Arterial Plasma Flow, mL/min}}$$

$$= \frac{\dot{V}_{In}}{\dot{V}_{PAH}/0.9} \quad (7.30)$$

The filtration fraction in man normally varies from 16% to 20%.

The filtration fraction as defined above may be regarded as the glomerular extraction ratio. The overall extraction ratio of any substance was defined earlier (Equation 7.2) as the fraction of the total quantity brought to the kidney which is excreted in the urine. The quantity brought to the kidney is equal to the product of the plasma concentration and the renal arterial plasma flow, i.e., $[x]_p \cdot \dot{V}_p$. The quantity removed by the kidney is equal to the quantity excreted in the urine, i.e., $[x]_u \cdot \dot{V}_u$. Therefore, the overall renal extraction ratio, assuming permeability ratio of unity, will be

$$\text{Renal Extraction Ratio} = E_x = \frac{[x]_u \cdot \dot{V}_u}{[x]_p \cdot \dot{V}_p} \quad (7.31)$$

Equation 7.31 may also be written as

$$E_x = \frac{[x]_u \cdot \dot{V}_u}{[x]_p} \cdot \frac{1}{\dot{V}_p} = \frac{\dot{V}_x}{\dot{V}_p} \quad (7.32)$$

This latter equation shows that the extraction ratio for a substance is equal to the fraction of the renal arterial plasma flow ($\dot{V}_p$) which is cleared of that substance ($\dot{V}_x$).

Knowing the filtration fraction and the extraction ratio, we can determine the manner by which the kidney processes a given substance. The ratio of the filtration fraction (FF) (Equation 7.30) to the extraction ratio (Equation 7.31) provides such information:

$$\frac{FF}{E_x} = \frac{\dfrac{\dot{V}_{In}}{\dot{V}_p}}{\dfrac{[x]_u \cdot \dot{V}_u}{[x]_p \cdot \dot{V}_p}} = \frac{\dot{V}_{In} \cdot [x]_p}{[x]_u \cdot \dot{V}_u} \quad (7.33)$$

This equation states that the ratio of the filtration fraction to the extraction ratio is equal to the ratio of the filtered load ($\dot{V}_{In} \cdot [x]_p$) to the excretion rate ($[x]_u \cdot \dot{V}_u$) of that substance. It may be used to estimate net rates of secretion or reabsorption of a substance. Therefore, a ratio of one implies that the net excretion rate is equal to the quantity filtered per minute. A ratio less than one implies that net secretion has occurred, and a ratio greater than one implies that net reabsorption has occured.

Note that Equation 7.33 gives only the net rates of secretion and reabsorption, respectively. The fact that there is net reabsorption does not rule out secretion, and vice versa.

Clearance Ratio

A closer look at Equation 7.33 reveals that the right-hand side of this equation, when inverted, is merely a ratio of the clearance of a substance ($\dot{V}_x$) to that of inulin ($\dot{V}_{In}$). Clearance of a substance relative to that of inulin can also provide an approximate index of the manner by which that substance is processed by the kidneys. If a substance is filtered and subsequently secreted by the tubules, its renal clearance is greater than that of inulin. However, if the substance is filtered and then is subsequently reabsorbed by the tubules, its renal clearance would be less than that of inulin. In either case, if the secretion and reabsorption processes involve a saturable type of mechanism, i.e., maximum tubular transport, then, as the plasma concentration of either substance is increased, the renal clearance of the substance approaches that of inulin (see Figure 7.3).

Partition of RBF

Normal renal function is dependent on normal renal circulation. Acute reduction in RBF is often accompanied by a marked alteration of function, which may lead to renal insufficiency and oliguria. The resulting acute derangements in function are due to abnormal changes in renal hemodynamics and maldistribution of blood flow within the kidney. In the normal kidney, changes in RBF and its distribution between the cortex and medulla have profound influence on the ability of the kidney to concentrate or dilute urine and hence on the rate of excretion of salt and water (see Chapter 12 for details).

To understand the magnitude of these changes in intrarenal circulation and the regions of the kidney affected by the redistribution of RBF, a knowledge of normal distribution of RBF is necessary. The conventional clearance techniques described earlier in this chapter provide an overall measure of GFR and RPF. Although PAH clearance is the most commonly used method in clinical medicine for measuring RBF, it provides little information about the distribution of blood between the cortex and medulla. In addition to clearance techniques, other methods have been used in animal models to measure total RBF. They include the use of noncannulating electromagnetic flowmeters and Doppler ultrasound detectors. Of the many different methods used to measure RBF and its distribution in various experimental circumstances,[13,15] we shall describe two techniques that have provided some insights into the intrarenal distribution of blood flow in response to a variety of experimental perturbations.

Inert Diffusable Gas Method

This method, introduced by Thorburn and associates,[21] is based on the assumption that the rate of accumulation (or washout) of an inert substance, such as krypton (^{85}Kr) or xenon (^{133}Xe) is proportional to the rate of blood flow. Briefly, the technique for measuring RBF consists of injecting a small bolus of saline saturated with radioactive krypton (^{85}Kr) or xenon (^{133}Xe) rapidly into the renal artery. The lipid-soluble gas diffuses so rapidly across the capillary membranes that the renal tissue is saturated almost instantaneously. The removal of the isotope from the tissue is thus limited by the rate of blood flow in the capillaries supplying these tissues. The higher the RBF, the faster the disappearance of the radioactive isotope. The disappearance of the radioactive isotope can be monitored by placing an external detector over the kidney.

The disappearance curve of radioactive krypton from the kidney, thus obtained, follows a complex (multicomponents) exponential, which has been analyzed into four mono-exponential components, using "tail subtraction" technique, representing blood flow in the following four regions of the kidney: (I) cortex, (II) outer medulla, (III) inner medulla, and (IV) perirenal and hilar fat. These components have been localized in the dog by autoradiography.

Briefly, the tail subtraction technique involves extrapolating the linear portion of the multicomponent exponential curve back to zero time. The resulting line gives the disappearance curve for component VI. Subtracting this curve from the original curve yields another exponential curve. Extrapolating the linear portion of this curve back to zero time yields the disappearance curve for component III. Repeating the subtraction and extrapolation processes once more yields a third exponential curve, which can be resolved into two linear lines, representing disappearance curves for components II and I, respectively. The blood flow for each component is determined by a method similar to that outlined in Chapter 2 (see also Appendix B) for measurement of body volume and composition. From such an analysis, Thorburn and co-workers[21] found that in dog the cortical RBF averages about 80% of the total RBF, whereas blood flow in the outer medulla and inner medulla represent about 10% and 5%, respectively.

In addition to partitioning the total RBF into four regions, calculation of the slopes of the mono-exponential curves yielded half-time ($t_{1/2}$) values, a useful index to estimate the rate of blood flow in each region. The results indicated that the blood flow in the cortex is nearly three times faster than that in the outer medulla and that the blood flow in the outer mdulla is nearly ten times faster than that in the inner meddula.

The differences in the rates of blood flow in the cortex and medulla, as indicated by differences in half-time values, are of more than academic interest. On the average, the cortical reabsorption accounts for nearly 60% to 80% of the filtered volume and its dissolved solutes. Since the cortical blood flow is fairly rapid, tubular reabsorption has little influence on the interstitial composition and osmolality. In contrast, the blood flow in the medulla is relatively slow. Consequently, tubular reabsorption profoundly affects the interstitial composition and osmolality.[3] The differences in the rates of blood flow between the cortex and the medulla and the volume of the filtrate normally reabsorbed in these two regions are largely responsible for the renal response to expansion or contraction of the extracellular fluid volume. Thus, expansion of the extracellular volume with intravenous infusion of a large volume of physiological saline results in an increase in urinary excretion of sodium (natriuresis). The increase in the extracellular volume causes an increase in both GFR and blood flow through the cortex. However, owing to elevated capillary hydrostatic pressure, reabsorption of sodium and water from the proximal tubule is depressed.[8] As a result, a greater volume of filtrate is delivered to the more "distal" segments of the nephron for final processing. The increased delivery of the filtrate to the "distal" segments of the nephron in excess of their reabsorptive capacity, coupled to an increase in the medullary blood flow, will lead to the solute washout observed in diuresis (Chapter 12). In fact, the effect of increase in tubular flow rate in the Henle's loop appears to be more important in determining the final medullary solute concentration than changes in the medullary blood flow.[5] Conversely, contraction of the extracellular volume, as in hemorrhage, will produce opposite changes and result in conservation of sodium and water and excretion of a concentrated urine.

Although the inert diffusable gas method has the advantage of being a noninvasive repeatable technique for measuring total RBF, its value in measuring regional blood flow has recently been questioned.[13] In part, this is because the four renal components of blood flow deduced from the tail subtraction analysis of the disappearance curve cannot be matched with any constancy to a particular anatomical area of the kidney.[1]

Radioactive Microsphere Method

This method, first used by McNay and Abe,[14] is the most widely used technique for measuring intrarenal blood flow distribution. It is based on the assumption that the fraction of RBF in a given region of the kidney is the same as the fraction of the injected microspheres trapped in the tissues of that region. It is further assumed that injection of microspheres does not alter renal function or systemic hemodynamics. Briefly, the procedure involves injecting radioactive microspheres into the left ventricle and measuring the fraction that is trapped in the glomerular capillaries. After sacrificing the animal, total RBF is calculated from the following formula:

$$\text{RBF} = \frac{\text{Total Renal Radioactivity (cpm)}}{\text{Total Radioactivity of Injected Microspheres (cpm)}} \cdot \text{CO} \quad (7.34)$$

where cpm = counts per minute, CO = cardiac output in milliliters per minute, and RBF is measured in milliliters per minute. The intrarenal blood flow distribution is then determined by measuring the fraction of the microspheres (radioactivity) trapped in different sections of the cortex, as well as the outer and the inner medullary zones.

The goal of this method is to use microspheres of the smallest diameter so that they will all be trapped in glomeruli, thereby providing an index of glomerular blood flow (GBF). The accuracy of this technique in measuring GBF is based on two critical assumptions. The first assumption is that all the injected microspheres are trapped in the glomeruli. This of course depends on the size of the microspheres used. Thus, microspheres of 80 μm diameter are generally trapped in interlobular arteries, whereas microspheres smaller than 10 μm diameter will pass through the kidney.[12] The second assumption is that the sequesteration of the microspheres in a given zone of the kidney represents the blood flow in that region. Recent studies have suggested that the accuracy of regional blood flow measurements may be limited. Thus, Bankir and co-workers showed that using microspheres of 15 ± 2 μm diameter greatly overestimated the outer cortical blood flow in rats and rabbits.[2] This has, in part, been attributed to the axial streaming of the microspheres in the larger arteries of the cortex.

Despite these shortcomings, the radioactive microsphere technique has provided useful information

TABLE 7.2. Changes in regional blood flow in response to selected forcings in dogs as determined by the radioactive microsphere technique.

Forcings	RBF (mL/min)	Renal resistance (mm Hg/mL/min)	Fractional cortical blood flow (%)	
			Outer cortical	Inner cortical
Acetylcholine	+	−	−	+
Bradykinin	+	−	−	+
Hemorrhagic hypotension	−	−	−	+
Norepinephrine	−	+	0	0
Angiotensin	−	+	0	0
Sympathetic nerve stimulation	−	+	0	0
Prostaglandin inhibition	−	+	+	−

Source: Stein.[19]

about normal intrarenal blood flow distribution and its alterations in a variety of conditions. Table 7.2 presents a summary of some of these findings. For each forcing, the deviation from normal is indicated by (+) for increase, (−) for decrease, and (0) for no change.

Problems

7.1. Complete the following table by making the appropriate calculations:

TABLE 7.a

	Substance	Total plasma concentration (mg/dL)	Permeability ratio (K_x)	Filtrate concentration (mg/dL)
a.	Inulin	80.0	1.0	
b.	Phenol red	1.0	0.2	
c.	Sodium PAH		0.8	4.0
d.	Diodrast iodine		2.0	1.5

7.2. Complete the following table by making the appropriate calculations:

TABLE 7.b

	Substance	$[x]_{pa}$ (mg/dL)	$[x]_{pv}$ (mg/dL)	$[x]_{pa} - [x]_{pv}$	E_x
a.	Inulin	50	40		
b.	Glucose	100			0
c.	Diodrast iodine	4	0		
d.	Sodium PAH	4			1.0
e.	Creatinine		15		0.25

7.3. What must the renal venous concentration be when the extraction ratio is 1? _____

7.4. What single concentration measurement will be equal to the renal arteriovenous difference for a substance with an extraction ratio of 1? _____

7.5. Complete the following table by making the appropriate calculations?

TABLE 7.c

	Plasma inulin concentration (mg/dL)	Urine inulin concentration (mg/mL)	Urine flow rate (mL/min)	Inulin filtered (mg/min)	GFR (mL/min)
a.	565	260	2.8		
b.	380	160	3.0		
c.		110	2.5		120
d.		112	2.0		125
e.	85		2.3	110	
f.	50		2.3	70	

7.6. Calculate the renal plasma flow from the following data obtained with sodium PAH:
  a. Plasma concentration = 5 mg/dL
  b. Urine concentration = 14.6 mg/mL
  c. Urine flow = 2.4 mL/min
  d. Assume PAH extraction ratio = 0.9

7.7. If hematocrit is 0.55, calculate the total RBF. _____

7.8. If the cardiac output is 4 L/min, calculate the percentage of the total cardiac output that passes through the kidney. _____

7.9. Using the average value for GFR from Table 7.c, calculate the filtration fraction. _____

7.10. Using the value for RPF that you calculated in problem 7.6, calculate the quantity of inulin (mg/min) brought to the kidney in each of the six experiments in Table 7.c. Record your results in Table 7.d below:

TABLE 7.d

Experiments	Total inulin brought to kidney (mg/min)	Extraction ratio
a.		
b.		
c.		
d.		
e.		
f.		

7.11. Calculate the extraction ratio of each of these six inulin experiments and enter the results in Table 7.d above.

7.12. The filtration fraction may be regarded as a glomerular extraction ratio for plasma, i.e., it is the ratio of the volume of plasma removed by the ultrafilter to the total volume brought to it.
  a. If a substance has a permeability ratio of 1 and is neither reabsorbed nor secreted by the tubules, which of the following relationships would hold? (Check correct one).
    (1) Extraction ratio = filtration fraction _____.
    (2) Extraction ratio > filtration fraction _____.
    (3) Extraction ratio < filtration fraction _____.
  b. If a substance has a permability ratio less than 1 and is neither reabsorbed nor secreted by the tubules, which of the relationships of 7.12a. would hold? _____
  c. If a substance has a permeability ratio of 1 and is partially reabsorbed by the tubules, which of these relationships would hold? _____
  d. If a substance has an extraction ratio greater than the filtration fraction, what process must be involved in its excretion? _____

7.13. What requirements must a substance meet for its plasma clearance to equal the GFR? _____

7.14. What requirements must a substance meet for its plasma clearance to equal the rate of RPF? _____

7.15. If a substance has a plasma clearance greater than the GFR, what process must be concerned in its excretion? _____

7.16. If a substance has a plasma clearance less than the GFR, what are two possible explanations in terms of fundamental renal excretory processes?
  a. _____
  b. _____

7.17. If RPF is 700 mL/min, calculate the extraction ratios corresponding to the following plasma clearances given in Table 7.e.

TABLE 7.e

Substance	Plasma clearance (mL/min)	Extraction ratio
a. Sodium PAH	700	
b. Phenol red	400	
c. Creatinine	180	
d. Inulin	130	
e. Glucose	0	

7.18. Construct a graph with extraction ratios (0 to 1.0) as abscissae and plasma clearance (0 to 700 mL/min) as ordinates. Plot the data from Table 7.e.
  a. From your graph, what plasma clearance would be found for substances having the following extraction ratios?
  $E_x = 0.8$; $\dot{V}_x = $ _____
  $E_x = 0.4$; $\dot{V}_x = $ _____
  $E_x = 0.2$; $\dot{V}_x = $ _____

7.19. The data listed below may be described by a double exponential curve of the form,

$$y(t) = A e^{-k_1 t} + B e^{-k_2 t}$$

Time, t (min)	$[x]_p$ (mg/dL)
5	50.0
10	41.9
15	36.0
20	32.0
25	28.0
30	24.8
35	22.0
40	19.5
45	17.5
50	16.1
60	13.2

Calculate (a) the parameters A, B, k_1, and k_2, and (b) the time constant for each exponential component (Appendix B).

7.20. Renal function studies in a patient produced the following data:

  Clearance of inulin = 80 mL/min
  Plasma concentration of Z = 25 mg/dL
  Urine concentration of Z = 12 mg/mL
  Urine flow rate = 5 mL/min

  Assume permeability ratio for substance Z is 1.0.
  a. Is substance Z absorbed or secreted? Compute it.
  b. Compute the filtered load of substance Z.

7.21. In a patient the GFR = 130 mL/min, RPF = 700 mL/min, and the permeability ratio for glucose =

1.0. Complete the tables below by calculating the following:
1. Quantity of glucose filtered, mg/min.
2. Quantity of glucose absorbed, mg/min.
3. Total quantity of glucose brought to the kidney, mg/min.
4. Plasma clearance of glucose, mL/min.
5. Extraction ratio for glucose.

Plasma glucose (mg/dL)	Urinary loss of glucose (mg/min)	Glucose filtered (mg/min)	Total glucose brought to kidney (mg/min)
a. 100	0		
b. 500	275		

Plasma clearance of glucose (mL/min)	Extraction ratio of glucose	Glucose reabsorption rate (mg/min)
a.		
b.		

7.22. Match the numbered items at the right to what is called for by each of the statements at the left. (A statement may call for more than one item.)

a. To calculate the GFR _____.
b. To calculate the rate of plasma flow of x to the peritubular capillaries _____.
c. To calculate $\dot{T}_m$ of substance x _____.
d. To calculate the clearance of substance x _____.
e. To determine whether substance x is secreted or reabsorbed _____.
d. To calculate the osmolar clearance _____ (see Chapter 12).

1. Urine flow rate.
2. RPF.
3. Urine inulin concentration.
4. Urine PAH concentration.
5. Urine concentration of x.
6. Plasma concentration of inulin.
7. Plasma PAH concentration.
8. Plasma concentration of x.
9. GFR.
10. Solute-free water clearance.

7.23. The following data were obtained in a patient subjected to renal function tests:

Plasma concentration of sodium = 140 mEq/L
Urine concentration of sodium = 120 mEq/L
Plasma concentration of glucose = 400 mg/dL
Urine concentration of glucose = 25 mg/mL
Urine flow rate = 4 mL/min
Inulin clearance = 100 mL/min
PAH clearance = 500 mL/min

Assume permeability ratio = 1.0 for sodium and glucose. Also assume extraction ratio of 1.0 for PAH. Using these data, calculate the following:
a. Filtered load of sodium _____
b. Rate of reabsorption of sodium _____
c. Maximum tubular transport capacity for glucose _____
d. Plasma concentration at which glucose transport is just maximal _____
e. Filtration fraction _____

7.24. What is the osmolar concentration (osmolality) of blood urea when blood urea nitrogen (BUN) is 100 mg/dL. Assume that nitrogen constitutes 47% of the urea molecule and that the molecular weight of urea is 60.

7.25. The intravenous infusion of a substance raised its plasma concentration to 145 mg/dL. If the concentration of the substance in the renal vein is 0.55 mg/mL, calculate (a) the amount of the substance removed by the kidney, and (b) the extraction ratio of this substance.

7.26. If the inulin clearance is 120 mL/min and the maximum tubular transport of glucose (T_{mG}) is 360 mg/min, calculate (a) the plasma glucose concentration at which glucose is absorbed maximally, and (b) the amount of glucose excreted in the urine when the plasma glucose concentration is 130 mg/dL.

7.27. Substances X, Y, and Z are freely permeable across the glomerular membrane. They are not stored, formed, or destroyed by the kidney tissue. The following data are obtained from a normal subject.

Substance	Plasma concentration (mg/mL)	Urine concentration (mg/mL)
X	1.0	300
Y	5.0	300
Z	10.0	300
Inulin	10.0	600

In the table below, indicate the renal processing of which of these three substances (X, Y, and Z) most closely resemble that for the substances listed below. Place the appropriate letter (X, Y, or Z) in the space provided.

Substance	Renal processing resembles that of:
a. _____	Glucose
b. _____	PAH
c. _____	Urea
d. _____	Inulin

7.28. Using clearance technique, describe explicitly how you can determine whether a substance is filtered, reabsorbed, or secreted by the kidney.

References

1. Aukland K: Methods for measuring renal blood flow: Total flow and regional distribution. *Ann Rev Physiol* 1980; 42:543–555.
2. Bankir L, Trinh Trang Tan MM, Grunfeld JP: Measurement of glomerular blood flow in rabbits and rats: Erroneous findings with 15 μm microspheres. *Kidney Int* 1979; 15:126–133.
3. Barger AC, Herd JA: Renal vascular anatomy and distribution of blood flow, in Orloff J, Berliner RW (eds): *Handbook of Physiology, Sec. 8, Renal Physiology*. Washington DC, American Physiological Society, 1973, pp 249–313.
4. Brod J, Sirota JH: Renal clearance of endogenous "creatinine" in man. *J Clin Invest* 1948; 27:645–654.
5. Chou SY, Spitalewitz S, Faubert PF, et al: Inner medullary hemodynamics in chronic salt-depleted dogs. *Am J Physiol* 1984; 246:F146–F154.
6. Davies DF, Shock NW: Age changes in glomerular filtration rate, effective renal plasma flow, and tubular excretory capacity in adult males. *J Clin Invest* 1950; 29:496–507.
7. De Wardener HE: *The Kidney: An Outline of Normal and Abnormal Structure and Function*. Boston, Little Brown & Co, 1958.
8. Earley LE, Friedler RM: Changes in renal blood flow and possibly the intrarenal distribution of blood during the natriuresis accompanying saline loading in the dog. *J Clin Invest* 1965; 44:929–941.
9. Goldstein MH, Lenz PR, Levitt MF: Effect of urine flow rate on urea reabsorption in man: Urea as a "tubular marker". *J Appl Physiol* 1969; 26:594–599.
10. Gutman Y, Gottschalk CW, Lassiter WE: Micropuncture study of inulin absorption in the rat kidney. *Science* 1965; 147:753–754.
11. Kampmann J, Siersbaek-Nielsen K, Kristensen M, Hansen JM: Rapid evaluation of creatinine clearance. *Acta Med Scand* 1974; 196:517–520.
12. Katz MA, Blantz RC, Rector FC Jr, Seldin DW: Measurement of intrarenal blood flow. I. Analysis of microsphere method. *Am J Physiol* 1971; 220:1903–1913.
13. Knox FG, Ritman EL, Romero JC: Intrarenal disribution of blood flow: Evaluation of a new approach to measurement. *Kidney Int* 1984; 25:473–479.
14. McNay JL, Abe Y: Pressure-dependent heterogeneity of renal cortical blood flow in dogs. *Circ Res* 1970; 27:571–587.
15. Moore CD, Gewertz BL: Measurement of renal blood flow. *J Surg Res* 1982; 32:85–95.
16. Pitts RF: *Physiology of Kidney and Body Fluids*, ed 3. Chicago, Year Book Medical Publishers, Inc, 1974.
17. Relman AS, Levinsky NG: Clinical examination of renal function, in Strauss MB, Welt LG (eds): *Diseases of the Kidney*. Boston, Little Brown & Co, 1963.
18. Smith HW: *The Kidney—Structure and Function in Health and Disease*. New York, Oxford University Press, 1951.
19. Stein J: The renal circulation, in Brenner BM, Rector FC Jr (eds): *The Kidney*. Philadelphia, WB Saunders Co, 1976, pp 215–250.
20. Steinitz K, Turkland H: Determination of glomerular filtration by endogeous creatinine clearance. *J Clin Invest* 1940; 19:285–298.
21. Thorburn GD, Kopald HH, Herd JA, et al: Intrarenal distribution of nutrient blood flow determined with Krypton[85] in the unanesthetized dog. *Circ Res* 1963; 13:290–307.
22. Tobias JG, McLaughlin RF Jr, Hopper J Jr: Endogenous creatinine clearance. *New Eng J Med* 1962; 266:317–323.
23. Van Slyke DD: Renal function tests. *New York State J Med* 1941; 41:825–833.
24. Winberg J: 24-hour true endogenous creatinine clearance in infants and children without renal disease. *Acta Paediat* 1959; 48:443–452.

8

Structural and Biophysical Basis of Tubular Transport

The ultrafiltrate as it traverses the various nephron segments undergoes considerable modifications both in volume and in composition. The small volume of excreted urine is largely due to reabsorption of nearly all the ultrafiltrate, both solutes and water, along with the addition of some important solutes by tubular secretion.

Although mechanisms of renal tubular transport differ, depending on the particular solutes involved, certain transport mechanisms are common to both reabsorption and secretion processes. In Chapter 5 those mechanisms related to the formation of the glomerular ultrafiltrate were described. In this chapter we examine the principles of transport across the cell membrane. These ideas should provide the basis for understanding the mechanisms of renal tubular reabsorption and secretion of solutes and water, which is discussed in Chapters 9 and 10.

We begin with a detailed description of the morphological basis of tubular transport.

Structural Basis

All epithelia along the uriniferous tubules are unilayered. They have a luminal (apical) surface that borders the tubule lumen. Basally, they rest on a basement membrane, which is already considered as part of the interstitium. Transports through the epithelia may follow two different routes: either between the cells, "paracellular," or through the cells, "transcellular" (Figure 8.1). The paracellular route comprises the lateral intercellular spaces and the tight junction. With respect to transport function, the other components of the junctional complex, namely, the intermediate junction and the desmosomes, are of minor importance; their function is to provide the mechanical connection of the cells. The transcellular route comprises the uptake of a substance across the luminal (or basolateral) membrane, its transport through the cytoplasm, and its secretion across the opposite cell membrane. Paracellular transports are passive in nature, whereas transcellular transport may be passive as well as active. Transports by both routes are not strictly independent from each other; within the lateral intercellular spaces both types of transports may compete with each other.

Paracellular Transport Route

Substances transported via this route have to pass through the tight junction and the lateral intercellular spaces. Since the transport by this route is passive, this means that the transport is a permselective barrier function. From this point of view the tight junction is the main determinant of transport by the paracellular pathway.

Tight Junctions

A tight junction may be compared to a belt that encircles the cells near their apical surface. It separates the luminal from the basolateral cell membrane. Within this belt the cell membranes of two opposite cells are "fused." This "fusion" does not comprise the entire area of the belt but is effected by "strands", that, in freeze-fracture replicas of unfixed material, appear to be composed of individual globular subunits anchored to both cell membranes.[17, 22] Whether they consist of proteins or of lipids is not clear.[15] Usual fixation tends to connect the individual subunits. The continuous strands, as usually seen in freeze-fracture replicas of fixed material, are considered to be fixation artifacts. The

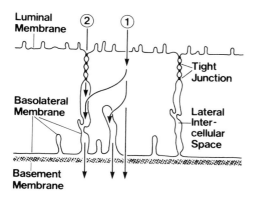

FIGURE 8.1. The transcellular transport (1) and paracellular transport (2) routes.

interspaces between the globular subunits are thought to be the structural correlate for the selective permeability in apico-basal direction.

Tight junctions differ in apico-basal depth and in the number of junctional strands (Figure 8.2). There is no strict correlation between depth and number of strands: the strands may be narrowly or loosely packed. Generally, epithelia with tight junctions consisting of many strands have a high electrical resistance; they are called "tight" epithelia. Epithelia with junctions of only one or two strands have a low electrical resistance; they are called "leaky" epithelia.

Tight junctions also differ in length per unit area of the epithelium. In a given epithelium consisting of large polygonal cells (e.g., amphibian urinary bladder), the length of the tight junction is short. In contrast, in an epithelium established by cells with meandering cell borders (e.g., an epithelium with interdigitating cells like the proximal tubule), the tight junction is comparably very long.[4] If we assume that the permeability of a tight junction results from the sum of channels between the globular subunits of the junctional strands, a long tight junction will provide many more channels per unit area of epithelium than a short one. This interpretation correlates with the observation that meandering cell borders only occur in "leaky" epithelia.[13] An elongation of a tight junction together with a tight epithelium makes no sense and has never been found.

The subdivision of epithelia into "tight" and "leaky" epithelia on the basis of the structural elaboration of their tight junction is crude. Individual differences with respect to certain substances and ions are not fully understood. Fixed negative charges associated

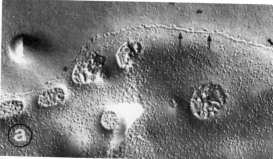
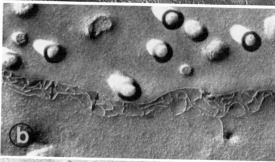
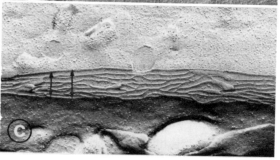

FIGURE 8.2. Freeze-fracture replicas of tight junctions. (a) Depicts a "leaky" tight junction established by one junctional strand (*arrows*). Circular structures are cross-fractured microvilli. (b) Depicts a tight junction with three to four anastomosing junctional strands (*two-headed arrows*). (c) Depicts a "tight" tight junction consisting of many junctional strands (*two-headed arrows*).

with a tight junction obviously restrict the permeability for anions when compared with cations.[6] Other functional differences may be based on differences in the chemical composition of the junction which are not known. In general, the water permeability of tight junctions appears to be very low, but the issue is not settled.[17]

Lateral Intercellular Spaces

Within the uriniferous tubule two types of lateral intercellular spaces are of relevance: those within epithelia with interdigitating cells like the proximal tubule and those between noninterdigitating cells like the collecting duct epithelium. The former are of constant narrow width; the latter may vary in width according to function. As known from the renal collecting duct epithelium and the gallbladder, the lateral intercellular spaces dilate when the overall fluid transport through the epithelium is high.[8, 19] The width of this type of intercellular spaces can be measured in micrometers; they are visible by the light microscope. Typical for these interspaces are lateral microfolds (microvilli) that protrude from the opposite cell membranes into the spaces and that frequently contact each other by desmosomes. This arrangement is interpreted as limiting the dilation of the interspaces. Where the space is very narrow, the lateral microfolds are found to be closely apposed to each other.

The lateral intercellular spaces between proximal and distal tubular epithelia are of uniform narrow width. The opposing cell membranes run strictly parallel to each other, and lateral microfolds are lacking. Their width is only moderately variable in response to a functional stimulus (about 30 angstroms).[21] It should, however, be mentioned that owing to the elaborate interdigitation of the cells, the total volume of interspaces increases tremendously in the apico-basal direction.

Transcellular Transport Route

The limiting steps of this route are the transports across the luminal and the basolateral membranes. Almost no details are known about the transport through the cytoplasm. For most ions, and small solutes like urea and water, a transport by diffusion is assumed through an intracellular but not clearly defined compartment that contacts the lumial and the basolateral membranes. The transports across both cell membranes are determined by the specific structural components of these membranes such as receptors, carriers, channels, enzymes, and other transport proteins. Qualitative features of these kinds will be presented later together with the discussion of particular functions of an epithelium. In this chapter, we only deal with the various forms of membrane amplifiction.

The quantity of transports through a cell membrane is dependent on the area of the membrane. Transport proteins (i.e., receptors, carriers, etc.) can obviously be accumulated within a membrane only up to a certain density.[10] If more transport proteins are necessary, the membrane area must be increased (amplification). Several forms of membrane amplification are typical for "transporting" epithelia. The type and extent of luminal and basolateral membrane amplification are characteristic for a certain epithelium and are used as structural criteria to distinguish the various epithelia.

Along the uriniferous tubule, amplification of the *luminal membrane* is achieved by microvilli, microfolds, or brush borders. Microvilli are a very common structure. They vary extensively in height and density in different epithelia. Even in a given type of epithelium the elaboration varies, possibly in response to functional stimuli.

Luminal microfolds are a less common structure. In the kidney they are typical for the intercalated cells of the connecting tubule and collecting duct. In these cells they are part of a membrane shuttle system, by which the area of the luminal membrane can be increased by the formation of new folds from pre-existing membrane reservoirs (flat vesicles) that are stored in the apical cytoplasm (see intercalated cells).

Brush borders are typical for absorbing epithelia (small intestine and proximal tubule). Brush borders are not simply assemblies of regularly arranged microvilli, but are highly organized formations of thin long-cell protrusions, which are also called microvilli. All microvilli of a brush border are of the same size, and they are very regularly packed together. Height and density of the microvilli of a brush border are typical for a given cell type (see differences between the segments of the proximal tubule). The "intervillous space" of a brush border must be considered functionally as a specific compartment. It provides a region of unstirred layer through which diffusion may take place. Brush border microvilli possess an axial system of contractile filaments by which they obviously can be decreased in length.[5]

In *basolateral membranes* three forms of membrane amplification are known (Figure 8.3): lateral microfolds (microvilli), basal infoldings, and cellular interdigitation. Along the uriniferous tubules lateral microfolds are typical for the collecting duct and the macula densa epithelium. Lateral microfolds only occur in epithelia with potentially wide intercellular spaces (see above).

Infoldings of the basal cell membrane (basal infoldings) are widespread among epithelia and are very variable in elaboration. Few and short infoldings are typical for the medullary collecting duct epithelium. Long and densely packed infoldings occur in the cells of the connecting tubule. In the cortical collecting duct it has been shown that basal infoldings increase in height and density in response to a chronic aldosterone stimulus.[9, 23] The "extracellular" spaces between the infolded membrane are not part of the lateral intercellular spaces since they terminate blindly and do not reach apically the tight junction; consequently, they do not belong to the paracellular transport route.

Amplification of the (baso-)lateral cell membrane by cellular interdigitation is the most typical feature of salt-transporting epithelia. Along the nephron all proximal and distal tubular segments exhibit this type of an epithelium. The interdigitating cell processes are densely packed with mitochondria. Thus, the transporting membrane and the energy providing mitochondria are closely associated with each other. A positive correlation has been found beween the membrane area and the quantity of mitochondria.[14] Such a close association between mitochondria and transporting membrane is lacking in epithelium with basal infoldings. The extracellular spaces between the interdigitating cell processes are all part of the paracellular transport route; they communicate with each other and extend apically to the tight junction.

Biophysical Basis

The transport mechanism of solutes and water across the cell membrane has been the subject of extensive theoretical and experimental studies. Without attempting to examine the details or resolve the controversies,[24] we shall consider those principles that have been both well documented and widely used by renal physiologists.

As mentioned in Chapter 2, translocation of solutes and water across the cell membrane may be classified into *three* categories: (1) transport by convection, bulk flow, or filtration; (2) transport by simple (passive) diffusion; and (3) carrier-mediated transport. The last category is further subdivided into (a) facilitated diffusion, (b) active transport, and (c) cotransport.

The characteristics and forces involved in solute and water transport by filtration have already been described (Chapter 5). This mechanism is suitable for transporting solutes and water over relatively long distances, such as movement of blood through the circulatory system, air across the lungs, and food through the gastrointestinal system. In this chapter, we focus on the last two mechanisms, which are

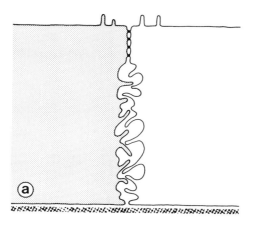
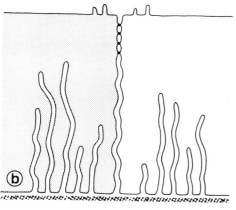
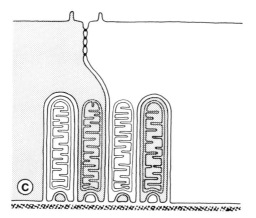

FIGURE 8.3. The three forms of amplification of the basolateral membrane: (a) lateral microfolds; (b) basal infoldings; and (c) amplification of lateral membranes by cellular interdigitation. (From Kriz; Der Bau transportierender Epithelien. VEB Gustav Fischer Verlag Jena. 1984. Verhandlungen der Anatomischen Gesellschaft 78.)

concerned with transport of solutes and water over relatively short distances. We begin with a description of the principles of simple or passive diffusion and then examine the characteristics and rules governing carrier-mediated transport and its subdivisions.

Simple or Passive Diffusion

Diffusion may be defined as the translocation of solutes and water across a porous membrane, permeable to both, along their respective chemical and electrical gradients until equilibrium is reached. The rate at which a solute molecule, such as glucose, diffuses across the membrane is called *flux*. It has a unit of amounts per second (e.g., millimole per second of glucose) and is designated by the letter J. Flux of a given solute molecule (S) across a membrane is directly proportional to its concentration [S], the absolute temperature T, the mobility m, or the ease with which the molecule can move through the membrane, and the membrane cross-sectional area A available for diffusion. Flux is inversely proportional to the size of the molecule and the distance over which diffusion must take place (membrane thickness, x). Thus, the flux of a solute molecule from one side of the membrane (side 1) to the other side (side 2) may be expressed by the following equation:

$$J_{1\to2} = -mRTA[S]_1/x \quad (8.1)$$

where R is the universal gas constant that converts temperature into units of energy per mole. The negative sign on the right-hand side of Equation 8.1 indicates that the flux is in a direction opposite to that of increasing concentration. Note that the smaller the molecular size, the greater is the mobility, and hence the higher would be the molecular flux across the membrane. Also, the higher the temperature, the greater the random molecular motion, and hence the higher would be the molecular flux across the membrane.

The flux in the opposite direction, from side 2 to side 1, may be described by an expression analogous to Equation 8.1:

$$J_{2\to1} = mRTA[S]_2/x \quad (8.2)$$

Note that as the concentration $[S]_2$ builds up on side 2, so does the flux from side 2 to side 1. Thus, $J_{2\to1}$ progressively increases as $J_{1\to2}$ decreases, until $J_{1\to2} = J_{2\to1}$, a condition that defines the equilibrium state.

$J_{1\to2}$ and $J_{2\to1}$ are called unidirectional fluxes so as to be distinguished from net flux, $J = J_{1\to2} - J_{2\to1}$. It is the net flux that determines the effective rate at which a molecule is transported across the cell membrane. Combining Equations 8.1 and 8.2 yields

$$J = -\frac{mRTA}{x}([S]_1 - [S]_2) \quad (8.3)$$

Equation 8.3 is known as Fick's law of diffusion. There are two forms of Fick's equation, depending on the grouping of various terms in Equation 8.3. Letting the product term, mRTA, equal D, the so-called diffusion coefficient yields the general form of Fick's equation:

$$J = -\frac{D}{x}([S]_1 - [S]_2) \quad (8.4)$$

Diffusion across many biological membranes involves flux through a finite distance, represented by the membrane thickness x. Incorporating this parameter into the diffusion coefficient yields a more meaningful and easily measurable index of the permeability of the membrane to a given solute. This new index is called the permeability coefficient K_p and is defined by Equation 8.5:

$$K_p = \frac{D}{x} \quad (8.5)$$

Incorporating K_p into Equation (8.4) yields:

$$\begin{aligned}J &= -K_p([S]_1 - [S]_2)\\ &= -K_p \cdot \Delta[S]\end{aligned} \quad (8.6)$$

where the terms inside the parentheses represent the concentration gradient, $\Delta[S]$.

Expanding Equation 8.6, we obtain an expression relating the net flux (J) across the membrane to the difference between the influx and the efflux:

$$\text{Net Flux (J)} = -K_p \cdot [S]_1 + K_p \cdot [S]_2 \quad (8.7)$$

where $-K_p \cdot [S]_1$ represents the flux into and $+K_p \cdot [S]_2$ represents the flux out of the membrane, as indicated by the minus and plus signs, respectively. Figure 8.4 illustrates schematically the operational significance of the various terms in Equation 8.4. The negative sign in the equation is represented by the arrow within the membrane, connecting the lines indicating the level of concentration of S on the two sides of the membrane, $[S]_1$ and $[S]_2$, respectively.

As depicted in Figure 8.4, solute transport by diffusion across the cell membrane depends on two factors: solute concentration gradient and membrane permeability to that solute. It is well recognized that

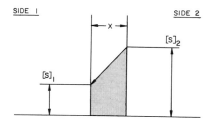

FIGURE 8.4. The general features of Fick's law of diffusion (Equation 8.4). $[S]_1$ and $[S]_2$ represent the concentrations of S on the two sides of the membrane with thickness x, respectively. (From Koushanpour.[11])

many solutes diffuse less rapidly across biological membranes than they do in the fluid surrounding the two sides of the membrane. Because of this permselective property, such a membrane is said to be semipermeable to the diffusion of the particular molecular solute. This permselectivity of the biological membrane serves important physiological functions, such as storage of different substances within the cell (e.g., hormones, enzymes, etc.) and their controlled release into the extracellular environment.

Chemical analyses of various membranes have revealed that the major constituents of many biological membranes are cholesterol, phospholipids, oligosaccharides, and proteins. This finding has led to the formulation of a fluid mosaic model of the cell membrane.[18] This model proposes that the membrane is made up of a skeleton of phospholipid bilayer with the polar heads (which are hydrophilic) facing the outside of the membrane and the nonpolar tails (which are hydrophobic and hence lipophilic) facing each other. In this model most of the proteins are assumed to be embedded in the lipid bilayer.

According to this model, polar molecules could pass through aqueous channels in the proteins, or they could be transported by a membrane-mediated carrier mechanism. The rate of diffusion is highly dependent on the molecular size, and the ease of diffusion diminishes as the molecular size exceeds 8 angstroms. Thus, permeability of the membrane to polar molecules varies inversely with the molecular size.

In contrast, diffusion of nonpolar molecules depends on their lipid solubility. This suggests that nonpolar molecules effectively dissolve in the membrane, diffuse through the membrane in the lipid phase, and then redissolve in water on reaching the other side of the membrane. Thus, the rate of diffusion depends on the effective concentration gradient, taking into account the relative lipid/water solubility ratio, referred to as the partition coefficient, for that substance. This effect may be incorporated into Fick's Equation 8.6 by simply multiplying the permeability coefficient K_p by the partition coefficient.

The transport of molecules across the capillary membrane involves passive filtration, a process described in Chapters 3 and 5. Such a transport process represents a special case of diffusion principles, with hydraulic and oncotic pressure gradients as the driving force, rather than the concentration gradient. Another special case is the transport of water. As described in Chapter 1, water is transported by osmosis, which is the diffusion of water along its concentration gradient.

In summary, it is evident that if simple diffusion is the mechanism by which a solute is transported across the membrane, then the net flux is linearly and directly proportional to the difference in the solute concentrations on the two sides. In such a case, solute concentration would be the only factor limiting the net flux. However, if the solute transport involves mechanisms other than or in addition to diffusion, then the net flux is not linearly related to the concentration difference. Such a nonlinear relationship exemplifies a carrier-mediated type of transport, in which case both solute and carrier concentrations on the two sides of the membrane will be the factors limiting the net flux.

Although many conceivable relationships between the net flux and the solute concentration exist, we have selected three for consideration. The choice was dictated by (a) the ease of handling the theoretical treatment of the concepts, and (b) more importantly, the wide usage of these relationships in studying transport characteristics of biological membranes. Figure 8.5 presents the relationship of reaction rate, a factor related to net flux, to solute concentration for two transport mechanisms: (a) simple diffusion and (b) carrier-mediated transport, represented by right-rectangular hyperbola and sigmoid (S-shape) curves. Having already characterized transport by simple diffusion, let us now consider the carrier-mediated transport.

Carrier-Mediated Transport

As noted earlier, the cell membrane is now believed to consist of a dynamic structure with the ability to regulate, through its internal machinery, transport of solutes across it. According to the most popular cur-

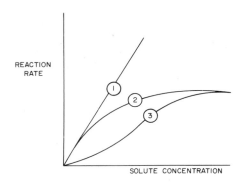

FIGURE 8.5. The relation of reaction rate to solute concentration for three types of transport: (1) simple diffusion; (2) carrier-mediated transport with fixed amount of carrier, represented by the right-rectangular hyperbola; and (3) carrier-mediated transport with variable amounts of carrier, represented by the sigmoid curve. (From Koushanpour.[11])

rent concept, translocation of solutes across the membrane involves a temporary binding to parts of the membrane. If this binding is not rigidly fixed within the membrane and the solute-membrane complex can in some way move through the membrane, the transport is called carrier-mediated. This concept, although still hypothetical, explains many qualitative and quantitative aspects of cellular transport. Carrier-mediated transport involves two types of chemical reactions, namely, unimolecular and bimolecular reactions. We shall begin this section with a quantitative description of the kinetics of these two types of reactions.

Unimolecular Reactions

The simplest chemical reaction involves the conversion of one molecule (A) into another (B). In chemical terminology, molecule A is called the reacting substance and B the product.

The rate of conversion of substance A into B, called rate constant (K), is determined by the quantity of substance A reacting or B produced by the reaction per unit of time. This idea is symbolized in Equation 8.8:

$$A \xrightarrow{K} B \qquad (8.8)$$

According to the principle of chemical-reaction kinetics, the rate of a chemical reaction (conversion of A to B in Equation 8.8) is proportional to the product of the concentrations of the reacting substances. Applying this principle to the above reaction, we obtain

$$[\dot{A}] = -K \qquad (8.9)$$

where $[\dot{A}]$ is the rate of change of concentration of A per unit time. The negative sign indicates that the quantity of reacting substance decreases as the product is formed.

The rate of change of the reacting substance $[\dot{A}]$ may or may not depend on the concentration of the reacting molecule [A]. If $[\dot{A}]$ is independent of [A], then the chemical reaction is called a zero-order reaction and is described by the rate Equation 8.9. However, if $[\dot{A}]$ depends on [A], then the chemical reaction is called a first or higher-order reaction, depending on the sum of exponents of the concentration factor which appears in the rate equation. For example, if the conversion of A into B is an irreversible first-order chemical reaction, then it can be described by the following rate equation:

$$[\dot{A}] = -K[A] \qquad (8.10)$$

Chemically, the rate constant K is a measure of the probability of conversion of A into B and is defined by the following equation:

$$K = Ze^{-\frac{E_a}{RT}} \qquad (8.11)$$

where Z is the frequency of collisions of molecules of A, and $e^{-E_a/RT}$, called Boltzman's constant, is the fraction of molecules of A with kinetic energy greater than the activation energy (E_a) required to initiate the conversion. Activation energy is measured in calories per mole. R is the molar gas constant in calories/degree/mole, and T is the absolute temperature in degrees Kelvin. Addition of a catalyst or an enzyme lowers the activation energy.

Often it is convenient to express the increase in reaction rate occurring as a function of temperature. The term Q_{10} has been employed to describe the relative increase in velocity for a rise in temperature of 10° C.

To get an insight into the meaning of rate constant K, we have plotted the change in concentration of the reacting substance with time (Figure 8.6) during a chemical reaction. As shown, the concentration changes exponentially with time ($[A]_t$) reaching a value equal to 37% of the original concentration ($[A]_{t=0}$) in one time-constant (τ). The slope of the exponential curve is equal to the rate constant K, which is the reciprocal of the time constant (see Appendices A and B for further meaning of τ).

In the case of reversible first-order chemical reactions, not only the reactant A is converted into the

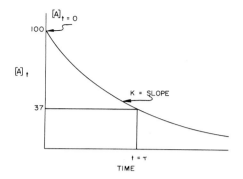

FIGURE 8.6. Time-course of change in the concentration of the reacting substance. $[A]_{t=0}$ is the original concentration, and $[A]_t$ is the concentration at any other time during the reaction. They are related by the exponential equation $[A]_t = [A]_{t=0}e^{-Kt}$. (From Koushanpour.[11])

product B, but also B is converted into A, as shown by the following reaction:

$$A \underset{K_2}{\overset{K_1}{\rightleftarrows}} B \qquad (8.12)$$

where K_1 is the rate constant for the conversion of A into B (forward reaction), and K_2 is the rate constant for the conversion of B into A (backward or reverse reaction).

According to the principles of chemical reaction kinetics, the net rate of change in reactant concentration with time ($[\dot{A}]$) is determined by the algebraic sum of the forward and backward reactions:

$$[\dot{A}] = -K_1[A] + K_2[B] \qquad (8.13)$$

where $-K_1[A]$ represents the fractional decrease and $+K_2[B]$ the fractional increase in $[\dot{A}]$. At equilibrium, $[\dot{A}] = 0$, and Equation 8.13 becomes

$$-K_1[A]_{eq} + K_2[B]_{eq} = 0 \qquad (8.14)$$

The subscript eq indicates that these are concentrations at equilibrium. From Equation 8.14 we get

$$\frac{[B]_{eq}}{[A]_{eq}} = \frac{K_1}{K_2} = K_{eq} \qquad (8.15)$$

where K_{eq} is the equilibrium constant of the reaction.

In any chemical reaction, the rate of change of the total free energy of the system (ΔF) depends on the change in the free energy of the reactants and products. Since K_{eq} is a measure of this latter quantity, it is related to ΔF by the following equation:

$$\Delta F = -RT \ln K_{eq} \qquad (8.16)$$

where ln is the logarithm to the base e or the natural logarithm, and R and T were already defined. Taking advantage of rules for logarithm ($\ln = \log_e = 2.3 \log_{10}$), Equation 8.16 may be written in a more convenient form:

$$\Delta F = -2.3\, RT \log K_{eq} \qquad (8.17)$$

where $\log_{10}$ is the logarithm to the base 10. The negative sign indicates that the reaction occurs spontaneously. This happens if the numerical value of K_{eq} is greater than zero. If the value of K_{eq} is less than zero, ΔF will be positive, which means that energy must be expended in converting the reactants into the product.

Having thus delineated the fundamental laws of chemical kinetics for unimolecular reactions, let us now apply them to some physiologically important bimolecular reactions.

Bimolecular Reactions

The simplest bimolecular reaction is one involving combination of two molecules to form a third one. Such a reaction may be illustrated by the following equation:

$$A + B \underset{K_2}{\overset{K_1}{\rightleftarrows}} AB \qquad (8.18)$$

where A and B are the reactants, AB is the product, K_1 is the rate constant for the forward reaction, and K_2 is the rate constant for the reverse reaction. At equilibrium, the rate constant for the formation of product is given by

$$K_{eq} = \frac{K_1}{K_2} = \frac{[AB]}{[A] \cdot [B]} \qquad (8.19)$$

A high value of K_{eq} indicates a high affinity (association) of A for B to form the product AB. Conversely, the equilibrium rate constant for the reverse reaction—conversion of product to reactants—is given by the equation:

$$K_m = \frac{K_2}{K_1} = \frac{[A] \cdot [B]}{[AB]} \qquad (8.20)$$

where K_m is the dissociation constant and is a measure of the extent of the binding of A to B to form AB. The value of K_m is given by the reciprocal of K_{eq}. Thus, the greater the binding of the reactants, the smaller is the K_m (but the greater is the K_{eq}), and the greater would be the formation of the product.

Bimolecular reactions are the most common steps

in any chain of complex chemical reactions occurring in biological systems. Because of this strategic position, they play a key role in any carrier-mediated transport. Basic to such a transport mechanism is the interaction of the so-called substrate—either a hormone, a neurochemical transmitter, or a drug—with a carrier substance, which in some circumstances may be replaced by a receptor or an enzyme. It is this carrier-substrate or enzyme-substrate interaction that constitutes the core of most chemical reactions in biological systems, including the carrier-mediated transport.

To understand the mechanism of such a transport scheme and the factors affecting its efficiency, a rigorous mathematical treatment of the kinetics of the enzyme-substrate interaction is necessary. Such a formulation was first described by Michaelis and Menten.[7] In this section, we first detail their formulation and then apply the concepts to the drug-receptor interactions, which have a profound physiological and pharmacological significance in renal tubular transport of solutes. An extensive treatment of this subject is given in the reviews by Bodansky,[1] Frieden,[7] and Christensen.[3]

The Michaelis-Menten Formulation of Enzyme-Catalyzed Reactions

Suppose we are interested in the effect of substrate concentration on the rate at which the substrate is converted into products in a chemical reaction involving an enzyme or carrier. For simplicity, let us assume that at the start of the reaction only the substrate (S) and enzyme (E) are present. As the reaction proceeds, a certain amount of S combines with E to form an enzyme-substrate complex (ES). The latter will then undergo appropriate chemical changes to yield the products (P) plus the enzyme to be reused in further conversion of the substrate. These two chemical reactions may be represented as follows:

$$S + E \underset{K_2}{\overset{K_1}{\rightleftarrows}} ES \underset{K_4}{\overset{K_3}{\rightleftarrows}} E + P \quad (8.21)$$

Applying the laws of chemical-reaction kinetics, the relationship between the rate of product formation ($[\dot{P}]$) and the substrate concentration ($[S]$) may be described by the following equations:

$$[\dot{ES}] = K_1[E]_f[S] + K_4[E]_f[P] - (K_2 + K_3)[ES] \quad (8.22)$$

$$[\dot{P}] = K_3[ES] - K_4[E]_f[P] \quad (8.23)$$

$$[E]_t = [E]_f + [ES] \quad (8.24)$$

where subscripts t and f stand for total and free, respectively.

Since at steady state, $[\dot{ES}] = 0$, and assuming that the rate constant $K_4 = 0$ (irreversible dissociation of ES), and the substrate concentration [S] is much greater than [ES], Equation 8.22 may be simplified to

$$K_1[E]_f[S] = (K_2 + K_3)[ES] \quad (8.25)$$

Substituting for $[E]_f$ its equivalent expression from Equation 8.24 and rearranging terms, [ES] may be expressed by the following equation:

$$[ES] = \frac{[E]_t \cdot [S]}{[S] + \dfrac{K_2 + K_3}{K_1}} \quad (8.26)$$

Chemically, the rate of formation of the product ($d[P]/dt = [\dot{P}]$) is given by the velocity of reaction yielding the product (v = dP/dt). Incorporating this fact into Equation 8.23 and assuming that $K_4 = 0$ (irreversible dissociation of ES), Equation 8.23 may be expressed as

$$v = K_3[ES] \quad (8.27)$$

If we assume further that at high substrate concentration all of the enzyme is bound to the substrate, then $[E]_f = 0$ and $[E]_t = [ES]$. Consequently, at high substrate concentration the rate of product formation is maximal and hence the velocity of reaction becomes maximum. Therefore, at high substrate concentration Equation 8.27 becomes

$$\left(\frac{dP}{dt}\right)_{max} = V_{max} = K_3[E]_t \quad (8.28)$$

where V_{max} is the maximum velocity. Now multiplying both sides of Equation 8.26 by K_3 and replacing equivalent terms by v (Equation 8.27) and V_{max} (Equation 8.28), we obtain the following important equation relating the reaction velocity to substrate concentration in an enzyme-catalyzed reaction:

$$v = \frac{V_{max} \cdot [S]}{[S] + \dfrac{K_2 + K_3}{K_1}} \quad (8.29)$$

where the rate constant terms in the denominator are collectively called Michaelis constant (K_m) and defined by the equation

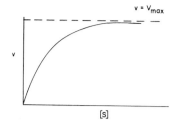

FIGURE 8.7. Hyperbolic relation of reaction velocity to substrate concentration. (From Koushanpour.[11])

$$K_m = \frac{K_2 + K_3}{K_1} \quad (8.30)$$

Hence, Equation 8.29 becomes

$$v = \frac{V_{max} \cdot [S]}{[S] + K_m} \quad (8.31)$$

Assuming that the rate constant K_2 is much greater than K_3, then Equation 8.30 gives the dissociation constant of the enzyme-substrate complex:

$$K_m = \frac{K_2}{K_1} = \frac{[S] \cdot [E]}{[ES]} \quad (8.32)$$

The reciprocal of Michaelis constant ($1/K_m$) is a measure of the affinity of the enzyme for the substrate.

Solving Equation 8.31 for K_m and rearranging, we get

$$K_m = [S] \left(\frac{V_{max}}{v} - 1 \right) \quad (8.33)$$

When the reaction is half completed, that is, $v = \frac{1}{2} V_{max}$, Equation 8.33 becomes

$$K_m = [S] \quad (8.34)$$

That is, K_m is numerically equal to the substrate concentration at which half-maximum velocity of reaction is obtained. K_m also has the dimension of concentration (moles/liter).

Equation 8.31 describes a rectangular hyperbola, where v is the dependent variable, [S] is the independent variable, K_m is the slope, and V_{max} is the asymptote. Figure 8.7 shows the hyperbolic relation of the reaction velocity to the substrate concentration. Note that at high substrate concentration the velocity approaches a maximum.

Experimentally, it is possible to estimate K_m by determining the concentration at half-maximum velocity. However, because of the hyperbolic rela-

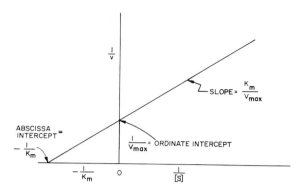

FIGURE 8.8. Lineweaver-Burk double-reciprocal plot of the reaction velocity against substrate concentration. (From Koushanpour.[11])

tionship between velocity and substrate concentration, the estimate is only approximate. The result can be improved enormously if we transform Equation 8.31 into a form that yields a straight line. This transformation is called Lineweaver-Burk double-reciprocal:

$$\frac{1}{v} = \frac{K_m}{V_{max}} \cdot \frac{1}{[S]} + \frac{1}{V_{max}} \quad (8.35)$$

Equation 8.35 is of the form $y = a + bx$, where $1/v = y$, $1/[S] = x$, $1/V_{max} = a$ or the y-intercept, and $K_m/V_{max} = b$ or the slope. Therefore, when $1/v$ is plotted against $1/[S]$, a straight line results (Figure 8.8). At infinite substrate concentration, when $1/[S] = 0$, from Equation 8.35 we get $1/v = 1/V_{max}$ [that is, the intercept on the velocity (ordinate) axis yields the value for the maximal velocity]. At infinite velocity, when $1/v = 0$, the intercept on the concentration (abscissa) axis gives the value for the dissociation constant of the enzyme-substrate complex, $1/K_m = -1/[S]$.

The enzyme-catalyzed reactions are influenced by a number of factors, including temperature, pH, and inhibitors. Let us briefly consider the effects of each.

Effects of Temperature. When enzyme-catalyzed reactions are studied over a range of temperatures, it is commonly observed that the velocity of reaction passes through a maximum (optimum). This temperature is not well defined and may vary with the pH, substrate concentration, purity of preparation, activators, or inhibitors present, etc. The failure to get proportionate increases in reaction rate at higher temperatures is due to inactivation (either reversible or irreversible) of the catalyst. As in other chemical reactions at low temperatures (before inactivation), a relationship between reaction velocity (v) and the temperature is given by the Arrhenius equation (Equation 8.36):

$$v = E_o e^{-\frac{E_a}{RT}} \qquad (8.36)$$

where E_o is the minimum or threshold activation energy, and E_a, R, and T have already been defined.

Effects of pH. When enzyme-catalyzed reactions are studied over a range of pH, under otherwise standardized conditions, it is also commonly observed that the rate passes through a maximum. This phenomenon, like the effects of temperature, can be adequately explained by inactivation of the enzyme. Usually the curve depicting the pH effects is monophasic, but like the optimum temperature, the optimum pH is not a fundamental constant of each enzyme but may vary with the temperature, concentration, and type of buffer cofactor and substrate concentration. The decrease from optimal enzyme activity often follows a bell-shaped curve—like an acid-base dissociation curve. In any particular case, however, the effects of pH on catalysis could be due to ionization of the substrate, or the enzyme.

Effects of Inhibitors. It is often observed that various compounds inhibit a particular enzyme-catalyzed reaction. It is possible from a knowledge of chemistry and kinetics to characterize the relative effectiveness of the inhibitor as well as to obtain information concerning the mechanism of inhibition.

Inhibitors generally fall into three categories:

1. Competitive inhibitors. They are often structurally related to the substrate and compete for it with the enzyme. Presumably inhibition occurs by combination of the inhibitor with the same enzyme site that binds with the substrate during the catalytic process. In this type of inhibition, K_m but not V_{max} is affected. Hence, increasing the substrate concentration reverses the inhibition. A Lineweaver-Burk plot of reaction velocity as a function of competitive inhibitor and substrate concentrations is shown in Figure 8.9. Note that increasing the inhibitor concentration affects both K_m and slope.

The general equation describing this type of inhibition is

$$\frac{1}{v} = \frac{K_m}{V_{max}} (1 + \frac{[I]}{K_I}) \frac{1}{[S]} + \frac{1}{V_{max}} \qquad (8.37)$$

where [I] is the inhibitor concentration and K_I is the

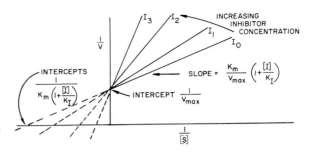

FIGURE 8.9. Lineweaver-Burk plot of the reaction velocity against substrate concentration in the presence of a competitive inhibitor. (From Koushanpour.[11])

rate constant for the dissociation of the enzyme-inhibitor complex.

2. Uncompetitive inhibitors. They influence V_{max} but not the slope, K_m/V_{max}. In this case an increase in substrate concentration does not overcome the inhibition. Presumably the mechanism of inhibition involves combination with the enzyme-substrate complex, although the same type of inhibition would be produced from an essentially irreversible combination of enzyme with inhibior, yielding a catalytically inactive complex. Uncompetitive inhibitors are very frequently specific for a particular enzyme or group of enzymes. A Lineweaver-Burk plot of uncompetitive type inhibition yields typical curves shown in Figure 8.10.

The general equation describing this type of inhibition is

$$\frac{1}{v} = \frac{K_m}{V_{max}} \cdot \frac{1}{[S]} + (1 + \frac{[I]}{K_I}) \frac{1}{V_{max}} \qquad (8.38)$$

3. Noncompetitive inhibitors. They affect V_{max} but not K_m. These inhibitors do not compete with the substrate for the enzyme, and at a constant inhibitor

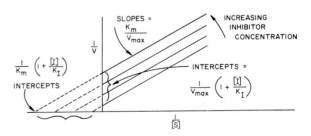

FIGURE 8.10. Lineweaver-Burk plot of the reaction velocity against substrate concentration in the presence of an uncompetitive inhibitor. (From Koushanpour.[11])

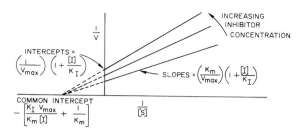

FIGURE 8.11. Lineweaver-Burk plot of the reaction velocity against substrate concentration in the presence of a noncompetitive inhibitor. (From Koushanpour.[11])

concentration, the extent of inhibition is the same at all substrate concentrations. Inhibitors that react with functional groups (SH, OH, COOH, NH_2, etc.) and many types of reagents that alter protein structure with consequent denaturation fall into this category. A Lineweaver-Burk plot of noncompetitive Inhibition is shown in Figure 8.11.

The general equation that describes this type of inhibition is

$$\frac{1}{v} = (1 + \frac{[I]}{K_I})(\frac{K_m}{V_{max}} \cdot \frac{1}{[S]} + \frac{1}{V_{max}}) \quad (8.39)$$

The Drug or Hormone-Receptor Interaction

The action of various drugs or hormones on target tissues is believed to be a consequence of a series of complex chemical reactions beginning with the interaction of, for example, the hormone (H) with a receptor site (R). The hormone-receptor complex (HR) thus formed will then initiate a series of complex chemical reactions within the target tissues, leading eventually to the emergence of the desired response.

Figure 8.12 presents a block diagram of some of the steps involved in drug or hormone-receptor interaction, as revealed by recent elaborate studies.[20] The receptor is represented by a box in the left, with hormone concentration ([H]) as the input and the hormone-receptor complex concentration ([HR]) as the output. In this scheme it is assumed that the hormone combines with one receptor and that all receptors are identical. The two arrows impinging at the bottom of the box are the indirect forcings, and more specifically may be called the properties of the receptor box. As such, for a given hormone concentration (input) the receptor properties determine the magnitude of the hormone-receptor complex concentration (output). The first property (K_{aff}) denotes the relative affinity of the hormone for the receptor. It is analogous to the reciprocal of the Michaelis constant (K_m). The second property (K_{sat}) denotes the maximum number of receptors occupied by the hormone when its concentration is infinite. It is a measure of the receptor binding capacity and is analogous to V_{max}. Accordingly, the concentration of the hormone-receptor complex ([HR]) depends not only on the hormone concentration but also on the two receptor properties (K_{aff}) and K_{sat}).

Analogous to the Michaelis-Menten formulation of the kinetics of the enzyme catalyzed reactions, we may write an equation relating the output of the receptor box ([HR]) to the input ([H]) and receptor properties (K_{aff}) and (K_{sat}):

$$[HR] = \frac{K_{sat} \cdot [H]}{K_{aff} + [H]} \quad (8.40)$$

The next box, called the coupler, represents the physical or chemical coupling of the hormone-receptor complex formed on the outside surface of the target cell membrane. The three arrows, designated as K_1, K_2, and K_3, impinging at the bottom of the coupler box, represent the properties of the coupler. They influence the reactions depicted by the next two boxes, as discussed below.

The enzyme involved in the chemical reaction represented by box three is adenylate cyclase; it breaks two phosphate groups from the adenosine triphosphate (ATP) yielding a cyclic form of the adenosine 3',5'-monophosphate (cAMP). In the block diagram

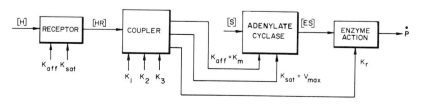

FIGURE 8.12. Possible steps involved in drug-receptor or hormone-receptor interactions. (From Koushanpour.[11])

the ATP concentration in the cell is represented by [S], which is the substrate on which adenylate cyclase acts, and the rate of production of cAMP is represented as $\dot{P}$, which is related to the final tissue response. Substantial experimental evidence indicate that the response induced by the hormone is directly related to the rate of formation of cAMP.

As shown in the diagram, the rate of production of cAMP ($\dot{P}$) is directly related to the concentration of the enzyme-substrate complex, [ES]. In accordance with the Michaelis-Menten concept of enzyme kinetics, we may relate [ES] to the substrate concentration [S] and the two enzyme properties (K_m and V_{max}) as follows:

$$[ES] = \frac{1}{K_r} \left(\frac{V_{max} \cdot [S]}{K_m + [S]} \right) \quad (8.41)$$

where K_r represents the rate of conversion of ES to the product and is analogous to K_3 in Equation 8.21.

The rate of formation of the product ($\dot{P}$) may be related to [ES] by the equation

$$\dot{P} = K_r[ES] \quad (8.42)$$

As indicated by the block diagram, through the coupler, the hormone-receptor complex influences the formation of both ES and cAMP. In the absence of precise information, these relationships may be represented by the following three linear equations:

$$K_m = K_1[HR] \quad (8.43)$$

$$V_{max} = K_2[HR] \quad (8.44)$$

$$K_r = K_3[HR] \quad (8.45)$$

Incorporating these relationships into Equation 8.41 and then substituting the results into Equation 8.42 yields an epression relating the rate of formation of the product (P) or the ultimate tissue response to the hormone concentration ([H]):

$$\dot{P} = \left(\frac{K_2[S]}{K_2 + [S]} \right) \left(\frac{K_{sat} \cdot [H]}{K_{aff} + [H]} \right) \quad (8.46)$$

Assuming that there is a large and constant supply of the substrate S (that is, ATP) within the cell, then, according to Equation 8.46, the rate of product or response formation is a hyperbolic function of the hormone or drug concentration, as depicted in Figure 8.13. Because the first portion of this graph rises very steeply, the effect of drug concentration on the rate of product formation may be better displayed by plotting $\dot{P}$ against the logarithm of the drug concentration. Such a plot yields a sigmoid or S-shape

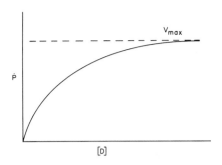

FIGURE 8.13. The dose-response relationship of a typical drug. (From Koushanpour.[11])

curve, as shown in Figure 8.14. In this plot the rate of product formation at maximal drug concentration is called the intrinsic activity. This is the conventional method used by pharmacologists to represent the dose-response relationship for a drug, which has also been widely adopted by biochemists.

Facilitated Transport

Simple diffusion provides for transport of those solutes that are either small or lipid soluble. There are many biologically important solutes, such as amino acids and sugars, which are polar molecules and generally are larger than 8 angstroms. These solutes are transported by carrier-mediated mechanisms that involve some membrane-mediated process. Facilitated transport (also called facilitated diffusion) represents a type of carrier-mediated transport. It differs from free diffusion in two respects: (a) facilitated diffusion leads to equilibration, rather than accumulation, of the substrate; and (b) its efficiency de-

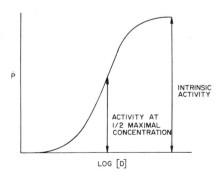

FIGURE 8.14. Conventional plot of the dose-response relationship of a typical drug. (From Koushanpour.[11])

pends on both substrate and carrier concentrations and their interactions.

Experimental analyses of substrate transport, particularly in red blood cells,[2] have revealed that the kinetics of facilitated diffusion strongly resemble enzyme-catalyzed chemical reactions. Keeping this in mind, certain characteristics common to carrier-mediated transport, and facilitated diffusion in particular, may be listed as follows:

1. The transport mechanism is saturated at high substrate concentration.
2. There is both structural specificity and affinity of the carrier for the substrate.
3. Presence of a second substrate (B) competitively inhibits the transport of the first (A), such that the two fluxes are not additive ($J_A + J_B > J_{A+B}$).
4. With near saturation of carrier-substrate interactions, the flux is proportional to the reciprocal of the difference in substrate concentration ($J_A \simeq 1/\Delta[A]$).
5. Under similar conditions, substrate flux is inversely related to carrier affinity ($J_A \simeq 1/K_m$). This relationship holds with either fixed or mobile receptor sites or carriers.
6. The concentration gradient of one substrate may cause uphill transport of another substrate. This is called counter transport (see below), not active transport, and can occur only with mobile carrier.
7. Addition of one substrate will increase the flux of another substrate, depending on their affinities for the carrier. This is called competitive acceleration.

The kinetics of carrier-mediated transport are much more complex than that for enzyme-catalyzed chemical reactions. This is, in part, due to the fact that conversion of the substrate to the products is not irreversible. Recognizing this, Rosenberg and Wilbrandt[16] formulated a quantitative description of carrier-mediated transport by applying the Michaelis-Menten concept of enzyme kinetics using the following assumptions:

1. Substrate (S) is transported through the membrane as a complex (SC) with the carrier (C).
2. Rates of substrate movement are proportional to the difference in the substrate-carrier complex at the two sides of the membrane ($J_s \simeq [SC_1] - [SC_2]$).
3. Carrier and carrier-substrate complex have the same diffusion coefficient in the membrane (though it may be greater than that for substrate alone).
4. There is a fixed amount of the carrier within the membrane, so that $[C]_t = [C]_f + [SC]$, where $[C]_t$ and $[C]_f$ are the total and free carrier concentrations, respectively.
5. There is a chemical equilibrium between the substrate and carrier at each membrane interface, as expressed by Equation 8.47:

$$K_{CS} = \frac{[S] \cdot [C]_f}{[SC]} \quad (8.47)$$

where K_{CS} is the dissociation constant for the substrate-carrier complex. Also, the equilibrium time is very rapid relative to the diffusion time.

Figure 8.15 incorporates these assumptions in a scheme for carrier-mediated transport. As shown, combination of the substrate and carrier on side 1 yields a certain concentration of carrier-substrate complex, determined by the equilibrium reaction shown on the left. The substrate then diffuses through the membrane as SC, according to the SC concentration gradient within the membrane. When SC reaches side 2, it dissociates into the substrate and carrier according to equilibrium equations shown on the right. Carrier regeneration on side 2 increases its concentration, resulting in the return of carrier to side 1 according to the carrier concentration gradient. In all these reactions, the diffusion of the carrier and the carrier-substrate complex is the rate-limiting reaction. The transport process depicted in Figure 8.15 continues as long as the substrate is available.

Having described those features of carrier-mediated transport that are common to both facilitated diffusion and active transport, let us now consider quantitatively the distinguishing characteristics of each.

In facilitated diffusion the affinity between the carrier and substrate is the same on both sides of the membrane. Thus, the dissociation constants for the carrier-substrate complex on both sides are equal: $K_{CS_1} = K_{CS_2}$. This results in rapid equilibration even if $[SC_1]$ is not equal to $[SC_2]$. Therefore,

$$\frac{[S_1] \cdot [C_1]}{[SC_1]} = \frac{[S_2] \cdot [C_2]}{[SC_2]} \quad (8.48)$$

Note that if there is no substrate concentration gradient across the membrane (that is, $[S_1] = [S_2]$), the above equation becomes

$$\frac{[C_1]}{[C_2]} = \frac{[SC_1]}{[SC_2]} \quad (8.49)$$

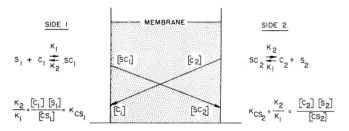

FIGURE 8.15. A schematic representation of carrier-mediated transport. See text for explanation of symbols. (From Koushanpour.[11])

which states that the substrate-carrier concentration gradient in one direction is equal to the free-carrier concentration gradient in the other direction. In such a condition, no substrate transport occurs. Therefore, to have any substrate flux, there must be a substrate concentration gradient.

To explore other factors affecting this type of transport, let us examine the kinetics of facilitated diffusion. For this analysis, we assume that the total carrier concentration will, in the steady state, be the same on the two sides and throughout the area within the membrane:

$$[C]_t = [C]_f + [SC] \quad (8.50)$$

Substituting for $[C]_f$ in Equation 8.47 its corresponding value from Equation 8.50 and solving for $[SC]$ yields

$$[SC] = [C]_t \left(\frac{[S]}{K_{CS} + [S]}\right) \quad (8.51)$$

The terms in parentheses are a measure of the fraction of total carrier bound to substrate within the membrane. Since substrate can only pass through the membrane in combination with the carrier, then its flux (J_S) is equal to the substrate-carrier complex flux (J_{SC}). Both substrate and complex fluxes are proportional to the substrate-carrier concentration gradient:

$$J_S = J_{SC} = \frac{D_{SC}}{x} ([SC_1] - [SC_2]) \quad (8.52)$$

where x is the membrane thickness, D_{SC}/x is the membrane permeability to the complex, and the term in parentheses is the concentration gradient of the complex. Expressing the substrate-carrier concentration on each side of the membrane in terms of total and free carrier concentrations (Equation 8.50), Equation 8.52 becomes

$$J_S = J_{SC} = \left(\frac{D_{SC}}{x} \cdot [C]_t\right) \left(\frac{[S_1]}{K_{CS_1} + [S_1]} - \frac{[S_2]}{K_{CS_2} + [S_2]}\right) \quad (8.53)$$

The product term in the first parentheses represents the maximum substrate flux, and the terms in the second parentheses represent the difference in the fraction of total carrier bound to the substrate on the two sides of the membrane.

Equation 8.53 may be used to explore the effects of four conditions on transport by facilitated diffusion.

1. Equal substrate concentration on both sides ($[S_1] = [S_2]$). In the absence of a substrate concentration gradient, the terms in the second parentheses in Equation 8.53 become zero and hence there will be no substrate flux ($J_S = J_{SC} = 0$).

2. Low carrier saturation. When substrate concentration is below saturating level, less substrate-carrier complex will be formed. This means that the value of K_C far exceeds $[S]$ in Equation 8.53. Hence, if $K_{CS_1} = K_{CS_2} = K_{CS}$, we get

$$J_S = \frac{D_{SC}}{x} \cdot [C]_t \cdot \frac{1}{K_{CS}} ([S_1] - [S_2]) \quad (8.54)$$

where $1/K_{CS} = K_m$ is a measure of the affinity between carrier and the substrate. In this case, substrate flux is largely determined by the product of total carrier concentration and affinity ($[C]_t \cdot 1/K_{CS}$). Thus, at low substrate concentration or low carrier saturation, substrate flux may be approximated by a simple diffusion.

3. High carrier saturation. At high substrate concentration, the carrier will be fully saturated, in which case the value of $[S]$ will be much greater than K_{CS}. Assuming that $[S_1]/(K_{CS_1} + [S_1]) = 1$ and $[S_2]/(K_{CS_2} + [S_2]) < 1$, Equation 8.51 becomes

$$J_s = \frac{D_{SC}}{x} \cdot [C]_t \left(1 - \frac{[S_2]}{K_{CS_2} + [S_2]}\right) \quad (8.55)$$

It is apparent from this equation that the substrate flux is limited by its concentration on side 2 of the membrane ($[S_2]$). Reducing $[S_2]$, such as might occur if the substrate is removed or metabolized, increases the carrier-substrate concentration gradient, thereby increasing substrate flux across the membrane.

4. Competitive inhibition. Suppose there are two substrates, R and S, that can combine with the carrier, C. Then Equation 8.50 will become

$$[C]_t = [C]_f + [CR] + [CS] \quad (8.56)$$

Assuming, as before, that the substrate transport is equal to the substrate-carrier complex flux, we can express the relative rate of complex formation in terms of dissociation constant:

$$K_{CS} = \frac{[S] \cdot [C]}{[SC]} \quad (8.57)$$

$$K_{CR} = \frac{[R] \cdot [C]}{[CR]} \quad (8.58)$$

Substituting for [CR] and [CS] in Equation 8.56 their corresponding expressions from Equations 8.57 and 8.58, and expressing the resulting equation in terms of [CS], we obtain

$$[CS] = [C]_t \left(\frac{[S]}{[S] + K_{CS} + R\frac{K_{CS}}{K_{CR}}}\right) \quad (8.59)$$

Comparison of Equations 8.51 and 8.59 shows the effect of the addition of a second substrate (R) on the relative concentration of the complex formed between the carrier and the first substrate ([CS]). This effect is determined by the term $R(K_{CS}/K_{CR})$ in Equation 8–57. If the carrier has a greater affinity for R than S, then $R(K_{CS}/K_{CR}) > 1$, in which case an increase in the concentration of R results in a proportional decrease in J_s.

Active Transport

In Chapter 2, active transport, which represents another type of carrier-mediated transport, was defined as the movement of solute against an electrochemical gradient. Unlike facilitated diffusion for which $K_{CS_1} = K_{CS_2}$ and for which there was equilibration of the carrier on both sides of the membrane, active transport is characterized by inequality of K_m on both sides and an uphill transport. This uphill flow of material (from lower to higher chemical potential) is believed to be coupled to a downhill transport of another material. The Na-K-ATPase (the biochemical equivalent of the so-called "sodium-potassium pump") found in mammalian cells provides the machinary for the active transport process.

Briefly, some important characteristics of the Na-K-ATPase pump, germane to the present discussion, are as follows[10]:

1. Electron microscopic studies have revealed that Na-K-ATPase consists of disk-shaped, triple-layered membrane fragments covered with clusters of particles, 30 to 50 angstroms in diameter, which protrude from the plane of the membrane. On purification these particles, which constitute the Na-K-ATPase, consist of two protein subunits in molar ratio of 1:1: the larger α-subunit has an approximate molecular weight of 100,000, and the smaller β-subunit has an approximate molecular weight of 40,000, respectively. The α-subunit contains the catalytic site of the enzyme, which is phosphorylated by ATP and binds with ouabain. Although the role of the β-subunit in not known, it is required for normal function of the enzyme.

2. Histochemical studies indicate that the Na-K-ATPase enzyme is confined to the basolateral cell membrane. ATP is the primary substrate for this enzyme, and the reaction requires the presence of magnesium. The energy released from the hydrolysis of ATP is coupled to a cyclic exchange of sodium and potassium across the basolateral cell membrane. Such a cycle consists of the following steps: (a) ATP binds to Na-K-ATPase and interacts with sodium on the inside (the cytoplasmic side) of the membrane, followed by the phosphorylation of the enzyme and hydrolysis of ATP to ADP; (b) conformational changes of the phosphorylated carrier complex, thereby changing its affinity for potassium instead of sodium, during which the sodium binding sites are turned inside out and release sodium into the extracellular fluid; (c) ATP-dephosphorylation-coupled binding of the extracellular potassium with the carrier; and (d) conformational change of the carrier and release of potassium into the intracellular fluid.

3. The stoichiometry of Na-K-ATPase activated Na-K exchange is such that there are two potassium ions transported into the cell for every three sodium ions transported out of the cell. However, an increase in intracellular sodium can cause the Na/K ratio to

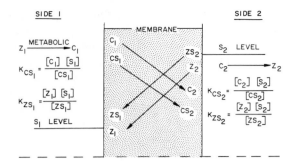

FIGURE 8.16. An uphill transporting carrier system representing active transport. See text for explanation of symbols. (From Koushanpour.[11])

exceed 1.5. This would result in a net outward movement of positive charge, thereby making the membrane potential more negative. Because of this characteristic, the sodium pump is said to be electrogenic.

4. An important function of the Na-K-ATPase pump is to create a potassium gradient across the cell membrane, thereby generating the membrane potential.

5. Another function of the sodium pump is to maintain an electrochemical gradient for sodium, which serves as a source of energy to fuel other carrier molecules used to transport sugars and amino acids against their concentration gradients. This is called coupled transport, which is described in the next section. Further discussion of Na-K-ATPase functions in kidney cells is presented in Chapter 9.

Analogous to facilitated diffusion, Rosenberg and Wilbrandt have made a quantitative treatment of active transport based on the assumption that there are "asymmetric chemical reactions of a fixed quantity of the carrier between two forms C and Z, differing in their affinities to the substrate.[16] The necessary gradient, then, is maintained by expenditure of metabolic energy (ATP hydrolysis), and the continuous carrier flow is replaced by a metabolically driven carrier cycle."

Figure 8.16 illustrates schematically the essential elements of an uphill transporting carrier system. The basic assumptions are (1) the membrane contains a fixed total quantity of carrier, cycling between the two forms C and Z, and (2) the two carriers have different affinities for the substrate S, characterized by the Michaelis-Menten constants K_{CS} and K_{ZS}, respectively. Furthermore, it is assumed that the carriers are in equilibrium with the substrate, as shown,

and can move through the membrane by diffusion, both in the free form (C and Z) and bound to the substrate (CS and ZS). Free substrate cannot pass through the membrane.

Assuming that the diffusion coefficients (D) for both free and bound carriers are identical, then the total carrier concentration will, in the steady state, be the same on the two sides of the membrane:

$$[C]_t = [C]_f + [Z]_f + [CS] + [ZS] \quad (8.60)$$

Since the substrate can only move through the membrane in combination with the carrier, its flux is given by

$$J_S = J_{CS+ZS}$$
$$= \frac{D}{x} [C]_t ([CS_1] + [ZS_1] - [CS_2] - [ZS_2]) \quad (8.61)$$

Expressing the various carrier-substrate concentration terms in the parentheses in terms of the dissociation equations given in Figure 8.16, Equation 8.61 becomes

$$J_S = \frac{D}{x} [C]_t \left[\frac{[S_1]}{[S_1] + K_{CS_1} \left(\frac{K_{ZS_1} + K_{ZS_1} \frac{[Z_1]}{[C_1]}}{K_{ZS_1} + K_{CS_1} \frac{[Z_1]}{[C_1]}} \right)} - \frac{[S_2]}{[S_2] + K_{CS_2} \left(\frac{K_{ZS_2} + K_{ZS_2} \frac{[Z_2]}{[C_2]}}{K_{ZS_2} + K_{CS_2} \frac{[Z_2]}{[C_2]}} \right)} \right] \quad (8.62)$$

The above equation can be simplified by replacing portions of the denominator, as shown, by the symbols Θ_1 and Θ_2, respectively:

$$\Theta_1 = K_{CS_1} \left(\frac{K_{ZS_1} + K_{ZS_1} \frac{[Z_1]}{[C_1]}}{K_{ZS_1} + K_{CS_1} \frac{[Z_1]}{[C_1]}} \right) \quad (8.63)$$

and

$$\Theta_2 = K_{CS_2} \left(\frac{K_{ZS_2} + K_{ZS_2} \frac{[Z_2]}{[C_2]}}{K_{ZS_2} + K_{CS_2} \frac{[Z_2]}{[C_2]}} \right) \quad (8.64)$$

Then,

$$J_s = \frac{D}{x}[C]_t \left\{ \frac{[S_1]}{[S_1] + \Theta_1} - \frac{[S_2]}{[S_2] + \Theta_2} \right\} \quad (8.65)$$

where Θ_1 and Θ_2 are equal to the concentration of S_1 and S_2 at half-saturation. They are the "steady-state" constants relating substrate-carrier reactions on the two sides of the membrane.

Equation 8.65, describing the kinetics of the C-Z carrier system, differs from Equation 8.53 for the equilibrating carrier system or facilitated diffusion in two respects. First, the Michaelis-Menten constants (K_{CS_1} and K_{CS_2}) in Equation 8.53 are replaced by two complex functions (Θ_1 and Θ_2) containing the Michaelis constants for the two carrier complexes CS and ZS and the concentration ratios of the two carriers on the two sides of the membrane. Second, owing to metabolic asymmetry ($[Z_1]/[C_1] \neq [Z_2]/[C_2]$), the functions Θ_1 and Θ_2 differ.

Finally, there are two conditions that make Θ_1 equal to Θ_2 — that is transforming the active transport Equation 8.65 into the equilibrating carrier system or facilitated diffusion. First, if $[Z_1]/[C_1] = [Z_2]/[C_2]$ (i.e., if they are symmetrical metabolic reactions), and second, if the two carriers have equal affinities for the substrate on both sides of the membrane ($K_{CS} = K_{ZS}$).

Coupled Transport

This type of transport (also called cotransport) uses the energy generated from the sodium concentration gradient to fuel the uphill transport of other substances, provided a mechanism exists for coupling the downhill movement of sodium to uphill movement of the other substance. The essence of this type of transport can best be described by considering the transport of glucose by the renal proximal tubule. As discussed in Chapter 7, glucose is reabsorbed in the proximal tubule by a T_m-limited transport mechanism. Glucose is reabsorbed uphill against its concentration gradient (active transport) from the tubular lumen into the proximal cell. Once inside the cell, glucose is reabsorbed into the blood downhill along its concentration gradient. As we shall see in Chapter 9, in the same nephron segment, sodium is reabsorbed downhill along its electrochemical gradient from the tubular lumen into the proximal cell. Sodium is then transported uphill against its electrochemical gradient from the cell into the blood. Kinetic studies have shown that the active transport of glucose is coupled to the active transport of sodium.

The coupling of these two separate transport processes is achieved by membrane-carrier molecules that have binding sites for both sodium and glucose:

Tubular Lumen   *Proximal Cell*

$$[Na^+]_l \rightleftarrows [Na^+]_c \quad (8.66)$$

$$[G]_l \rightleftarrows [G]_c \quad (8.67)$$

where $[Na^+]_l$ and $[G]_l$ are the sodium and glucose concentrations in the tubular lumen, $[Na^+]_c$ and $[G]_c$ are the sodium and glucose concentrations inside the proximal cell, respectively, and the lengths of the arrows depict the magnitude of the respective gradients.

Coupling the above two reactions by the carrier yields

$$[Na^+]_l + [G]_l \rightleftarrows [Na^+]_c + [G]_c \quad (8.68)$$

The coupled transport of sodium and glucose will proceed in a direction determined by the combined gradients for sodium and glucose. The direction of transport will be into the cell if

$$E_{Na} - E_m > \frac{RT}{F} \ln \frac{[G]_c}{[G]_l} = 60 \log \frac{[G]_c}{[G]_l} \quad (8.69)$$

where the term on the right converts the concentration gradient of the transported glucose into an equivalent electrical unit. The Faraday constant, F, is defined as the charge upon one mole of univalent ions (96,500 coulombs/mole), and R and T were already defined.

Kinetic studies have shown that the carrier molecules when occupied by only one substance are not mobile within the membrane, whereas unoccupied and dually occupied carriers are mobile and undergo passive translocation within the membrane. This has led to the following concept for the coupled transport cycle for sodium and glucose: (a) binding of sodium and glucose to the carrier, (b) translocation of sodium-glucose-carrier complex within the membrane, (c) release of sodium and glucose into the cell, (d) translocation of carrier, and (e) repeat of step (a). It should be noted that the special carriers capable of dual binding are present only on the mucosal membrane (luminal side of the renal proximal tubule) of the cell, where the uphill transport of glucose is coupled to downhill transport of sodium. The Na-K-ATPase (sodium-potassium pump) located in the serosal membrane (peritubular blood side of the renal proximal tubule) restores the sodium concentration gradient and the standard glucose carrier located there mediates downhill glucose transport.

Transport of amino acids from intestinal mucosa and their renal reabsorption involves special carriers capable of coupling uphill transport of these molecules with that of sodium. Kinetic studies have revealed that unlike the glucose coupled transport mechanism, the carrier must bind amino acids before it can bind sodium and that all forms of the carriers are equally mobile and translocated within the membrane. Except for these differences, the cycle of coupled transport of amino acids and sodium is similar to that outlined above for cotransport of glucose and sodium.

Neutral amino acids are transported by the same special carrier. However, the order of binding affinity appears to decrease with increasing size of the side chain, giving the following binding order: glycine > alanine > valine > leucine > phenylalanine.

Anionic amino acids are transported by a different carrier system that resembles the glucose-sodium carrier system. The cationic amino acids are transported by another carrier system that appears to be even less sodium-dependent. Further details of amino acid transport systems are given in Chapter 10.

With these materials serving as the background, we are now ready to consider in Chapters 9 and 10 the mechanisms of renal transport for various solutes present in the ultrafiltrate as well as a quantitative account of those that are finally excreted in the urine.

Problems

8.1. What is the main barrier along the paracellular transport pathway through an epithelium?

8.2. What is the difference between a "leaky" and a "tight" tight junction?

8.3. What are the distinguishing features between the lateral intercellular spaces of the proximal tubule and the collecting duct?

8.4. What are the different kinds of basolateral membrane amplification?

8.5. Describe the structural organization of a brush border.

8.6. The following table lists the dose-response data for the action of a drug with and without the presence of an inhibitor.

Drug concentration (mg/mL)	Response (arbitrary units)	
	without inhibitor	with inhibitor
20	25	16
22	26	18
25	28	19
29	29	21
33	32	23
40	34	26
50	37	28
67	40	32
100	45	38
200	50	45

a. Calculate the dissociation constant for the drug-receptor complex.
b. Calculate the dissociation constant for the inhibitor-receptor complex.
c. Indicate the type of inhibitor used.

8.7. On a Lineweaver-Burk transformation of a rectangular hyperbolic process,
a. What is the ordinate intercept a measure of?
b. What is the abscissa a measure of?

8.8. For carrier-mediated transport, the net flux of substrate is considered to be *directly proportional* to what?

8.9. Give a descriptive explanation for how a carrier-mediated transport could have substrate flux inversely related to substrate-carrier affinity.

8.10. What are the two conditions under which active transport is reduced to facilitated diffusion?
a. _____
b. _____

8.11. The dissociation equilibrium constant between D-mannose and human red blood cells (RBCs) has been estimated as 0.02 M. If the RBCs were allowed to equilibrate in a 0.2 M D-mannose solution, what would be the expected initial flux, expressed as a fraction of the maximum possible flux, of that sugar?

References

1. Bodansky O: Diagnostic applications of enzymes in medicine. *Am J Med* 1959; 27:861–874.
2. Britton HG: Permeability of the human red cell to labelled glucose. *J Physiol* 1964; 170:1–20.
3. Christensen HN: *Biological Transport*, ed 2. Reading, Mass, Addison-Wesley Publishing Co, 1975.
4. DiBona DR, Mills JW: Distribution of Na^+-pump sites in transporting epithelia. *Fed Proc* 1979; 38:134–143.
5. Drenckhahn D, Groschel-Stewart U: Localization of myosin, actin, and tropomyosin in rat intestinal epithelium: Immunohistochemical studies at the light and

electron microscope levels. *J Cell Biol* 1980; 86:475–482.
6. Erlij D, Martinez-Palomo A: Role of tight junctions in epithelial function, in Giebisch G, Tosteson DC, Ussing HH (eds): *Membrane Transport in Biology.* vol 3. Berlin, Springer-Verlag, 1978, pp 27–53.
7. Frieden C: Treatment of enzyme kinetic data. *J Biol Chem* 1964; 239(19):3522–3531.
8. Ganote CE, Grantham JJ, Moses HL, et al: Ultrastructural studies of vasopressin effect on isolated perfused renal collecting tubules of the rabbit. *J Cell Biol* 1968; 36:355–367.
9. Kaissling B: Structural aspects of adaptive changes in renal electrolyte excretion. *Am J Physiol* 1982; 243:F211–F226.
10. Katz A: Renal Na-K-ATPase: its role in tubular sodium and potassium transport. *Am J Physiol* 1982; 242:F207–F219.
11. Koushanpour E: *Renal Physiology: Principles and Functions*, ed 1. Philadelphia, WB Saunders Co, 1976.
12. Kriz W: Der Bau transportierender Epithelien. *Verh Anat Ges* 1984; 78:25–40.
13. Kriz W, Schiller A, Taugner R: Freeze-fracture studies on the thin limbs of Henle's loop in Psammomys obesus. *Am J Anat* 1981; 162:23–34.
14. Pfaller W: Structure function correlation on rat kidney. *Adv Anat Embryol Cell Biol* 1981; 70:1–100.
15. Pinto da Silva P, Kachar B: On tight junction structure. *Cell* 1982; 28:441–450.
16. Rosenberg T, Wilbrandt W: Carrier transport uphill. I. General. *J Theoret Biol* 1963; 5:288–305.
17. Schiller A, Taugner R: Freeze-fracturing and deep-etching with the volatile cryoprotectant ethanol reveals true membrane surfaces of kidney structures. *Cell Tissue Res* 1980; 210:57–69.
18. Singer SJ, Nicolson GL: The fluid mosaic model of the structure of cell membranes. *Science* 1972; 175:720–731.
19. Spring KR, Hope A: Dimensions of cells and lateral intercellular spaces in living Necturus gallbladder. *Fed Proc* 1979; 38:128–133.
20. Sutherland EW, Robison CA, Butcher RW: Some aspects of biological role of adenosine 3',5'-monophosphate (cyclic AMP). *Circulation* 1968; 37:279–306.
21. Tisher CC, Kokko JP: Relationship between peritubular oncotic pressure gradients and morphology in isolated proximal tubules. *Kidney Int* 1974; 6:146–156.
22. Van Deurs B, Luft JH: Effects of glutaraldehyde fixation on the structure of tight junctions. A quantitative freeze-fracture analysis. *J Ultrastruct Res* 1979; 68:160–172.
23. Wade JB, O'Neil RG, Pryor JL, et al: Modulation of cell membrane area in renal collecting tubules by corticosteroid hormones. *J Cell Biol* 1979; 81:439–445.
24. Wilbrandt W, Rosenberg T: The concept of carrier transport and its corollaries in pharmacology. *Pharmacol Rev* 1961; 13:109–183.

9

Tubular Processing of Glomerular Ultrafiltrate: Mechanisms of Electrolyte and Water Transport

Once the glomerular filtrate is formed, the next major steps in the renal regulation of body fluids occur in the tubular exchanger. Here, as the filtrate flows under a hydrostatic pressure head along the various nephron segments, almost all the filtered water and solutes are reabsorbed from the tubules, while some organic compounds are secreted into the tubules, yielding a small, hypertonic volume of urine. The final volume and composition of this excreted urine will largely depend on the efficiency of these two tubular transport processes and their intrarenal and extrarenal regulation.

Our current knowledge of the mechanisms of tubular processing of the filtrate is primarily due to the development of numerous quantitative methods for the study of renal function. Hence, we begin this chapter with a brief description of the various methods used to characterize the renal transport mechanisms involved in the sequential processing of the ultrafiltrate and its components.

Methods of Studying Tubular Transport

Although the clearance techniques described in Chapter 7 provided much useful information about the overall performance of the kidney, they have limited value in assessing the sequential processing of the filtrate and its constituents along the various nephron segments. Direct information about the mechanisms of tubular processing of the filtrate and its constituents has been obtained from the application of several widely used methods:

Stop-Flow Analysis of Tubular Urine in Whole Kidney

This modified clearance technique, often referred to as "the poor man's micropuncture," developed by Malvin and his associates[172] has provided useful information about tubular transport function and possible sites of transport along the nephron for a variety of solutes. The method is based on the idea that stopping the urine flow allows the various tubular processes to function at maximum efficiency. Thus, during acute ureteral occlusion, any substance that is normally reabsorbed is continuously reabsorbed, whereas any substance that is secreted will be continuously added to the trapped urine. In this manner an exaggerated concentration profile, for a given solute, along the various nephron segments will be established. After approximately 10 minutes the occlusion is removed and small (0.5 mL) serial urine samples are collected. Each sample removed represents urine trapped in a successively more proximal segment of the nephron.

The approximate tubular location of each collected sample will be verified from the analysis of urine-to-plasma concentration (U/P) ratio of inulin administered prior to occlusion. The tubular transport will be determined from the comparison of the U/P ratio for the solute under study relative to that of inulin.

Figure 9.1 shows the tubular concentration profiles for sodium, glucose, and p-aminohippuric acid (PAH) as obtained by this method. To eliminate the effect of variable flow rates during collection, the various solute concentrations are plotted against the accumulated volume of urine that has appeared since the removal of the clamp. Note that the concentration of glucose—a substance maximally reabsorbed—is

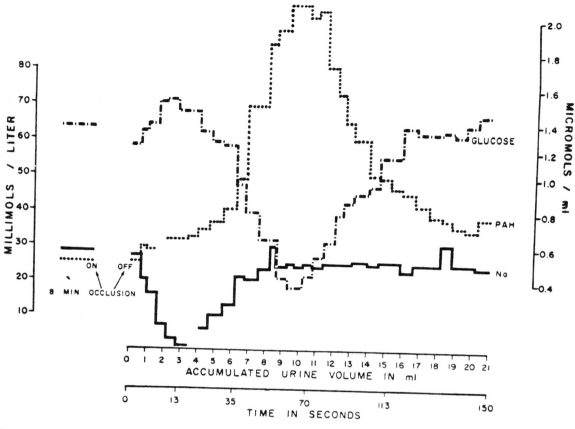

FIGURE 9.1. Concentration profiles developed for PAH, sodium, and glucose along the nephron as determined by stop-flow technique. (From Malvin.[172])

low in the more proximal samples, corresponding to 9 to 11 mL on the volume scale. On the other hand, sodium, which is actively reabsorbed all along the nephron, shows an initial decrease in concentration, followed by an increase, reaching a plateau at about the same time on the volume scale. In the case of PAH, the maximal urinary PAH concentration occurs in the same tubular location where glucose concentration is minimal, indicating that PAH is secreted by the proximal tubule.

Accurate localization of tubular transport by the stop-flow method in whole kidney is limited by some "smearing" of the solute concentration profiles. This is attributed to (1) the mixing of the tubular urine with urine already in the catheter when occlusion is removed, (2) the heterogeneity of the nephron population in the whole kidney, and (3) the glomerular filtration, which proceeds (even though to a small extent) during the complete tubular blockade. In addition, the composition of the filtrate issuing from the proximal tubule may be greatly altered as it passes through the distal parts of the nephron. However, this effect is assumed to be small owing to rapid tubular flow rate immediately after the clamp is released. Because of these shortcomings, this method is currently used less frequently.

Free-Flow Micropuncture Collection of Tubular Fluid

The free-flow micropuncture collection and analysis of fluid from single mammalian nephrons was first introduced by Richards and Walker.[198] Since then, this technique has been successfully applied to collect and analyze fluids from the proximal and distal convoluted tubules of the superficial nephrons as well as the medullary segments of the loops of Henle and collecting duct.[94] Refinements of these techniques have made it possible to isolate and analyze fluids collected from segments of the nephron previously inaccessible to direct study.[23, 151]

Briefly, the technique involves puncturing, under a stereomicroscope and quartz rod illumination, the desired tubular segment with micropipette, having an outer diameter of 6 to 10 μm. A small volume of tubular fluid (about 1 μL) is then collected under free-flow conditions and analyzed. After collecting the tubular fluid, a dye solution is then injected to mark the punctured site for subsequent determination of its position relative to the glomerulus by microdissection. Such a procedure permits quantitative analysis of the tubular transport process along the length of a given nephron segment.

The tubular concentration of a given solute, thus obtained, at any position along the nephron depends not only on how much of the solute has been reabsorbed, but also on how much water has left the tubular segment. Since we are interested in solute transport, the effect of water movement can easily be eliminated by comparing the ratio of the tubular fluid-to-plasma solute concentration, or simply TF/P, to that of a nonabsorbable solute that remains in the tubule, such as inulin. Since for a given collected sample, the tubular volume flow is the same, the TF/P ratios are identical to clearance ratios described in Chapter 7. The rationale for comparing the TF/P ratio of a solute to that of inulin to assess the nature of tubular transport of that solute is as follows.

The quantity of a substance x filtered per minute is given by the product of the glomerular filtration rate (GFR) and the plasma concentration of the substance ($[x]_p$), assuming a permeability ratio of unity,

$$\text{quantity of } x \text{ filtered} = \text{GFR} \cdot [x]_p \quad (9.1)$$

Similarly, the quantity of a substance x remaining in any tubular segment of the nephron is given by the product of the volume of the filtrate remaining in that segment ($\dot{V}_f$) and the tubular concentration of substance x in that segment ($[x]_f$):

$$\text{quantity of } x \text{ remaining} = \dot{V}_f \cdot [x]_f \quad (9.2)$$

Thus, at any point along the nephron the difference between the quantity filtered and the quantity remaining in the tubule is the amount of substance x which has been transferred across the tubular cells ($\dot{T}_x$). Therefore, for any nephron segment, we can write a material balance equation relating the quantity filtered to that remaining in the tubule:

$$\text{GFR} \cdot [x]_p = \dot{V}_f \cdot [x]_f \pm \dot{T}_x \quad (9.3)$$

where the ± signs indicate whether the solute is secreted ($+\dot{T}_x$) into or reabsorbed ($-\dot{T}_x$) from the tubule. If a substance is neither reabsorbed nor secreted, then $\dot{T}_x = 0$, and Equation 9.3 becomes

$$\text{GFR} \cdot [x]_p = \dot{V}_f \cdot [x]_f \quad (9.4)$$

Writing Equation 9.4 in terms of the fluid-to-plasma concentration ratios, we get

$$[x]_f/[x]_p = \text{GFR}/\dot{V}_f \quad (9.5)$$

Analogous equation for inulin yields

$$[\text{In}]_f/[\text{In}]_p = \text{GFR}/\dot{V}_f \quad (9.6)$$

where the right hand side of Equation 9.6 is now recognized as the reciprocal fraction of fluid remaining in the tubule, or the so-called rejection ratio.

Now, from Equation 9.4, the fraction of a substance x remaining in any tubular segment of the nephron may be expressed as

$$\text{fraction of } x \text{ remaining} = \dot{V}_f \cdot [x]_f / \text{GFR} \cdot [x]_p \quad (9.7)$$

Note that for inulin the value of this fraction is unity. Rearranging Equation 9.7 in terms of the fluid-to-plasma concentration ratios, we obtain

$$\text{fraction of } x \text{ remaining} = \frac{[x]_f/[x]_p}{\text{GFR}/\dot{V}_f} \quad (9.8)$$

Comparing this equation with Equation 9.6 reveals that the denominator of Equation 9.8 is the same as the fluid-to-plasma concentration ratio for inulin ($[\text{In}]_f/[\text{In}]_p$). Substituting this ratio in Equation 9.8, the fraction of substance x remaining in the tubular fluid may now be expressed as a ratio of the fluid-to-plasma concentration ratio of the substance under study to that for inulin:

$$\text{fraction of } x \text{ remaining} = \frac{[x]_f/[x]_p}{[\text{In}]_f/[\text{In}]_p} = \frac{(\text{TF}/\text{P})_x}{(\text{TF}/\text{P})_{\text{In}}} \quad (9.9)$$

If the TF/P ratio of the solute in question is greater than that for inulin, then the solute must have been secreted into the tubule. If the TF/P ratio for the solute is less than that of inulin, then the solute must have been reabsorbed.

The successful application of micropuncture technique over the past few decades has greatly contributed to the evolution of our present concepts of renal function. Specifically, it has elucidated the renal transport mechanisms for a variety of plasma solutes,[93] the renal action of drugs, particularly the di-

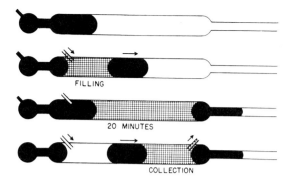

FIGURE 9.2. Technique of stop-flow microperfusion. (From Giebisch.[77])

uretics, and the mechanism of concentration and dilution of urine and its modulation by hormones.[258] The knowledge thus gained has provided the basis for the present integrated analysis of the regulation of the extracellular volume and osmolality by the renal-body fluid regulating system.

In-Situ Stop-Flow Microperfusion Techniques

Another powerful method for direct determination of tubular transport is the in-situ stop-flow microperfusion of a small tubular segment with a known solution. This procedure, first introduced by Richards and Walker[198] and later modified by Shipp and associates[220] and Gertz,[74] is a modification of the free-flow micropuncture collection of tubular fluid as described above.

Briefly, the method involves injecting a small volume of colored mineral oil into the lumen of a renal tubule, and then splitting the oil column in two by the injection of the test perfusate. In this way, the injected test fluid is completely sealed off by oil on both sides, and any subsequent observed changes in the volume and composition of the perfusate are attributed to the activity of the tubular cells to which the fluid was exposed. The injected fluid is usually left in contact with the tubular cells for a known period of time and then collected and analyzed for any changes in composition. Figure 9.2 illustrates the essence of this technique.

The reabsorptive rate of the injected fluid can be estimated as follows. As the fluid is reabsorbed, the length (L) of the injected fluid column between the oil decreases without visible changes in the tubular diameter. Assuming that the tubular radius does not change, then the fractional change in length (L/L_0) corresponds to the fraction of fluid volume remaining (V/V_0), where L_0 and V_0 are the initial length and volume, and L and V are the length and volume of the injected fluid at the time of collection. This fractional volume declines exponentially as a function of time:

$$\ln \frac{V}{V_0} = - k \cdot t \qquad (9.10)$$

where k is the reabsorptive rate constant per second and t is the time in seconds.

Another way of describing this relationship is the following (adapted from Arrizurieto-Muchnik and his associates[3]).

For a given tubular segment of length (L) and cross-sectional area (πr^2), the reabsorptive rate constant (k) may be expressed as the ratio of reabsorptive rate per unit tubular length ($\dot{R}$) and cross-sectional area (πr^2). Expressed mathematically,

$$k = \frac{\dot{R}}{\pi r^2} \qquad (9.11)$$

If $\dot{R}$ and r remain constant along the length of the tubular segment, then the change in fluid volume per unit time (dV/dt) should be inversely proportional to the initial volume injected.

$$\frac{dV}{dt} = - \frac{\dot{R}}{\pi r^2} \cdot V \qquad (9.12)$$

or

$$\frac{dV}{V} = - \frac{\dot{R}}{\pi r^2} \cdot dt \qquad (9.13)$$

Integrating Equation 9.13 from t = 0 to t = T, where T is the time required for fluid reabsorption in the split-oil drop experiments, we get

$$\ln \frac{V_t}{V_0} = - \frac{\dot{R}}{\pi r^2} \cdot T \qquad (9.14)$$

This equation also holds for the free-flow experiments, where T is defined as the tubular transit time of lissamine green dye, except that $V_0/V_t = [In]_f/[In]_p$ as defined by Equation 9.6, and hence,

$$\ln \frac{[In]_f}{[In]_p} = \frac{\dot{R}}{\pi r^2} \cdot T \qquad (9.15)$$

or

$$\frac{\ln \dfrac{[In]_f}{[In]_p}}{T} = \frac{\dot{R}}{\pi r} \qquad (9.16)$$

A modification of this method is the flow-through microperfusion of a punctured tubular segment. The procedure involves introducing the perfusate into a tubular lumen through one glass micropipette and recollecting it at a more distal site with a second micropipette. To prevent nonperfusing luminal fluid from contaminating the perfusate, the tubular segments not involved in perfusion are sealed off with oil columns. These methods have yielded extensive quantitative information about the tubular activity on the perfused fluid, such as bidirectional rate of ion movements, and the effects of tubular geometry and volume flow on fluid reabsorption rate.

In-Vitro Perfusion of Isolated Tubules

This method, first introduced by Burg and associates,[23] involves in-vitro perfusion of small segments of tubules microdissected from rabbit kidney. It has been successfully applied to study tubular transport in different nephron segments, including those segments inaccessible to direct micropuncture techniques, in a variety of mammalian species such as human, rabbit, rat, mouse, and hamster. Thus, with this method it has been possible to study tubular transport in the proximal and distal tubules, the loops of Henle, the connecting tubule, and the cortical and medullary collecting ducts. The experimental setup employed in these studies varied from a simple system consisting of a perfusion-holding pipette, a perfusion pipette, and a collecting pipette[23] to the most complex system, which incorporates the above setup with a dual-channel perfusion pipette with two fluid-exchange pipettes, and a Sylgard pipette for the perfusion side.[98-100] Incorporation of a recently developed pH-sensing electrode[212] allows for continuous monitoring of the pH in the collected fluid.

The application of this method has yielded significant information about water transport, unidirectional fluxes of radioactive tracers, and net fluxes of electrolytes and organic substances along the nephron. Additionally, this method has the unique advantage of allowing the investigator to vary the composition of both lumen and bath perfusion solutions as desired. Thus, with this method, it is possible to study essentially the transport properties of the entire nephron, as well as provide an important tool for defining the effects of hormones and drugs on various nephron segments. At present, the drawbacks of the technique are that it is difficult to master and that only one tubule can be studied per day.[98, 99]

Short-Circuit Current Measurement

This method, first applied to toad bladder and frog skin,[241] provides a direct quantitative measure of the net active transport of an ion across the cell membrane. Since active transport involves the transport of a solute, such as sodium ion, against an electrochemical gradient, this method provides a direct measure of the magnitude of this transport.

Briefly, the technique involves measuring the electrical current that must be applied to reduce the membrane potential to zero. This applied current, which abolishes the electrical gradient, is called the short-circuit current. It has been shown to be a direct measure of the active transport of an ion by the membrane.

Combining the short-circuit current with in-situ stop-flow microperfusion and in-vitro tubular perfusion techniques, several investigators have studied the characteristics of active transport in isolated nephron segments. The experimental procedure involves splitting the previously injected oil column with small volume of Ringer's solution. Then, a microelectrode, filled with 3 mol/L potassium chloride solution, with tip potential less than 2.5 mV, is inserted into the previously punctured tubule. The potential difference is measured between the luminal electrode and the indifferent electrode placed on the surface. Figure 9.3 shows schematically the technique of electrode placement and measurement of short-circuit current. The interpretation of experimental data thus obtained is analogous to that described for frog skin.

Methodological Basis of Active Reabsorption of an Electrolyte

As mentioned in Chapter 8, passive transepithelial transport of an electrolyte may result from the action of three physical forces: (1) chemical concentration or activity gradients, (2) electrical potential gradients, and (3) convection or solvent drag. The transepithelial transport of the electrolyte not accounted for by these forces was classified as active transport. Since an electrical potential gradient always coexists with the active transport of an electrolyte, its measurement is a key to assessing whether the electrolyte is transported actively or not.

To determine the site and magnitude of the active transport of an electrolyte, three types of electrical

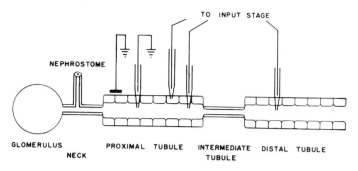

FIGURE 9.3. Single tubule of Necturus nephron showing relation of indifferent and recording microelectrodes to various tubular structures. (From Giebisch.[78] By permission of the American Heart Association, Inc.)

potential gradients have been measured on single nephrons:

1. The transepithelial potential gradient, obtained by recording the electrical potential difference between the tubular lumen and the peritubular interstitial fluid. Comparison of this measured electrical potential gradient (PD) with the theoretical potential gradient (E) for that electrolyte calculated from the Nernst equation allows one to decide whether an ion is transported actively or not. According to the Nernst equation, if a membrane were permeable to only one type of ion, such as sodium, in the steady state the theoretical transmembrane electrical potential gradient would be a function of the chemical concentration gradient of that ion:

$$E = - \frac{RT}{zF} \ln \frac{[x]_l}{[x]_b} \qquad (9.17)$$

where z is the valence of the electrolyte, F is the Faraday constant, defined as the charge upon one mole of univalent ions (96,500 C/mole), $[x]_l$ and $[x]_b$ are the concentrations of the particular electrolyte in the tubular lumen and blood, respectively, and R and T have already been defined. If measured transepithelial potential gradient is equal to the calculated theoretical gradient (PD = E), then the ion is transported passively, otherwise active transport may be involved.

2. The transluminal and transbasolaterl electrical potential gradients, obtained by measuring the electrical potential differences across the luminal and basolateral membranes of single renal tubular cells. Such electrical measurements, in conjunction with ionic concentration gradients across the same membranes, permit one to locate the site of active transport within the cell.

3. The measurement of short-circuit current, which, along with the knowledge of net movement (flux) of a given ion, allows one to determine the fraction of the total ion movement that is due to active transport.

Whether an ion is actively transported or not can be determined experimentally by comparing the observed bidirectional fluxes for that ion, and hence their ratio, with the theoretical ratio calculated from Ussing's[240] flux equation:

$$\frac{J_{l \rightarrow b}}{J_{b \rightarrow l}} = \frac{[x]_l}{[x]_b} \exp - \frac{zFE}{RT} \qquad (9.18)$$

where $J_{l \rightarrow b}$ is the ion flux from lumen to blood, $J_{b \rightarrow l}$ is the ion flux from blood to lumen, and the other terms have already been defined. Note that the value of transepithelial electrical potential gradient (E) used in this equation is calculated from the Nernst equation. Also, in this formulation the effect of solvent drag on ion movement is assumed to be small and hence is not included.

To evaluate Equation 9.18, and hence to determine the mechanism of ion transport, we need to measure simultaneously three quantities: (1) the bidirectional ion flux ratio, $J_{l \rightarrow b}/J_{b \rightarrow l}$; (2) the chemical concentration gradients for the ion across the cell membranes; and (3) the electrical potential gradient across the cell membranes. If the measured ion flux ratio is equal to the theoretical ratio calculated from Equation 9.18, the ion must be transported passively. If the measured ion flux ratio exceeds the theoretical ratio, the ion is transported actively. A measured flux ratio less than the calculated ratio indicates that some of the ions do not diffuse freely under the influence of the prevailing transmembrane electrochemical gradients. Such an ion movement, independent of the electrochemical gradients, is called exchange diffusion. It is believed to be due to the competition for the carrier by the same ion species present on both sides of the membrane. Hence, exchange diffusion does not contribute to the net ion transport, but results in increased turnover rate.

As mentioned above, the Nernst equation defines the dependency of the transmembrane electrical potential gradient on the transmembrane chemical concentration gradient of an ion, assuming that the membrane is permeable only to that ion. As such, the Nernst equation represents a special case of the more general Ussing's flux equation. It is derived by assuming that $J_{l \to b} = J_{b \to l}$, and then rewriting Equation 9.18 in terms of E in logarithmic form:

$$E = -\frac{RT}{zF} \ln \frac{[x]_l}{[x]_b} \qquad (9.19)$$

Replacing the natural logarithm ($\ln = \log_e$) by the logarithm to the base 10 ($\ln = 2.3 \log$) yields

$$E = -2.3 \frac{RT}{zF} \log \frac{[x]_l}{[x]_b} \qquad (9.20)$$

At an ambient temperature of 36° C and for a univalent ion, the term (2.3 RT/zF) has a value of 60 and is expressed in millivolts. This substitution simplifies Equation 9.20 to:

$$E = -60 \log \frac{[x]_l}{[x]_b} \qquad (9.21)$$

The fact that the net ion flux across the membrane depends on the prevailing electrochemical gradients (Ussing's equation) and that the generated electrical gradient is a function of the chemical concentration gradient (Nernst equation) has provided the basis for determining the relative permeability of the cell membrane to a given ion. The procedure involves measuring the bidirectional flux of that ion from the disappearance rate of the isotope of that ion added to the tubular fluid at zero transepithelial electrical potential gradient (PD) induced by short-circuit current. The ion flux is measured in the direction opposite to that of the net ion transport—that is, from the backflux—on the assumption that it is passive.

Recent modifications and improvements of electrical measurements in the nephron, including introduction of ion-sensitive electrodes filled with liquid ion exchangers, have made it possible to measure intracellular activities of potassium, sodium, chloride, and calcium ions, as well as the pH in different tubular segments along the nephron. Application of these and other recently developed methods—such as patch-clamp technique,[106] isolation of glomeruli[188] and nephron segments with different cell types[63]—and preparation of membrane vesicles from either luminal or basolateral cell side[113] for biochemical and transport studies not only have confirmed the existence of the various transport mechanisms described in Chapter 8, but also have extended our understanding of the underlying complex mechanisms involved in tubular processing of the ultrafiltrate along the different segments of the nephron.

Sequential Processing of the Filtrate Along the Nephron

Glomerular filtrate undergoes considerable modification of both volume and composition as it flows along the various nephron segments. However, depending on the type of solute present, the tubular processes involved are quite different in each nephron segment, in terms of both the mechanisms of transport and the quantity of the solute and fluid transported. Therefore, to facilitate presentation, we shall consider the tubular processing of the major constituents of the filtrate along three sequential nephron segments: (1) the proximal tubule, (2) the Henle's loop, and (3) the distal tubule and collecting duct. Where appropriate, any similarities and differences in the filtrate processing in these segments will be mentioned, and their known physiological significance in relation to body fluid homeostasis will be discussed. Furthermore, to emphasize the importance of correlation between structure and function, we begin each section with a detailed description of the morphology of that nephron segment.

The Proximal Tubule

Morphology

The proximal tubule is not an homogeneous segment. With respect to its epithelial organization, the proximal tubule exhibits an intrasegmental axial heterogeneity, as is found in several nephron segments (thin descending limb of a long loop, thick ascending limb, connecting tubule, and collecting duct). Such an intrasegmental axial heterogeneity is characterized by having the specific structural features best developed in the beginning portions and decreasing in complexity toward the end of the segment; but the basic epithelial organization is maintained. In parallel, other features may well increase in quantity and elaboration; sometimes those secondary features change rather abruptly and are then used as criteria for subdivisions.

Microanatomically, the proximal tubule is divided into an extensively convoluted part (pars convoluta) followed by a straight part (pars recta), which extends into the outer stripe of the medulla. The subdivision into three subsegments (P1, P2, and P3; other often used abbreviations are S1, S2, and S3) is based on ultrastructural criteria and reflects the intrasegmental axial heterogeneity (Figure 9.4). P1 transforms gradually into P2 within the second half of the convoluted part. The change from P2 and P3 occurs along the straight part; it is gradual in rabbit and human, but

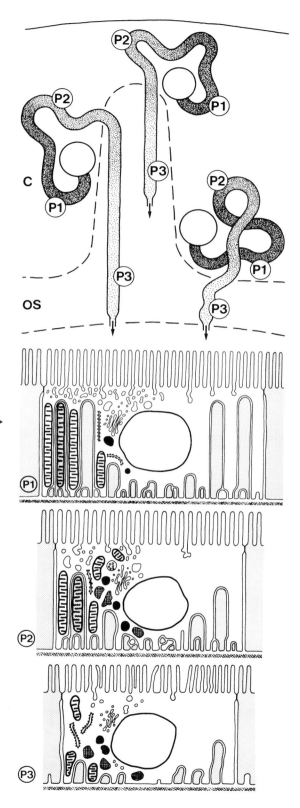

FIGURE 9.4. Schematics of the proximal tubule of three nephrons (*top panel*) showing the subdivision of the proximal tubule into three segments (P1, P2, and P3) and their cellular structures (*bottom panel*). In the top panel, the gradual transition of the three segments is shown by the intensity of the shading. As can be seen, the P1 and most of the P2 are located in the cortical labyrinth, and the rest of P2 and P3 are located in the medullary ray or in the outer stripe (OS). The schematics of the epithelia in the bottom panel illustrate the relationship of a middle cell (shown as white with a centrally located nucleus) with two parts of neighboring cells (*shaded*). A basement membrane underlies the epithelium. For clarity of presentation, the basic organization of the cells are shown on the right, whereas the cytoplasmic details are added on the left. In all three epithelia the cells interdigitate by large processes with each other. The shaded processes of the neighboring cells are found within the body of the middle cell. The degree of interdigitation decreases from P1 to P3. As shown on the left, the interdigitating processes are filled with mitochondria. The quantity of mitochondria correlates with the degree of interdigitation. Note the differences in cell size and the density of the brush border. The apical cytoplasm contains the vacuolar apparatus. Peroxisomes (*cross-hatched* profiles) are only found in considerable amounts in P2 and P3. (Modified from Kaissling and Kriz.[133])

is commonly described to be rather abrupt in rat and dog.[22, 133, 175]

The proximal tubule epithelium exhibits all characteristics typical for salt absorbing epithelia,[174, 235] and is most perfectly developed in its beginning part (P1). The tall epithelial cells extensively interdigitate with each other by large complexly shaped lateral cell processes that extend from the luminal surface to the base of the cell. Along their way, they split into secondary processes, which, near the base of the epithelium, fall apart into slender folds (Figures 9.5 and 9.6). All these various types of processes interdigitate with the same type of processes from the adjacent cell. In this manner, the surface area of the lateral cell membrane is extensively amplified; the amplification increases in the apicobasal direction.[248, 249] Apart from the slender basal folds, the lateral processes are occupied by large mitochondria with densely arranged cristae.

The amplified basolateral membrane contains sodium-potassium-adenosine triphosphatase (Na-K-ATPase), which is a key enzyme of the proximal tubule function. This enzyme, together with the associated mitochondria, represents the active machinery not only for sodium transport, but also for all contransport mechanisms that occur across the apical membrane. Although adenosine triphosphate (ATP) synthesis occurs in mitochondria and is tightly coupled to metabolic oxygen consumption, the substrate for this process is derived from fatty acids and not from sugars.[104] This is consistent with the findings that the proximal tubule has little or no enzymes for the glycolytic pathway but is rich with enzymes necessary for gluconeogenesis.[209] The luminal cell

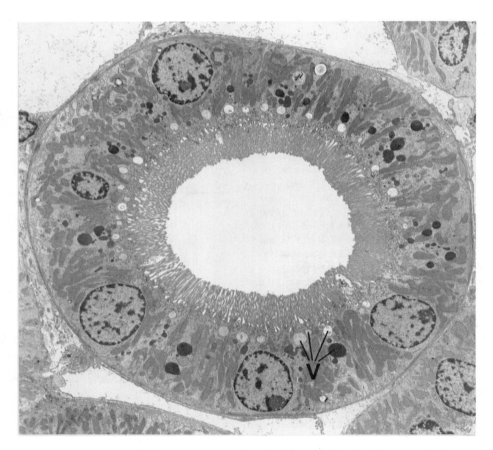

FIGURE 9.5. An electron micrograph of a cross-section of a rat proximal tubule (P1 segment) surrounded by capillaries. Note the dense and high brush border on the luminal side. The interdigitating cell processes can not be seen with this magnification. However, the dense stuffing with mitochondria is visible. Note the many vesicles and vacuoles (V) that comprise the vacuolar apparatus.

membrane bears a high and densely packed brush border (Figure 9.7); the individual microvilli are covered by a prominent cell coat. The luminal membrane contains numerous specific carrier systems that are responsible for the sodium-driven reabsorption of a variety of organic substances, such as sugars and amino acids.[141]

The paracellular pathway is fitted with "leaky" tight junctions that consist of only one junctional strand; even this single strand appears to be discontinuous at many sites.[203] The lateral intercellular spaces are of a constant narrow width. They are as complexly shaped as the lateral cell membranes. Consequently, they increase toward the base of the epithelium. In contrast to all other nephron segments, proximal tubular cells are electrically coupled by gap junctions.

Prominent constituents of the proximal tubule cytoplasm are the large mitochondria within the lateral cell processes (already mentioned above), the many peroxisomes whose specific function within the proximal tubule is poorly understood,[7] and the so called "vacuolar apparatus," which is a structural correlate for the endocytic absorption and digestion of various kinds of proteins (see Chapter 10). Other less specific constituents of the proximal tubule epithelium are the Golgi apparatus, considerable amounts of the rough and the smooth variety of the endoplasmic reticulum, and polysomes.

The subdivision of the proximal tubule into three subsegments must be, as already stated, regarded as an intrasegmental axial heterogeneity (Figure 9.4). Thus, the transition from P1 via P2 to P3 consists of gradual decease in structural complexity along with maintenance of the basic epithelial organization. The degree of basolateral interdigitation by large lateral processes is gradually decreased and finally greatly reduced. The associated mitochondria are diminished in parallel. In most species, including rabbit, dog, and man, the brush border decreases in height and density from P1 via P2 to P3. In rat, however, the brush border reaches its greatest height in the P3 segment. In all species so far investigated, peroxisomes are most numerous in P3. The tight junctions of P1 and P2 are shallow (one single junctional strand) but increase in depth and number of strands in P3 (in rat, dog, and cat, but not in rabbit). Considerable intrasegmental differences are also obvious in the elaboration of the "vacuolar apparatus," in addition to differences between the sexes[264] and to tremendous interspecies differences. It is not yet possible to correlate these structural differences with

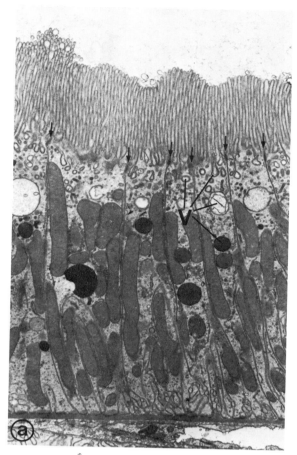

FIGURE 9.6. (a) An electron micrograph of a proximal tubular epithelium (P1 segment). The interdigitation of the epithelium is apparent by many longitudinally running profiles separated by tight junctions (*arrows*). Note the close association of the lateral cell membrane with the mitochondria. The different structures comprising the vacuolar apparatus (V) are prominently shown. (b) Three-dimensional model of a rabbit proximal tubule cell. (From Welling and Welling.[248])

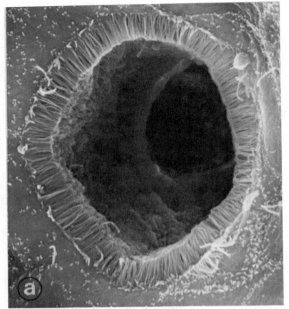

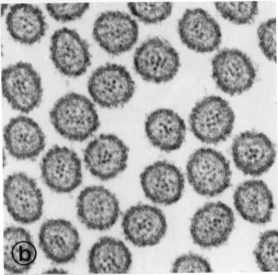

FIGURE 9.7. (a) A scanning electron micrograph of a urinary pole of renal corpuscle allowing a view into the proximal tubule. Note the abrupt beginning of the brush border. (b) A transmission electron micrograph of a cross-section through microvilli of brush border. The centrally located filaments that are part of the contractile apparatus are visible. The microvilli surface is covered by a cell coat.

differences in function, and therefore they are not described in detail in this book.

General Characteristics of Filtrate Transport

As described in Chapter 8, transepithelial transport in the proximal tubule involves two distinct routes arranged in parallel (Figure 8.1): transcellular and paracellular pathways. The transcellular pathway consists of the apical (luminal) cell membrane, the cell cytoplasm, and the peritubular (basolateral) cell membrane arranged in series. All active solute transport occurs via this pathway. The paracellular pathway consists of junctional complexes (tight or leaky) and the intercellular spaces arranged also in series. Although the paracellular pathway comprises less than 1% of the epithelial surface area,[13] it appears to be the major route for passive diffusion of ions and large polar nonelectrolytes. Estimates of the osmotic water flow permeability of the paracellular pathway indicate that this pathway accounts for only 5% to 10% of transepithelial transport and that the remaining 90% to 95% of water must be transported via the transcellular pathway. In the proximal tubule the paracellular pathways have high electric and hydraulic conductances, and lower reflection coefficients (or high permeability) for sodium and chloride than for bicarbonate. Consequently, a relatively large fraction of the iso-osmotic fluid reabsorption in the proximal tubule occurs via this pathway. The substantial fluid reabsorption by the paracellular pathway might explain the relatively high ratio of sodium reabsorbed per oxygen consumed (28 Na ions per one O_2) in the kidney, as compared with the expected 18 from the Na-K-ATPase stoichiometry.[136]

The apical membrane contains a number of specific carrier systems that are responsible for the cotransport of glucose and amino acids with sodium.[141, 239] As described in Chapter 8, the cotransport mechanism couples the downhill transport of sodium, along its electrochemical gradient, with the uphill active transport of glucose and amino acids into the cell across the luminal membrane. The resulting accumulation of glucose and amino acids inside the cell increases their respective concentrations so that they are transported by passive facilitated diffusion across the basolateral membrane. A similar mechanism governs phosphate transport in the proximal tubule.[112] The low sodium and high potassium activities inside the cell, which are required for the sodium-driven cotransport of the organic and nonorganic substances across the luminal membrane, are maintained by the activity of the Na-K-ATPase enzyme system situated in the basolateral membrane. A more detailed description of the sodium-dependent cotransport systems for renal transport of organic solutes and organic acids is presented in Chapter 10.

Reabsorptive mechanisms for sodium and its accompanying anions (bicarbonate and chloride) are different in the three segments of the proximal tubule. In the early proximal convolution, most of the sodium reabsorption is coupled to that of bicarbonate and only 10% to 20% of sodium reabsorption is coupled to that of chloride. The reabsorption of chloride occurs mainly by solvent drag and electrodiffusion driven by the lumen-negative transepithelial electrical potential difference (PD), generated by the electrogenic sodium-dependent cotransport of glucose and amino acids.[152] Bicarbonate reabsorption occurs partly as a result of its dissociation into CO_2 and H_2O secondary to H^+ secretion at the apical membrane by an electroneutral $Na^+ - H^+$ countertransport mechanism.[185] H^+ secretion accounts for the gradual decrease in pH in the proximal tubular fluid. In the late proximal convolution, as the bicarbonate, glucose, and amino acids concentrations fall and chloride concentration rises, the transepithelial PD becomes lumen-positive. This change in PD is generated by the bicarbonate and chloride concentration gradients and the greater permeability of the luminal membrane to chloride. Thus, sodium chloride is reabsorbed by both an active electroneutral and a passive process. Chloride is reabsorbed passively down its concentration gradient, and sodium is driven by the lumen-positive transepithelial PD. The available evidence to date indicates that approximately 60% of sodium chloride is reabsorbed by an active electroneutral process, via the transcellular pathway, whereas the remaining 40% is reabsorbed by a passive mechanism, presumably via the paracellular shunt pathway.[15] Currently the mechanisms of the electroneutral transport process and the chloride transport across the basolateral membrane are poorly understood. Finally, in the proximal straight segment, sodium chloride is reabsorbed by both active and passive components. The active component involves simple rheogenic (sodium transported down its electrochemical gradient, not coupled to any other substance, driven by the action of Na-K-ATPase in the basolateral membrane) process across the luminal membrane (transcellular pathway) with chloride driven through the paracellular pathway by the lumen-negative voltage. The passive component involves sodium chloride transport via the paracellular pathway.

Reabsorptive capacity is heterogeneous within the proximal tubule, and it appears to be correlated with the known morphological differences along this segment. However, these differences are more quantitative than qualitative in nature. Thus, the P1 segment of the proximal tubule has a relatively high Na-K-ATPase activity, allowing for an high active sodium transport and an efficient sodium-coupled net reabsorption of glucose, amino acids, and phosphate, and net secretion of H^+ ions. This results in iso-osmotic reabsorption of a large volume of filtrate in this segment, leading to a decrease in the luminal concentration of glucose, amino acids, phosphate, and bicarbonate, and a corresponding increase in chloride concentration.

In the P2 segment, the reabsorption of glucose, amino acids, and phosphate and H^+ secretion is somewhat reduced. However, because of high lumen-to-plasma chloride concentration gradient, there is enhanced passive reabsorption of sodium and chloride along with appreciable amounts of volume reabsorption in this segment. In the P3 segment, the Na-K-ATPase activity is low, which accounts for relatively less active sodium transport and volume reabsorption. The main transport function of this segment is that of organic acids and bases, such as PAH.

To sum up, numerous in-vivo micropuncture and in-vitro microperfusion studies of the proximal tubule in the rat, dog, and monkey, under a variety of experimental conditions, have established that as the filtrate flows through this segment: (1) Water is always reabsorbed in direct proportion to the sum of the osmolar solutes, so that the filtrate-to-plasma osmolality ratio ($[TF/P]_{Osm}$) does not change regardless of whether the final urine is more or less concentrated. This phenomenon, shown in Figure 9.8, indicates that the filtrate is reabsorbed iso-osmotically along this segment, meaning that the tubular fluid and reabsorbate have the same osmotic pressure. (2) In the absence of significant amounts of nonreabsorbable solute (such as mannitol):

(a) Sodium is reabsorbed in direct proportion to water, so that its tubular fluid-to-plasma molal concentration ratio does not change. This is illustrated in Figure 9.9, where TF/P ratio for sodium and sodium-to-inulin TF/P ratios are plotted, on a logarithmic scale, as a function of the tubular length.

(b) The osmolar sum of nonsodium solutes is also reabsorbed in direct proportion to water, although individual solutes may be reabsorbed in greater (glu-

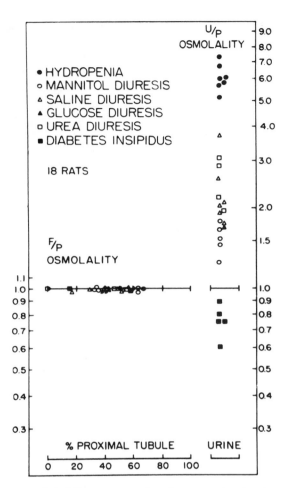

FIGURE 9.8. Micropuncture data from rats showing iso-osmotic reabsorption of fluid along the proximal convoluted tubule. Note that proximal tubular fluid osmolality relative to plasma water (F/P) remains constant at unity whether the final urine is more concentrated than plasma (hydropenia) or less concentrated than plasma (diuresis). (From Gottschalk.[92])

cose, for example), or lesser (urea, for example) amounts than water.

Figure 9.10 summarizes the changes in the composition of the luminal fluid for several substances, expressed as tubular fluid-to-plasma (TF/P) concentration ratios (upper panel) and the corresponding changes in the transepithelial PD (lower panel) as a function of length of the proximal tubule. Note that TF/P for inulin, a substance that is filtered but neither reabsorbed nor secreted, increases to a value of 2, suggesting that about 50% of the filtered water is reabsorbed along the proximal tubule. TF/P for both glucose and amino acids fall to a value of about 0.1,

indicating that about 90% of the filtered glucose and amino acids are reabsorbed in the first 25% of the length of the proximal tubule. Since bicarbonate reabsorption is linked to Na^+/H^+ antiporter at the luminal membrane, the continuous fall in the TF/P ratio for bicarbonate indicates that acidification takes place along the entire length of the proximal tubule, lowering the luminal bicarbonate to about 5 mEq/L or less. In the early portion of the proximal convolution, where transepithelial PD is lumen negative, a small amount of chloride is reabsorbed. However, preferential reabsorption of bicarbonate along with iso-osmotic reabsorption of water in this segment leads to an increase in luminal chloride concentration and the change in transepithelial PD from lumen-negative to lumen-positive voltage. This lumen-positive voltage is generated by the prevailing (reciprocal) bicarbonate and chloride concentration gradient and the higher permeability of the tubule to chloride.

Since sodium is the most abundant constituent of the filtrate, its tubular reabsorption is the key to the overall filtrate processing. Thus, any factor that interferes with sodium reabsorption has an adverse effect on filtrate reabsorption. Indeed, as we shall see later, interference with sodium reabsorption is the most common basis for inducing diuresis. Hence, it is important to have a clear understanding of the mechanism of sodium reabsorption.

In the following sections, we shall detail the salient features of specific mechanisms involved in sodium, potassium, bicarbonate, and chloride reabsorption in the proximal tubule. The sodium-dependent cotransport of organic solutes is described in Chapter 10.

Mechanisms of Sodium Reabsorption

Recent techniques for isolation of cell membranes, coupled with electrophsiological methods, have provided powerful tools for characterizing the mechanisms of ion transport across both the luminal and basolateral membranes in various segments of the mammalian nephron. Application of these methods have made it possible to study directly the transport properties of individual membranes involved in the transcellular transport in electrically "leaky" epithelia, such as the proximal tubule, without the complicating effects of the paracellular shunt pathways.[113] These studies have revealed that all forms of coupled sodium chloride transport are driven by the sodium gradient across the luminal membrane and, therefore, are examples of secondary active transport.

Sodium enters the proximal cell via the luminal

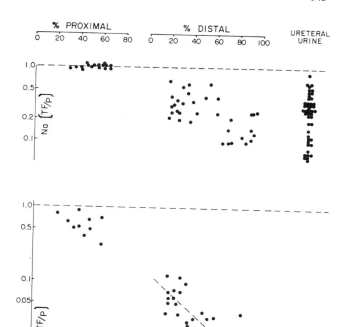

FIGURE 9.9. Tubular fluid-to-plasma concentration ratios (TF/P) for sodium (*upper panel*) and sodium-to-inulin (Na/In) (*lower panel*), plotted on logarithmic scale, as a function of proximal and distal tubule lengths and in ureteral urine from rats under control conditions. Note that in accordance with recent findings, the term distal tubule used in this figure denotes a heterogeneous segment consisting of the distal convoluted tubule, the connecting tubule, and part of the cortical collecting duct. (From Malnic, et al.[170])

membrane by at least four mechanisms[247]: (1) Simple rheogenic process that involves passive entry of sodium across the luminal membrane down its electrochemical gradient, and not coupled to any other solute. The electrochemical gradient for luminal sodium entry is maintained by the action of the Na-K-ATPase. (2) Directly coupled electroneutral Na^+ and Cl^- transport. This form of sodium chloride transport has been demonstrated in the Necturus proximal tubule, but not in the mammalian proximal tubule.[197] (3) Indirectly coupled Na^+ and HCO_3^- transport via the operation of Na^+/H^+ antiporter (exchange process) across the luminal membrane. Murer and associates[185] were the first to demonstrate the coupling between hydrogen secretion and sodium reabsorption across the rat renal brush-border membrane vesicles. The rate of Na^+/H^+ exchange has been shown to depend on the transluminal and transepithelial pH gradients. And (4) sodium-dependent cotransport of organic solutes. This mechanism will be detailed in Chapter 10.

Taken together, the results of these as well as numerous in-vivo micropuncture and in-vitro microperfusion of isolated tubular segments, and electrophysiological studies in the mammalian nephron[80, 90, 119, 120, 156, 214] have revealed the following patterns of electrochemical potential gradients for sodium transport across the proximal tubule epithelial cell and its membrane components:

1. In the early proximal convolution of both superficial and juxtamedullary nephrons the *transepithelial* PD is lumen negative and is about 4 mV. The lumen-negative voltage is due to electrogenic Na^+-coupled cotransport of glucose and amino acids across the luminal membrane. However, in the late proximal convolution and pars recta of the superficial nephrons, owing to the higher chloride/bicarbonate permeability ratio, the lumen-negative voltage changes to a lumen-positive voltage, secondary to the chloride diffusion potential. In contrast, in the juxtamedullary nephrons, owing to a bicarbonate/

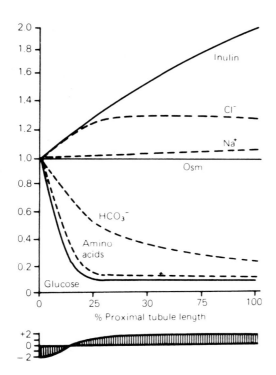

FIGURE 9.10. Changes in the composition of the luminal fluid (*upper panel*) and transepithelial PD (*lower panel*) as a function of the length of the proximal tubule. (From Rector.[197])

chloride permeability ratio of unity, the lumen-negative voltage, though somewhat reduced, is maintained throughout the proximal tubule.

2. The cell interior is electronegative (about 70 mV) with respect to both the tubular and peritubular fluids. However, the *transluminal* PD is somewhat smaller (by about 4 mV, cell interior negative) than the *tranbasolateral* PD.

3. The intracellular sodium concentration, $[Na]_c$, is significantly lower (about 40 mEq/L) than that in the tubular, $[Na]_f$, and peritubular fluids, $[Na]_p$, (both concentrations being about 145 mEq/L).

Synthesis of the above electrochemical potential gradients yields the following patterns for sodium transport across the two membrane components of the proximal cell. At the luminal membrane there is an electrical potential gradient of some 60 mV (cell interior negative) that favors sodium (a cation) entry into the cell. This electrical force is further complemented by a concomitant concentration gradient for sodium from lumen to cell interior ($\Delta[Na] = 105$ mEq/L). Consequently, sodium enters the cell *down* its electrochemical potential gradients by *passive* diffusion. However, this rheogenic process accounts for only part of the transluminal sodium transport. The remaining sodium enters the cell by the other two mechanisms outlined above. On the other hand, since the cell interior has a lower sodium concentration than the peritubular fluid, and is also electronegative, it follows that sodium must leave the cell across the basolateral membrane *against* its electrochemical gradient by *active* transport. This conclusion is confirmed by the finding that the basolateral membrane contains the sodium-potassium stimulated adenosine triphosphatase (ATPase) enzyme system,[136, 138] a requirement for active transport. Finally, it has been found that transepithelial reabsorption of sodium is a function of the magnitude of the intracellular sodium pool. The size of this pool is determined by the rates at which sodium enters the proximal cell by passive or facilitated diffusion and leaves the cell by active transport. Of these two processes, the active transport component at the peritubular membrane appears to be the rate-limiting process.[32]

Figure 9.11 summarizes the various components of the electrochemical potential gradients involved in sodium transport across the mammalian proximal tubule cell. As shown, sodium enters the cell by passive (or facilitated) diffusion (depicted by broken arrow) along its electrochemical gradients. Once inside, sodium is then actively extruded against its electrochemical gradient (depicted by solid arrow) across the peritubular (basolateral) membrane into the surrounding interstitial fluid. From there, sodium, along with an anion (such as bicarbonate or chloride) and water, passes through the basement and capillary membranes as a result of the prevailing regional hydrostatic and protein oncotic pressure gradients.[163] In short, as a result of both passive and active transport processes and peritubular hydrostatic and oncotic pressure gradients, sodium ions are reabsorbed *iso-osmotically* along the proximal tubule. A more complete description of the mechanism of this iso-osmotic transepithelial sodium transport is given later in this chapter.

Besides partitioning the ion transport process into active and passive components, measurement of short-circuit current in single nephrons has provided the basis for correlating the proximal tubule sodium reabsorption with renal oxygen consumption. These measurements have revealed a stoichiometric relationship between sodium reabsorption and renal oxygen consumption, such that about 28 sodium ions are reabsorbed per molecule of oxygen.[136] Further-

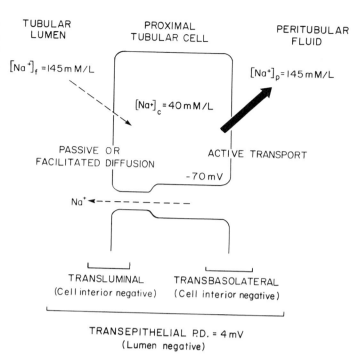

FIGURE 9.11. The electrical potential difference (P.D.) across a mammalian early proximal tubule cell and the localization of sites of passive and active sodium transport. See text for details.

more, it has been established that both renal oxygen consumption and proximal tubule sodium reabsorption are positive functions of glomerular filtration rate.

Mechanism of Potassium Reabsorption

In the 1960s, numerous in-vivo micropuncture studies in rat delineated the magnitude and direction of potassium transport along the nephron.[170, 171] These studies showed that most of the filtered potassium is reabsorbed in the proximal tubule.

Figure 9.12 summarizes the axial profile of potassium transport along the proximal convoluted tubules of superficial nephrons. The upper left panel of this figure shows a plot, on logarithmic scale, of the tubular fluid-to-plasma concentration ratios (TF/P) for potassium as a function of tubular length in rat proximal tubule under control conditions. Note that potassium TF/P ratios fall consistently below unity, suggesting net transepithelial potassium reabsorption.[171] To eliminate the effect of water reabsorption on potassium TF/P ratios and to determine the relative amount of potassium remaining in the tubule and hence not reabsorbed, the potassium TF/P ratios are divided by inulin TF/P ratios, yielding clearance ratios in the corresponding micropuncture samples.

The resulting potassium-to-inulin (K/In) TF/P ratios plotted, on logarithmic scale, as a function of tubular length are shown in the lower left panel of Figure 9.12. It is evident that by the time the filtrate reaches 65% of the proximal tubular length, about 70% of the filtered potassium has been reabsorbed (K/In TF/P = 0.3).

Subsequent studies, using a variety of experimental maneuvers, have revealed the following important characteristics of potassium transport in the proximal tubule[17, 72, 139, 236]:

1. Potassium reabsorption in the proximal tubule is dependent on fluid reabsorption. Thus, an increase in fluid reabsorption, and hence a decrease in tubular flow rate, enhances potassium reabsorption, whereas a decrease in fluid reabsorption, and hence an increase in tubular flow rate, reduces potassium reabsorption. This coupling of potassium reabsorption with fluid reabsorption suggests solvent-drag as a possible mechanism for potassium reabsorption. The paracellular pathway would be a likely route for this mode of transport.

2. As mentioned earlier and depicted in Figure 9.11, in the early proximal tubule the transepithelial PD is lumen-negative. This finding, coupled with the fact that the tubular concentration of potassium

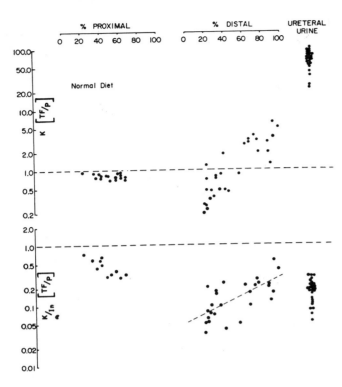

FIGURE 9.12. Tubular fluid-to-plasma concentration ratios (TF/P) for potassium (*upper panel*) and potassium-to-inulin (K/In) (*lower panel*), plotted on logarithmic scale, as a function of proximal and distal tubule lengths and in ureteral urine from rats under control conditions. Note that in accordance with recent findings, the term distal tubule used in this figure denotes a heterogeneous segment consisting of the distal convoluted tubule, the connecting tubule, and part of the cortical collecting duct. (From Malnic, et al.[171])

($[K]_f = 4$ mEq/L) is significantly lower than that in the cell interior ($[K]_c = 150$ mEq/L), suggests the possibilty of an active component for potassium reabsorption and suggests that the luminal membrane may be the site of active transport.

3. Cellular concentration of potassium, $[K]_c$, is determined by a balance between active potassium uptake from the peritubular interstitium by the ATP-driven Na-K-ATPase enzyme at the basolateral membrane and passive leakage from the cytosol to the interstitium via the potassium conductance pathway, which provides a route for passive diffusion of potassium driven by the electrical gradient (*electrodiffusion*). Thus, spontaneous changes in Na-K-ATPase activity are compensated for by directional changes in potassium electrodiffusion, thereby minimizing fluctuations in $[K]_c$ and cell volume.

Taken together, the above findings suggest three possible mechanisms for potassium transport across the proximal tubule. First, in the early segment, where there is lumen-negative voltage, potassium is reabsorbed actively across the luminal membrane. Second, in the middle and late portions of the proximal tubule, where transition from lumen-negative to lumen-positive voltage takes place, the resulting transepithelial voltage provides a favorable driving force for net passive potassium reabsorption via the paracellular pathway. Finally, presence of potassium conductance pathways in both the luminal and basolateral membranes provide routes for leakage of potassium out of the cell by electrodiffusion. In short, the net reabsorptive flux of potassium at any point along the proximal tubule appears to be determined by the dynamic interaction of these three mechanisms. The adaptive response of these mechanisms to various physiological and pathological perturbations is described later in this chapter and in Chapter 11.

Mechanism of Bicarbonate Reabsorption

Reabsorption of filtered bicarbonate in the proximal tubule is coupled to $Na^+ - H^+$ exchange across the luminal membrane, a process controlled by the enzyme carbonic ahnydrase. The H^+ ion secreted into the lumen combines with filtered bicarbonate (HCO_3^-) to form carbonic acid (H_2CO_3). The latter is readily dissociated into carbon dioxide and water. Carbon dioxide then diffuses into the proximal cell where it combines with OH^- to form HCO_3^-. The HCO_3^- ions thus formed are transported across the basolateral membrane along a voltage-dependent conductive pathway for bicarbonate. Stated another way, the bicarbonate generated inside the cell exits across the basolateral membrane along its electrochemical gradient. The rate of H^+ secretion is controlled in part by the acid-base composition (pH) of the luminal and peritubular fluids.

Histochemical and cell fractionation studies have shown that the enzyme carbonic anhydrase (CA) is present in the luminal brush border and basolateral membrane as well as in the cytoplasm.[47] The function of the luminal CA is to catalyze the breakdown of carbonic acid (H_2CO_3) formed in the lumen. The function of the cytosolic CA is to maintain an adequate supply of H^+ for secretion across the luminal membrane. The function of basolateral CA is not known.

Two models have been proposed to account for secretion of hydrogen ions across the luminal membrane. The first model postulates an *electroneutral sodium-hydrogen antiporter* across the luminal membrane. This process is driven by the sodium concentration gradient across the luminal membrane maintained by the activity of Na-K-ATPase in the basolateral membrane.[185] The second model postulates an *electrogenic sodium-independent* active hydrogen ion secretion in the luminal membrane. Most of the available evidence support the first model, namely, the Na^+/H^+ antiporter as the primary mechanism for acidification in the proximal tubule.[197]

Evidence from in-vivo micropuncture[79] and renal microvillus membrane vesicles[145] indicate that H^+ ion is actively secreted by the luminal cell membrane and that it is coupled to passive sodium reabsorption at the same site by an electroneutral countertransport mechanism. More recent studies have shown that the Na^+-H^+ exchange mechanism may be modified by a number of factors.[1, 118, 144] Thus, amiloride inhibits Na^+-H^+ countertransport, whereas dexamethasone (an anti-inflammatory adrenocortical glucogenic steroid), metabolic acidosis, and hypercapnia stimulate H^+-Na^+ exchange. A change in intracellular H^+ ion concentration has recently been suggested as the common factor modifying Na^+-H^+ exchange.[2]

It should be apparent that bicarbonate reabsorption via the Na^+-H^+ exchange process raises the hydrogen ion concentration (low pH) and acidifies the proximal tubular fluid. This role of Na^+-H^+ countertransport in bicarbonate reabsorption and acid-base regulation is presented in Chapter 11.

Mechanism of Chloride Reabsorption

The conventional concept of transepithelial chloride transport consists of a passive movement of Cl^- ions

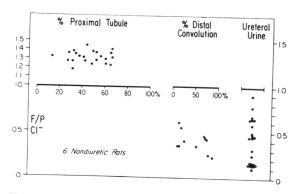

FIGURE 9.13. Tubular fluid-to-plasma concentration ratios (F/P) for chloride along the proximal and distal convoluted tubules and in ureteral urine from nondiuretic rats. (From Gottschalk.[91])

via the paracellular shunt pathway accompanied by rheogenic Na^+ transport across the transcellular pathway. However, available evidence suggests two possible mechanisms for chloride entry into the proximal tubule: (1) Rheogenic Cl^- reabsorption, and (2) Cl^- reabsorption coupled to H^+ gradient. The following is a brief description of each of these two mechanisms.

Rheogenic Chloride Reabsorption. The existence of sodium-dependent organic solute cotransport in the proximal tubule[239] has raised the interesting possibility that a rheogenic (via an electrodiffusion conductive pathway) route for anion transport may also exist in this nephron segment. The transcellular anionic current flow necessary for such a transport requires the existence of anion conductive pathways in both the luminal and basolateral membranes. Evidence exists that HCO_3^- is transported across the basolateral membrane by an electrodiffusive mechanism.[16] A possible Cl^-/HCO_3^- exchange mechanism across the basolateral membrane could provide the necessary transcellular anionic current flow, the chloride transport, and the recycling of HCO_3^- across the basolateral membrane. Such a Cl^-/HCO_3^- exchanger has been found in Necturus proximal tubule.[247]

Chloride Reabsorption Coupled to Hydrogen Ion Gradient. Although 40% to 50% of the filtered chloride is reabsorbed along the proximal convoluted tubule in nondiuretic rats, the tubular fluid concentration is found to be considerably higher than that in the plasma (Figure 9.13). This higher tubular fluid chloride concentration is not due to tubular secretion, but rather is postulated to be the result of early and preferential reabsorption of bicarbonate in the P1 segment with concomitant fall in the intratubular pH.[151] This leads to a transition of lumen-negative to a lumen-positive voltage, which along with the prevailing chloride concentration gradient favors electroneutral chloride transport presumably via the transcellular pathway. Additionally, this preferential bicarbonate reabsorption, as suggested by Windhager,[254] may provide a favorable chloride concentration gradient, thereby making it possible for sodium and water to be reabsorbed mainly by a passive process in the late proximal tubule.

The chloride concentration in the proximal tubule may be lowered in the presence of poorly reabsorbable anions, such as sulfate,[169] and after infusion of carbonic anhydrase inhibitors and mannitol.[30] Furthermore, hypokalemia (as might be developed in potassium deprivation or depletion) elevates tubular chloride concentration while increasing bicarbonate reabsorption.[6]

As described above, a large fraction of renal bicarbonte reabsorption is linked to the $Na^+ - H^+$ exchange process. A recent study suggests that chloride reabsorption may also be linked to Na^+ transport.[215] They have proposed a double ion ($Na^+ - H^+$ and $Na^+ - Cl^-$) exchange mechanism to account for chloride reabsorption across the luminal membrane. According to this model, active Na^+ reabsorption across the basolateral membrane generates an inwardly directed Na^+ gradient across the luminal membrane. Secondary active H^+ secretion via $Na^+ - H^+$ exchange mechanism leads to an increase in OH^- activity inside the cell as a result of dissociation of water. The latter is generated from the CA-catalyzed breakdown of intracellular H_2CO_3 into carbon dioxide and water. Intracellular carbon dioxide combines with OH^- to form HCO_3^-, favoring its outward diffusion into the lumen and Cl^- into the cell across the luminal membrane. In short, primary active Na^+ transport across the basolateral membrane leads to secondary $Na^+ - H^+$ exchange and tertiary active Cl^- reabsorption via Cl^-/HCO_3^- exchange across the luminal membrane. This mechanism may also explain the inverse relationship between bicarbonate and chloride reabsorption in the kidney. The role of chloride reabsorption in acid-base regulation is described in Chapter 11.

Mechanism of Coupled Sodium and Water Transport

Analysis of tubular fluid-to-plasma concentration ratios for inulin ($[TF/P]_{In}$) along the proximal tubule

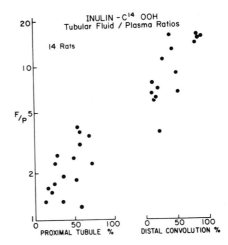

FIGURE 9.14. Tubular fluid-to-plasma concentration ratios (F/P) for inulin along the proximal and distal convoluted tubules of nondiuretic rats. (From Gottschalk.[92])

of nondiuretic rats has revealed a progressive increase in this ratio, reaching a value of about 3 by the end of the convoluted segment (Figure 9.14). Since inulin is a nonreabsorbable solute, the rise in its tubular concentration relative to that in plasma ($[TF/P]_{In} > 1$) is taken as evidence of water reabsorption along the tubule. Accordingly, it has been found that in rats, about 60% of the filtered water is reabsorbed by the time the filtrate reaches the end of the proximal convoluted tubule,[81] whereas in dogs and monkeys only 45% and 30% is reabsorbed, respectively.[10, 11] The percentage of the filtrate volume reabsorbed along the proximal tubule has been found to be relatively constant, under normal physiological conditions. Consequently, fluid reabsorption along this segment is called *obligatory*.

Because the filtrate (solutes and water) is reabsorbed iso-osmotically along the proximal tubule, as is evident from $[TF/P]_{Osm} = 1.0$ in Figure 9.8, then one would expect a large osmotic force (estimated as 20 mOsm/L) to exist between the tubular lumen and the peritubular capillary blood to account for the observed reabsorption of such a large fraction of the filtered water. However, osmotic force of such a magnitude has never been found experimentally.[75]

To resolve this discrepancy between the expected and actually observed osmotic driving force, Giebisch[76] proposed a model of the renal proximal tubule cell to account for the observed reabsorption of the large quantity of sodium and water along the proximal tubule. Actually, this model is based on a scheme proposed by Curran and MacIntosh[33] for isotonic transport across the epithelial membrane. Diamond and Bossert[43] provided both functional and anatomical evidence in support of this scheme for isotonic reabsorption across the mammalian gallbladder wall.

Figure 9.15 presents schematically the salient features of Giebisch's model. Each renal tubule cell is depicted with two distinct membranes: a *luminal* (apical) and a *peritubular* (basolateral) membrane. The adjacent cells are connected by the *tight junction* near the apical surface. Between the lateral cell membranes an *intercellular space* is present (see Chapter 8, Figure 8.1) that constitutes an extracellular compartment within the renal tubular epithelium. As envisioned by this model, this extracellular space is believed to play a key role in the transepithelial reabsorption of sodium and water. The essence of this transepithelial transport may be summarized as follows.

Luminal sodium enters the renal cell along its electrochemical gradient. Once inside, sodium ions are then actively transported into the intercellular space across luminal membrane. This leads to accumulation of sodium in the intercellular space, thereby generating a local region of hyperosmolality within the tubular epithelium. The osmotic concentration gradient thus established induces osmotic flow of water into the intercellular space either through the tight junction or across the cell compartment, or both. The accumulated iso-osmotic fluid, owing to the limited expandability of the intercellular space, develops a small but transient hydrostatic pressure gradient along this space, which forces the fluid to move from the apical to the basal side of the intercellular space. This reabsorbate will then cross the basement membrane and capillary endothelium by a force determined by the balance of the prevailing regional hydrostatic and protein oncotic pressure gradients (Starling forces).

Giebisch[76] presented three arguments in support of the important role played by the intercellular space in the proposed coupled transport of sodium and water:

1. Electrophysiological studies in rat proximal tubule have shown that the specific electrical resistance

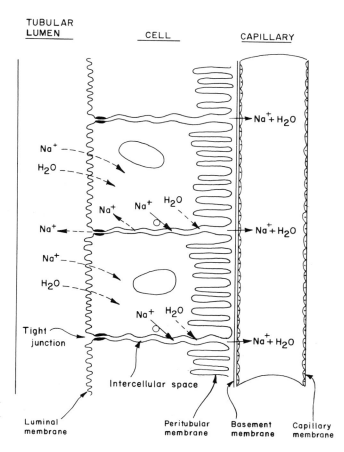

FIGURE 9.15. Renal tubule cell showing transepithelial movement of sodium and water. (Modified from Giebisch and Windhager.[80])

of this tubule is quite low compared with that measured in other epithelial tissues as well as the more distally located nephron segments. This low electrical resistance may be due to one of two properties of the tubular epithelium: (a) the tubular cell membranes have an inherently low electrical resistance; and (b) there are extracellular channels within the tight junction (*paracellular shunt pathway*) connecting the tubular lumen to the peritubular fluid. Recent experimental evidence indicates that the "leaky" (low electrical resistance) nature of the proximal tubule is due to the latter possibility.[197]

2. Measurement of ion permeability of the luminal and basolateral membranes of the proximal tubule cell has revealed marked ion selectivity for these membranes.[255] For example, it has been found that the peritubular membrane is some 25 times more permeable to potassium than to sodium ions, and some 2.5 times more permeable to potassium than to chloride ions. In contrast, when the permeability properties of the whole proximal tubule cell are measured, such an ion selectivity is not observed. This marked difference between the ion permeability of the proximal tubule cell as a whole and of its individual membranes strongly suggests that other structures besides the cell membranes, namely the paracellular shunt pathways, must account for the ease of sodium and water transport across the proximal epithelium.

3. The above proposed intercellular shunt model can readily explain a number of observations in which experimentally induced changes in the peritubular capillary hemodynamics (namely, changes in peritubular capillary hydrostatic and protein oncotic pressures) have been shown to profoundly influence the net reabsorption of sodium and water in the proximal tubule.[14, 52, 225, 256] A more detailed description of these experiments and their physiological importance is given in the next section.

Mechanism of Sodium-Potassium Exchange at the Basolateral Membrane

It is generally agreed that active sodium extrusion from the proximal tubule cell into the peritubular fluid (intercellular space) is partly coupled to potassium uptake into the tubular cell from the peritubular fluid and that this process is mediated by a sodium-potassium-stimulated ATPase enzyme system. The stoichiometry of the coupled sodium-potassium transport is such that 3 sodium ions leave the cell in exchange for 2 potassium ions that enter the cell. Furthermore, this enzymatic action requires the presence of ATP, sodium, and potassium as transport substrates. Of these, the intracellular sodium concentration appears to be the major determinant of the Na-K-ATPase activity under normal physiological conditions.[136]

Evidence indicates that the Na-K-ATPase behaves as an *adaptive* enzyme whose activity is altered in response to changes in sodium and potassium transport under a variety of physiologic and experimental conditions. Thus, any factor that increases the filtered load of sodium, such as a chronic increase in glomerular filtration rate, enhances the net transepithelial sodium reabsorption and causes a parallel increase in Na-K-ATPase activity. This is followed by the *de novo* synthesis of the enzyme and the amplification of the basolateral membrane.[136]

The activity of the Na-K-ATPase is also influenced by the action of several hormones, including aldosterone, thyroid hormone, insulin, and catecholamines. *Aldosterone* influences reabsorption of only 1% to 2% of the filtered sodium in the collecting duct; it has no effect on sodium reabsorption in the proximal tubule. In the collecting duct the effect of aldosterone on Na-K-ATPase activity depends on the duration of hormone treatment. Short-term (*acute*) administration of aldosterone stimulates Na-K-ATPase secondary to its enhancement of passive sodium influx into the cell across the luminal membrane. However, long-term (*chronic*) administration of aldosterone appears to directly stimulate Na-K-ATPase, followed by *de novo* synthesis of the enzyme and basolateral amplification. The net effect is an increase in transepithelial sodium transport. Regardless of the duration of treatment, current evidence indicate that the mechanism of the cellular action of aldosterone involves intracellular formation of a hormone-receptor complex which initiates DNA-dependent RNA synthesis of a specific protein which enhances transepithelial sodium transport (see Figure 9.40 for further details). Analogous to the mechanisms described for the cellular action of aldosterone, *thyroid hormone* enhances sodium reabsorption in the proximal tubule via stimulation of Na-K-ATPase and/or an increase in sodium permeability of the luminal membrane. Clearance studies in dogs and humans have shown that *insulin* decreases renal excretion of sodium and phosphate.[38, 39] Micropuncture studies have localized the proximal tubule as the site of action of this hormone.[39] However, the exact cellular mechanism whereby insulin increases sodium reabsorption in the proximal tubule is not known.

Finally, *catecholamines* stimulate Na-K-ATPase activity by blocking the inhibitory effect of *vanadate*, a naturally occuring endogenous substance that reversibly binds to a phosphate site on the enzyme.[136] Unlike ouabain, which acts from the basolateral side, vanadate acts from the cytoplasmic side, and its inhbition is potentiated by extracellular potassium. Recent studies provide strong circumstantial evidence in favor of vanadate as a potential endogenous regulator of the Na-K-ATPase activity.

Regulation of Filtrate Processing Along the Proximal Tubule

Renal regulation of volume and osmolality of body fluids ultimately depends on the capacity of the tubular epithelium to adjust its rate of sodium reabsorption in response to changes in glomerular load. It is well established that changes in glomerular filtration rate (GFR) are usually compensated for by parallel changes in tubular reabsorption along the nephron. More explicitly, it has been found that in response to normal changes in GFR, the fraction of filtered sodium reabsorbed by the tubules and that which is not reabsorbed (and therefore excreted) remain constant. This implies that the absolute amounts of both sodium reabsorbed and sodium not reabsorbed vary in the same direction as the changes in GFR. This parallel relationship between GFR and tubular reabsorption is commonly referred to as the *glomerular-tubular balance* (see Chapter 6). Since the proximal tubule accounts for the reabsorption of nearly two thirds of the filtered sodium, this nephron segment must play a major role in the maintenance of such a balance, and it is, therefore, the most prominent regulator of body sodium. Furthermore, inasmuch as renal transport of sodium appears to be a saturable process (see below), the existence of a glomerular-tubular balance for sodium serves to conserve body stores of sodium and to minimize excessive loss in conditions that produce sodium imbalance.

Numerous studies, reviewed by Earley and Schrier[53] and De Wardener,[41] have demonstrated that the glomerular-tubular balance is influenced by several factors. Chief among them are (a) changes in plasma concentration of sodium, (b) presence of poorly reabsorbable solutes, such as mannitol, urea, and sulfate, in the filtrate, (c) changes in hemodynamics of systemic circulation following expansion of intravascular or extracellular volume by fluid infusion, and (d) alterations in Starling forces in the blood perfusing the kidney, particularly at the postglomerular capillary vasculature. Since the first two exert their effects primarily at the luminal side and the last two at the peritubular side of the renal cell membrane, we shall describe the modes of their actions in accordance with whether they exert their effects on the luminal or peritubular side of the renal cell membrane.

Luminal Control of Filtrate Reabsorption

Proximal reabsorption of sodium and water may be modified by altering the luminal concentration of sodium. This may be done either directly, by changing plasma concentration of sodium, or indirectly, by intravenous infusion of poorly reabsorbable solutes. Let us examine the effects of each of these forcings on filtrate reabsorption in the proximal tubule.

Effect of Changes in Plasma Sodium Concentration. It is well established that changes in the filtered load of sodium result in parallel changes in sodium reabsorption, both at the level of the whole kidney and the single nephron. The filtered load of sodium is the product of the GFR and plasma sodium concentration, $[Na]_p$. Therefore, it is of interest to know whether changes in tubular sodium reabsorption occur in response to changes in $[Na]_p$ or to changes in GFR. This question has been addressed in anesthetized dogs in which $[Na]_p$ was varied, by changing sodium concentration of saline administered into one renal artery, while allowing GFR to vary spontaneously.[127, 187] The results showed that the fraction of filtered sodium reabsorbed remained constant when the filtered load was varied by changing the GFR. However, when the filtered load was varied by changing $[Na]_p$ at constant GFR, the resulting hypernatremia depressed tubular sodium reabsorption and disrupted the level of sodium reabsorption at which glomerular-tubular balance occurred. These studies suggest that at the whole kidney level, changes in tubular sodium reabsorption occur in response to spontaneous changes in GFR and not to acute changes in the filtered load of sodium. It should be pointed out that acute changes in plasma sodium concentration, sufficient to raise $[Na]_p$ to between 140 and 180 mEq/L, have been found to produce vasoconstriction, which in turn reduced GFR.[187] Although the increase in $[Na]_p$ is counterbalanced by the decrease in GFR, resulting in a relatively constant filtered load of sodium, the resulting hypernatremia increased sodium excretion rate and reduced the absolute amount of sodium reabsorbed.

A recent study has identified the various factors that modulate variations in tubular sodium reabsorption in response to changes in [Na]$_p$ produced by saline infusion.[89] These include a reduction in plasma protein concentration, hematocrit, filtration fraction, and plasma pH. However, when the combined effects of these variables were taken into account, the increase in [Na]$_p$ still resulted in a decrease in tubular sodium reabsorption.

Micropuncture studies have confirmed the inhibitory effect of the increased [Na]$_p$ on proximal sodium reabsorption and have further shown that the luminal transport of sodium is a saturable function of luminal sodium concentration.[105, 224] Furthermore, these studies have demonstrated that at the single nephron level, hypernatremia also disrupts the glomerular-tubular balance.

Taken together, these studies indicate that at the whole kidney and single-nephron levels, the adaptive changes in tubular reabsorption of sodium are correlated to spontaneous variations in GFR and not to changes in the filtered load of sodium. Furthermore, an abnormal increase in plasma sodium concentration may lead to increased urinary sodium excretion (natriuresis), secondary to inhibition of tubular sodium reabsorption and disruption of the normal operation of the glomerular-tubular balance.

The mechanisms mediating the natriuretic response of the kidney following saline infusion have been the focus of intensive investigations. De Wardener and co-workers were the first to demonstrate the possibility that the natriuretic effect is in part due to the dilution of a nonrenal, nonadrenal circulating hormone.[42] They showed that infusion of isotonic saline in dogs resulted in natriuresis that could not be prevented by either administration of aldosterone or experimental reduction in GFR. Since that study, many investigators have used a variety of experimental models to substantiate the effect of such a natriuretic hormone.

The experimental model that has afforded the best insight into this question is the cross-circulation technique. The procedure involves transfusing blood from the aorta of a donor dog to the aorta of a recipient dog. The blood is then returned from the recipient's inferior vena cava to the inferior vena cava of the donor dog. In this way hemodynamics and renal response of both dogs to intravenous saline infusion into the donor dog can be studied. The response of the recipient dog to such an infusion has been analyzed to detect the effect of concentration changes of the natriuretic hormone in question.[126]

The results of numerous cross-circulation experiments have strongly implied that changes in concentration of a natriuretic hormone are responsible for natriuresis observed after saline infusion or blood volume expansion with iso-oncotic solutions or blood. This conclusion has been confirmed by numerous variations of the original whole animal cross-circulation experiments, including perfusion of isolated kidney with blood from intact donor animal in combination with micropuncture analysis of sodium transport along the nephron.[237]

The presence and potency of the natriuretic factor in urine was tested by Kruck in fasted, sodium-depleted rats injected with a urine extract from an orally hydrated normal man.[157] He found a dose-dependent increase in renal sodium excretion without a change in potassium excretion. Since urine extract from patients with congestive heart failure failed to produce natriuresis in rats, the author concluded that the absence of a normally present humoral factor is responsible for sodium retention and edema in these patients.

Much research has been done to elucidate the organ(s) that might be the source of this natriuretic hormone. It is now clear that the kidney is not the source of this hormone. Some studies have implicated the brain[167] and liver[34] as the possible organs producing this natriuretic hormone.

Recent studies have shown that the mammalian atria contain *granules* that secrete a *natriuretic factor*. The atrial natriuretic factor (ANF) has no effect on sodium reabsorption in the proximal tubule, but affects renal circulation and sodium transport in the collecting duct. Therefore, the function of ANF is considered later in this chapter.

Effect of Poorly Reabsorbable Solutes. Numerous micropuncture studies have established that the presence of poorly reabsorbable solutes, such as mannitol, urea, and sulfate, in the filtrate alter the tubular concentration of sodium and hence its proximal reabsorption. For example, Giebisch and Windhager found that during hypertonic mannitol diuresis in rats, TF/P for sodium decreased along the proximal tubule, leading to development of a blood-lumen sodium concentration gradient of 30 to 50 mEq/L by the end of this segment.[81] The final urine was found to be iso-osmotic, with urinary sodium concentration below that in the plasma.

Since tubular fluid remains iso-osmotic relative to

plasma during mannitol diuresis, the presence of poorly reabsorbable mannitol in the tubular fluid obligates osmotic retention of water within the tubule. This would reduce obligatory osmotic flow of water despite active sodium reabsorption along the proximal tubule. As a result, excess tubular water dilutes tubular sodium, decreasing its concentration along this segment, a factor responsible for the observed diminished net transepithelial sodium reabsorption.

Recent micropuncture studies in dogs have shown that the major effect of mannitol diuresis is the reduction in sodium and water reabsorption in Henle's loop.[260] Thus, it has been found that although 10% to 20% more of the filtered water leaves the proximal tubule during vigorous mannitol diuresis, only 4% to 5% of filtered sodium leaves this segment. In contrast, as much as 40% more of the filtered water and 25% more of the filtered sodium enter the distal tubule. These results are consistent with the view that the mannitol acts as an important loop diuretic. The underlying mechanism is thought to be a reduction in medullary tonicity, thereby reducing water reabsorption from the thin descending limb of Henle's loop. This would lead to less concentration of the tubular fluid, which, in turn, reduces passive sodium reabsorption from the thin ascending limb (long-loop nephrons only), a step necessary for the maintenance of the hypertonicity of the renal medulla. As we shall learn in Chapter 12, this will lead to medullary solute "washout" and eventually diuresis and natriuresis.

The diuretic effect of urea in the proximal tubule is different from that of mannitol. Urea appears to act directly or indirectly to reduce net proximal sodium reabsorption. This could occur by either a reduction of sodium efflux or an increased backflux of sodium into the lumen. There is no available evidence to suggest any effect of urea on the active component of sodium transport across the proximal epithelium. In the rat, the permeability of the distal tubule to urea is found to be only about one tenth that of the proximal tubule. Furthermore, the urea-induced diuresis and natriuresis appear to depend on the amount of urea administered and occur mainly within the distally located nephron segments. In the dog, only a small fraction of the filtered urea is reabsorbed in the distal tubule.[58] It is possible that even less urea is reabsorbed under the conditions of high-urea load. Thus, the magnitude of urea-induced diuresis and natriuresis was found to be a function of reabsorptive response of the distally located nephron segments.

Because normally the intratubular sodium concentration in the proximal tubule is the same as that in the plasma, its control is not considered to be the most important mechanism regulating proximal sodium reabsorption. However, most investigators concur that the most important mechanism is the proportionality that exists between the rate of delivery of fluid into the tubule and the rate of its proximal reabsorption, in the absence of changes in tubular fluid sodium concentration.[254] Experimental evidence supporting this relationship and the factors that modify it are described next.

Peritubular Control of Fluid Reabsorption

Numerous studies have shown that the "peritubular blood environment" and the factors modifying it exert profound influence on the net proximal reabsorption of sodium and water, and hence on the glomerular-tubular balance. To elucidate the mechanisms involved and to dissociate the effects of systemic and intrarenal hemodynamic factors, the peritubular environment has been modified using two classes of experimental forcings: (a) alterations in systemic hemodynamics induced by intravenous fluid expansion of extracellular or intravascular volumes; (b) alterations in intrarenal hemodynamics induced by changes in GFR and peritubular Starling forces. Let us now examine the effects of each forcing and the underlying physiological mechanisms.

Effect of Extracellular Volume Expansion. It is well established that changes in the extracellular volume produce inverse changes in ADH secretion and hence parallel changes in urine flow (see Chapter 13). On the other hand, changes in plasma sodium concentration and therefore plasma osmolality produce parallel changes in ADH secretion and hence an inverse change in urine flow. It follows that for the volume and osmolality of the extracellular fluid to remain constant, changes in the extracellular volume should influence not only the rate of water excretion, but also the rate of sodium excretion. Thus, as we have seen before, a decrease in the volume of the extracellular fluid, as in hemorrhage, reduces urinary sodium excretion; whereas urinary sodium excretion is increased after infusion of saline or plasma, or blood infusion.

Mechanisms underlying the relationship between the extracellular fluid or blood volume expansion and the urinary sodium excretion have been the subject of considerable investigations. Dirks and associates were the first to study, by micropuncture, the effect of saline infusion on the proximal tubule sodium reabsorption.[46] They found that the observed natriuresis following saline infusion was the result of depression of proximal sodium reabsorption. The disruption of the glomerular-tubular balance after saline infusion was found to be due, in part, to a decrease in the *tubular transit time* subsequent to elevation of GFR. It should be noted that a decrease in the tubular transit time tends to reduce the contact time between the epithelial cells and the filtrate, as the latter flows along the tubule. The net effect is that less time would be available for complete equilibration of tubular fluid with surrounding epithelial cells, and hence there is reduced fluid reabsorption.

Subsequent micropuncture studies, using split-oil drop and microperfusion techniques, have not only confirmed these findings but have also demonstrated that the depression of tubular reabsorption, following the expansion of the extracellular volume, was limited to the proximal tubule and had no effect on the more distally located nephron segments—namely, Henle's loop, the distal tubule, and the collecting duct. For example, micropuncture studies of superficial nephrons in both rats and dogs have shown that although saline infusion decreases the fraction of filtered sodium reabsorbed as well as the absolute sodium reabsorption rate in the proximal tubule by about 50%, the urinary excretion is only increased to about 10% of the filtered load.[53, 210] If the response of these nephrons is representative of the nephron population within the kidney, such findings imply that a major fraction of the sodium escaping proximal reabsorption is reabsorbed by the distally located nephron segments. Thus, urinary sodium excretion represents that fraction of the filtered load that has escaped reabsorption by the distal nephron segments.

Landwehr and associates studied, by micropuncture technique, sodium reabsorption in various segments of rat nephron in response to saline infusion.[158] They found that both the fraction of filtered sodium reabsorbed by Henle's loop and the absolute rate of sodium reabsorption in this segment were increased. However, saline infusion did not alter the absolute rate of sodium reabsorption in the distal convoluted tubule and collecting duct. Thus, it appears that the extent of urinary excretion of sodium in saline diuresis depends on how well the increase in sodium reabsorption (both fractional and absolute) by Henle's loop compensates for the decrease in sodium reabsorption by the proximal tubule.

That the proximal fluid reabsorption is highly sensitive to the degree of extracellular fluid volume expansion was demonstrated by Brenner and Berliner.[19] Using re-collection micropuncture technique (a procedure whereby fluid is collected from the same tubule under both control and experimental conditions), they found an inverse relationship between the fraction of sodium reabsorbed by the proximal tubule and the degree of expansion of the extracellular fluid volume. However, they observed that the fraction of filtered sodium excreted in urine changes little over a wide range of changes in the extracellular fluid volume. Thus, this substantiates the findings cited earlier that the compensatory increase in sodium reabsorption by the distally located nephron segments determines the extent of natriuresis, despite the volume expansion sufficient to depress the proximal sodium reabsorption.

The effect of extracellular volume expansion on proximal reabsorption explains in part the observed normal range in fluid reabsorption (50% to 75%) by this segment, as determined from TF/P inulin ratios. The results, however, are not specific to saline loading. Similar depression of the proximal fluid reabsorption has been found after an expansion of the intravascular volume by hyperoncotic albumin solution.[114]

Although both extracellular and intravascular volume expansion depress proximal sodium reabsorption, only the former leads to significant natriuresis. Thus, only saline loading appears to compromise the ability of the distally located nephron segments to compensate for the reduced proximal sodium reabsorption. The probable mechanisms for this difference will be discussed later.

The various studies cited above have all emphasized the important role of the peritubular Starling forces in explaining the inhibition of filtrate reabsorption in the proximal tubule following acute volume expansion. More recent studies have provided further insight into the possible mechanisms involved. Thus, volume expansion inhibited both volume and bicarbonate reabsorption when the rat proximal convoluted tubule was perfused with a high chloride and low bicarbonate solution.[197] Considering the normal composition of tubular fluid in the late proximal convolution, the inhibition of volume reabsorption was attributed to an increased back diffusion of bicarbonate into the lumen caused by an increase in the permeability of the paracellular pathway. However, inhibition of sodium chloride reabsorption by volume expansion was attributed to the inhibition of active transcellular sodium chloride transport rather than an increase in permeability of the paracellular pathway. Furthermore, studies in isolated rabbit proximal convoluted tubule have shown that a decrease in protein concentration in the bath reduced fluid reabsorption without altering transepithelial PD or paracellular resistance.[14] Also, reduction in protein oncotic pressure inhibited only sodium chloride reabsorption and had no effect on glucose or bicarbonate transport. Moreover, reduction of protein oncotic pressure completely inhibited active electroneutral transcellular transport of sodium chloride, but had no effect on the passive transport via the paracellular pathway.

Taken together, these studies indicate that the effect of Starling forces on the permeability of the paracellular pathway plays a minor role in proximal fluid reabsorption. This mechanism is important only for those substances whose concentration in the tubular lumen falls considerably below that in the plasma. In contrast, the major effect of alterations in the peritubular Starling forces is on the active electroneutral transcellular transport of sodium chloride. At present, the precise mechanism mediating this effect remains unclear.

Effect of Changes in GFR. That the extracellular fluid or intravascular volume expansion produces its natriuretic effects, at least in part, via induced parallel changes in the systemic and renal hemodynamics is well documented. Of particular interest here are those studies that have specifically modified the delivery rate of sodium to the proximal tubule by changing GFR, thereby altering the balance of Starling forces in the postglomerular peritubular capillary vasculatures. Experimentally, GFR may be reduced by partial constriction of the abdominal aorta or renal artery, partial occlusion of the renal vein, or partial obstruction of the ureter. The GFR may be increased by acute expansion of extracellular volume by intravenous infusion of saline, iso-oncotic solutions, or blood.

As mentioned earlier, an acute increase in GFR, produced by expansion of the extracellular fluid volume, depresses proximal sodium reabsorption and hence promotes natriuresis. On the other hand, an acute reduction of GFR by the above three maneuvers has not always produced expected renal responses. Several factors may contribute to the heterogeneity of the response.

Because reduction of GFR induced by partial constriction of renal artery, renal vein, or ureter is mediated by different mechanisms, these maneuvers produce different urinary sodium excretion patterns. To understand the reasons, a brief description of the mechanisms whereby GFR is reduced by these procedures is in order.

Partial constriction of the abdominal aorta or renal artery reduces renal perfusion pressure, thereby reducing the hydrostatic blood pressure at the glomerulus and hence reducing GFR (see Chapter 5). Partial constriction of the renal vein leads to a retrograde increase in the peritubular hydrostatic pressure, thereby causing partial decrease in tubular lumen, which increases the intratubular hydrostatic pressure and reduces GFR. Finally, partial ureteral obstruction increases ureteral pressure, which, in turn, leads to an increase in intratubular hydrostatic pressure, thereby reducing GFR. However, as we pointed out in Chapter 5, micropuncture studies in rats have shown that the changes in GFR produced by ureteral obstruction depend on the fluid balance of the animal under study. Thus, during hydropenia the fall in single-nephron GFR (SNGFR) is due to a reduction in glomerular filtration coefficient, K_g. In contrast, in volume-expanded rats, the fall in SNGFR is secondary to a rise in intratubular pressure. Finally, it should be noted that all three procedures are known to produce intrarenal vasodilation. They differ, however, in that aortic constriction and venous occlusion tend to reduce cortical blood flow, whereas ureteral obstruction elevates total renal blood flow.

Regardless of which procedure was employed, Rodicio and associates found that in hydropenic rats acute reduction of GFR increased fractional sodium and water reabsorption in the proximal tubule, suggesting excellent glomerular-tubular balance.[201] This was associated with an induced increase in tubular transit time, a factor that promotes filtrate reabsorption by increasing the contact and equilibration time of the tubular fluid with the surrounding epithelial cells and the peritubular blood. In contrast, intravenous saline infusion superimposed on the same maneuvers disrupted the glomerular-tubular balance. The mechanism whereby salt load does this is not known. However, other studies[41] have shown that although reduction of GFR by partial constriction of the renal artery increased proximal sodium reabsorption and decreased urinary sodium excretion, reduction of GFR by occlusion of renal vein and ureteral obstruction produced opposite effects: it reduced proximal sodium reabsorption and increased urinary sodium excretion.

The apparent conflict in these findings may be resolved if we remember the following points:

1. Changes in GFR produced by these three man-

uevers are brought about by different mechanisms. Constriction of the renal artery reduces hydrostatic pressure at the glomerular as well as postglomerular capillaries. The first effect reduces GFR and increases tubular transit time, a factor that favors increased luminal sodium reabsorption. The second effect reduces the balance of hydrostatic-oncotic pressure gradients across the peritubular membrane, thereby increasing peritubular uptake of reabsorbate. The net result of both effects is increased proximal sodium reabsorption.

2. When GFR is reduced by constriction of the renal vein, although tubular transit time is increased, the hydrostatic-oncotic pressure gradient across the peritubular membrane is also increased. The former effect tends to increase reabsorption, whereas the latter effect tends to decrease reabsorption. The results of the available experiments suggest that the effect of peritubular pressure gradient on proximal sodium reabsoprtion is more important than changes in tubular transit time. Thus, renal vein occlusion was found to decrease proximal sodium reabsorption. However, the compensatory increase in sodium reabsorption by the distally located nephron segments somewhat blunts this effect and reduces urinary sodium excretion.

3. It should be noted that the effect of changes in GFR may be mediated by both a change in filtration fraction (GFR/renal plasma flow, RPF) and a change in the balance of hydrostatic-oncotic pressure gradients across the peritubular capillaries.

In summary, an increase in GFR produces a parallel increase in filtration fraction and a secondary increase in the protein oncotic pressure in the postglomerular capillary vasculature. The former decreases tubular transit time, thereby reducing tubular reabsorption; the latter increases capillary uptake of reabsorbate, thereby masking the extent of the effect of increase in GFR on tubular reabsorption. The compensatory adjustments in renal sodium excretion in response to changes in GFR or sodium load appear to be mediated by changes in reabsorption in both proximal and distal tubules. The mechanism of such a compensatory adjustment involves induced changes in reabsorption characteristics at the luminal and peritubular membranes.

Effect of Changes in Renal Perfusion Pressure. Changes in renal perfusion pressure bring about parallel changes in sodium excretion rate. However, as described in Chapter 6, changes in mean renal perfusion pressure between 80 and 180 mm Hg (the autoregulatory range) are not accompanied by changes in GFR and RPF.[176]

The mechanism whereby changes in mean arterial pressure bring about parallel changes in sodium excretion (*pressure-diuresis*) has been examined by a number of experiments.[53] The results of these studies suggest that the natriuretic effect of the increase in mean arterial pressure is due to the transmission of the elevated pressure to the postglomerular capillary vasculature. At this site, the relative increase in hydrostatic pressure over the protein oncotic pressure decreases the capillary uptake of the reabsorbate by the proximal tubule, thereby increasing urinary sodium excretion.[148] That an increase in renal perfusion pressure, without an equivalent rise in GFR, depresses tubular reabsorption and produces marked natriuresis has also been well established.[216] Finally, as we pointed out in Chapter 6, recent studies have implicated increased renal prostaglandin synthesis and elevation of renal interstitial pressure as possible mediators of the phenomenon of pressure natriuresis.

Influence of Intrarenal Hemodynamics on Sodium Transport. The results of numerous recent experiments strongly favor the concept of intrarenal transcapillary hydrostatic-oncotic pressure gradients as an important determinant of capillary uptake of reabsorbate along the proximal tubule and hence sodium transport along the nephron. This concept best explains a number of experimental results, although it is based on the assumption that the capillary uptake of the reabsorbate is the limiting step in the overall filtrate reabsorption along the nephron.

The major intrarenal hemodynamic variables in such a control of capillary uptake of reabsorbate are capillary resistance, capillary hydrostatic, and capillary oncotic pressures. Thus, any factor that alters these variables will have a profound influence on the capillary reabsorption of the filtrate. For example, vasodilation reduces capillary uptake, whereas a rise in protein concentration increases the uptake. Micropuncture studies have localized the site of action of these intrarenal hemodynamic variables as the proximal tubule. Furthermore, their effects on tubular reabsorption do not appear to be mediated by changes in tubular volume. Thus, in keeping with the proximal cell model depicted in Figure 9.15, changes in intrarenal hemodynamic variables exert their influence on the uptake of reabsorbate from the intercellular spaces into the capillary. Hence, an increase in protein oncotic pressure speeds the rate of such a fluid uptake, whereas an increase in hydrostatic pressure retards it.

An important factor modulating the effect of transcapillary hydrostatic-oncotic pressure gradient on tubular reabsorption is the GFR. Of more practical interest is the filtration fraction (GFR/RPF), a factor determining the protein concentration in the postglomerular plasma and hence peritubular capillary reabsorption.

The results of several experiments have shown that expansion of blood volume with hyperoncotic solutions depresses proximal sodium reabsorption, with little change in urinary excretion. This implies that the increased plasma oncotic pressure, at the peritubular capillary level, has a lesser effect on reabsorption than the parallel increase in the hydrostatic pressure induced by expansion of blood volume. However, when blood volume was expanded to the same degree by infusion of hypo-oncotic or iso-oncotic solutions, despite the same degree of depression of the proximal tubule, the urinary sodium excretion was increased. This difference in urinary sodium excretion is because whereas blood volume expansion depresses proximal sodium reabsorption, the effect of plasma oncotic pressure is more pronounced in the distal tubule than in the proximal tubule.

Windhager and co-workers have provided the most direct evidence for the effect of peritubular plasma oncotic pressure on proximal sodium reabsorption.[256] Microperfusing the neighboring peritubular capillaries with hyperoncotic dextran, they found a significant increase in proximal sodium reabsorption as measured by split-oil drop technique. They also found that proximal sodium reabsorption increased as filtration fraction increased, further emphasizing the effect of plasma oncotic pressure.

Perhaps the strongest evidence implicating the importance of plasma oncotic pressure on proximal sodium reabsorption is the experiments of Brenner and associates.[20] They found that the proximal sodium reabsorption in rat varies directly with protein concentration of the postglomerular capillary, whether the latter was altered by bolus infusion of saline, iso-oncotic or hyperoncotic solutions.

We conclude this section with a simplified model of renal vasculature (Figure 9.16) so that we may synthesize quantitatively the influence of the intrarenal hemodynamics on sodium transport in the proximal tubule. This model is an extension of the scheme used to analyze the pressure-flow-resistance characteristics of the glomerular capillaries in Chapter 3 (Figure 3.5).

In this scheme we have lumped all the intrarenal resistances as the linear combination of three resistances: preglomerular resistance (R_a), postglomerular resistance (R_e), and venular resistance (R_v). P_c and π_c are the hydrostatic and protein oncotic pressures of the peritubular capillary blood. The other terms have already been defined.

Using this model, let us write defining equations for the pressure-flow relationships for the entire kidney and the two components relevant to the present discussion—namely, the glomerulus and the peritubular capillary network.

In accordance with the notions introduced in Chapters 3 and 5, the renal blood flow (RBF) may be defined as the ratio of the hydrostatic pressure drop across the kidney vasculature ($P_{AS} - P_{VS}$) to the linear sum of the three intrarenal resistances. Expressed algebraically:

$$\text{RBF} = \frac{P_{AS} - P_{VS}}{R_a + R_e + R_v} \quad (9.22)$$

Since RPF equals RBF(1 − Hct), where Hct is the hematocrit, we can rewrite the above equation in terms of RPF:

$$\text{RPF} = \frac{P_{AS} - P_{VS}}{R_a + R_e + R_v} (1 - \text{Hct}) \quad (9.23)$$

For a given RPF, the hemodynamics of the postglomerular peritubular capillary blood are determined by the fraction of RPF, which is filtered at the glomerulus, or GFR. The latter was defined earlier by the equation:

$$\text{GFR} = \frac{1}{2} \left(\frac{P_a - P_e}{R_a} - \frac{P_a - P_e}{R_e} \right) \quad (9.24)$$

The filtration fraction, FF, is then given by dividing Equation 9.24 by Equation 9.23:

$$\text{FF} = \frac{\text{GFR}}{\text{RPF}}$$
$$= \frac{(1/2)\,[(P_a - P_e)/R_a - (P_a - P_e)/R_e]}{[(P_{AS} - P_{VS})/(R_a + R_e + R_v)](1 - \text{Hct})} \quad (9.25)$$

The hydrostatic blood pressure in the peritubular capillary (P_c) is given by

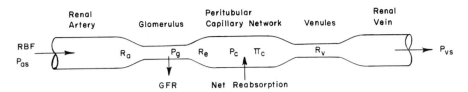

FIGURE 9.16. A schematic model of renal vasculature. (From Koushanpour.[155])

$$P_c = (\frac{RPF}{1 - Hct} - GFR) R_v + P_{vs} \quad (9.26)$$

where the term in parentheses is the fraction of RBF, which has escaped filtration. The protein oncotic pressure in this blood (π_c) is given by

$$\pi_c = (\frac{RPF}{RPF - GFR}) \pi_p \quad (9.27)$$

Finally, the net rate of fluid reabsorption ($\dot{T}_f$) by the peritubular capillary network is given by the difference between the hydrostatic and protein oncotic pressures, with the latter as the major determining force:

$$\dot{T}_f = k_f(\pi_c - P_c) \quad (9.28)$$

where k_f is the permeability coefficient of the capillary membrane. Thus, an increase in π_c relative to P_c increases $\dot{T}_f$, whereas a relative decrease in π_c reduces $\dot{T}_f$.

Because changes in π_c and P_c are ultimately determined by the renal and extrarenal factors defined by Equations 9.23 and 9.24, in the final analysis the magnitude of $\dot{T}_f$ depends on the influence of the systemic hemodynamics and their effective regulation.

From Equation 9.27 it can be seen that the major factor influencing π_c is the GFR and π_p, independently of changes in RPF. Thus, if π_p and RPF remain constant, changes in π_c and hence capillary uptake of reabsorbate ($\dot{T}_f$) are related to changes in GFR. Furthermore, as shown by Equation 9.26, the effect of P_c on $\dot{T}_f$ will be evident only in the absence of large changes in GFR.

Table 9.1 summarizes the effects of several maneuvers on the intrarenal hemodynamics and the net capillary uptake of the reabsorbate, both predicted by Equations 9.26 to 9.28 and observed by experiments discussed earlier.

Hormonal Regulation of Filtrate Reabsorption Along the Proximal Tubule

In-vitro measurement of hormone-dependent adenylate cyclase activity has been used to determine the presence of receptors specific to a given hormone in samples of defined tubular fragments.[177] Using this criteria, it has been shown that the epithelium of the proximal tubule contains receptors for a number of hormones. They include parathyroid hormone (PTH), thyroid hormone, and catecholamines. The following is a brief review of their actions.

Parathyroid Hormone

The principal function of PTH is the regulation of calcium concentration in the extracellular fluid compartment. This is achieved by the action of PTH on the bone and the kidney via activation of the adenylate cyclase enzyme system (see Chapter 8). In the kidney, PTH stimulates calcium reabsorption and inhibits sodium-dependent phosphate, bicarbonate, and fluid reabsorption in the proximal tubule. The mechanism involves stimulation of gluconeogenesis via stimulation of the adenylate cyclase enzyme system situated at the basolateral membrane, and an increase in intracellular synthesis of cyclic adenosine monophosphate (cAMP).[102] However, the precise details of the cellular steps involved are presently not known.

The exchange of intravesicular H^+ for intravesicular Na^+ across the apical membrane of the proximal tubule (the so-called electroneutral $Na^+ - H^+$ countertransport) occurs via an amiloride-sensitive carrier system.[143] Recent studies have demonstrated that PTH and cAMP directly inhibit countertransport,[128] whereas glucocorticoids, but not mineralocorticoids, stimulate $Na^+ - H^+$ countertransport.[142]

Finally, several recent studies have shown that PTH stimulates the enzyme responsible for synthesis of 1,25-dihydroxycholecalciferol, the active form of

TABLE 9.1. Directional changes in some intrarenal hemodynamics in response to some selected forcings.

Forcings	Response						
	RPF	GFR	FF	P_c	π_c	$\dot{T}_f$	Tubular-transit time
Step increase in P_{VS} (venous occlusion)	−	−	0	+	0	−	+
Step increase in R_v	−	−	0	+	0	−	+
Step increase in R_a	−	−	0	−	0	+	+
Step increase in R_e	−	+	+	−	+	+	−
Step increase in ureteral pressure (ureteral obstruction)	+	−	−	+	−	−	+
Step decrease in P_{AS} (aortic constriction)	−	−	0	−	0	+	+

Source: Koushanpour.[155]

vitamin D_3, which is produced mainly in the proximal tubule.[140] Furthermore, these studies have shown that the effect of PTH is mediated via the adenylate cyclase system.

Thyroid Hormone

Recent studies have shown that triiodothyronine (T_3), the biologically active form of the thyroid hormone thyroxine (T_4), stimulates sodium reabsorption in the proximal tubule.[166] The mechanism of action of thyroid hormone in the kidney is similar to that of aldosterone.[54] Briefly (see Figure 9.40 for more details), both T_4 and T_3 forms of the hormone enter the proximal cell through the basolateral membrane. Once inside, T_4 is converted to T_3, and the latter combines with high-affinity cytoplasmic receptors, forming a hormone-receptor complex. The cytoplasmic hormone-receptor complex is then transported into the nucleus, where it binds with the nuclear chromatin. The formation of the latter complex somehow activates DNA-dependent RNA synthesis, a process referred to as *gene transcription*, meaning that the information in the DNA is "transcribed" into RNA. The RNA, thus synthesized, is called the messenger RNA (mRNA), because it leaves the nucleus and enters the cytoplasm where it induces *de novo* synthesis of specific proteins. This step is called *translation*, meaning that the nucleotide of mRNA is "translated" into the amino acid sequence of the synthesized protein. The thyroid hormone-induced protein thus synthesized, through a series of poorly understood steps, stimulates Na-K-ATPase and/or increases the permeability of the luminal memrane, thereby increasing sodium reabsorption in the proximal tubule.

Catecholamines

As mentioned in Chapter 4, postganglionic sympathetic nerves that enter the kidney innervate the smooth muscles of the afferent and efferent arterioles and of the descending vasa recta. Currently, there is no agreement on the direct innervation of the renal tubules. Consequently, any effect on tubular function following experimental manipulations of renal nerves must be secondary to their effects on the renal vasculature. Moreover, tubular catecholamines receptors have so far been demonstrated at connecting tubules, which are located alongside of the interlobular arteries. With this in mind, numerous studies have shown that renal denervation increases and low-frequency renal nerve stimulation decreases urinary sodium excretion.[182] Micropuncture studies have suggested that these effects are due to alterations in sodium and water reabsorption in the proximal tubule. However, a recent micropuncture study in Munich-Wistar rats demonstrated that the antinatriuresis effect of the low-frequency stimulation is due to increases in afferent and efferent arteriolar resistances and decreases in glomerular capillary filtration (K_g) and peritubular capillary reabsorption coefficients.[153]

Clearance studies have shown that β-adrenergic agonists stimulate and α-adrenergic agonists inhibit sodium reabsorption in the proximal tubule.[8] A recent micropuncture-microperfusion study has shown that norepinephrine (an α-adrenergic receptor agonist) increased fluid reabsorption in the rat proximal convoluted tubule.[27] Additionally, it has been shown that dopamine causes vasodilation and natriuresis.[83] These findings are consistent with the concept that α- and β-adrenergic and dopaminergic receptors are present in the renal vasculature and presumably the prximal tubule. However, at present, definitive evidence for the presence of adrenergic receptors in renal tubules is not available.

Henle's Loop

Morphology

There are three types of loops of Henle: cortical, short, and long loops. Cortical loops, which are found only in a few species (pig and man), do not enter the renal medulla but turn back within the medullary rays of the cortex. If present in a species, they originate from the most superficially located glomeruli. Short loops usually originate from superficial and midcortical glomeruli and descend to comparable medullary levels roughly at the junction between the inner and outer medulla. Hence, the medullary portions are all of about the same length. Long loops originate from juxtamedullary and deep midcortical glomeruli and turn back at successive levels in the inner medulla. They characteristically differ in the length of the inner medullary portions. There are "short" long loops, "intermediate" long loops, and "long" long loops (Figure 9.17).

A short loop of Henle consists of a thick descending limb (i.e., the straight part of the proximal tubule), a thin descending limb, and a thick ascending limb (i.e., the straight part of the distal tubule). The terminal portion of the thick ascending limb contains the macula densa. A long loop, in addition, has a thin ascending limb.

The epithelial organization of the thick descending limb has already been presented as a part of the proximal tubule (pars recta). Those parts of the pars recta which, are situated in the outer stripe, are all established by P3 segments (see above).

The Thin Limbs of Henle's Loop

The thin limbs of Henle's loop (intermediate tubule) are composed of four segments that have different epithelia: the descending thin limbs of short loops, the descending thin limbs of long loops subdivided

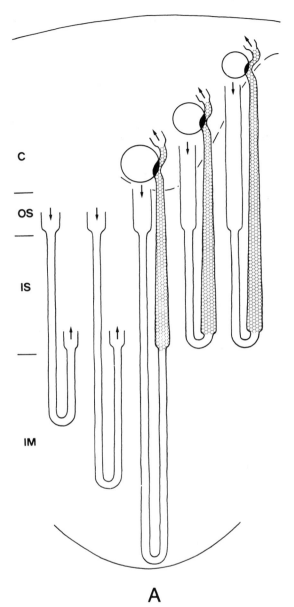

FIGURE 9.17. *Panel A*: Schematic diagram showing two types of loops. The short loops on the right belong to superficial and midcortical glomeruli and they descend to the same level within the inner stripe of the medulla. Note that these loops only have a descending thin limb. The thick ascending limb passes over into the distal convoluted tubule a short distance past the macula densa (shown in *black*). The long loop in the middle belongs to a juxtamedullary glomerulus. It contains a descending thin limb and an ascending thin limb; the bend of the loop is located within the tip of the papilla. The two additional long loops on the left demonstrate the heterogeneity among long loops turning back at different levels within the inner medulla.

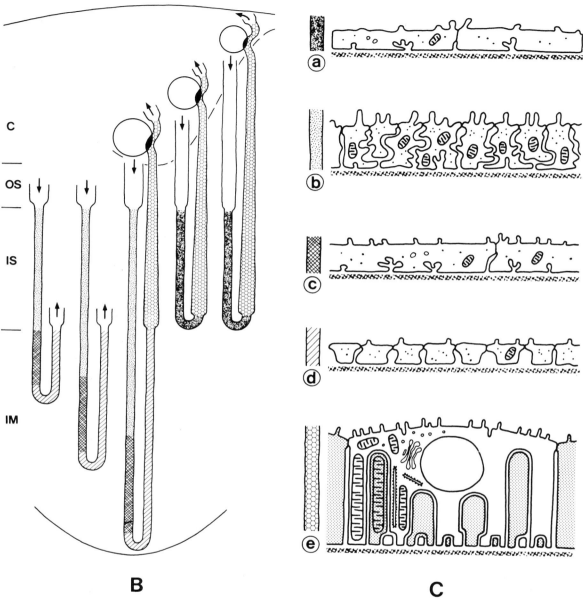

B

C

Panels B and C: Schematic diagrams showing the various epithelia (*panel C*) and their distribution along the loops of Henle (*panel B*). There are four types of thin limb epithelia (a-d). The thin descending limbs of short loops are made up of flat, simply organized epithelium (a). In contrast, the descending thin limbs of long loops are heterogeneous and are made up of two types of epithelia belonging to an upper (b) part and a lower (c) part. Note that the upper parts extend down into the inner medulla for variable distances according to actual length of a loop. The upper part epithelium (b) is made up of a very complexly organized epithelium with prominent paracellular pathways and amplification of the basolateral cell membrane, whereas the lower part epithelium (c) is very simply constructed. The thin ascending limb epithelium (d) is made up of heavily interdigitated epithelial cells exhibiting paracellular pathways. Note that this epithelium begins a short distance before the loop bend. In the thick ascending limb epithelium (e) the basolateral membrane is extensively amplified by cellular interdigitation with large mitochondria embedded into the interdigitating cell processes.

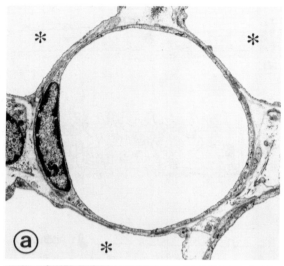

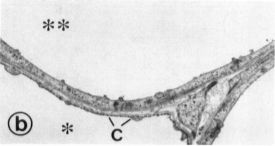

FIGURE 9.18. (a) An electron micrograph showing an overview of a descending thin limb of the short loop of the rat kidney. Note the tubule is surrounded by three ascending vasa recta depicted by asterisks (*). (b) The simplicity of the thin epithelium of such a loop limb is shown. The tubular lumen is marked by two asterisks (**), and the underlying vas rectum is marked by one asterisk (*).

in turn into an upper part and a lower part, and the ascending thin limbs.[5, 133, 213]

The *descending thin limbs of short loops* have a very simply organized epithelium. It consists of flat noninterdigitating cells connected by tight junctions that are composed of several anastomosing junctional strands (Figure 9.18).

Descending thin limbs of long loops are generally much larger in diameter and have a thicker epithelium than those of short loops.[44] Moreover, descending thin limbs of long loops are heterogeneous. Obviously, those of the longest long loops begin in the inner stripe as much thicker tubules than those of shorter long loops. The epithelium exhibits an axial heterogeneity that gradually simplifies as these limbs descend toward the inner medulla. Therefore, an upper part is distinguished from a lower part. Moreover, this process of epithelial simplification appears to be related to the length of each loop. It occurs earlier and more quickly in the shorter long loops and is delayed in the longest. Hence, upper parts are still found even deep in the inner medulla[213] (Figure 9.19).

The upper part of long descending thin limbs is established by an epithelium with an extremely high degree of cellular interdigitation. The tight junctions are shallow, usually consisting of only one junctional strand. Hence, the junctions appear leaky and the amount of junctional area available per unit area of epithelial surface is increased several fold by the tortuousity of the junction because of cellular interdigitation. Moreover, the basolateral membrane area is extremely amplified and forms an elaborate labyrinth that may span the entire thickness of the epithelium. Remarkably similar features have been found in salt secreting epithelia of birds and fish.[65, 206] In addition, cytochemical studies have demonstrated a membrane-bound Na-K-ATPase activity[66] and a carbonic anhydrase activity[48] in this epithelium. This is in contrast to other thin limb segments. From the morphology, therefore, one may conclude that this epithelium should be able to perform active salt transport, most likely to secrete salts into the tubular fluid.

The epithelium of the lower part of long descending thin limbs is comparably undifferentiated. The epithelium consists of relatively flat, noninterdigitatng cells with tight junctions of an intermediate apico-basal depth. The basal membrane is amplified by regularly distributed basal infoldings.

The epithelium of the *ascending thin limbs* is very flat, but consists of heavily interdigitating cells (Figure 9.20). Tight junctions are very shallow and usually made up of not more than one junctional strand. This combination of shallow junctions and lengthening of the junctional area by cellular interdigitation correlates with functional studies, which have demonstrated that the ascending thin limbs are highly permeable to ions but almost impermeable to water.[116, 173] The transition from the descending type of epithelium to the ascending type occurs some short distance before the bend. The bend of the loop already contains the ascending type epithelium.

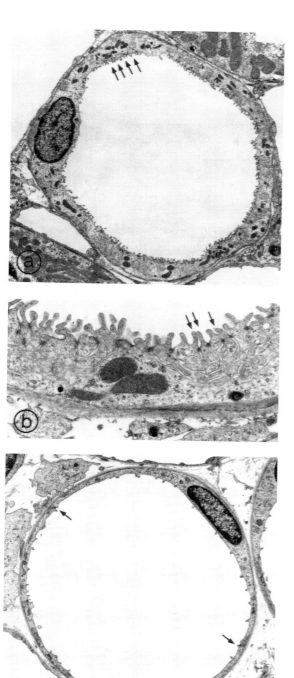

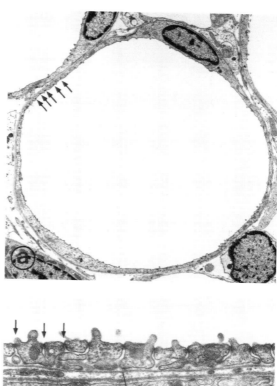

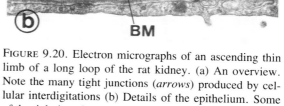

FIGURE 9.20. Electron micrographs of an ascending thin limb of a long loop of the rat kidney. (a) An overview. Note the many tight junctions (*arrows*) produced by cellular interdigitations (b) Details of the epithelium. Some of the tight junctions are marked by *arrows*. BM = Basement membrane.

The Thick Ascending Limb

The thick ascending limb exhibits an intrasegmental axial heterogeneity. Its epithelium gradually simplifies from the beginning of the medulla toward its termination in the cortex. Therefore, a medullary part is usually distinguished from a cortical part. This subdivision is substantiated by functional differences (see later). The cortical parts are much shorter in juxtamedullary nephrons (generally the long loops)

◁FIGURE 9.19. Electron micrographs of a descending thin limb of a long loop of the rat kidney. (a) An overview of the upper part. Note the many tight junctions, some of which are marked by *arrows*. (b) Details of the epithelium. Some of the tight junctions are marked by *arrows*. (c) An overview of the lower part established by simple epithelium. Only four tight junctions are visible (*arrows*). (d) Details of the epithelium. BM = basement membrane.

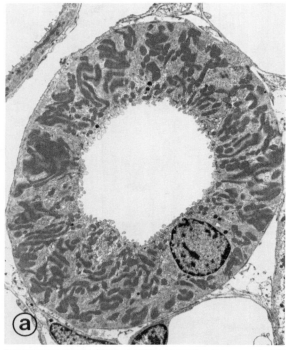

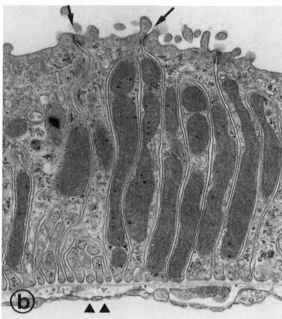

FIGURE 9.21. Electron micrographs of a thick ascending limb of the rat kidney typical of both short and long loops. (a) An overview. (b) Details of the epithelium. Note the cell processes are filled with large mitochondria. The cell junctions (*arrows*) are of intermediate apico-basal depth. Note the fenestrated endothelium of a capillary (*arrowheads*).

than in midcortical and superficial nephrons (generally the short loops).

At the beginning of the medulla the epithelium of thick ascending limbs is established by tall, heavily interdigitating cells[133, 234, 250] (Figure 9.21). The large, but narrow lateral cell processes are filled with plate-like mitochondria. Toward the base of the epithelium they further ramify, splitting finally into short, slender ridges. The tight junctional belt is of an intermediate apico-basal depth, but consists of several narrowly packed junctional strands. The apical surface has only a few microvilli. Many small round vesicles of the uncoated variety are encountered in the apical cytoplasm.

Consistent with the appearance of this epithelium, Na-K-ATPase activity has been found to be high in this nephron segment.[137, 208] The energy for ATP synthesis is provided by glycolysis, as is evident by a high glycolytic enzyme activity.[103] The major portion of oxygen consumption is associated with active sodium transport.

Toward and within the cortex the epithelium gradually decreases in cell height.[133, 154] In the rabbit the cortical thick ascending limb finally is a "thin tubule." The microvilli covering the luminal surface increase in density. The basic character of the epithelium is, however, maintained throughout. The thick ascending limb terminates a variable short distance (from a few cells up to 500 μm) beyond the macula densa (see Chapter 6, Figure 6.3).

General Characteristics of Filtrate Transport

Until recently, the inaccessibility of the medullary portions of the nephron to direct micropuncture made it virtually impossible to study the mechanism of sequential processing of the filtrate along the different segments of Henle's loop. Our current understanding of the processing of the tubular fluid along Henle's loop has proceeded in two successive stages. The first was the extension of the micropuncture techniques to collection and analysis of fluids obtained from the tip of the loop and adjacent structures in hamsters and young rats. The resulting information, in conjunction with that previously obtained about filtrate processing in the late proximal and early distal convoluted tubules of rats, provided the basic information regarding the overall processing of the tubular fluid along Henle's loop. The second was the application of microperfusion techniques to isolated segments of rabbit loop of Henle. This made it possible to study directly the mechanism of sequential processing of the tubular fluid along the inaccessible segments of Henle's loop.

To facilitate presentation, we begin this section with a description of the overall processing of the tubular fluid along the entire loop of Henle. And, where appropriate, we incorporate recent findings to provide insight into the mechanism of sequential processing of tubular fluid within the loop. Next, we present a detailed analysis of the processing of tubular fluid along different segments of Henle's loop. A fuller discussion of these findings and their significance in urine concentration and dilution is given in Chapter 12.

In 1951 Wirz and co-workers,[257] measuring osmolality of renal tissue slices in rat by direct cryoscopy, found a progressive increase in osmolality from the corticomedullary junction to the tip of the renal papilla. Their findings have subsequently been confirmed by several micropuncture studies showing that the osmolality of fluid collected from the loop bend was significantly higher than that collected from the late proximal convoluted tubule.[121, 122, 161, 173] From this evidence emerged the concept that the filtrate becomes progressively concentrated as it flows down the descending limb. The probable mechanism for the increase in osmolality was thought to be net water reabsorption from or net solute addition to the tubular fluid as it flows along the descending limb.

In a series of in-vitro microperfusion studies of rabbit descending limb, Kokko[151] not only confirmed the progressive increase in tubular fluid osmolality, but also found that the descending limb cell membranes have a much lower passive permeability to sodium and urea as compared with that for water. This finding provided strong evidence that, at least in the rabbit, the increase in osmolality of the tubular fluid as it flows down the descending limb is due primarily to water abstraction and not to solute addition.

In contrast, as shown in Figure 9.22, micropuncture analysis of the fluid collected from the early distal convoluted tubule, and hence that emerging from the thick ascending limb, found it to be always hypo-osmotic, whether the final urine was concentrated (U/P osmolality > 1.0) or not. Since the fluid collected at the loop bend was always hyperosmotic, the hypotonicity of the early distal fluid implied that the tubular fluid must become progressively dilute as it flows up the ascending limb. Direct measurement of permeability properties of the ascending limb

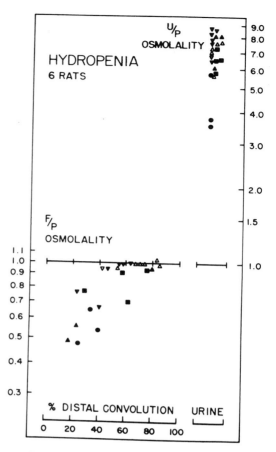

FIGURE 9.22. Tubular fluid-to-plasma (F/P) osmolality ratios along the distal convolution and urine-to-plasma osmolality ratios (U/P) from hydropenic rats. Different symbols refer to different rats. Note that in accordance with recent findings the term distal convolution used in this figure denotes a heterogeneous segment consisting of the distal convoluted tubule, the connecting tubule, and part of the cortical collecting duct. (From Gottschalk.[92])

in rats[179] and rabbits[151] have revealed that in both species the ascending limb membranes are highly impermeable to osmotic flow of water, but not to passive diffusion (thin segments) and passive and active transport (thick segment) of solutes. Taken together, these studies firmly established that the mechanism for the decrease in osmolality in the ascending limb is net solute reabsorption.

To sum up, the overall operation of the sequential processing of the filtrate along Henle's loop leads to a progressive concentration of the tubular fluid in the descending limb followed by its gradual dilution in the ascending limb. The dynamic interaction between this concentration difference and the surrounding interstitium leads to the generation of a concentration gradient within the renal medulla, which ultimately determines the final osmolality of the excreted urine. A number of recent studies have further clarified the above stated function of Henle's loop by measuring the permeability characteristics for water and several solutes and their modes of transport in the thin and thick segments of Henle's loop (Table 9.2). The following is a brief review of these findings.

The Thin Segments

As is evident from the data summarized in Table 9.2, there are striking differences in the modes of electrolytes, urea, and water transport along Henle's loop compared with that in the proximal and distal tubules and the collecting duct.

Compared with the proximal tubule, the thin descending limb is characterized by a relatively lower permeability to sodium and urea, a higher permeability to osmotic flow of water (hydraulic conductance), and by having a transepithelial electrical potential gradient (PD) of zero. These findings provide the electrochemical evidence for the proposed passive mode of sodium and urea transport and osmotic reabsorption of water secondary to the osmotic force generated by passive solute movement along this segment.

Recent in-vitro microperfusion studies[115, 117] have revealed a significant intranephron (axial), internephron (differences between short and long loops), and interspecies heterogeneity in sodium chloride and urea permeabilities of the descending limb of Henle's loop. Thus, in both hamster and rat, the upper portion of the thin descending limb of the long-loop nephron is highly permeable to water, sodium, and chloride, but is less permeable to urea. However, in rabbit, this segment is less permeable to sodium and chloride, but is highly permeable to water. In contrast, the thin descending limb of the short-loop nephron is highly permeable to water, moderately permeable to urea, and even less permeable to sodium and chloride in all these species of animals. Since the structural organization of Henle's loop in the human kidney is similar to that in rat and hamster, it is reasonable to assume that the permeability char-

TABLE 9.2. Transport characteristics of various segments of rabbit nephrons perfused in-vitro.

Nephron segments	Substance	Passive permeability (x 10^{-5} cm/s)	Measured transepithelial electrical potential, PD (mV)	Mode of transport
1. Proximal convoluted tubule	Sodium	2.1–5.0	−5.8 (lumen negative)	Active
	Chloride	2.4–5.5		Passive
	Urea	2.5–6.6		Passive
	Water	29–63		Osmosis
2. Thin descending limb	Sodium	0.17	0	Passive
	Urea	1.0–1.5		Passive
	Water	171		Osmosis
3. Thin ascending limb	Sodium	25.5	0	Passive
	Chloride	117		Passive
	Urea	6.7		Passive
	Water	0		None
4. Thick ascending limb	Sodium	6.3	+10 (lumen positive)	Active
	Chloride	1.1		Active
	Urea	0.9		Passive
	Water	0		None
5. Distal convoluted tubule	Sodium	low	−8 to −45 (lumen negative)	Active
	Chloride	low		Passive
	Water	0		None
6. Cortical collecting duct	Sodium	0.08	−25 (lumen negative)	Active
	Potassium	1.0		Active (secretion)
	Chloride	4.7		Passive

From data reported by Kokko,[151] Burg and Stoner,[24] Jacobson,[119] and Greger and Schlatter.[98, 99]

acteristics from these species are more relevant to the function of the human kidney than those obtained from the rabbit kidney.

In contrast, the membrane characteristics of the thin ascending limb are qualitatively similar among these species.[116] This segment is characterized by being highly impermeable to osmotic flow of water, highly permeable to sodium and chloride, but less permeable to urea, and by having a transepithelial PD of zero.

The asymmetric permeability properties for sodium, urea, and water in the thin limbs account for the efficient osmotic equilibration, primarily by water abstraction (depending on species), of the fluid in the thin descending limb as it encounters progressively hypertonic interstitium in the medulla. Consequently, the hypertonic fluid entering the thin ascending limb contains a higher sodium concentration and a lower urea concentration than the surrounding medullary interstitium, even though both fluids are in osmotic equilibrium. In contrast, owing to a very high permeability to sodium and somewhat lower permeability to urea of the thin ascending limb, as the tubular fluid traverses this segment a large amount of sodium chloride is reabsorbed into the interstitium by passive diffusion because of its concentration gradient. Because of low hydraulic conductance of the thin ascending limb, the diffusion of sodium chloride leads to substantial dilution of the luminal fluid relative to the surrounding interstitium.

The Thick Ascending Limb

As is evident from the data in Table 9.2, the thick ascending limb is highly impermeable to osmotic flow of water, it is relatively impermeable to solute, and it has a transepithelial PD that is both steeper than that for the proximal tubule (10 m V as compared with 5.8 m V) and opposite in sign (lumen positive). These findings have been taken as evidence for active chloride transport along this segment.[151] However, recent studies have provided strong evidence, at least in the rabbit, for an electroneutral active sodium chloride cotransport along this segment.[98,99] This conclusion is further supported by the finding that the epithelial cells comprising this segment are rich in mitochondria and Na-K-ATPase enzyme.[136]

The ability of the epithelial cells of the thick ascending limb to generate and maintain the hypotonicity of the luminal fluid is attributed to two important properties. First, the permeability to urea and the hydraulic conductance for water are very low in both the luminal membrane and the cell junctions, allowing the epithelial cells to maintain a large osmotic pressure difference between the tubular fluid and the surrounding medullary interstitium.[119] Second, an electroneutral sodium chloride cotransport system is present in the thick ascending limb, which allows net reabsorption of Na^+ and Cl^- ions against their respective concentration gradients. Recent studies have further clarified the mechanism of sodium chloride cotransport in this nephron segment. Greger,[97] using isolated microperfused rabbit cortical thick ascending limb, found that replacement of Na^+ or Cl^-, or addition of furosemide to the lumen or ouabain to the bathing media abolished the lumen-positive transepithelial voltage. He suggested that the sodium chloride cotransport was electrogenic, 2 Cl^- ions were transported per Na^+ ion, thereby accounting for the observed lumen-positive voltage. In a subsequent study, using the same type of preparations, Greger and Schlatter[98,99] found that removal of K^+ from the luminal fluid reduced potassium short-circuit current. This finding, coupled with their previous observations, led them to propose that an electroneutral 1 Na^+ − 1K^+ − 2 Cl^- cotransport system accounts for active chloride transport across the luminal membrane. Since this cotransport system is considered to be electroneutral, the mechanism responsible for generating lumen-positive voltage is thought to be due to K^+ recycling from cell to the lumen through a K^+ conductive pathway in the luminal membrane. Taken together, these findings have led to the concept that the active sodium chloride cotransport mechanism is driven primarily by the Na-K-ATPase enzyme, with 1 Na^+, 2 Cl^-, and 1 K^+ cotransported across the luminal membrane and K^+ being recycled across the luminal membrane, thereby generating the lumen-positive transepithelial PD.[98-100] On the basolateral side, some Cl^- ions leave the cell along with K^+, utilizing the chemical gradient for K^+, and the remaining Cl^- ions leave through Cl^- conductive pathways. Because the net Cl^- flux proceeds against both electrical (lumen-positive transepithelial PD) and chemical gradients, it has been concluded that Cl^- rather than Na^+ is actively transported across the entire epithelium. Since the net flux of Na^+ and Cl^- are blocked by ouabain, a glycoside inhibiting Na-K-ATPase, this suggests that the primary active transport is the continuous extrusion of Na^+ across the basolateral membrane by the

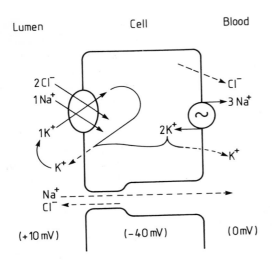

FIGURE 9.23. The sodium-chloride-potassium cotransport system in the thick ascending limb of Henle's loop.

Na-K-ATPase enzyme. Figure 9.23 summarizes the salient features of the Na^+-Cl^--K^+ cotransport mechanism in the thick ascending limb.

Since the tubular fluid in the thick ascending limb is hypotonic, this requires that equal quantities of Na^+ and Cl^- ions must be reabsorbed in this segment. In view of a stoichiometry of 1 Na^+ and 2 Cl^- for the transcellular cotransport mechanism, it follows that half of the net Na^+ reabsorption must occur via the paracellular pathway. This conclusion is supported by both anatomical and electrical data that have shown the thick ascending limb to be a "leaky" epithelium.[108]

In addition to Na^+ and Cl^-, it has been shown that Ca^{++} and Mg^{++} are also reabsorbed from the thick ascending limb. However, there is no evidence for net transepithelial transport of sugar, amino acids, phosphate, or other organic substances.

As described earlier, the epithelial structure in the thick ascending limb is not uniform throughout its length. This intrasegmental axial heterogeneity is the basis for subdividing the thick ascending limb into a medullary and a cortical part. Recent microperfusion studies in isolated tubules have demonstrated that associated with this structural heterogeneity, there is also a functional heterogeneity within the thick ascending limb. These functional differences are described in the following sections.

Medullary Thick Ascending Limb. The major solute reabsorbed in the medullary thick ascending limb is sodium chloride. Recent studies, using isolated single cells from the medullary ascending limb of rabbit kidney,[62, 63] have shown that sodium chloride is reabsorbed via a sodium chloride cotransport system and that furosemide directly inhibits this cotransport. A previous study in microperfused isolated rabbit tubule had shown that furosemide inhibited the sodium chloride transport in the thick ascending limb only when it was added to the luminal fluid.[25] Taken together, these findings suggest that furosemide may act on the carrier for the sodium chloride cotransport, which is located in the luminal membrane.

Several lines of evidence suggest that the epithelium of the medullary thick ascending limb constitutes a high-capacity transport system capable of reabsorbing the bulk of sodium chloride entering the thick ascending limb. Thus, it has been found that sodium chloride cotransport in the medullary thick ascending limb can only lower luminal sodium chloride concentration from 140 to 117 mM/L, whereas sodium chloride cotransport in the cortical thick ascending limb can lower luminal sodium chloride concentration to 65 mM/L.[199] These findings suggest that despite the fact that the medullary thick ascending limb reabsorbs the bulk of filtrate entering this segment, it can only generate a modest transepithelial sodium chloride gradient, as compared with the cortical thick ascending limb.

Sodium chloride transport in the medullary thick ascending limb is influenced by antidiuretic hormone (ADH) in some species. Thus, a recent study, using isolated medullary thick ascending limb from the mouse kidney, has shown that ADH, at physiological concentration, added to the bathing media doubled the lumen-positive voltage and increased the net chloride transport by fourfold to fivefold.[109] The ADH effect was found to be both Na^+ and Cl^- dependent and was inhibited by luminal furosemide and peritubular ouabain. This ADH effect was found to be species dependent and was demonstrated in the rat and the mouse kidney. It was found to be highly variable in the rabbit and was not found in the human kidney. The ADH-induced increase in sodium chloride transport in the medullary thick ascending limb contributes further to the ability of this nephron segment to concentrate urine.

Taken together, the above studies suggest that the bulk of sodium chloride entering the thick ascending limb is reabsorbed in the medullary segment via an active sodium chloride cotransport system across the luminal membrane. The luminal cotransport is, in turn, driven by the favorable transbasolateral electrochemical gradient for Na^+, which is maintained by the action of Na-K-ATPase localized in the basolateral membrane. However, this high sodium chloride transport capacity, even though enhanced by ADH in some species, leads to generation of only a modest transepithelial sodium chloride gradient along this segment. The large transepithelial sodium chloride gradient, responsible for creating the dilution of the tubular fluid at the end of the thick ascending limb, is generated by the epithelium of the cortical thick ascending limb.

Cortical Thick Ascending Limb. Some of the data that have defined the mechanism of sodium chloride transport in the thick ascending limb have come from studies in the isolated cortical thick ascending limb. As noted earlier, these studies have shown that sodium chloride transport in this segment involves an electroneutral sodium chloride cotransport whereby 2 Cl^- ions are cotransported with 1 Na^+ and 1 K^+ ion (Figure 9.23), with K^+ recycling across the luminal membrane accounting for the observed lumen-positive voltage.[98-100] Moreover, it has been shown that net sodium chloride transport in the cortical thick ascending limb is significantly lower than that in the medullary thick ascending limb.[229] However, in contrast to the medullary thick ascending limb, sodium chloride transport in the cortical thick ascending limb can lower sodium chloride concentration to 65 mM/L.[199] Taken together, these findings suggest that the cortical thick ascending limb has a lower capacity for sodium chloride transport but is a better generator of the transepithelial sodium chloride gradient, as compared with the medullary thick ascending limb.

In summary, sodium and chloride are transported by an active sodium chloride cotransport mechanism in both medullary and cortical portions of the thick ascending limb. However, owing to differences in epithelial transport capacity in these two segments, sodium chloride transport serves two distinct functions. In the medullary thick ascending limb, the epithelium has a high capacity for sodium chloride transport but can only generate a modest transepithelial sodium chloride gradient. In contrast, the cortical thick ascending limb epithelium has a low capacity for sodium chloride transport but can generate a large transepithelial sodium chloride gradient.

Function of Henle's Loop

The asymmetric permeability characteristics of the descending and ascending limbs to electrolytes, urea, and osmotic flow of water, described above, have provided the basis for numerous theories advanced to explain the function of Henle's loop in concentration and dilution of urine. The following is a brief description of the overall processing of the filtrate by this nephron segment. Details of these theories and our current understanding of the mechanisms of urine concentration and dilution are discussed in Chapter 12.

To facilitate presentation, we begin by considering the filtrate processing along the thick ascending segment. Here, active sodium chloride reabsorption without osmotic flow of water leads to the development of a local region of hyperosmolality in the surrounding medullary interstitium at each horizontal level along the ascending limb. The tubular fluid, in turn, becomes hypo-osmotic, at each level, relative to the surrounding medullary interstitium. To maintain tissue isotonicity, the tubular fluid in the descending limb comes to osmotic equilibrium at each level, primarily owing to abstraction of water, with the surrounding hyperosmotic interstitium. In so doing, the descending limb fluid becomes progressively hyperosmotic, reaching its maximum value at about the loop bend. The net result is that the fluid emerging from Henle's loop will be slightly hypo-osmotic relative to that entering the loop. This would lead to a net accumulation of solute within the medullary interstitium, making it hyperosmotic and thereby inducing osmotic absorption of water from the adjacent structures, including the collecting duct. It is this latter effect that determines the final osmolality of the excreted urine.

From the foregoing considerations, it is evident that any factor that alters solute and water reabsorption along Henle's loop will have a profound effect on the magnitude of the osmolality gradient established within the medullary interstitium and hence on the final concentration of the excreted urine. Let us now examine some of these factors.

Regulation of Filtrate Processing Along Henle's Loop

Effects of Changes in GFR

Like the proximal tubule, the epithelial cells of Henle's loop have the ability to adjust the rate of sodium and chloride reabsorption to the load delivered to this nephron segment. However, in contrast to the proximal tubule, because of the very low permeability of the ascending limb epithelium to osmotic flow of water mentioned above, there is considerable lag between sodium chloride and water

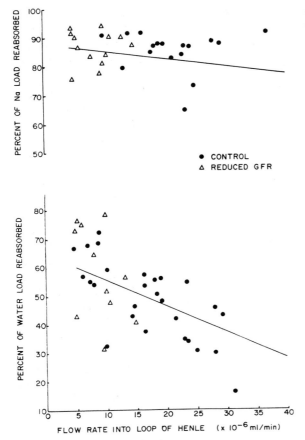

FIGURE 9.24. Fractional reabsorption of sodium and water by short loops of Henle as a function of flow rate into Henle's loop during control conditions and after acute reduction of GFR by partial clamping of the renal artery. *Upper panel*: Relation between fractional reabsorption of sodium load from the Henle's loop and flow rate entering the loop. *Lower panel*: Relation between fractional reabsorption of water load from Henle's loop and flow rate entering the loop. (From Landwehr et al.[158])

acute reduction in GFR, and hence in the fluid load entering the loop, will result in an increase in the transit time of fluid through the loop. Hence, a reduction in flow rate into the loop will result in a more complete sodium chloride reabsorption from the tubular fluid into the surrounding medullary interstitium. Thus, within low physiological flow rates (10 to 15×10^{-6} mL/min), the sodium concentration in the tubular fluid leaving the ascending limb will be low because of increased sodium reabsorption. Conversely, an increase in GFR above normal will increase sodium concentration in the tubular fluid leaving the ascending limb because of reduced sodium reabsorption.

Effects of Hormones and Diuretics

Analogous to the proximal tubule, measurement of hormone-dependent adenylate cyclase stimulation has been used to determine the responsiveness of the thin and thick segments of Henle's loop to various hormones.[177] Thus, it has been shown that the thin descending limb is not responsive to any hormone tested. In contrast, both the thin and the thick segments of the ascending limb respond to different hormones. The thin ascending limb was found to be only responsive to ADH (vasopressin). However, this effect was species dependent. Thus, the thin ascending limb was found to be highly responsive to ADH in rat and mouse, less responsive in rabbit, and completely unresponsive in human.[26] Although ADH exerts a well-established antidiuretic effect in the collecting duct, its stimulation of the adenylate cyclase in the thin ascending limb was not associated with any effect on osmotic water permeability of this segment. The physiological effect of ADH on the thin ascending limb remains to be clarified.

The thick ascending limb contains the adenylate cyclase system, which is responsive to PTH, ADH, calcitonin, and prostaglandin. Insofar as the hormonal regulation of ion transport is concerned, there are major qualitative differences between the medullary and cortical segments of the thick ascending limb.[40, 59, 119] Thus, in rabbit, PTH stimulates adenylate cyclase and calcium transport in the cortical but not in the medullary portion of thick ascending limb. In contrast, vasopressin stimulates sodium chloride transport (mouse and rat) and calcitonin stimulates calcium transport (rabbit) in the medullary but not in the cortical portion of the thick ascending limb. In the rat, calcitonin acts on the entire thick ascending limb.

reabsorption along the loop. In fact, it has been found that both sodium chloride and water reabsorption along the loop are flow-rate dependent, with the flow dependency of water reabsorption being much greater.[158] These findings are shown in Figure 9.24, which depicts an inverse relationship between the fractional reabsorption of sodium and water loads by the Henle's loop and the flow rate into the loop.

The sequence of events thought to be responsible for this inverse relationship may best be explained by considering the effect of reduced GFR on sodium and water reabsorption rate along Henle's loop. An

Although osmotic flow of water along the collecting duct is known to be affected by blood levels of ADH, ADH has no effect on water transport along Henle's loop. However, ADH has been shown to stimulate the transepithelial PD and sodium chloride transport in the thick ascending limb in mice and rat.[40] This effect of ADH on salt transport provides a steeper concentration gradient between the thick ascending limb and the surrounding medullary intertitium, an effect that leads to the development of a steeper longitudinal gradient and to the formation of more concentrated urine. Similar effects have been observed with calcitonin.[59] Thus, it has been shown that calcitonin stimulates the same cells in the thick ascending limb that are responsive to ADH. Hence, calcitonin also stimulates sodium chloride reabsorption in the thick ascending limb, thereby contributing to the development of the medullary concentration gradient and increased urine concentration.

Both the medullary and cortical portions of the thick ascending limb were responsive to calcitonin.[177] In contrast, prostaglandin (PGE_2) decreased lumen-positive voltage and net chloride transport in the medullary but not in the cortical portion of the thick ascending limb.[229]

Finally, the net water and sodium chloride reabsorption along the loop can also be modified by selective diuretics. For example, furosemide[29] and ethacrynic acid[84] have been found to reduce, by about 50%, the net sodium chloride and water reabsorption in Henle's loop. A recent in-vitro microperfusion study has shown that both of these diuretics inhibit sodium chloride reabsorption in the thick ascending limb.[207] Furosemide, acting from the luminal side, inhibits the 1 Na^+ − 2 Cl^- − 1 K^+ cotransport system in the luminal membrane. However, ethacrynic acid inhibits sodium chloride transport by a mechanism that does not involve the 1 Na^+ − 2 Cl^- −1 K^+ cotransport system. The physiological significance of the action of these diuretics in relation to the urine concentrating function of Henle's loop is considered in Chapter 12.

The Distal Tubule and the Collecting Duct

Morphology

The term "distal tubule" has somewhat different meanings when used by morphologists and physiologists. Morphologically, the distal tubule is composed of a straight part, a segment that is commonly called the thick ascending limb, and a convoluted part (the distal convoluted tubule), which transforms from the straight part shortly after the macula densa. It is much shorter than the convoluted part of the proximal tubule; after two or three coils it passes over into the connecting tubule. Physiologists often refer to all of the cortical tubular segments beyond the macula densa as the "distal tubule," including the distal convoluted tubule, the connecting tubule, and the cortical collecting duct. From a functional point of view, there are reasons to separate the straight part of the distal tubule from the subsequent convoluted part and to present the distal convoluted tubule, the connecting tubule, and the cortical collecting duct together. Functionally, these latter three are intimately involved in adjusting the final electrolyte composition of the urine. This correlates with the fact that the borders between these three tubular segments are not clearly discernable in most species, including man (Figure 9.25). In this section, we present the structural organization of all tubular segments beyond the macula densa, including the distal convoluted tubule (DCT), the connecting tubule (CNT), the cortical collecting duct (CCD), and the medullary collecting duct (MCD).

Distal Convoluted Tubule

The DCT (Figures 9.25, 9.26, and 9.27) begins a short distance beyond the macula densa with an abrupt increase in epithelial height[134, 193]; the short postmacula segment of the thick ascending limb has a rather low epithelium. The epithelium of the DCT is composed of tall extensively interdigitating cells.[133, 233, 234] The nucleus is characteristically situated in the apical cell portion almost apposed to the luminal membrane. In contrast to proximal tubule cells, the cellular interdigitation is restricted to the basal two thirds of the cells; the apical cell portion has a simple polygonal outline and is smoothly apposed to the adjacent cell. The tight junctions are much deeper than in the proximal tubule and consist of several narrowly arranged junctional strands. The luminal membrane bears short, stubby microvilli in varying amounts. The cellular interdigitation comprises the middle and basal parts of the epithelium and is similar to that of the proximal tubule. Large but rather narrow cell processes arise laterally from the middle part of the cells. Toward the base of the epithelium they divide into large ridges and finally split into short, slender ramifications. The interdi-

174   9. Tubular Processing of Glomerular Ultrafiltrate: Mechanisms of Electrolyte and Water Transport

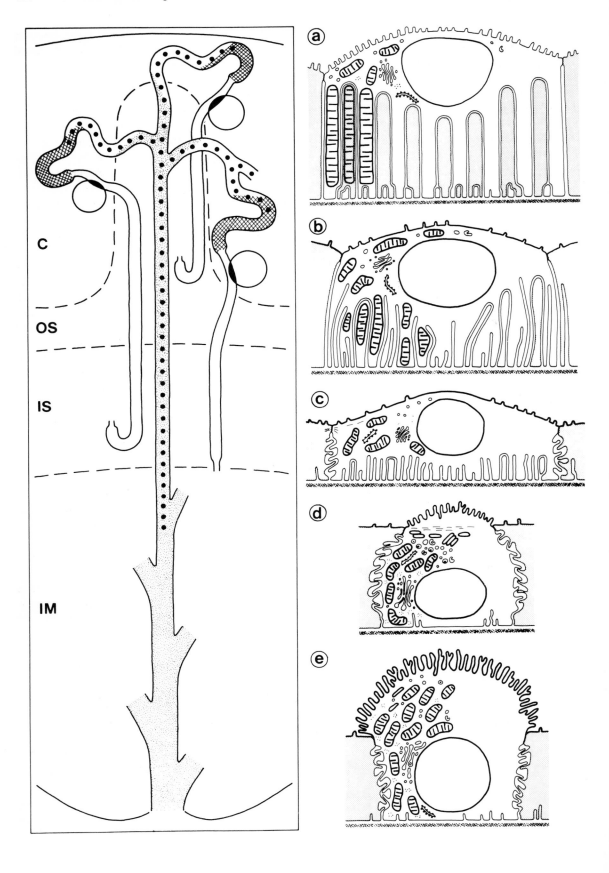

gitating cell processes are densely furnished with large mitochondria, closely apposed to the cell membranes. The narrow association of mitochondria and transporting membranes, a general characteristic for salt-transporting epithelia, is best developed in this nephron segment. This is consistent with a very high Na-K-ATPase activity found in this segment.[137] In contrast to the proximal tubule (see above), the energy for ATP synthesis and hence for tubular transport is derived from glycolysis as is evident by a high activity of glycolytic enzymes found in this nephron segment.[103]

Connecting Tubule

Microanatomically, the CNT consists of the arcades, i.e., branched tubular portions established by the fusion of deep nephrons (see Chapter 4) and the unbranched CNT of midcortical and superficial nephrons (Figure 9.25). Ultrastructurally, the epithelium is the same in both, being heterogeneously composed of CNT cells and intercalated cells (IC cells) (Figure 9.28.

CNT cells are rather large cells (Figure 9.28). Generally, the luminal membrane bears only few microvilli but is covered by a distinct cell coat. The basolateral membrane is extensively amplified. In contrast to the preceding DCT, the amplification is accomplished by infoldings of the basal membrane rather than by cellular interdigitation. The basal infoldings are deep, frequently extending up into the apical cytoplasm. The cells contain abundant mitochondria, but the infolded membranes are never as closely related to mitochondria as interdigitating epithelia. CNT cells in the beginning of this segment are generally larger and more complexly developed than those toward the end.[133]

Intercalated cells (Figures 9.29 and 9.32) are not specific for the CNT. In addition, they are found in cortical and outer MCDs. In some species, including man, they are also found within the terminal portions of the DCT.[234] Their name refers to the fact that intercalated cells never occur in clusters but always as individuals dispersed among other cells.

Intercalated cells are different in many respects from all other cell types along the nephron. One of the ways in which they are unique is the variability of the apical cell pole[129, 130] (Figure 9.25). The apical cell portion, densely covered by microfolds and microvilli, may be found to be prominently exposed into the tubule lumen. In other situations the cells may be retracted behind the surrounding cells (CNT cells or the principal cells of the collecting duct) and may be almost fully devoid of microfolds. Changes from one manifestation into the other may occur within a short span of time. These changes are possibly based on a membrane shuttle system.[132, 227] The so-called "flat vesicles" found in the apical cytoplasm when the cells are retracted are considered to be reservoirs of the luminal membrane that can be incorporated into the luminal surface. There is evidence that like mitochondrial cells of the turtle bladder, the luminal membrane of intercalated cells contains a specific H^+ transport system by which the tubular urine can be acidified.[82, 150] A very recent study suggests that the two manifestations of intercalated cells correlate with H^+ − secreting (exposed manifestation) and HCO_3^--secreting/H^+-reabsorbing (retracted manifestation) functions.[211] In the CCD the HCO_3^--secreting type cells appear to be more numerous, whereas in the MCD the H^+-secreting type cells appear to be more prominent.

Intercalated cells have a conspicuously dark-staining cytoplasm, and they have therefore frequently been called dark cells. This dark-staining cytoplasm is due to a densely meshed smooth endoplasmic re-

---

◁ FIGURE 9.25. The DCT, CNT, and collecting duct system of three nephrons (*left panel*) and their cellular structures (*right panel*). In the left panel it can be seen that the DCT (*cross-hatched*) begins abruptly a short distance beyond the macula densa (*black*), and transforms gradually into the CNT (*white*). The CNT gradually transforms into the CCD (*stippled*), which then descends into the medulla. The intercalated cells (*dark dots*) begin to occur in the terminal portion of the DCT and end in the early portion of the inner MCD. The schematics of the epithelia in the right panel illustrate the relationship of the middle cell (*white with a nucleus*) with the two neighboring cells (*shaded*). A basement membrane underlies the epithelium. For clarity of presentation, the basic organization of the cells is drawn on the right, whereas the cytoplasmic details are added on the left. (a) Cells of the distal convoluted tubule (DCT cells), which are heavily interdigitated. (b) Cells of the connecting tubule (CNT cells) in which the amplication of the basolateral cell membrane are effected by basal infolding. (c) Principal cells of a CCD in which the basal infoldings are found only in the basal third of the cell. (d and e) Intercalated cells in retracted and exposed manifestations, respectively. (Modified from Kaissling and Kriz.[133])

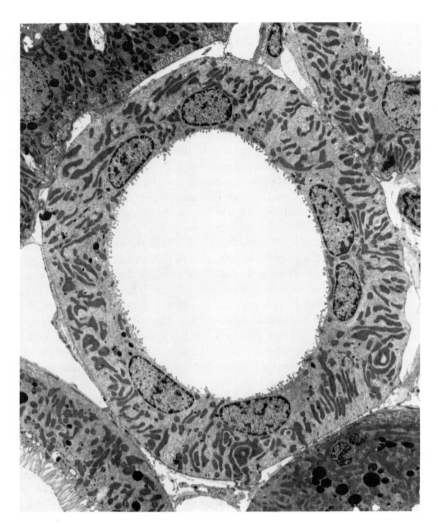

FIGURE 9.26. An electron micrograph of a rat distal convoluted tubule. Note the typical location of cell nuclei in the apical cytoplasm, the many mitochondria, and the short and stubby microvilli of the luminal membrane, which do not form a brush border.

ticulum, a large Golgi apparatus, many polysomes, and numerous small mitochondria with narrow mitochondrial cristae. The basolateral cell membrane is amplified by basal infoldings, variable in amount among individual cells, individuals, and species. Surprisingly, intercalated cells have little or almost no Na-K-ATPase activity.[64] In histochemical reactions they reveal a strong carbonic anhydrase activity,[21,48] a finding that correlates with their H^+-secreting function.

Cortical Collecting Duct

The collecting ducts are subdivided into cortical, outer medullary, and inner medullary collecting ducts (Figure 9.25). The epithelium of the CCD is similar to that of the connecting tubule, because it is composed of two cell types: principal cells and intercalated cells (Figures 9.30–9.32). The latter appear to be identical to those occurring in the CNT. Even if similar in some features, the principal cells of CCD are different from the CNT cells.

Principal cells have a simple polygonal outline and they do not interdigitate laterally with each other.[133] They are joined to each other or to adjacent IC cells by deep tight junctions.[196] The collecting duct epithelium is a tight epithelium. The lateral intercellular spaces run almost straight from the apical to the basal side of the epithelium. They have been found to be dilated in the presence of ADH (i.e., when water is transported through the epithelium) and narrowed when the hormone is ab-

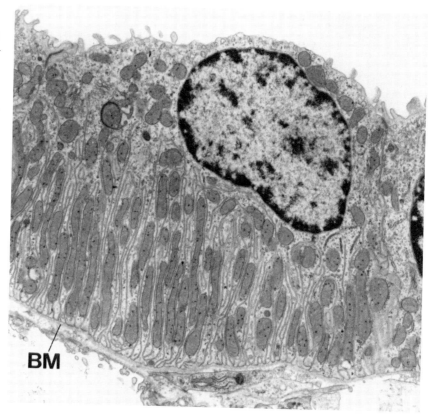

FIGURE 9.27. An electron micrograph of a distal convoluted tubule cell of the rat kidney showing the close association of mitochondria and the lateral membrane of the interdigitating processes. BM = basement membrane.

sent.[73, 146, 259] In the dilated state, one may most clearly observe lateral microfolds and microvilli, which protrude into the interspaces and are connected to microfolds from the opposite cell membrane by desmosomes (Figure 9.33). At their base the lateral intercellular spaces open through basal slits (which are narrow and do not change in width) to the intersititial compartment.

The luminal cell membrane of principal cells bear some stubby microvilli and a prominent cilium (the function of which is unknown) and is covered by a distinct cell coat. The basolateral membrane is amplified by the lateral microvilli (mentioned above) and by basal infoldings arranged in a typical pattern. The infoldings are all of the same height extending for about one third of the cell height into the basal cytoplasm.[133] Since there are no other major cell organelles (no mitochondria) found in this part of the cytoplasm, the infoldings appear like a fence occupying a basal rim of the cells. The nucleus and all other usual cell organelles, including many small mitochondria, lie in the cytoplasm above this fence.

In summary, the cortical tubular segments beyond the macula densa, which include the DCT, CNT, and CCD, are formed by four cell types: the DCT cell, the CNT cell, the principal cell of the CCD, and the IC cells. In most species, including man, the borders between these segments are not clear-cut but gradual in the sense that the respective cell types intermingle with each other for some distance. Hence, at the transition from the DCT to the CNT there is a transitional tubular portion, which contains three cell types: DCT cells, CNT cells, and IC cells. The same happens at the transition from the CNT to the CCD where there are CNT cells, principal cells, and IC cells.

Medullary Collecting Duct

Outer medullary and inner medullary collecting ducts are distinguished within the MCD. Outer MCDs traverse as unbranched tubes the entire length of the outer medulla. Their epithelium is made up of principal cells and IC cells. Inner MCDs fuse succes-

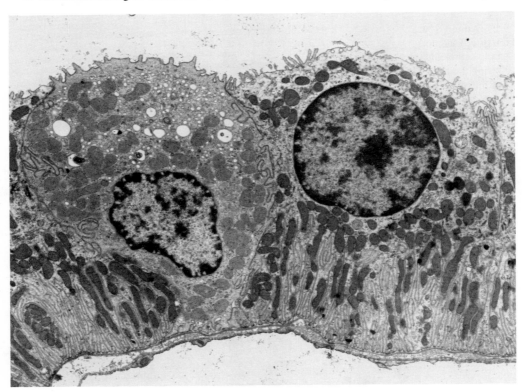

FIGURE 9.28. An electron micrograph of a CNT epithelium of the rat kidney. An intercalated cell (*left*) and a CNT cell (*right*) are shown. Note that the basal infoldings of the CNT cell do not exhibit a close association to mitochondria as seen in the DCT (Figure 9.27). The IC cell is in a retracted manifestation. Also note here the many mitochondria not related to any cell membrane and the many vesicles in the apical cytoplasm. Note, in addition, the fenestrated endothelium of the underlying capillary.

sively; after eight fusions they empty into the pelvis at the papillary tip. In the very beginning, the epithelium is still heterogeneous. For the major part of the inner MCD the epithelium is established by principal cells only; IC cells are absent (Figure 9.25).

The IC cells in outer MCDs are identical to those found in the CCDs. Basically, the principal cells along the entire collecting duct are of the same cell type. They are polygonal cells with a pale cytoplasm, basal infoldings, and some apical microvilli, as described above. Within the medulla toward the papillary tip, however, the cells increase considerably in cell height up to tall columnar cells,[186] and the amount of basal infoldings decreases. As in the cortex, the tight junctions are deep, and the lateral intercellular spaces are of variable width, with many microfolds protruding from the lateral cell membrane (Figure 9.34). Along the entire collecting duct, principal cells are sensitive to ADH, being highly water permeable when the hormone is present. The outer MCDs are virtually impermeable to urea. However, the permeability of the inner MCDs is high, and this difference can not, so far, be correlated to a structural difference.

General Characteristics of Filtrate Transport

Micropuncture analysis of fluids collected from the accessible portions of the "distal tubule" (DCT, CNT, and CCD) and collecting duct in rats, dogs, and monkeys, and in-vitro microperfusion of these tubular segments as well as of the inaccessible segments isolated from rabbits, have established the following: (1) The fluid in the early distal convolution is always hypo-osmotic, becoming iso-osmotic as it reaches the early portion of the CCD. (2) Sodium is reabsorbed actively in the DCT, CNT, and CCD. Sodium reabsorption is adversely affected by changes in tubular flow rate and by the presence of poorly reabsorbable solutes in the tubular fluid. Sodium reabsorption in the collecting duct is enhanced in the

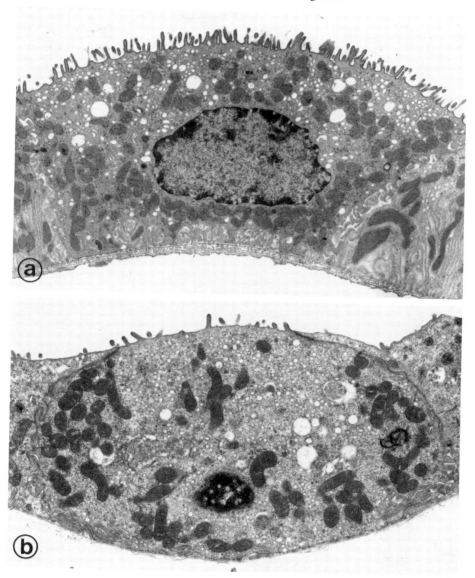

FIGURE 9.29. Electron micrographs of two IC cells of the rat kidney showing the exposed (a) and retracted (b) manifestations, respectively. Note the many microvilli in the exposed and the almost bold-head cell in the retracted manifestations.

presence of aldosterone. (3) Potassium is secreted actively into the CNT and the CCD, and reabsorbed passively from the outer MCD. The potassium reabsorbed in this segment is added to the descending limb of Henle's loop, and hence is recycled in the medulla. Potassium secretion in the collecting duct is stimulated by an increase in tubular flow rate, by chronic administration of aldosterone, and by low tubular chloride concentration, secondary to the presence of poorly reabsorbable solutes in the tubular fluid. (4) Chloride is reabsorbed actively in the DCT and passively in the CCD. It is not known how chloride is transported in the MCD. However, in the papillary collecting duct, chloride is actively reabsorbed. (5) Water is always reabsorbed by osmosis, as a result of the osmotic force developed by active and passive solute reabsorption. Osmotic reabsorption of water in the CCD is enhanced in the presence of ADH, and is reduced when ADH is absent.

Let us now examine some of the evidence that has

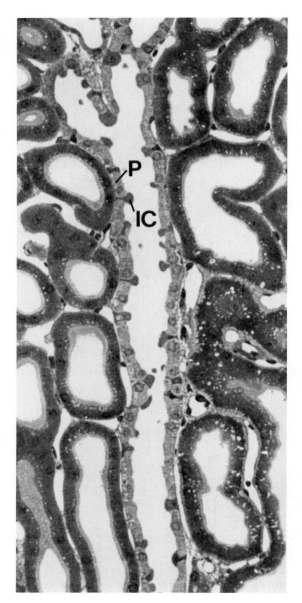

FIGURE 9.30. A light micrograph of the rabbit kidney cortex showing a CCD (*center*). Note the heterogeneous epithelium consisting of principal (P) and the prominent intercalated (IC) cells. (Courtesy of Dr. Brigitte Kaissling).

delineated the above characteristics of the sequential processing of the filtrate along the "distal tubule" and the collecting duct.

Filtrate Osmolality

As shown in Figure 9.22, the fluid emerging from the ascending limb of Henle's loop and hence entering the distal convoluted tubule is hypo-osmotic (F/P osmolality ratio less than unity). It remains hypotonic during dehydration, during induced osmotic diuresis, and whether ADH is present or not. However, as this hypotonic fluid passes through the cortical collecting duct, it gradually becomes iso-osmotic (F/P osmolality ratio approaches unity). But if ADH is absent, the fluid remains hypo-osmotic as it reaches the latter portion of the cortical collecting duct. ADH enhances the osmotic water permeability of the principal cells in this nephron segment. As decribed below, the hormone activates adenylate cyclase, which in turn stimulates intracellular cAMP synthesis, which eventually leads to an increase in water permeability of the apical membrane of the principal cell. During its passage through the medullary collecting duct, the normally iso-osmotic tubular fluid becomes progressively hyperosmotic owing to osmotic efflux of water into the surrounding hypertonic medullary interstitium. In the absence of ADH, the urine will be dilute.

In the final analysis, the observed osmolality profile is determined by the different rates of solute and water transport along these distally located nephron segments. Since the major constituents of the filtrate delivered to these segments are sodium, potassium, chloride, and urea, the extent of their tubular processing will largely determine the final volume and osmolality of the excreted urine. Therefore, the remainder of this section is devoted to a closer examination of the tubular processing of these filtrate components and the factors that modify their transport along these segments.

Sodium Transport

As depicted in Figure 9.9, sodium is reabsorbed continuously in the "distal tubule," thereby lowering its TF/P ratios below unity all along the length of this segment. Moreover, comparison of TF/P ratios for sodium along the distal and proximal tubules reveals a steep blood-to-lumen concentration gradient for this ion in the "distal tubule" (TF/P < 1), compared with no gradient in the proximal tubule (TF/P = 1). Furthermore, split-oil drop experiments have shown that sodium reabsorptive rate per unit tubular length in the "distal tubule" is only one fourth of that in the proximal tubule. Also, water permeability of the distal epithelial cells was found to be only 40% of that for the proximal tubule, a factor that accounts for the smaller fraction of water load being reabsorbed in the distal segment. These striking differences in reabsorptive rates for sodium and water between the proximal and distal tubules sug-

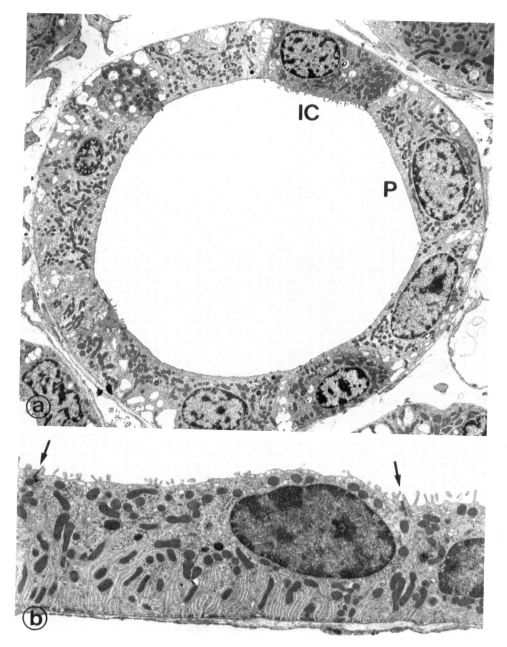

FIGURE 9.31. Electron micrographs of the rat kidney. (a) An entire cross-section of a CCD showing the heterogeneity of the epithelium consisting of principal (P) and intercalated (IC) cells (From Kaissling and Kriz.[131]) (b) A detailed view of a principal cell. Note the infoldings of the basal cell membrane, which only penetrate a short distance into the cytoplasm. Note the deep tight junctions (*arrows*).

gest that active sodium and osmotic water reabsorption along the "distal tubule" require much steeper electrochemical potential gradients (see Table 9.2).[24,170]

Inspection of the sodium-to-inulin TF/P ratios in the lower panel of Figure 9.9 shows that nearly two thirds of the filtered sodium is reabsorbed by the time the filtrate reaches the end of the proximal convoluted tubule. Furthermore, sodium reabsorption is about 90% complete by the time the tubular fluid reaches the first 15% of the distal tubule past the macula densa.

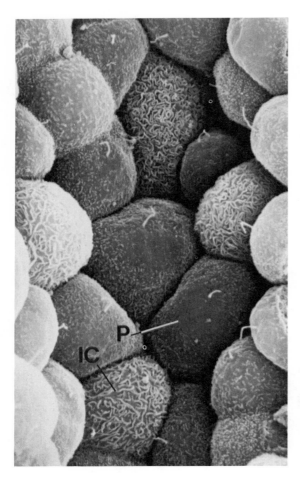

FIGURE 9.32. A scanning electron micrograph of the luminal aspect of the rat CCD showing the intercalated (IC) cells bearing many microfolds and the principal (P) cells, which characteristically have one centrally located cilium.

Similar comparison of sodium-to-inulin TF/P ratios between the early and late "distal tubule" and ureteral urine reveals that most of the remaining 10% of the filtered sodium is reabsorbed along the "distal tubule," with only 1% being reabsorbed along the collecting duct. Quantitatively, similar reabsorption patterns for sodium have been observed in dogs and monkeys.[80]

In-vitro perfusion of isolated rabbit distal convoluted tubule has revealed a lumen-negative transepithelial PD (Table 9.2), which was abolished when ouabain (an inhibitor of Na-K-ATPase) was added to the basolateral side or amiloride (a potassium-sparing and luminal sodium-channel blocker) to the luminal side, suggesting the presence of an electrogenic active sodium transport process.[101] Thus, sodium is reabsorbed against a very high electrochemical gradient, which together with the low osmotic water permeability of this segment helps to maintain the hypotoncity of the tubular fluid. This conclusion is supported by the very high Na-K-ATPase[136] and glycolytic enzyme activity[243] found in this tubular segment. The latter finding suggests that glucose oxidation is the main source of energy for ATP synthesis. Furthermore, this segment was found to be unresponsive to vasopressin and PTH, but was highly responsive to calcitonin.[119]

Like the distal convoluted tubule, the connecting tubule is characterized by high Na-K-ATPase and glycolytic enzyme activity.[243] In microperfused rabbit connecting tubule, addition of PTH to the peritubular side enhanced calcium reabsorption.[119] This effect was mimicked by cAMP derivatives. Furthermore, it has been shown that isoproterenol stimulated adenylate cyclase via β-adrenergic receptors. Taken together, these findings suggest that PTH and β-adrenergic agonists stimulation of cAMP synthesis evokes different physiological responses from the different cell types of the connecting tubule.

The magnitude of sodium reabsorption along the "distal tubule" may be modified markedly by the presence of poorly reabsorbable solutes in the tubular fluid and by the rate of tubular volume flow. Lassiter and associates[160] observed an inverse relationship between the tubular concentration of sodium and urea along the "distal tubule" in nondiuretic rats, concentration of sodium being low and that of urea being high. They attributed this to passive *recirculation* (by back-diffusion) of the urea reabsorbed from the collecting duct into Henle's loop, thereby increasing the urea concentration in the fluid entering the distal convoluted tubule. In contrast, in saline diuresis (induced by intravenous infusion of hypertonic sodium chloride), urea concentration in the "distal tubule" was found to be lower and sodium concentration higher, compared with the nondiuretic conditions. These latter findings were attributed to (1) a reduction in sodium reabsorption in the proximal tubule, thereby increasing its concentration in the "distal tubule," and (2) a reduction in the recirculation of urea from the collecting duct into Henle's loop. Both of these factors tend to reduce urea concentration emerging from Henle's loop and hence in the "distal tubule." The rationale for this conclusion may be stated as follows.

Numerous clearance studies have established that *osmotic diuresis*, induced by intravenous infusion of saline or solutions containing poorly reabsorbable electrolytes or nonelectrolytes, increases urinary excretion of sodium. Micropuncture studies have re-

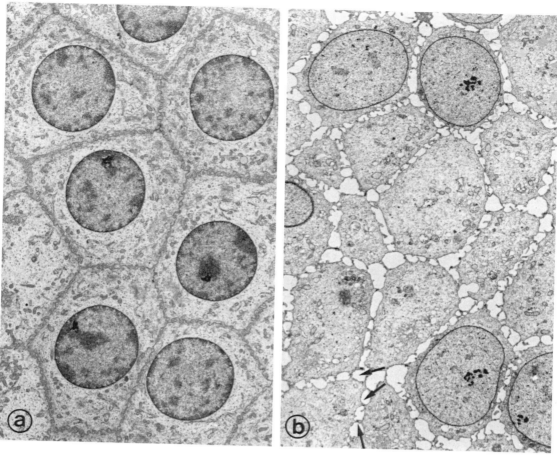

FIGURE 9.33. Electron micrographs of cross-sections through the inner MCD epithelium. In (a) the intercellular spaces are narrow and in (b) they are dilated. The dilated intercellular spaces are probably the result of an ADH-induced water flow through the epithelium. Note that the extent of intercellular space dilation is limited by the lateral microfolds that are connected by desmosomes (*arrows*).

vealed that the natriuresis results from a suppression of proximal reabsorption of sodium, despite a compensatory increase in sodium reabsorption by Henle's loop and to a minor extent by the "distal tubule" and collecting duct. Consequently, a larger than normal fraction of the glomerular filtrate would reach the "distal tubule" and collecting duct. Since urea is reabsorbed by passive diffusion, the high rates of volume flow through Henle's loop, "distal tubule," and collecting duct, caused by diuresis, would reduce the tubular transit time, a factor that prevents the development of a steep urea concentration gradient required for its passive recirculation. As a result, urea concentration in the fluid emerging from the ascending limb of Henle's loop and hence entering the distal convoluted tubule will be lowered. The reduced tubular transit time will also decrease tubular contact time for sodium, a factor that reduces active sodium reabsorption in the "distal tubule," thereby increasing its tubular concentration. Accordingly, in saline diuresis, an increase in sodium concentration in the "distal tubule" is accompanied by a simultaneous decrease in urea concentration. Thus, the extent of normal iso-osmotic reabsorption of the filtrate along the late "distal tubule" depends on the amount of poorly reabsorbable solute present in the tubular fluid. Presence of such a solute will induce osmotic retention of water, which along with continuous active sodium reabsorption, leads to the development of a significant blood-to-lumen concentration gradient for sodium and the opposite for urea.

Both the cortical and the medullary collecting ducts in rabbit exhibit a low osmotic permeability to water and a capacity to reabsorb sodium.[119] Sodium reab-

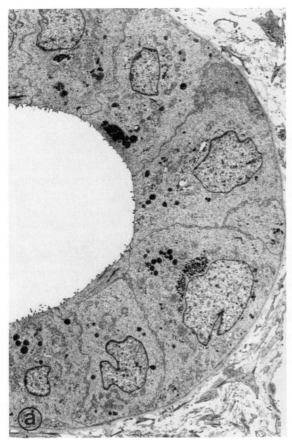 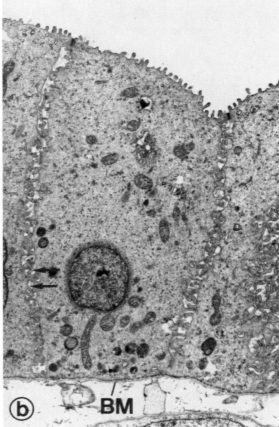

FIGURE 9.34. Electron micrographs of an inner MCD. (a) An overview of the collecting duct in the human kidney. The epithelium is composed of only the principal cells. (b) Details of the epithelium of the inner MCD of the rat kidney. The cells are of an high columnar type. Note the many lateral microfolds (*arrows*) protruding into the lateral intercellular space. BM = basement membrane.

sorption involves active transport because it occurs against an electrochemical gradient and is blocked by ouabain. Sodium reabsorption rate is low, as is evident by the low Na-K-ATPase activity. Additionally, it has been shown that the cortical and outer medullary collecting ducts are relatively impermeable to urea, whereas the inner medullary and papillary collecting ducts have a much higher permeability to urea.[119] This difference in urea permeability along the collecting duct is the basis for urea recycling by diffusion, in the inner medulla, between the collecting duct and the thin limbs of Henle's loop. Its role in concentration and dilution of urine is detailed in Chapter 12.

Another factor that influences the rate of sodium reabsorption by the collecting duct is the blood levels of the adrenal hormone aldosterone.[168] The fraction of sodium reabsorbed under the influence of this hormone is very small, amounting to only 2% of the filtered sodium. Nevertheless, the continuous loss of this amount of sodium, in the absence of aldosterone, would be fatal if not replaced. The available evidence (see below) indicate that aldosterone enhances passive reabsorption of sodium across the luminal principal cell membrane of the cortical collecting duct.[61] Finally, several lines of evidence suggest that the intercalated cells are involved in a catecholamine-dependent active electrogenic secretion of hydrogen ions and/or reabsorption of potassium ions.[119]

From the foregoing discussion it is clear that in contrast to the proximal tubule, the distal convoluted tubule, connecting tubule, and collecting ducts can reabsorb variable amounts of filtrate and its constituents, subject to hormonal intervention. Thus, fluid reabsorption in these segments is nonobligatory and hence is properly called *facultative*. As we will see later, it is the smooth coordination of the adaptive reabsorptive capacity of the "distal tubule" and collecting duct along with hormonal control of fluid and salt reabsorption in these nephron segments that eventually determines the final volume and osmolality of the excreted urine.

Potassium Transport

Numerous micropunture studies in most mammalian nephrons have revealed that potassium is the only plasma electrolyte that is both reabsorbed from and secreted into the renal tubules. Furthermore, these studies have led to the generalization that most of the filtered potassium is actively reabsorbed in the proximal tubule, whereas the excreted potassium is derived from potassium secretion in the distally located nephron segments. However, recent studies[123, 261] have revealed that the renal potassium transport is more complex and that the relatively constant fraction of filtered potassium (10% to 20%) collected from the early distal convolution is the result of potassium reabsorption in the proximal tubule, followed by potassium secretion and reabsorption along Henle's loop. Thus, it appears that analogous to urea, potassium also undergoes a *recycling* in the medulla by being added to the descending limb of Henle's loop and reabsorbed from the ascending limb and/or medullary collecting duct. The following represents a brief review of the major mechanisms involved in potassium transport along the "distal tubule" and collecting duct. Further aspects of renal potassium excretion and the factors affecting it is given in a subsequent section in this chapter.

The marked difference in potassium transport pattern between the proximal and distal tubules is well illustrated in Figure 9.12. In the proximal tubule we see that both the tubular fluid-to-plasma concentration ratios (TF/P) for potassium and the potassium-to-inulin TF/P ratios are clustered below unity, suggesting a net reabsorption of this ion along this nephron segment. This is consistent with the net reabsorption of a large fraction of filtered sodium and water in this segment.

In contrast, in the "distal tubule" we see that the TF/P ratios for potassium increase from an initial low value in the early portion to a high value in the late portion, indicating a progressive increase in tubular potassium concentration. This is due to either a net potassium secretion or net reabsorption of water in excess of solute, or both. However, inspection of the potassium-to-inulin TF/P ratios reveals that although these ratios increase along the "distal tubule," they all fall below unity, suggesting a net potassium reabsorption. Despite this, some potassium must have been secreted into the tubules and eventually excreted in the urine, as is evident by the measurable TF/P ratios for potassium in the ureteral samples.

It appears that potassium secretion is confined only to the late segments of the "distal tubule" (connecting tubule) and collecting duct. This is strongly suggested by the relative decrease in the potassium-to-inulin TF/P ratios in the ureteral urine, compared with those at the end of the "distal tubule." This decline in TF/P ratios indicates net potassium reabsorption in the nephron segments beyond the "distal tubule." This conclusion has been confirmed by recent in-vitro microperfusion studies, which have shown that potassium is secreted into the cortical collecting duct and reabsorbed from the medullary collecting duct.[123, 261] Furthermore, it is well to remember that the distal samples, yielding the data plotted in Figure 9.12, were collected from the *superficial* nephrons which have short loops of Henle, while the ureteral urine samples are an admixture of fluids issued from the superficial nephrons, *midcortical* and *juxtamedullary* nephrons which have long loops of Henle.

A recent quantitative analysis of the potassium mass flow data obtained from in-vitro microperfusion and in-vivo micropuncture studies along the superficial and juxtamedullary nephrons has provided compelling evidence that potassium, like urea, is recycled within the renal medulla.[123] Accordingly, potassium is added to the descending limb of the juxtamedullary nephrons as a result of increased potassium concentration in the medullary interstitium. In the ascending limb, the relatively high ionic permeability of the thin segment and the lumen-positive transepithelial voltage in the thick segment provide favorable electrochemical gradients for reabsorption of potassium by these segments. By re-entry into descending limbs potassium is trapped within the medulla by Henle's loop. In addition, potassium reabsorption from medullary collecting ducts (subsequent to secretion into cortical collecting ducts) will elevate potassium concentration in the medullary interstitium and contribute to the process of potassium trapping.

Finally, morphological studies have shown that potassium secretion is associated with the principal cells,[246] whereas potassium reabsorption is associated with the intercalated cells.[132] As mentioned earlier, the principal cells are found in the collecting duct; however, the connecting tubule cells have many structural and functional similarities with the principal cells. The intercalated cells are found in the connecting tubule and collecting duct. Although the mechanism of potassium secretion has received the most attention, little is known about the mechanism of potassium reabsorption.

Chloride Transport

As shown in Figure 9.13, and similar to sodium, chloride concentration decreases along the "distal tubule," reaching a value of about one fifth of that in the plasma by the end of this segment. Furthermore, comparison of the early and late "distal" tubular F/P ratios indicates that about 6% of the filtered chloride is reabsorbed along this nephron segment.

Recent in-vitro microperfusion studies of the isolated rabbit "distal tubule" have revealed that chloride is reabsorbed actively against its electrochemical gradients,[24] presumably coupled to the energy provided by passive entry of sodium into the cell. This is consistent with the findings of a recent study[244] that showed that the net sodium reabsorption doubled in the presence of luminal chloride, as compared with the absence of chloride. Also like sodium, reabsorption of chloride along the "distal tubule" was adversely affected by the presence of poorly reabsorbable anions, such as sulfate. In addition to active reabsorption, a portion of chloride may also be transported by passive diffusion.

In-vitro microperfusion studies in rabbit have shown that the permeability of the cortical collecting duct to chloride is much higher than that to sodium or potassium.[190] This suggests that at least a portion of chloride is reabsorbed passively down its electrochemical gradient in this nephron segment. However, it is not known what fraction of chloride reabsorption occurs by transcellular and paracellular pathways. Furthermore, the question of active chloride reabsorption in this segment has not been completely resolved.

Measurement of transepithelial PD in the medullary collecting duct shows that the lumen-negative voltage becomes positive as one proceeds from the corticomedullary border toward the papilla.[230] However, currently, the mechanism of chloride transport in this nephron segment is incompletely understood.

Finally, in-vivo micropuncture and in-vitro microperfusion studies have shown that chloride is reabsorbed in the papillary collecting duct.[111] Isotopic studies have shown that the permeability of this segment to chloride is rather low.[200] This finding, coupled with the very low transepithelial PD (0 to -10 mV) measured across this segment,[200] suggests that most of the chloride is actively reabsorbed along the papillary collecting duct. Although the possibility that a small fraction of chloride may be reabsorbed by passive diffusion can not be excluded. The precise mechanism of active chloride transport in this nephron segment is currently not known.

Mechanism of Potassium Secretion

It is now well established that among the distally located nephron segments the connecting tubule and the collecting duct are the sites of potassium secretion. However, in older micropuncture studies, these segments were considered part of the "distal tubule." Therefore, in this section we have placed the term "distal tubule" in quotation marks to remind the reader of this difference in nomenclature (see also Appendix C).

Results of numerous micropuncture studies of the type illustrated in Figure 9.12 have revealed that net secretion and net reabsorption of potassium occur in the connecting tubule and collecting duct. Moreover, these studies have shown that potassium secretion by the principal cells (and probably the connecting tubule cells) is influenced by several factors. These include (a) changes in tubular concentration of potassium, sodium, and chloride, (b) changes in tubular flow rate, (c) administration of luminally acting inhibitors of sodium and potassium transport, (d) changes in plasma concentrations of potassium, hydrogen, and bicarbonate ions, and (e) changes in plasma levels of aldosterone and ADH. Since the first three exert their effects primarily at the luminal side and the last two at the peritubular side of the principal cell membrane, we describe the mechanism of their action in accordance with the sites of their actions.

To better understand how the luminal and peritubular factors exert their effects on potassium secretion, we begin with an overview of the factors that govern filtrate transport along the distal tubule.

Direct measurements of electrochemical potential gradient along the rat "distal tubule" have revealed that (1) Both the concentration and the percentage of potassium remaining in the tubular fluid increase,

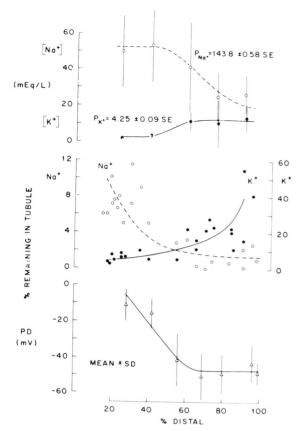

FIGURE 9.35. Tubular fluid sodium and potassium concentrations, relative rates of sodium reabsorption and potassium secretion (expressed as % remaining), and transepithelial electrical potential difference (PD) along the "distal tubule" of rat nephron. P_{Na}^+ and P_K^+ are the measured plasma concentrations of sodium and potassium, respectively. Vertical bars depict standard deviation (SD) of the plotted data. (From Giebisch and Windhager.[80])

while those for sodium decrease (Figure 9.35, upper and middle panels). (2) There is a progressive increase in the luminal negativity along the "distal tubule" (Figure 9.35, lower panel). This would account for the relative increase in potassium permeability in the second half of the "distal tubule," as is evident by the increased percentage of this ion appearing in that segment (Table 9.2).

This observed inverse relationship between the tubular potassium and sodium concentrations, together with the reduced reabsorptive rate for sodium along the "distal tubule," alluded to earlier, would imply a lowering of the intracellular concentration of sodium. If so, to preserve isotonicity within the tubular cell, there must be a compensatory increase in the intracellular potassium concentration.

The above arguments suggest the possibility of a one-to-one exchange as the possible mechanism for sodium reabsorption and potassium secretion at the luminal membrane in the connecting tubule and collecting duct. That this is not the only mechanism of potassium secretion is supported by several studies, notably those by Malnic and associates.[170]

Malnic and co-workers,[170] using in-vivo micropuncture techniques, made a comprehensive study of sodium and potassium transport along the rat proximal and distal tubules in response to two classes of forcings: (1) Those that modify the intake of either potassium or sodium, thereby affecting the cellular uptake of these ions from blood as well as their filtered loads and hence their net tubular transport. (2) Those that alter the volume flow rate in the tubules, thereby modifying net tubular reabsorption of sodium and hence secretion of potassium. Their findings, summarized in Figures 9.36 and 9.37, reveal several important characteristics of potassium transport along the rat nephron.

Gross inspection of these figures shows that despite a wide variation in the urinary excretion rates of potassium (Figure 9.37), the fraction of filtered potassium entering the distal convoluted tubule varies but little (Figure 9.36). This indicates that regardless of the type of forcing used, the fractions of filtered potassium reabsorbed by the proximal tubule and by Henle's loop are quite comparable. In other words, even in conditions that caused the urinary excretion rate of potassium to exceed its filtered load, virtually all of the filtered potassium is reabsorbed, and the excreted potassium is primarily of secretory origin.

Using the data presented in Figures 9.36 and 9.37, let us now examine the effects of the luminal and peritubular factors on potassium transport capacity of the connecting tubule and collecting duct and analyze their possible mechanistic significance.

Luminal Factors Affecting Potassium Secretion

Effects of Changes in "Distal" Tubular Fluid Composition

It is well established that changes in the concentration of potassium, sodium, and chloride in the fluid entering the distal convoluted tubule exert profound effects on potassium secretion in the distal nephron

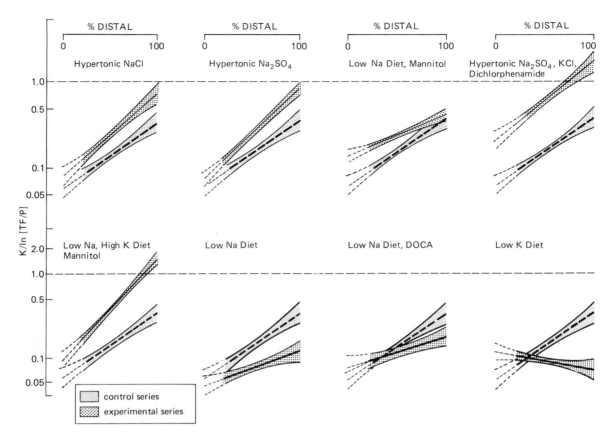

FIGURE 9.36. Comparison of potassium-to-inulin TF/P ratios along the "distal tubule" in rats under various dietary intake regimes and fluid loading. Regression lines were obtained by least-squares method. The width of the area around each regression line represents ± 1 standard error. Values between 0% and 20% distal tubular length (*dashed lines*) were extrapolated. (From Malnic et al.[170])

(the distal convoluted tubule, the connecting tubule, and the entire collecting duct). The tubular concentrations of these ions may be changed indirectly by changing their dietary intake or directly by perfusing the tubules with known concentrations.

The effects of reduced dietary intake of potassium and sodium on potassium secretion by the late "distal tubule" are shown in the lower panel of Figure 9.36. In both cases there is a marked suppression of potassium secretion compared with normal. In the case of low-potassium diet, reduced plasma concentration of potassium would tend to decrease cellular uptake of potassium, thereby lowering the intracellular concentration of this ion and hence its net secretion. A reduced filtered load of potassium secondary to a reduction in GFR, on the other hand, would tend to increase reabsorption of potassium along the nephron due to an increase in the tubular transit time and hence decrease its urinary excretion rate. The combined effect of these two mechanisms is to reduce urinary excretion rate of potassium (Figure 9.37) by maximizing net reabsorption and minimizing net secretion.

In the case of low-sodium diet, reduced filtered load of sodium coupled with its continuous active reabsorption in the proximal tubule would tend to lower sodium concentration in the fluid entering the distal convoluted tubule. The lowered sodium concentration in the tubular fluid will tend to decrease net potassium secretion. Conversely, any factor that increases "distal" tubular concentration of sodium, such as administration of poorly reabsorbable salts, like sodium sulfate, enhances potassium secretion and suppresses sodium reabsorption by this nephron segment. Since sulfate is a poorly reabsorbable anion, it tends to increase luminal negativity and hence transepithelial PD, a factor favoring net potassium secretion. A recent study proposed that the effect of

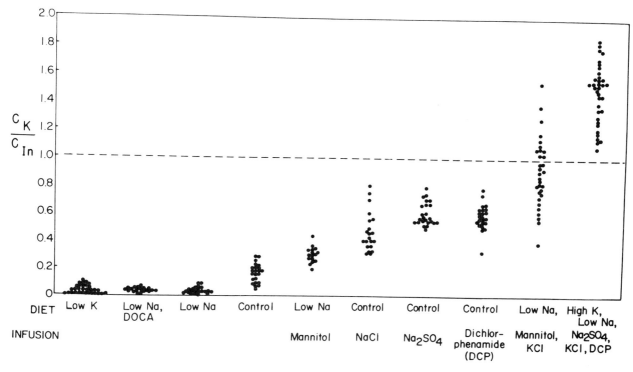

FIGURE 9.37. Comparison of potassium-to-inulin clearance ratios (C_K/C_{In}) along the "distal tubule" in rats under various dietary intake regimes and fluid loading. (From Malnic et al.[170])

a poorly reabsorbable anion on potassium secretion is secondary to its effect on enhancing chloride reabsorption in the predistal nephron segments, thereby lowering luminal chloride concentration and reducing coupled chloride-potassium cotransport across the luminal membrane.[244] This may explain why sodium sulfate is a more powerful kaliuretic agent than sodium chloride.

Recent in-vivo microperfusion studies in rat[85, 86] have further clarified the mechanism whereby changes in luminal sodium and potassium cncentrations modulate potassium secretion in the "disal tubule." Thus, a decrease in "distal" tubular potassium concentration increased potassium secretion. An increase in the "distal" tubular flow rate, without increasing sodium concentration, markedly increased potassium secretion while reducing the transepithelial PD. This was attributed to the flow-dependent decrease in luminal potassium concentration, thereby favoring potassium secretion from the cell into the lumen. However, an increase in "distal" tubular sodium, while keeping the flow rate unchanged, had no effect on potassium secretion or the transepithelial PD. Since luminal sodium concentration in the "distal tubule" is normally about 35 mM/L, it is unlikely that an increase in sodium concentration has a stimulatory effect on potassium secretion. In a subsequent study, Good and associates showed that lowering the tubular chloride concentration increased potassium secretion even at very low tubular sodium concentration.[87] This observation provides further evidence for the existence of a chloride-potassium cotransport system in the luminal membrane of the rat "distal tubule." Taking these findings together, they suggest that, similar to the chloride-dependent increase in potassium secretion in the presence of a poorly reabsorbable anion, the increase in potassium secretion secondary to changes in sodium concentration and transepithelial PD is also chloride-dependent. These findings further suggest that the effect of changes in luminal sodium concentration on potassium secretion is secondary to changes in the tubular flow rate at this site.

That electrical coupling of sodium-potassium rather than a one-to-one exchange may be the mechanism for potassium secretion is implied by those experiments that have used aldosterone and its inhibitors to dissociate sodium reabsorption from potassium secretion. Administration of deoxycorticosterone-acetate (DOCA), a synthetic mineralocorticoid known to induce hypokalemic alkalosis, to rats fed low-sodium diet (Figure 9.36), *lower panel*) had surprisingly little effect on potassium secretion. Thus, it appears that a low-sodium diet tends to protect the animal from the kaliuretic effect of DOCA.

A more striking finding in this respect is the observation of Wiederholt.[251] He found that in adrenalectomized rats treated with aldosterone, administration of actinomycin D inhibited the antinatriuretic effect of aldosterone but did not impair the ability of the "distal tubule" to secrete potassium. It appears that actinomycin D inhibits DNA-dependent RNA synthesis (the so-called gene transcription), an intermediate step in aldosterone-induced stimulation of sodium reabsorption in the connecting tubule and collecting duct (see subsequent sections in this chapter for further details). These findings are consistent with the view that a one-to-one sodium-potassium exchange can not account for potassium secretion.[180]

Effect of Changes in Tubular Flow Rate

As stated above, changes in tubular flow rate, independent of changes in tubular composition and transepithelial PD, produce directional changes in potassium secretion. Thus, any factor that increases tubular flow rate will increase potassium secretion. Conversely, any factor that decreases tubular flow rate will decrease potassium secretion. For example, expansion of the extracellular fluid volume with hypertonic sodium chloride, Na_2SO_4, or mannitol will have varying effects on "distal" tubular flow rate and potassium secretion, depending on the mechanisms involved in their renal transport. As depicted in Figure 9.36 (upper panel), intravenous infusion of either hypertonic sodium chloride or Na_2SO_4 increased the fraction of filtered potassium remaining at the end of the "distal tubule." This indicates that *osmotic diuresis* (as the renal effect of intravenous infusion of these and similar solutions is called) induces kaliuresis as a result of increased net potassium secretion. The explanation for this finding is as follows. As described earlier, osmotic diuresis depresses proximal tubular reabsorption of sodium and water, thereby increasing the volume of filtrate delivered to the "distal tubule". This high volume flow rate reduces potassium concentration in the luminal fluid, a factor that stimulates net potassium secretion. As is evident from the data plotted in Figure 9.36, kaliuresis can be markedly enhanced if the increase in tubular flow rate is induced by the presence of a poorly reabsorbable anion, such as sulfate.

In contrast, osmotic diuresis induced by hypertonic mannitol (a poorly reabsorbable nonelectrolyte) in rats fed low-sodium diet markedly reduced the contribution of the "distal" tubular potassium secretion to the total potassium excreted in urine, as compared with hypertonic sodium chloride or Na_2SO_4 loading (Figure 9.37). Also, it was found, but not shown in Figure 9.37, that mannitol diuresis reduced potassium reabsorption in the proximal tubule, a factor that contributed to the extent of urinary excretion of potassium.

The diminished urinary excretion of potassium is not an inherent characteristic of the "distal tubule." This is demonstrated by the plots shown in the lower left corner of Figure 9.36. These data are from rats fed low-sodium, high-potassium diet and receiving an isotonic potassium chloride solution mixed with hypertonic mannitol solution. The steep positive slope of the potassium-to-inulin TF/P ratios, including some values greater than unity, are indicative of the secretory capacity of the "distal tubule" for potassium. These results indicate further that sodium deprivation did not compromise the capacity of these tubular segments for potassium secretion.

Effect of Inhibitors of Sodium and Potassium Transport

Potassium secretion is affected by diuretic agents that inhibit luminal transport of sodium and potassium. Thus, when rats fed a high-potassium, low-sodium diet received hypertonic Na_2SO_4 and isotonic potassium chloride solutions along with dichlorophenamide, a diuretic agent, potassium secretion was maximally stimulated. This diuretic is a potent carbonic anhydrase inhibitor, the effect of which is to interfere with sodium reabsorption via the Na/H exchange mechanism in the proximal tubule, and hence to induce natriuresis. The results, plotted in the upper corner of Figure 9.36, show a marked increase in net potassium secretion and urinary excretion (Figure 9.37), which greatly exceeded those observed during the osmotic diuresis alone.

Two other agents are known to affect potassium secretion when present in the tubular lumen. The first is amiloride, a potassium-sparing diuretic, which inhibits both sodium reabsorption and potassium secretion by the "distal tubule" and collecting duct. It is believed that amiloride blocks the channels through which sodium enters into the cell across the luminal membrane.[189] The blockage of sodium diffusion into the cell reduces the lumen-negative transepithelial PD, thereby reducing the driving force favoring potassium diffusion from the cell into the lumen. The second is barium which, in small concentrations, also inhibits both sodium reabsorption and potassium secretion in the connecting tubule and collecting duct.[60] However, in contrast to amiloride, barium blocks the channels through which potassium diffuses out of the cell. This effect leads to hyperpolarization of the luminal membrane and to an increase in the lumen-negative transepithelial PD. The resulting change in the transepithelial PD opposes passive diffusion of sodium from the lumen into the cell.

Peritubular Factors Affecting Potassium Secretion

Effect of Changes in Plasma Concentration of Potassium, Hydrogen, and Bicarbonate Ions

As noted earlier, a chronic increase in potassium intake leads to an increase in plasma potassium concentration, which increases potassium secretion by activation of three separate mechanisms. First, the increase in plasma potassium concentration increases Na-K-ATPase stimulated cellular uptake of potassium. Second, the elevation of plasma potassium concentration stimulates the zona glomerulosa of the adrenal cortex (see below) to increase synthesis of aldosterone. The latter stimulates Na-K-ATPase activity and increases the number of sodium pumps, as is evident by amplification of the basolateral membrane.[226, 227] The net result is an increase in cellular potassium uptake. Finally, an increase in plasma potassium concentration is known to inhibit sodium and water reabsorption in the proximal tubule, thereby increasing the "distal" tubular flow rate, a factor stimulating potassium secretion.

As we shall see in Chapter 11, changes in bicarbonate reabsorption are accompanied by opposite changes in plasma chloride concentration. Furthermore, plasma concentration of bicarbonate is determined by fortuitous gain or losses of acid or alkali. As noted earlier, a large fraction of bicarbonate reabsorption is associated with sodium reabsorption via the Na/H exchange mechanism in the proximal tubule. Thus, any factor that interferes with the Na/H exchange mechanism will lead to a decrease in sodium and fluid reabsorption, thereby increasing "distal" tubular flow rate and potassium secretion.

Plasma hydrogen ion concentration can be altered by fortuitous gain or loss of acid or alkali. Thus, both hyperventilation (respiratory in origin) and excess alkali influx into the body (metabolic in origin) would lower the plasma hydrogen ion concentration, resulting in states of respiratory and metabolic alkalosis, respectively. The decrease in plasma hydrogen ion concentration causes hydrogen ions to be shifted from the cell into the plasma. To maintain electrical neutrality, there will be a concomitant increase in the cellular uptake of potassium. The resulting increase in the intracellular potassium concentration leads to an increase in the net secretion of this ion. Conversely, in acute respiratory and metabolic acidosis, conditions manifested by an increase in plasma hydrogen ion concentration, changes opposite to those mentioned above, will lead to a decrease in the cellular uptake of potassium and hence the net secretion of this ion. In contrast, in chronic acidosis there is an increase in net potassium secretion. The resulting condition, called *dissociated metabolic acidosis*, is characterized by excretion of alkaline urine (secondary to intracellular alkalosis) despite the extracellular acidosis (see Chapter 11). It is apparent that under the condition of acute hydrogen ion imbalance, potassium secretion is inversely related to hydrogen ion secretion.

Alteration in body water balance also influences potassium secretion by the connecting tubule and collecting duct. Excessive loss of hypotonic fluid from the body, such as may occur in severe sweating, will lead to extracellular hypertonic dehydration (see Chapters 1 and 3). This will induce a shift of water from the intracellular to the extracellular compartment, thereby increasing cellular concentration of potassium and hence net secretion of this ion. Conversely, excessive water intake results in extracellular hypotonic hydration. This will induce a shift of water from the extracellular to the intracellular compartment. The net effect is to reduce cellular concentration of potassium and hence net secretion of this ion.

Effect of Aldosterone, ADH, and Diuretics

Acute (short-term) administration of aldosterone causes an increase in sodium permeability of the luminal membrane, which in turn increases Na-K-ATPase activity and sodium transport. This conclusion is based on the observation that in the isolated microperfused rabbit cortical collecting duct, amiloride inhibited the acute increase in Na-K-ATPase activity after exposure to aldosterone.[194] However, chronic (long-term) administration of aldosterone stimulates sodium reabsorption and potassium secretion. As noted earlier, aldosterone stimulates Na-K-ATPase and induces amplification of the basolateral membrane of the principal cells, which reflects a possible increase in the number of Na-K-ATPase pumps. The net effect is an increase in sodium reabsorption and potassium secretion.

Recent studies have demonstrated that changes in plasma ADH concentration also affect renal potassium secretion.[70] It appears that ADH, by increasing the osmotic water permeability of the luminal membrane of the principal cells in the collecting duct, also increases permeability to potassium. The net result is an increase in potassium diffusion out of the cell into the lumen. They have suggested that this effect of ADH may serve to prevent potassium retention during periods of antidiuresis.

Finally, chronic administration of some diuretics, such as chlorothiazide (a carbonic anhydrase inhibitor), furosemide, and ethacrynic acid increases net potassium secretion. Although they differ in their sites of action, they all interfere with sodium reabsorption, thereby increasing both tubular volume flow rate and sodium concentration. As mentioned earlier, both of these factors stimulate net potassium secretion. In contrast, amiloride depresses potassium secretion by reducing the transepithelial PD.

Hormonal Regulation of Tubular Transport Along the Distal Tubule and Collecting Duct

In the preceding sections, while considering the filtrate transport along Henle's loop, we stated that in the mammalian kidney the final volume and osmolality of the excreted urine ultimately depends on the extent of osmotic equilibration of the fluids in the collecting duct with the surrounding hypertonic medullary interstitium. Briefly, there are three factors that contribute to this osmotic equilibration: (1) Osmotic reabsorption of water from the collecting duct, a process ensured by the presence of ADH; (2) passive reabsorption of urea from the cortical and outer medullary collecting duct, as well as ADH-enhanced passive urea reabsorption from the inner medullary collecting duct, which initiates the recycling of urea within the renal medulla (see Chapter 12); and (3) active reabsorption of sodium, which had escaped reabsorption in the proximal tubule, Henle's loop, and the distal convoluted tubule along with its associated anions from the collecting duct. This latter process is profoundly influenced by the circulating level of another hormone, namely, the adrenocortical hormone aldosterone. It therefore follows that any factor that influences the plasma concentration of ADH and aldosterone must exert a marked effect on the final volume and osmolality of the excreted urine.

To better understand the roles of these two hormones in the concentration and dilution of urine, and hence in the regulation of the final volume and osmolality of the excreted urine, it is necessary to closely examine the sites and mechanisms of their cellular actions as well as the factors governing their effects on the target tissues. To facilitate this presentation and subsequent discussion, we include a brief survey of the biosynthesis, secretion, and metabolism of each hormone and the factors affecting them.

Mechanism of Action of Antidiuretic Hormone

Biosynthesis, Secretion, and Metabolism

In a landmark study, Kamm and associates[135] showed that injection of crude extracts of the posterior lobe of the pituitary gland (neurohypophysis) into mammals produces three distinct physiological responses: (1) It causes uterine contraction, an action referred to as the *oxytocic* effect; (2) it raises the systemic arterial blood pressure, an action referred to as the *vasopressor* effect; and (3) it decreases the volume of excreted urine, an action referred to as the *antidiuretic* effect. Subsequent purification and separation of the crude extracts led to the successful isolation and synthesis of three chemically distinct hormones, each exhibiting the oxytocic, vasopressor, and antidiuretic effects to varying degrees.[51] Structurally, all three hormones are nanopeptides with molecular weights of slightly more than 1,000 and each consists of a cyclic unit formed by a disulfide-bridged pentapeptide and a tripeptide side chain unit. The three hormones are named oxytocin, [8-arginine]-vasopressin and [8-lysine]-vasopressin. Of these, [8-arginine]-vasopressin (AVP), which has an arginine residue in the 8 position in the side chain, is the most prevalent form of the posterior pituitary hormone found in mammals. It has been isolated as a natural hormone in man, dog, rat, rabbit, and sheep. In contrast, [8-lysine]-vasopressin (LVP), in which a lysine residue replaces the arginine residue in the 8 position in the side chain, has been isolated only from the hog pituitary. Although all three hormones cause antidiuresis, the antidiuretic effects of AVP and LVP are some 100 times that of oxytocin. For this reason, the two vasopressins are customarily referred to as the antidiuretic hormones or ADH.

It is generally accepted that vasopressin is primarily synthesized in the supraoptic nucleus, whereas oxytocin is synthesized in the paraventricular nucleus of the hypothalamus. The synthesized hormones are then carried by neurosecretory granules through the hypothalamo-hypophyseal nerve tracts into the neurohypophysis, where they are stored for subsequent release into the circulation.[204] Subsequent studies have confirmed this concept and have further clarified the role of the neurosecretory granules in the biosynthesis and transport of the neurohypophyseal hormones. According to these new findings, the neurosecretory granules contain a group of small molecular weight proteins, called neurophysin, which bind oxytocin and vasopressins. As such, neurophysin serves as both "precursor" and "carrier" of the neurohypophyseal hormones, thereby providing a vehicle for their transport from the hypothalamus to the posterior lobe of the pituitary gland, where they are stored. Subsequent to proper stimulation, the appropriate hormone is then separated from the neurophysin protein and is released as a free polypeptide into the circulation.[263]

The plasma concentration of ADH, based on bioassay of pressor activity of pure [8-arginine]-vasopressin (400 pressor units per milligram) is about 5 microunits (μU), or 5×10^{-5} μg/mL plasma, in normal subjects.[217] The plasma concentration of the hormone is maintained by a dynamic balance between the rates of its biosynthesis and secretion into the circulation and the rate of its removal from the blood. Although the rates of biosynthesis and secretion of vasopressins are at present not well understood, they appear to be largely dependent on the blood volume and its osmolality, with the latter being the major stimulus (see Chapter 13 for further details). However, there is little disagreement about the rates and major routes of removal of ADH from the circulation. Results of numerous clearance studies have shown that ADH is removed from the blood by three major routes: (1) hepatic clearance and metabolic inactivation by the liver, (2) renal clearance and excretion in the urine, and (3) utilization by the renal tubular target tissues.[162] Furthermore, clearance studies in dogs and rats have shown that only half of the injected ADH reaches the kidney, while the other half is cleared and inactivated by the liver. Of the amount of ADH that reaches the kidney, approximately half is excreted in the urine. The circulatory half-life of injected ADH averaged about 6.5 min in these studies. Recent clearance studies have shown that a fraction of ADH is bound to protein and that the extent of protein binding varies with its plasma concentration.[218] Thus, only 10% is bound to protein when plasma ADH concentration is less than 5 μU/mL, and nearly 40% is bound when the plasma ADH concentration exceeds 20 μU/mL. At normal plasma concentration, renal clearance of ADH is the result of glomerular filtration, degradation, or reabsorption in the proximal nephron, and secretion in the distal nephron.[218] Hence, its renal clearance and urinary excretion represent that fraction of the filtered hormone that has escaped tubular reabsorption.

Modes and Sites of Cellular Action

The classic experiments of Verney[245] were the first to define the nature of the ADH-releasing stimulus, the target organ, and its response characteristics. Verney found that intracarotid injection of hypertonic saline or sucrose solution reduced previously established water diuresis in conscious dogs. Moreover, a sustained 2% increase in the osmolality of blood was sufficient to produce a 90% reduction in maximal water diuresis. This response was completely abolished following removal of the posterior lobe of the pituitary gland. In a later study, Jewell and Verney[125] showed that oliguria (a marked reduction of urine volume) resulted whenever the hypertonic solution reached the supraoptic nucleus. On the basis of these findings, Verney and associates postulated that the cells of the supraoptic nucleus act as the *"osmoreceptors"* responding to alterations in the osmotic pressure or osmolality of the blood perfusing them. Since these cells were known to produce ADH, they postulated further that the release of this hormone increased the osmotic reabsorption of water by the kidney, a factor accounting for the inhibition of water diuresis and concentration of the excreted urine. Although subsequent studies have confirmed the above mechanism for the release of ADH and its overall renal action[217] (see also Chapter 13), localization of ADH-responsive target tissues within the nephron and the mode of its cellular action had to await the application of micropuncture and microperfusion techniques to the kidney tubules.

Micropuncture and microperfusion studies of the

"distal tubule" and collecting duct, cited earlier, have clearly established that (a) the collecting duct is the only ADH-responsive target tissues within the kidney, and (b) ADH not only increases the osmotic water permeability of these tubular segments, but it also enhances the passive urea permeability of the inner medullary collecting duct. Furthermore, it has been shown that ADH increases the osmotic water permeability of the luminal membrane of the renal cells in these tubular segments, but has no effect on the water permeability of the peritubular membrane.[178] However, to have any effect on the permeability of the luminal membrane, ADH must be added to the surface of the peritubular membrane, which is the blood side of the collecting duct. Thus, the antidiuretic effect of ADH is obtained only when the hormone reaches the collecting duct by way of the blood vessels. Consequently, although the ADH, which is filtered at the glomerulus, is exposed to the luminal side of the tubular cell, insofar as water reabsorption is concerned, it has no physiological effect.

In summary, ADH-enhanced osmotic reabsorption of water in the collecting duct appears to be a consequence of some complex intracellular biochemical processes that are presumed to be initiated by exposure of the surface of the basolateral membrane to ADH. Exactly how the presence of ADH at the basolateral surface initiates these complex cellular processes, and how these processes alter osmotic water permeability of the luminal surface, has been the subject of intensive investigation.

Our present knowledge of the mechanism of cellular action of ADH has largely been derived from an extension of studies in anuran membrane (toad urinary bladder) to mammalian renal tubules.[191, 192] In toad bladder it has been found that when ADH is added to the fluid in contact with the serosal (peritubular or blood) side, it causes an increase in both diffusional and osmotic reabsorption of water from the mucosal (luminal or urinary) side of the epithelial membrane. However, placing the ADH in the fluid bathing the mucosal side produces no effect, suggesting that ADH cannot enter the cell. In contrast, osmotic transfer of water from the mucosal side to the serosal side was increased when cAMP was placed in the fluid bathing either the serosal or mucosal side, suggesting that cAMP can enter the cell. From these studies, the following general inferences have been made: (1) The barrier for the osmotic reabsorption of water lies within the mucosal side of the epithelial membrane; (2) ADH, acting on the serosal side, somehow causes an increase in the osmotic water permeability of the mucosal side; and (3) since cAMP mimics the action of ADH, its cellular concentration somehow might mediate the ADH-enhanced water reabsorption. This latter inference is based on the postulated role of cAMP[231] as the "second messenger," mediating the hormone-stimulated biological response of a variety of target tissues (see Chapter 8, Figure 8.13).

In a series of elegant experiments, Grantham and associates[95, 96] extended these observations to the mammalian nephron, thereby providing the most direct experimental evidence for the site of action and the possible cellular mechanism of ADH-enhanced water reabsorption. Using isolated fragments of rabbit cortical collecting duct, they found that in the absence of ADH, when the peritubular surface of the tubular segment was exposed to hypo-osmotic, iso-osmotic, and hyperosmotic solutions, the tubular cells behaved like an osmometer: shrinking in hyperosmotic and swelling in hypo-osmotic solutions. However, no such response was observed when the luminal surface of the tubular segments was exposed to the same solutions. In contrast, when ADH was added to the fluid bathing the peritubular surface, the previously unresponsive luminal surface became "active" and once again the tubular cells behaved like an osmometer. However, no osmometer-like effect was observed when ADH was added to the fluid bathing the luminal surface. In contrast, when cAMP was added to the fluid bathing either the peritubular or luminal surfaces, the tubular cells once again behaved as an osmometer. Electron microscopic examination of both ADH-stimulated and cAMP-stimulated tubular fragments revealed an increase in the size of the existing aqueous channels (intercellular space with tight junction; see Figure 9.15) within the tubular cells, an observation that, in part, accounted for the altered osmotic water permeability. From these findings, analogous to those for the anuran membrane, the following general conclusions were drawn: (1) The luminal membrane of the tubular cell constitutes the barrier for the osmotic reabsorption of water; (2) ADH stimulation of the peritubular membrane of the tubular cell somehow increases the water permeability of the luminal membrane; and (3) since cAMP mimics the action of ADH but, unlike the latter, it can enter the renal cells, ADH-enhanced water reabsorption is somehow mediated by the intracellular production of cAMP, which, in an unknown manner, alters the water permeability of the luminal membrane.

The mystery of the intracellular processes initiated by ADH stimulation has now been partly resolved by additional experiments. The results of these experiments have led to the formulation of the hypothesis shown in Figure 9.38, depicting the cellular acion of ADH. From the functional point of view, we may divide the various biochemical reactions involved into two sequential steps: (1) An ADH-dependent formation of cAMP (cAMP breakdown within the cell is independent of the hormone); and (2) cAMP-dependent phosphorylation of the proteins in the luminal membrane, thereby bringing about altered water permeability (dephosphorylation of the luminal membrane is independent of the hormone).

Let us now consider the details of these two steps, starting with the interaction of ADH with the basolateral membrane.

Experiments with the isolated perfused collecting ducts, cited above, have shown that ADH induces osmotic reabsorption of water from these epithelia only when the hormone is present in the fluid bathing the basolateral membrane. From this, it has been inferred that the receptor for ADH must be located on the outer surface of the basolateral membrane of the tubular cell. Moreover, since cAMP has been found to mimic the cellular action of ADH, it is postulated that the stereospecific binding of ADH with the receptor somehow activates enzymatic production of cAMP. It has been shown that the enzyme adenylate cyclase, situated in the inner surface of

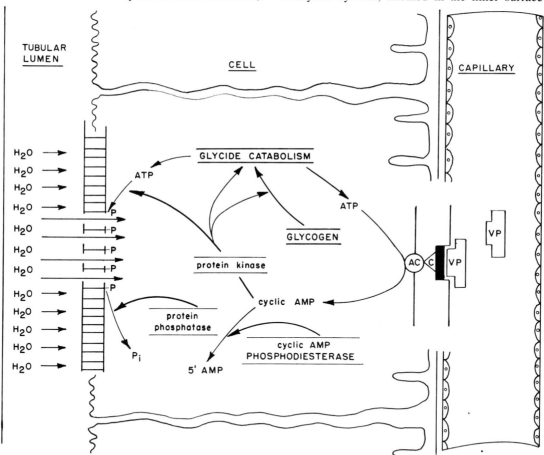

FIGURE 9.38. The proposed mechanism of cellular action of ADH. VP designates a molecule of vasopressin or ADH, and the triangle C represents the hypothetical coupler that connects the membrane-bound receptor (*dark rectangle*) to the catalytic site of the adenylate cyclase (AC). (Reprinted with permission from Life Sciences, 13:1033-1040. Modified from Dousa, TP., role of cyclic AMP in the action of antidiuretic hormone on the kidney. Copyright 1973, Pergamon Press, Ltd.)

the cell membrane, catalyzes the formation of cAMP from adenosine triphosphate (ATP).[231] The latter is supplied by cellular breakdown of glycogen, a reaction stimulated by both ADH and cAMP (Figure 9.38).

The concentration of cAMP thus formed within the cell is regulated by a cytoplasmic enzyme called cAMP-dephosphodiesterase. This enzyme, which is not influenced by ADH, converts cAMP into 5'-AMP, an inactive degradation product. Therefore, cAMP-phosphodiesterase plays an important role in the cellular metabolism of cAMP and the extent to which it mediates ADH-induced water reabsorption. Inhibition of this enzyme by reagents such theophylline, aminophylline, or chlorpropamide can potentiate the effect of ADH or exogenous cAMP on osmotic water reabsorption.[12]

The precise mechanism whereby the intracellular formation of cAMP induces altered water permeability of the luminal membrane is currently not well understood. However, on the basis of several in-vitro microperfusion as well as isolated cell experiments, the following picture has emerged. Intracellular formation of cAMP stimulates an intracellular enzyme called protein kinase, which catalyzes the transfer of γ-phosphate of ATP to serine or threonine residues of the protein phosphatase. The latter is located on the inner surface of the luminal membrane of the renal cell. Thus, cAMP, by activating protein kinase, may bring about phosphorylation of the luminal membrane, the postulated consequence of which is the alteration of the structure of the luminal membrane, and thereby an increase in its osmotic water permeability.

The scheme depicted in Figure 9.38 represents a mixture of both experimentally verified and postulated steps that may be involved in the cellular action of ADH. As such, it has served a useful purpose not only in synthesizing the various experimental and theoretical information, but also in guiding future research. Thus, it has helped to elucidate the nature of pathogenesis of nephrogenic diabetes insipidus. This is a disease characterized by polyuria and a decrease or loss of the ability of the kidney to increase the osmolality of urine in response to exogenous ADH. Although the exact mechanism underlying its pathogenesis is currently not well understood, results of a number of experiments suggest that the defect may be due to a decrease in ADH-stimulated cAMP formation within the renal cell.[50]

Recent studies, using vasopressin analogs, agonists, and antagonists, have characterized two types of vasopressin isoreceptors: antidiuretic and vascular.[124] The antidiuretic isoreceptor (V_2) exerts its biological action through cAMP, whereas the vascular isoreceptor (V_1) exerts its biological action by increasing cytosolic cell calcium. Arginine vasopressin, the naturally occurring hormone in humans and most other mammals, stimulates both types of receptors, whereas the synthetic analog dDAVP [1-desamino–8-D-arginine vasopressin] stimulates only the antidiuretic receptor. Other studies have shown that vasopressin-sensitive adenylate cyclase is localized in the collecting duct and the thick ascending limb of Henle's loop.[177] Furthermore, it has been shown that ADH stimulates prostaglandin (PGE_2) synthesis in three loci within the kidney.[181] First, ADH stimulates PGE_2 synthesis in the glomerular mesangial cells, thereby reducing GFR (see Chapter 5, Figure 5.11). Second, ADH stimulates PGE_2 synthesis in the renal medullary interstitial cells, thereby increasing the medullary blood flow and decreasing the medullary solute gradient. Both of these PGE effects are mediated via the vascular receptor V_1. Third, ADH stimulates PGE_2 synthesis in the cortical collecting duct epithelium, an effect mediated via the antidiuretic receptor V_2. In turn, the PGE_2 produced inhibits the antidiuretic effect of ADH in this nephron segment by supressing the ADH-dependent cAMP accumulation, possibly by stimulating cAMP phosphodiesterase.[238] In this manner, PGE_2 serves as a negative feedback signal to modulate the antidiuretic effect of ADH in the collecting duct.

Mechanism of Action of Aldosterone

Production and Site of Action

The modern history of the renal action of aldosterone began in the late 1940s, when Roemmelt and associates found that in adrenalectomized dogs only 2% to 2.5% of the total filtered sodium escaped tubular reabsorption, a deficiency that was corrected by administering the crude extracts of the adrenal gland.[202] Subsequent purification and isolation of the crude extracts identified the sodium-retaining substance as an 18-aldehyde derivative of corticosterone, and hence it was named aldosterone.[222]

Aldosterone is synthesized from cholesterol in the zona glomerulosa of the adrenal cortex. Its biosynthesis is regulated by several factors, of which dietary intake of sodium is the most prominent. In normal subjects whose sodium intake varies between 50 and 200 mEq/day, 50 to 250 μg of aldosterone is produced per day, of which about 10% is excreted in urine.[147] Clearance studies have shown that the renal clearance of aldosterone is about 14% that of inulin, suggesting that about 86% of the filtered aldosterone is reabsorbed by the renal tubules.[221]

In normal subjects kept in supine position for three consecutive days and given a daily dose of 10 mEq sodium/100 mEq potassium, plasma concentration of aldosterone showed a distinct diurnal rhythm. It varied from 55 ± 7 ng/100 mL plasma at 8:00 am to 33 ± 5 ng/100 mL plasma at 11:00 pm.[252] Although the physiological significance of this diurnal variation in plasma concentration is at present not known, it is believed to be mediated by variations in the secretion of the anterior pituitary adrenocorticotrophic hormone (ACTH).

The sites of renal action of aldosterone have been localized by both stop-flow and micropuncture experiments. Stop-flow studies have shown that administration of aldosterone to adrenalectomized dogs lowers the high sodium concentration found in the fluids emerging from the distally located nephron segments.[242] This suggested that the renal action of aldosterone is confined to the distal nephron, a conclusion that was subsequently confirmed by free-flow micropuncture studies.[110] They found that in adrenalectomized rats the high $[TF/P]_{Na}$ in the "distal tubule" was reduced after administration of aldosterone. However, no difference in the $[TF/P]_{Na}$ in the proximal tubule was noted between the normal and adrenalectomized animals, a finding that has also been confirmed in dogs.[262] Other evidence, however, has implicated the action of aldosterone in the proximal tubule. Thus, when tritiated aldosterone was injected into the rats, it appeared in both proximal and distal tubules. Moreover, actinomycin D (see discussion below for mechanism) inhibited a portion of proximal tubular sodium reabsorption in adrenalectomized rats treated with aldosterone.[184] Despite this, the proximal tubular action of aldosterone remains controversial and has yet to be verified by free-flow micropuncture experiments. Recent studies have localized the biological action of aldosterone to the principal cells of the cortical collecting duct, where it enhances the reabsorption of about 2% of the filtered sodium.[61, 228]

To summarize, reabsorption of sodium in the "distal tubule" and collecting duct occurs as a consequence of two mechanisms: One mechanism consists of aldosterone-independent active reabsorption of most of the sodium by a process analogous to that described for the proximal tubule. The second mechanism consists of aldosterone-dependent active reabsorption of a small fraction of the remaining sodium in the collecting duct and probably the connecting tubule (about 2% of the filtered load). When considered in a short-term situation, the quantity of sodium reabsorbed by the second mechanism is very small. However, the long-term loss of this amount of sodium from the body consequent to aldosterone deficiency is indeed very serious and detrimental to life.

Factors Regulating Biosynthesis

A number of studies (see reviews by Muller,[183] and Williams and Dluhy[253]) have shown that the biosynthesis of aldosterone from cholesterol is influenced by three major factors, each acting on different sites along the biosynthetic pathway shown in Figure 9.39. This scheme assumes that there is only one type of aldosterone-producing cell and that there is only one biosynthetic pathway for aldosterone. To facilitate presentation, we have divided this pathway into three stages: (1) The conversion of cholesterol into pregnenolone, designated as the early pathway; (2) the convrsion of pregnenolone into corticosterone, designated as the middle pathway, and (3) the conversion of corticosterone into aldosterone, designated as the late pathway. On the basis of this division, the various stimuli influencing aldosterone biosynthesis may be classified into those that affect the early, middle, or late pathway.

Results of extensive studies in both animals and man have shown that stimuli that influence the early pathway are all of acute nature, whereas those that act on the late pathway are all of chronic nature. Included in the first category are the renin-angiotensin system, acute changes in sodium and potassium intake, and acute changes in blood levels of ACTH. Of these, the first two are the most important; ACTH plays only a minor role in aldosterone biosynthesis. The second category includes chronic sodium and potassium loading and/or depletion, and their combinations. It should be remembered that increased production of aldosterone caused by acute stimuli is a consequence of increased activity in both early and late pathways. The increase in the activity of the late pathway is due to a secondary increase in corticosterone production consequent to increased activity in the early pathway. Thus, acute infusion of renin or angiotensin stimulates the early pathway and hence increases the production of aldosterone. Similarly, acute infusion or oral ingestion of potassium as well

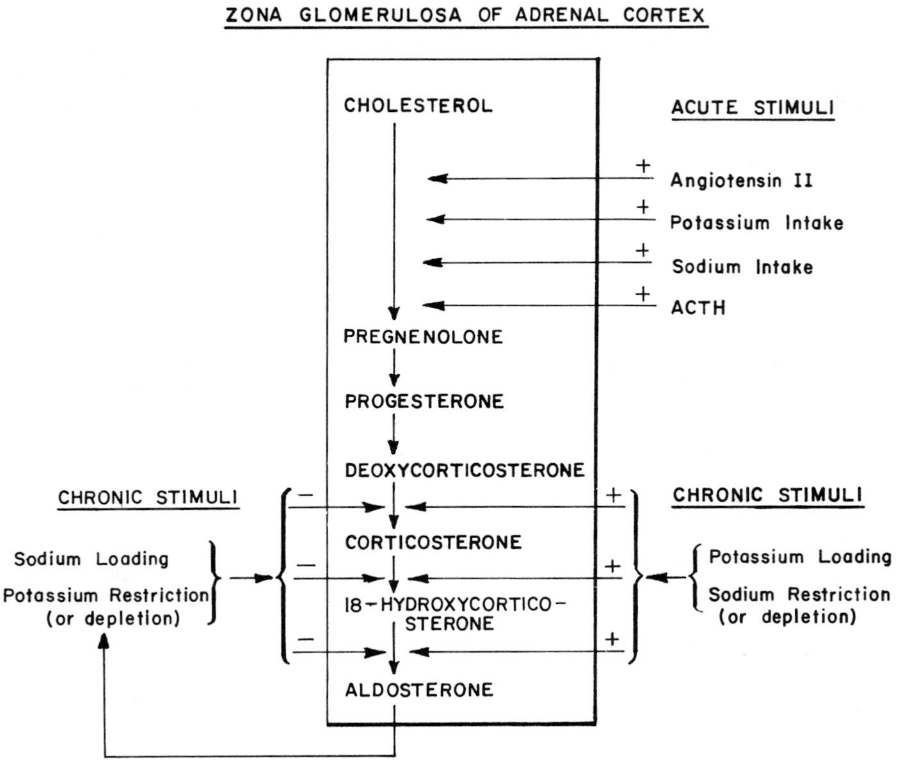

ZONA GLOMERULOSA OF ADRENAL CORTEX

```
                    CHOLESTEROL           ACUTE STIMULI
                         │         +
                         │◄────────── Angiotensin II
                         │         +
                         │◄────────── Potassium Intake
                         │         +
                         │◄────────── Sodium Intake
                         │         +
                         │◄────────── ACTH
                         ▼
                    PREGNENOLONE
                         │
                         ▼
                    PROGESTERONE
                         │
                         ▼
                 DEOXYCORTICOSTERONE
CHRONIC STIMULI    ─         │        +    CHRONIC STIMULI
Sodium Loading    ────►  CORTICOSTERONE  ◄──── Potassium Loading
Potassium          ─         │        +       Sodium Restriction
Restriction       ────► 18-HYDROXYCORTICO- ◄──── (or depletion)
(or depletion)              STERONE
                   ─         │        +
                  ────►      ▼       ◄────
                         ALDOSTERONE
```

FIGURE 9.39. Scheme summarizing the sites of action of acute and chronic stimuli on the biosynthesis of aldosterone. A plus sign (+) above the *arrows* indicates stimulation, and a minus sign (−) indicates inhibition. (Modified from Muller.[183])

as chronic potassium loading enhances aldosterone production, whereas chronic potassium depletion or sodium loading inhibits its production. The potassium effect on aldosterone production has been shown to be independent of the renin-angiotensin system. Let us now consider in more detail the effect of each of these stimuli on aldosterone biosynthesis.

Renin-Angiotensin System. As mentioned in Chapter 6, renin is an enzyme produced by the granular cells of the juxtaglomerular apparatus in response to (a) changes in mean renal perfusion pressure (a direct consequence of alterations in blood volume) and (b) flow-dependent changes in the chloride concentration in the distal tubular fluid passing by the macula densa cells. For the present, we focus only on the changes in blood volume and their effects on renin production and hence aldosterone biosynthesis.

Briefly, a decrease in the circulating blood volume leads to a reduction in the cardiac output, which in turn reduces the renal blood flow and the mean renal perfusion pressure. This is monitored by the granular cells of the juxtaglomerular apparatus, causing an increase in renin production by these cells. This renin acts on its circulating substrate, angiotensinogen (found chiefly in the α–2-globulin fraction of plasma) to split off the decapeptide angiotensin I, which is biologically inactive. However, on passing through the lung, which is rich in the converting enzyme, angiotensin I is converted into the biologically active octapeptide angiotensin II. This substance has at least three known effects: (1) It is a potent vasoconstrictor; (2) it has a direct effect on sodium transport by the kidney; and (3) it stimulates the zona glomerulosa cells of the adrenal cortex to secrete aldosterone.[35] The latter effect enhances plasma levels of aldosterone, which in turn increases sodium reabsorption, accompanied by osmotic reabsorption of water, from the collecting duct. This would lead to an increase in the circulating blood volume, thereby closing the negative feedback loop. Most of the animal studies have shown that renin-angiotensin levels and aldosterone secretion change in parallel fashion following alterations in blood volume and/or sodium intake. Further details of this feedback loop, as well as that initiated by the macula densa cells, and their roles in the regulation of volume and osmolality of body fluids are discussed in Chapter 13.

Changes in Sodium-Potassium Balance. The second category of stimuli affecting aldosterone biosynthesis consists of acute changes in dietary intake of sodium and potassium as well as chronic loading and/or restriction of these ions. The former influences the early pathway, whereas the latter acts on the late pathway.

Like angiotensin II, acute alterations of sodium and potassium intake stimulate the early pathway, thereby increasing the conversion of cholesterol to pregnenolone and eventually the secretion of aldosterone. In contrast, chronic loading and/or restriction of sodium and potassium alter aldsterone secretion by directly modifying the activity of the late pathway. Thus, either chronic potassium loading or sodium restriction causes an increase in aldosterone production, whereas sodium loading or potassium restriction produces a marked decrease in aldosterone production. An additional effect of these chronic stimuli is that by modifying the activity of the late pathway, they modulate the rate of response or the "sensitivity" of the aldosterone secretion to various acute stimuli. Furthermore, as depicted in Figure 9.39, restriction of sodium intake and potassium loading has an additive stimulatory effect on aldosterone production. In contrast, restriction of potassium intake and sodium loading has an additive inhibitory effect on the rate of biosynthesis of aldosterone. However, if severely sodium-depleted subjects are also potassium-restricted, aldosterone secretion is found to decrease to subnormal levels, suggesting the predominant effect of potassium.[253] In short, in normal man, it is the dynamic interrelationship between sodium and potassium balance that ultimatey determines the effect of these ions on aldosterone secretion and their potentiating effect in response to various acute stimuli.

Role of ACTH. As depicted in Figure 9.39, ACTH enhances aldosterone production by stimulating the early biosynthetic pathway. Its stimulatory effect is enhanced when the zona glomerulosa is sensitized by the action of chronic stimuli on the late biosynthetic pathway. Under normal conditions, ACTH plays a minor role in the biosynthesis of aldosterone, except for mediating its diurnal secretion.

Recent studies have shown that the effects of angiotensin II, potassium and sodium intake, and ACTH on the biosynthesis of aldosterone are all highly dependent on the extracellular calcium concentration.[67] Thus, an increase in these stimuli causes an increase in the intracellular calcium concentration, which in turn stimulates aldosterone synthesis. These studies are consistent with the concept that calcium is a second intracellular messenger, since its mobilization appears to be critical for stimulation of aldosterone secretion in response to the above stimuli.

Other Factors. Since changes in aldosterone secretion are often associated with parallel changes in water and electrolyte balance and hence edema formation, factors controlling its biosynthesis have received considerable attention in clinical medicine. Of particular interest are the factors controlling aldosterone secretion in anephric man and the consequences of the loss of sodium-retaining effect of aldosterone after prolonged administration.

In anephric man, in which the renin-angiotensin system is absent, it has been found that the increase in plasma aldosterone concentration that normally occurs on assuming upright posture is abolished. It should be remembered that in a normal subject, assuming upright posture decreases central venous volume and hence the circulating blood volume (owing to pooling of the blood in the lower extremities by gravity), thereby stimulating aldosterone production by the renin-angiotensin mechanism described above. Numerous studies in man and animals, referred to above, have shown a direct effect of alteration in potassium balance as well as plasma potassium levels on aldosterone secretion. Moreover, it has been found that aldosterone secretion in anephric man increases in response to stressful stimuli, a possible consequence of parallel changes in ACTH. On the basis of this evidence, it has been suggested that in anephric man, the most important factor controlling aldosterone secretion is potassium balance, with ACTH playing a secondary role. Another factor controlling aldosterone secretion in anephric man is the prolonged administration of the anticoagulant heparin, which is required for hemodialysis therapy. Prolonged administration of this substance produces hypoaldosteronism, the mechanism of which is currently not known.[253]

The other condition of clinical interest is the loss of sodium-retaining effect of aldosterone following prolonged administration. When aldosterone is administered for several days to normal subjects, the kidney escapes from the sodium-retaining, but not the potassium-losing effect of the hormone.[4] The onset of the escape is related to the amount of dietary sodium intake. Although the mechanism of this escape phenomenon is currently not known, it is believed to be responsible for the absence of edema in patients with primary aldosteronism. The failure of the escape mechanism is thought to contribute to the edematous states in patients with congestive heart failure, liver cirrhosis with ascites, and nephrosis.

In summary, it is evident that under normal conditions of dietary salt intake, the basal rate of aldosterone secretion is controlled by the additive effects of the renin-angiotensin system and the plasma levels of potassium. This is complimented by the additional effect of dietary sodium and potassium intake, which potentiates aldosterone secretion in response to acute stimuli. In short, changes in dietary intake of sodium and potassium, by virtue of the effect of these ions on the late biosynthetic pathway, sensitize (potentiate) the zona glomerulosa to acute stimuli, which act on the early biosynthetic pathway. Thus, with identical plasma levels of angiotensinn II, aldosterone production is found to be greater in sodium-restricted than in sodium-loaded subjects. However, potassium restriction has an adverse effect on angiotensin II-stimulated aldosterone seretion. As depicted in Figure 9.39, there exists a negative feedback loop between potassium restriction (or depletion) and aldosterone secretion. Since potassium loss from the body via the kidney is itself a consequence of increased aldosterone secretion, stimulation of aldosterone by angiotensin II is self-limited by the potassium loss it causes.

Modes and Sites of Cellular Action

Our current understanding of the mechanism of action of aldosterone and its effect on sodium transport in the mammalian kidney has largely been derived from extensive studies of sodium transport in isolated toad urinary bladder. Results of numerous studies[219] employing both short-circuit and isotope dilution techniques have revealed the following general characteristics for sodium transport across isolated toad urinary bladder. The direction of net transbladder sodium transport is from the mucosal (urinary) side to the serosal (blood) side of the bladder. Sodium is transported across the mucosal barrier by passive diffusion down its electrochemical gradient. Once inside the epithelial cells, the accumulated sodium is then actively transported by the membrane-bound Na^+-K^+ activated ATPase pump across the serosal barrier against its electrochemical gradient. This active transport is "electrogenic"; it occurs without obligatory coupled movement of an anion in the same direction or a cation in the opposite direction. The extrusion of sodium out of the cell leads to the separation of charges (sodium ions), which gives rise to the measured electrical potential gradient.

The effect of aldosterone on sodium transport was first demonstrated in the isolated toad bladder.[31] When 3×10^{-10} M of aldosterone (an amount considerably less than plasma aldosterone concentration, which is about 10^{-8} M) was added to the fluid bathing the serosal surface, it caused an increase in the short-circuit current, which is equivalent to an increase in active sodium transport, after a latent period of 45 to 90 minutes. This increase in active sodium transport was independent of the concentration of aldosterone in the bathing fluid. The response also occurred after the withdrawal of the hormone from the bathing fluid during the latent period. Subsequent studies showed that this latent period was not due to slow accumulation of the hormone in the tissue, since maximal tissue hormone concentration could be achieved only 30 minutes after the hormone exposure.[219] Nevertheless, one to four hours were required before a detectable change in the active sodium transport could be measured.

Similar results have been obtained in the intact mammalian kidney.[219] When aldosterone is injected into the renal artery of dogs and rats, a latent period of 45 to 120 minutes elapses before antinatriuresis is detected. In the rat the peak of aldosterone effect on sodium transport after a single injection occurred when the hormone was virtually absent from the circulation.

In both the isolated toad urinary bladder and the intact mammalian kidney, the antinatriuretic effect of aldosterone is inhibited by actinomycin D, puromycin, and cycloheximide, which are all well-known inhibitors of protein synthesis. In the mammalian kidney, however, these inhibitors failed to block the kaliuretic effect of aldosterone. This finding indicates that the aldosterone-induced renal reabsorption of sodium and secretion of potassium may occur by two separate pathways.

Exactly how aldosterone stimulates sodium transport has been the subject of intensive investigation.

Several types of studies, using toad urinary bladder, adrenalectomized rat, and isolated rabbit kidney tubules, have led to the formulation of a currently accepted hypothesis depicting the cellular action of aldosterone and the mechanism of its stimulation of sodium transport.

The observed latent period before the onset of aldosterone effect led Crabbe[31] to suggest that stimulation of sodium transport by aldosterone involves intracellular synthesis of an active intermediate substance. Since then, in a series of intricate experiments, Edelman and his associates[26, 54, 55, 57, 69] identified the nature of this active intermediate substance and consequently proposed the *induction theory* of aldosterone action. According to this theory, as depicted in Figure 9.40, aldosterone present in the blood crosses the serosal (peritubular) membrane by chemical diffusion and enters the cytoplasm of the cortical collecting duct epithelial cell, where it stereospecifically binds noncovalently with an aldosterone-receptor protein. Studies of the time-course of uptake of tritiated aldosterone by the four conventional cell fractions (nucleus, cytosol, mitochondria, and microsomes) have identified cytosol as the fraction containing the initial cytoplasmic aldosterone-receptor protein. This cytoplasmic binding is characterized by a latent period of 30 to 45 minutes, exclusive of the time required for the onset of the hormone effect. The cytoplasmic aldosterone-receptor complex is then translocated into the nucleus, where it binds noncovalently with the acceptor sites on the nuclear chromatin. The formation of this complex somehow activates deoxyribonucleic acid (DNA)-dependent synthesis of ribonucleic acid

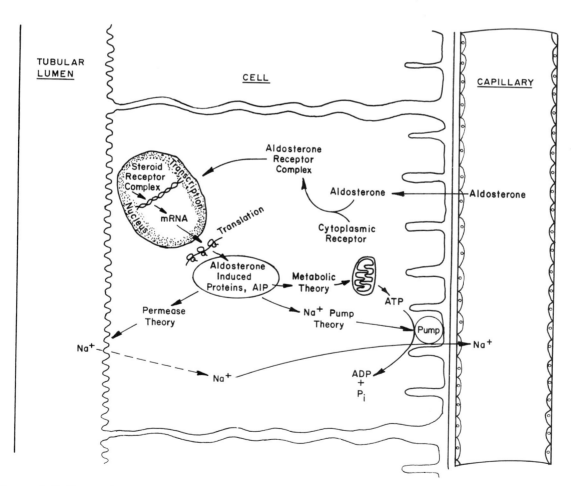

FIGURE 9.40. Proposed mechanism of cellular steps involved in aldosterone-stimulated sodium transport. (Modified from Feldman.[69])

(RNA). This is called *gene transcription*, meaning that the information in the DNA is "transcribed" into the RNA. The RNA, thus synthesized, is called the messenger RNA (mRNA), because it leaves the nucleus and enters the cytoplasm where it induces *de novo* synthesis of protein. This step is called *translation*, meaning that the nucleotide of mRNA is "translated" into the amino acid sequence of the synthesized protein. The *aldosterone-induced protein* (AIP) thus synthesized, through a series of poorly understood reations, mediates the increase in the transepithelial sodium transport.

Support for the induction theory has come from a number of experiments.[54] It has been shown that actinomycin D and cycloheximide inhibit aldosterone-induced sodium transport by inhibiting the DNA-dependent RNA synthesis of AIP. Actinomycin inhibits transcription by binding to the guanosine residue of DNA, and cycloheximide inhibits translation. In the rat, inhibition of DNA-synthesis of RNA by actinomycin D abolished the antinatriuretic effect of aldosterone but not its kaliuretic effect. Studies in the dog have confirmed these findings and have shown further that actinomycin D failed to abolish the aldosterone-enhanced hydrogen ion secretion.[164] These findings suggest that the effect of aldosterone on potassium and hydrogen ion secretion is independent of DNA-dependent RNA synthesis. Furthermore, the effect of aldosterone on increasing both the hydrogen and potassium secretion appears to occur by a pathway different from that involved in the transport of sodium and may involve a common biochemical pathway for hydrogen and potassium (see Chapter 11).

Spironolactone, a competitive aldosterone antagonist, inhibits binding of aldosterone to both cytsol and nuclear receptor proteins. However, it has no effect on the uptake of tritiated aldosterone by the other two cell fractions or the plasma concentration of the hormone.[56] This is in contrast to inhibition of the aldosterone effect by actinomycin D (inhibits transcription) and ouabain (inhibits Na^+-K^+ activated ATPase), which do not displace the hormone from the receptor sites.

The underlying assumption of the induction hypothesis of the cellular mechanism of aldosterone-stimulated sodium transport is the DNA-dependent synthesis of RNA, which in turn induces the synthesis of a new protein, the AIP. The only direct experimental proof for the synthesis of this protein, which may be a permease, a membrane component, or a carrier molecule, is the report of Benjamin and Singer.[9] Using electrophysiological and biochemical techniques, they demonstrated aldosterone-inducd synthesis of a low-molecular-weight protein (about 12,000) in isolated toad urinary bladder. The synthesized protein had a high specificity for aldosterone, as revealed by the inhibitory effects of actinomycin D and spironolactone.

How does AIP increase sodium transport? Based on the accepted model of transepithelial sodium transport described earlier (Figure 9.11) and as depicted in Figure 9.40, three theories have been advanced to explain the action of AIP: (1) The sodium pump theory—AIP increases the activity of Na^+-K^+ activated ATPase at the serosal (peritubular) membrane, and hence increases sodium transport; (2) the permease theory—AIP facilitates the entry of sodium across the mucosal (luminal) membrane, thereby providing more sodium for the pump to work on; and (3) the metabolic theory—AIP stimulates mitochondrial synthesis of ATP, hence raising the cellular ATP/ADP ratio. Let us consider the consequence of each theory and its supporting evidence.

The sodium pump theory, advanced by Goodman and associates,[88] is based on the finding that aldosterone potentiated the effect of vasopressin on sodium transport under anaerobic conditions in the isolated toad urinary bladder. This finding implicates aldosterone as activating the membrane-bound Na^+-K^+ ATPase pump and hence increasing sodium transport. Time-course studies of single injection of aldosterone have shown that its effect on sodium transport peaks at three to four hours and disappears after six hours. In contrast, multiple injections of the hormone are required to detect its effect on Na^+-K^+ ATPase activity, which begins six hours after hormone expression and peaks at 24 to 36 hours.[69] Moreover, the response is not specific to aldosterone and is seen after administration of corticosterone and dexamethasone, a steroid with predominant glucocorticoid activity and no effect on sodium transport. Both nonspecificity and delayed increase in the Na^+-K^+ ATPase activity may be due to noncovalent binding of aldosterone with glucocorticoids receptors within the cytoplasm. Recent studies have shown that both short-term and long-term administration of aldosterone enhances the Na-K-ATPase activity. Thus, El Mernissi and Doucet showed that adrenalectomy markedly reduced Na-K-ATPase activity in the cortical collecting duct.[61] The enzyme activity was restored to normal by short-term administration

(10 μg/kg body weight, three hours prior to experiment) of aldosterone. Other studies have found that short-term administration of aldosterone influences the transport porperties of the luminal membrane, whereas long-term administration of aldosterone affects the basolateral properties of the principal cells of the cortical collecting tubule.[68, 149, 205, 226] Both effects lead to enhanced Na-K-ATPase activity. Thus, short-term administration of aldosterone increases the permeability of the luminal membrane to sodium and thereby increases its intracellular concentration, which in turn increases Na-K-ATPase activity. In contrast, long-term administration of aldosterone increases the luminal membrane sodium and potassium conductances, as well as increasing amplification of the basolateral membrane area of the principal cells. The latter effect has been shown to be associated with an increase in Na-K-ATPase activity and an increase in the number of the sodium pump.[226] It is not known whether short-term administration of aldosterone has a similar effect on the basolateral membrane area.

The permease theory, advanced by Sharp and Leaf,[219] is based on the concept that AIP increases the passive influx of sodium across the luminal membrane, thereby indirectly increasing the active extrusion of sodium across the peritubular membrane. If AIP acts on the rate-limiting membrane at the urinary surface, there should be an increase in the intracellular sodium concentration. Alternatively, if AIP acts on the rate-limiting membrane at the blood surface, there should be a decrease in the intracellular concentration of sodium. Measurements of intracellular sodium concentration of epithelial cells isolated from toad urinary bladder have shown that both aldosterone and vasopressin increased intracellular sodium concentration.[107] Since vasopressin increases the permeability of the luminal membrane, aldosterone must have produced similar effects with respect to sodium. Similar increase in intracellular sodium concentration was obtained after exposure of the cells to ouabain. In contrast, amiloride, a diuretic agent that inhibits sodium transport to the same extent as ouabain, decreased intracellular sodium concentration, suggesting that amiloride inhibited the passive influx of sodium across the luminal membrane. These results, however, are in variance with those of Lipton and Edelman[165] who, using similar preparations, found no detectable change in the intracellular sodium concentration. They suggested that AIP has a bipolar effect; it enhances simultaneously plasma sodium influx at the luminal membrane and active sodium efflux at the peritubular memrane. As noted above, these discrepancies may have been caused by the length of time the cells were exposed to aldosterone.

Further support for the permease theory comes from direct measurement of transepithelial electrical resistance. If AIP acts on the sodium influx step, it should reduce transepithelial electrical resistance (Figure 9.11). However, if AIP acts on the sodium efflux step, there should be no change in resistance. Electrophysiological measurements have shown that aldosterone caused a small decrease in transepithelial electrical resistance in toad urinary bladder,[28] thereby increasing sodium conductance. This conclusion has been confirmed by a number of recent studies. Thus, Sansom and O'Neil,[205] using microelectrode techniques in microperfused isolated rabbit cortical collecting duct, have shown that short-term (after one day) administration of deoxycorticosterone acetate (DOCA), a mineralocorticoid, increased sodium conductance and current (an index of sodium influx) but had no influence on potassium current (an index of potassium secretion). However, after four days or more of DOCA treatment, there was a marked increase in potassium conductance. These studies suggest that acute effect of the mineralocorticoids is to enhance sodium transport across the luminal membrane, whereas the chronic effect of the hormone is to increase luminal conductance of potassium.

The metabolic theory, advanced by Edelman and Fimognari,[57] is based on the concept that AIP stimulates oxidative metabolism, thereby increasing the rate of supply of high-energy intermediates, which would act as a pacemaker for the sodium transport system. Therefore, the increase in sodium influx at the luminal membrane is secondary to enhanced active sodium efflux at the peritubular membrane. This hypothesis was prompted by the striking dependence of aldosterone-stimulated sodium transport on the presence of exogenous oxidative substrates of tricarboxylic acid cycle in isolated toad urinary bladder preparation.

Experimental evidence supporting the metabolic theory comes from several types of studies in isolated toad urinary bladder preparation[69]: (1) In bladders depleted of oxidative substrates the normal aldosterone-stimulated increase in sodium transport does not occur. However, addition of substrate to aldosterone-treated bladders causes an immediate increase in sodium transport (no latent period). This suggests that synthesis of AIP may occur in the absence of oxidative substrates, but that energy metabolism is required for the expression of its effects. (2) The aldosterone effect is maintained by any substance that leads to the formation of pyruvate or oxaloacetate, as well as glucose. (3) Several inhibitors of oxidative phosphorylation in mitochondria (such as rotenone and amobarbital) inhibit the aldosterone-induced effect in toad bladder, but do not block its response to vasopressin or amphotericin B. In the absence of aldosterone, amphotericin B increases net transbladder sodium transport, presumably by increasing the rate of sodium influx at the mucosal barrier. The results obtained with the inhibitors imply that the mitochondrial nicotinamide adenine dinucleotide dehydrogenase (NADH) pathway is specifically involved in the mediation of aldosterone-induced increase in sodium transport. (4) Renal citrate synthase activity decreases after adrenalectomy. It increases after administration of physiological concentratons of aldosterone with a time-course paralleling the urinary sodium-potassium response to the hormone. Furthermore, stimulation of citrate synthase could be blocked by spironolactone at concentrations that inhibit urinary electrolyte changes. In contrast, glucocorticoids had no inhibitory effect on renal citrate synthase activity.

In summary, the current available evidence is consistent with the view that the effect of aldosterone on sodium reabsorption and potassium secretion in the cortical collecting duct depends on the length of time the principal cells are exposed to the hormone. Thus, the short-term effect of the hormone acts via the permease mechanism, whereas the long-term effect of the hormone is mediated by the sodium pump and metabolic mechanisms, or a combination of all three mechanisms. However, there are many unresolved questions about the nature of action of AIP. There is as yet no direct experimental verification of the role of AIP on the observed increase in the net transport of sodium in aldosterone-sensitive tissues. Moreover, the stoichiometric relationship, if any, between the synthesis of AIP and net sodium transport is yet to be established.

Effects of Other Hormones

In addition to ADH and aldosterone, numerous recent studies have demonstrated that other hormones, notably PTH and calcitonin, influence electrolyte transport in the distal convoluted tubule and collecting duct system.[119] Thus, it has been shown that PTH stimulates adenylate cyclase activity in the collecting duct, but not in the distal convoluted tubule. On the other hand, calcitonin produces a greater effect in the distal convoluted tubule as compared with the collecting duct. The physiological effects of these two hormones on phosphate and calcium transport are described further in Chapter 10.

Finally, recent studies have shown that the mammalian atria contain secretory granules whose density is increased by sodium or water deficiency and decreased by saline infusion.[37, 232] Administration of the atrial extract produces an immediate, short-acting natriuresis,[195] accompanied by an increase in GFR and renal blood flow, particularly the blood flow in the inner cortex and medulla, and a slight hypotension.[18] Furthermore, micropuncture studies have localized the medullary collecting duct as the site of action of the atrial natriuretic factor (ANF).[223] Since this nephron segment plays an important regulatory role in the urine concentrating mechanism (see Chapter 12), these findings might explain the rapid onset of natriuresis after intravenous ANF administration. Recently, Flynn and co-workers[71] reported the amino acid sequence of a 28 residue peptide isolated from rat atria with potent diuretic and natriuretic properties. Subsequent studies,[223] using antibodies against synthetic ANF, have shown that ANF is released into the circulation; therefore, ANF can be considered as a natriuretic hormone. Further studies have shown that the mechanism of ANF release from the intracellular stores into the circulation is via receptor-mediated activation of the cellular polyphospho-inositide system. A recent study, using an isolated rat heart-lung preparation, indicates that stretch of the right atrium per se might be the stimulus for ANF release.[45] Thus, it has been suggested that ANF may be involved in the regulation of blood volume in response to acute expansion of the extracellular fluid volume.[223]

Problems

9.1. What significant information has been obtained from the application of micropuncture, stop-flow microperfusion, and short-circuit current techniques to single nephrons and stop-flow experiments in the whole kidney?

9.2. What experimental measurements do we need to decide whether an ion is transported actively or not?

9.3. Define the terms internephron heterogeneity and axial heterogeneity. Give an example for both terms.

9.4. What are the main structural features of the proximal tubule epithelium?

9.5. What is the main distinguishing feature between the epithelium of the proximal convoluted tubule and the distal convoluted tubule?

9.6. What are the major distinguishing features between proximal tubule cells from the beginning of the convoluted part and from the end of the straight part?

9.7. What is the major substrate for ATP synthesis in the proximal tubule cell?

9.8. What kind of tight junctions are characteristic of the proximal tubule?

9.9. What is the functional relevance of the "vacuolar apparatus" of the proximal tubule?

9.10. Micropuncture studies yielded the values of 10 and 0.5, respectively, for the tubular fluid-to-plasma concentration ratios of inulin and sodium at the end of the proximal tubule. If GFR was 120 mL/min and plasma sodium concentration was 140 mEq/L, calculate the fraction of filtered sodium reabsorbed along the proximal tubule.

9.11. Briefly describe the extrarenal and intrarenal factors affecting iso-osmotic transport of sodium and water along the proximal tubule.

9.12. Briefly explain the mechanisms whereby extracellular fluid expansion by intravenous mannitol and saline infusion induce natriuresis and diuresis.

9.13. Describe the histological differences between a short and a long loop of Henle.

9.14. What are the distinguishing ultrastructural features between the descending thin limb of a short and a long loop?

9.15. What is the most characteristic ultrastructural feature of an ascending thin limb?

9.16. What types of tight junctions are found in the ascending thin limb epithelium?

9.17. Describe the ultrastructural organization of the epithelium of the thick ascending limb.

9.18. What are the structural differences between the epithelium of the medullary and the cortical thick ascending limb?

9.19. What is Tamm-Horsfall protein and what is its presumed function?

9.20. How do changes in GFR modify tubular transport of sodium and water (a) in the proximal tubule and (b) in Henle's loop?

9.21. The term "distal tubule" when used by physiologists includes three structurally different segments. What are these segments?

9.22. What are the structural differences between the distal convoluted tubule and the connecting tubule?

9.23. Four cell types are found within the "distal tubule." What are these cell types?

9.24. What are the ultrastructural characteristics of an intercalated cell?

9.25. Along which nephron segments are intercalated cells found?

9.26. What are the differences between a connecting tubule cell and a principal cell of the cortical collecting duct?

9.27. Intercalated cells may be found to be either exposed into the tubular lumen or retracted behind the surrounding cells. Explain the functional mechanism that may cause these changes.

9.28. What are the histological differences between outer medullary and inner medullary collecting ducts?

9.29. What structural changes are produced in the collecting duct epithelium by the action of ADH in the presence of a transepithelial osmotic gradient?

9.30. What types of tight junctions are found in the collecting duct epithelium?

9.31. State the currently accepted concept of K^+ secretion in the collecting duct and briefly outline the fundamental experiments that led to its development.

9.32. In the table given below, indicate whether each forcing (during steady state) increases ($+$), decreases ($-$), or has no effect (0) on potassium secretion in the collecting duct.

Forcings	Urinary potassium excretion
1. Respiratory acidosis	
2. Respiratory alkalosis	
3. Metabolic acidosis	
4. Metabolic alkalosis	
5. Sodium loading	
6. Sodium depletion	
7. Potassium loading	
8. Potassium depletion	
9. Water loading	
10. Administration of aldosterone	
11. Administration of spironolactone	
12. Administration of acetazolamide	
13. During mannitol or urea diuresis	

9.33. Indicate the sites of action of ADH and aldosterone

within the nephron and briefly outline the cellular mechanisms mediating their effects.

9.34. List the major factors that influence the biosynthesis of aldosterone in the zona glomerulosa of the adrenal cortex and their proposed sites of action along the biosynthesis pathway.

9.35. What is meant by "aldosterone escape" phenomenon? Does it have any clinical significance?

9.36. What factors regulate aldosterone biosynthesis and secretion rate in anephric patients?

9.37. What are the functional differences between central diabetes insipidus and nephrogenic diabetes insipidus? Could you distinguish between these two types of diabetes by administration of ADH, and why?

9.38. What is the mechanism by which change in posture may affect aldosterone biosynthesis and secretion rate?

References

1. Alpern RJ, Cogan MG, Rector FC Jr: Effects of extracellular fluid volume and plasma bicarbonate concentration of proximal acidification in the rat. *J Clin Invest* 1983; 71:736–746.
2. Aronson PS, Nee J, Suhm MA: Modifier role of internal H in activating Na^+-H^+ exchanger in renal microvillus membrane vesicles. *Nature* 1982; 299:161–163.
3. Arrizurieto-Muchnik EE, Lassiter WE, Lipham EM, Gottschalk CW: Micropuncture study of glomerulotubular balance in the rat kidney. *Nephron* 1969; 6:418–439.
4. August JT, Nelson DH, Thorn GW: Response of normal subjects to large amounts of aldosterone. *J Clin Invest* 1958; 37:1549–1555.
5. Bachmann S, Kriz W: The thin limbs of Henle's loop in the golden hamster kidney: A TEM study. *Cell Tissue Res* 1982; 225:111–127.
6. Bank N, Aynedjian HS: A micropuncture study of renal bicarbonate and chloride reabsorption in hypokalemic alkalosis. *Clin Sci* 1965; 29:159–170.
7. Barrett JM, Heidger PM: Microbodies of the rat renal proximal tubule: Ultrastructural and cytochemical investigations. *Cell Tissue Res* 1975; 157:283–305.
8. Bello-Reuss E: Effect of catecholamines on fluid reabsorption by the isolated proximal convoluted tubule. *Am J Physiol* 1980; 238:F347–F352.
9. Benjamin WB, Singer I: Aldosterone-induced protein in toad urinary bladder. *Science* 1974; 186:269–272.
10. Bennett CM, Brenner BM, Berliner RW: Micropuncture study of nephron function in the Rhesus monkey. *J Clin Invest* 1968; 47:203–216.
11. Bennett CM, Clapp JR, Berliner RW: Micropuncture study of the proximal and distal tubule in the dog. *Am J Physiol* 1967; 213:1254–1262.
12. Berndt WO, Miller M, Kettyle WM, Valtin H: Potentiation of the antidiuretic effect of vasopressin by chlorpropamide. *Endocrinology* 1970; 86:1028–1032.
13. Berry CA: Water permeability and pathways in the proximal tubule. *Am J Physiol* 1983a; 245:F279–F294.
14. Berry CA: A lack of effect of peritubular protein on passive NaCl transport in the rabbit proximal convoluted tubule. *J Clin Invest* 1983b; 71:268–281.
15. Berry CA, Rector FC Jr: Active and passive sodium transport in the proximal tubule. *Miner Electrolyte Metab* 1980; 4:149–160.
16. Berry CA, Warnock DG, Rector FC Jr: Ion selectivity and proximal salt reabsorption. *Am J Physiol* 1978; 235:F234–F245.
17. Biagi B, Sohtell M, Giebisch G: Intracellular potassium activity in the rabbit proximal straight tubule. *Am J Physiol* 1981; 241:F677-F686.
18. Borenstein HB, Cupples WA, Sonnenberg H, Veress AT: The effect of natriuretic atrial extract on renal haemodynamics and urinary excretion in anesthetized rats. *J Physiol (Lond)* 1983; 334:133–140.
19. Brenner BM, Berliner RW: Relationship between extracellular volume and fluid reabsorption by the rat nephron. *Am J Physiol* 1969; 217:6–12.
20. Brenner BM, Falchuk KH, Keimowitz RI, Berliner RW: The relationship between peritubular capillary protein concentration and fluid reabsorption by the renal proximal tubule. *J Clin Invest* 1969; 48:1519–1531.
21. Brown D, Roth J, Kumpulainen TO: Ultrastructural immunocytochemical localization of carbonic anhydrase. *Histochem J* 1982; 75:209–213.
22. Bulger RE, Cronin RE, Dobyan DC: Survey of the morphology of the dog kidney. *Anat Rec* 1979; 194:41–66.
23. Burg MB, Grantham J, Abramow M, Orloff J: Preparation and study of fragments of single rabbit nephron. *Am J Physiol* 1966; 210:1293–1298.
24. Burg MB, Stoner L: Sodium transport in the distal nephron. *Fed Proc* 1974; 33:31–36.
25. Burg MB, Stoner L, Cardinal J, Green N: Furosemide effect on isolated perfused tubules. *Am J Physiol* 1973; 225:119–124.
26. Chabardes D, Gagnan-Brunette M, Imbert-Teboul M, et al: Adenylate cyclase responsiveness to hormones in various portions of the human nephron. *J Clin Invest* 1980; 65:439-448.
27. Chan YL: The role of norepinephrine in the regulation of fluid absorption in the rat proximal tubule. *J Pharmacol Exp Ther* 1980; 215:65–70.
28. Civan MM, Hoffman RE: Effect of aldosterone on electrical resistance of toad bladder. *Am J Physiol* 1971; 220:324–328.

29. Clapp RJ, Robinson RR: Distal sites of action of diuretic drugs in the dog nephron. *Am J Physiol* 1968; 215:228–235.
30. Cortney MA: Renal tubular transfer of water and electrolytes in adrenalectomized rats. *Am J Physiol* 1969; 216:589–598.
31. Crabbe J: Stimulation of active sodium transprt by the isolated toad bladder with aldosterone in vitro. *J Clin Invest* 1961; 40:2103–2110.
32. Crabbe J, De Weer P: Relevance of transport pool measurements in toad bladder tissue for the elucidation of the mechanism whereby hormones stimulate active sodium transport. *Arch Ges Physiol* 1969; 313:197–221.
33. Curran PF, MacIntosh JR: A model system for biological water transport. *Nature* 1962; 193:347–348.
34. Daly JJ, Roe JW, Horrocks P: A comparison of sodium excretion following the infusion of saline into systemic and portal veins in the dog: evidence for a hepatic role in the control of sodium excretion. *Clin Sci* 1967; 33:481–487.
35. Davis JO: The control of renin release. *Am J Med* 1973; 55:333–350.
36. Day H, Middendorf D, Lukert B, et al: The renal response to intravenous vanadate in rats. *J Lab Clin Med* 1980; 96:382–395.
37. De Bold AJ: Atrial natriuretic factor of the rat heart. Studies on isolation and properties. *Proc Soc Exp Biol Med* 1982; 170:133–138.
38. De Fronzo RA, Cooke CR, Andres R, et al: The effect of insulin on renal handling of sodium, potassium, calcium, and phosphate in man. *J Clin Invest* 1975; 55:845–855.
39. De Fronzo RA, Goldberg M, Agus ZS: The effects of glucose and insulin on renal electrolyte transport. *J Clin Invest* 1976; 58:83–90.
40. De Rouffignac C, Corman B, Roinel N: Stimulation by antidiuretic hormone of electrolyte. Tubular reabsorption in rat kidney. *Am J Physiol* 1984; 244:F156-F164.
41. De Wardener HE: The control of sodium excrtetion. *Am J Physiol* 1978; 235:F163–F173.
42. De Wardener HE, Mills IH, Clapham WF, Hayter CJ: Studies on the efferent mechanism of the sodium diuresis which follows the administration of intravenous saline in dog. *Clin Sci* 1961; 21:249–258.
43. Diamond JM, Bossert WH: Standing gradient osmotic flow. A mechanism for coupling of water and solute transport in epithelia. *J Gen Physiol* 1967; 50:2061–2083.
44. Dieterich HJ, Barrett JM, Kriz W, Bulhoff JP: The ultrastructure of the thin loop limbs of the mouse kidney. *Anat Embryol* 1975; 147:1–18.
45. Dietz JR: Release of natriuretic factor from rat heart-lung preparation by atrial distension. *Am J Physiol* 1984; 247:R1093–R1096.
46. Dirks JH, Cirksena WJ, Berliner RW: The effect of saline infusion on sodium reabsorption by the proximal tubule of the dog. *J Clin Invest* 1965; 44:1160–1170.
47. Dobyan DC, Bulger RE: Renal carbonic anhydrase. *Am J Physiol* 1982; 243:F311–FF324.
48. Dobyan DC, Magill LS, Friedman PA, et al: Carbonic anhydrase histochemistry in rabbit and mouse kidneys. *Anat Rec* 1982; 204:185–197.
49. Dousa TP: Role of cyclic AMP in the action of antidiuretic hormone on the kidney. *Life Sci* 1973; 13:1033–1040.
50. Dousa TP: Cellular action of antidiuretic hormone in nephrogenic diabetes insipidus. *Mayo Clin Proc* 1974; 49:188–199.
51. Du Vigneaud V: Trail of sulfur research: From insulin to oxytocin. *Science* 1956; 123:967–974.
52. Earley LE, Friedler RM: The effects of combined renal vasodilation and pressor agents on renal hemodynamics and the tubular reabsorption of sodium. *J Clin Invest* 1966; 45:542–551.
53. Earley LE, Schrier RW: Intrarenal control of sodium excretion by hemodynamic and physical factors, in Orloff J, Berliner RW (eds): *Handbook of Physiology, Section 8, Renal Physiology*. Washington DC, American Physiological Society, 1973, pp 721–762.
54. Edelman IS: Receptors and effectors in hormone action on the kidney. *Am J Physiol* 1981; 241:F333-F339.
55. Edelman IS: Mechanism of action of steroid hormones. *J Steroid Biochem* 1975; 6:147–159.
56. Edelman IS: The initiation mechanism in the action of aldosterone on sodium transport. *J Steroid Biochem* 1972; 3:167–172.
57. Edelman IS, Fimognari GM: On the biochemical mechanism of action of aldosterone. *Recent Prog Horm Res* 1968; 24:1–44.
58. Edwards BR, Novakova A, Sutton RAL, Dirks JH: Effects of acute urea infusion on proximal tubular reabsorption in the dog kidney. *Am J Physiol* 1973; 224:73–79.
59. Elalouf JM, Roinel N, De Rouffinac C: ADH-like effects of calcitonin on electrolyte. Transport by Henle's loop of rat kidney. *Am J Physiol* 1984; 246:F213–F220.
60. Ellison DH, Velazquez H, Wright FS: Stimulation of distal potassium secretion by low lumen chloride in the presence of barium. *Am J Physiol* 1985; 248:F638–F649.
61. El Mernissi G, Doucet A: Short-term effects of aldosterone and dexamethasone on Na-K-ATPase along the rabbit nephron. *Pfluegers Arch* 1983; 399:147–151.
62. Eveloff J, Bayerdorffer E, Silva P, Kinne R: Sodium-chloride transport in the thick ascending limb of Henle's loop. Oxygen consumption studies in isolated cells. *Pfluegers Arch* 1981; 389:263–270.

63. Eveloff J, Haase W, Kinne R: Separation of renal medullary cells: Isolation of cells from the thick ascending limb of Henle's loop. *J Cell Biol* 1980; 87:672–681.
64. Ernst SA: Transport ATPase cytochemistry: Ultrastructural localization of potassium-dependent and potassium-independent activities in rat kidney cortex. *J Cell Biol* 1975; 66:586–608.
65. Ernst SA, Mills JE: Basolateral plasma membrane localization of ouabain-sensitive sodium transport sites in the secretory epithelium of the avian salt gland. *J Cell Biol* 1977; 75:74–94.
66. Ernst SA, Schreiber JH: Ultrastructural localization of Na^+,K^+-ATPase in rat and rabbit kidney medulla. *J Cell Biol* 1981; 91:803–813.
67. Fakunding JL, Chow R, Catt KJ: The role of calcium in the stimulation of aldosterone production by adrenocorticotropin, angiotensin II, and potassium in isolated glomerulosa cells. *Endocrinology* 1979; 105:327-333.
68. Fanestil DD, Park CS: Steroid hormones and the kidney. *Ann Rev Physiol* 1981; 43:637–649.
69. Feldman D, Funder JW, Edelman IS: Subcellular mechanisms in the action of adrenal steroids. *Am J Med* 1972; 53:545–560.
70. Field MJ, Stanton BA, Giebisch GH: Influence of ADH on renal potassium handling: A micropuncture and microperfusion study. *Kidney Int* 1984; 25:502–511.
71. Flynn TG, de Bold ML, de Bold AJ: The amino acid sequence of an atrial peptide with potent diuretic and natriuretic properties. *Biochem Biophys Res Comm* 1983; 117:859–865.
72. Fromter E, Gessner K: Free flow potential profile along rat kidney proximal tubule. *Pfluegers Arch* 1974; 351:69–84.
73. Ganote CE, Grantham JJ, Moses HL, et al: Ultrastructural studies of vasopressin effect on isolated perfused renal collecting tubules of the rabbit. *J Cell Biol* 1968; 36:355–367.
74. Gertz KH: Transtubulare Natriumchloridflusse und permeabilitat fur Nichteletrolyte im proximalen und distalen Konvolut der Rattenniere. *Arch Ges Physiol* 1963; 276:336–356.
75. Giebisch G: Coupled ion and fluid transport in the kidney. *N Engl J Med* 1972; 287:913–919.
76. Giebisch G: Functional organization of proximal and distal tubular electrolyte transport. *Nephron* 1969; 6:260–281.
77. Giebisch G: The contribution of measurements of electrical phenomena to our knowledge of renal electrolyte transport. *Prog Cardiovas Dis* 1961; 3:463–482.
78. Giebisch G: Measurements of electrical potentials and ion fluxes on single renal tubules. *Circulation* 1960; 21:879–891.
79. Giebisch G, Malnic G: Some aspects of renal tubular hydrogen ion transport. *Proc Intern Congr Nephrol 4th Stockholm* 1970; 1:181–194.
80. Giebisch G, Windhager EE: Electrolyte transport across renal tubular membranes, in Orloff J, Berliner RW (eds): *Handbook of Physiology, Section 8, Renal Physiology*. Washington DC, American Physiological Society, 1973, pp 315–376.
81. Giebisch G, Windhager EE: Renal tubular transfer of sodium chloride and potassium. *Am J Med* 1964; 36:643–669.
82. Gluck ST, Cannon CH, Al-Awquati Q: Exocytosis regulates urinary acidification in turtle bladder by rapid insertion of H^+ pumps into the luminal membrane. *Proc Natl Acad Sci USA* 1982; 79:4327–4331.
83. Goldberg LI, Volkman PH, Kohli JD: A comparison of the vascular dopamine receptor with other hormone receptors. *Ann Rev Pharmacol Toxicol* 1978; 18:57–79.
84. Goldberg M, McCurdy DK, Foltz EL, Bluemle LW: Effects of ethacrynic acid (a new slauretic agent) on renal diluting and concentrating mechanisms: Evidence for site of action in the loop of Henle. *J Clin Invest* 1964; 43:201–216.
85. Good DW, Wright FS: Luminal influences on potassium secretion: Sodium concentration and fluid flow rate. *Am J Physiol* 1979; 236:F192–F205.
86. Good DW, Wright FS: Luminal influences on potassium secretion: Transepithelial voltage. *Am J Physiol* 1980; 239:F289–F298.
87. Good DW, Velazquez H, Wright FS: Luminal influences on potassium secretion: Low sodium concentration. *Am J Physiol* 1984; 246:F609–F619.
88. Goodman DD, Allen JE, Rasmussen H: On the mechanism of action of aldosterone. *Proc Natl Acad Sci USA* 1969; 64:330–337.
89. Gordon D, Nashat FS, Wilcox CS: An analysis of the regulation of sodium excretion during induced changes in plasma sodium concentration in anesthetized dogs. *J Physiol (Lond)* 1981; 1981; 314:531–545.
90. Gottschalk CW: Osmotic concentration and dilution of the urine. *Am J Med* 1964; 36:670–685.
91. Gottschalk CW: Renal tubular function: lessons from micropuncture. *Harvey Lecture Ser* 1962–1963; 58:99–123. Academic Press, Inc.
92. Gottschalk CW: Micropuncture studies of tubular function in the mammalian kidney. *Physiologist* 1961; 4:33–55.
93. Gottschalk CW, Lassiter WE: Micropuncture methodology, in Orloff J, Berliner RW (eds): *Handbook of Physiology, Section 8, Renal Physiology*. Washington DC, American Physiological Society, 1973, pp 129–143.
94. Gottschalk CW, Lassiter WE: A review of micropuncture studies of salt and water reabsorption in the mammalian nephron. *Proc Intern Congr Nephrol,*

Washington DC, Handler JS (ed). New York, Karger, 1967, vol 1, pp 357–373.
95. Grantham JJ, Ganote CE, Burg MB, Orloff J: Paths of transtubular water flow in isolated renal collecting tubules. *J Cell Biol* 1969; 41:562–576.
96. Grantham JJ: Mode of water transport in mammalian renal collecting tubules. *Fed Proc* 1971; 30:14–21.
97. Greger R: Chloride reabsorption in the rabbit cortical thick ascending limb of the loop of Henle. *Pfluegers Arch* 1981; 390:38–43.
98. Greger R, Schlatter E: Properties of the lumen membrane of the cortical thick ascending limb of Henle's loop of rabbit kidney. *Pfluegers Arch* 1983a; 396:315–324.
99. Greger R, Schlatter E: Properties of the basolateral membrane of the cortical thick ascending limb of Henle's loop of rabbit kidney. *Pfluegers Arch* 1983b; 396:325–334.
100. Greger R, Schlatter E, Lang F: Evidence for electroneutral sodium chloride cotransport in the cortical thick ascending limb of Henle's loop of rabbit kidney. *Pfluegers Arch* 1983; 396:308–314.
101. Gross JB, Imai M, Kokko JP: A functional comparison of the cortical collecting tubule and the distal convoluted tubule. *J Clin Invest* 1975; 55:1284–1294.
102. Guder WG: Stimulation of renal gluconeogenesis by angiotensin II. *Bichmim Biophys Acta* 1979; 584:507–519.
103. Guder WG, Ross BD: Enzyme distribution along the nephron. *Kidney Int* 1984; 26:101–111.
104. Guder WG, Wirthensohn G: Metabolism of isolated kidney tubules: Interaction between lactate, glutamine and oleate metabolism. *Eur J Biochem* 1979; 99:577–584.
105. Gyory AZ, Lingard JM: Kinetics of active sodium transport in rat proximal tubules and its variation by cardiac glycosides at zero net volume and ion fluxes. Evidence for a multisite sodium transport system. *J Physiol (Lond)* 1976; 257:257–274.
106. Hamill OP, Marty A, Neher E, et al: Improved patch-clamp techniques for high-resolution current recording from cells and cell-free membrane patches. *Pfluegers Arch* 1981; 391:85–100.
107. Handler JS, Preston AS, Orloff J: Effect of aldosterone on the sodium content and energy metabolism of epithelial cells of the toad urinary bladder. *J Steroid Biochem* 1972; 3:137–141.
108. Hebert SC, Andreoli TE: Control of NaCl transport in the thick ascending limb. *Am J Physiol* 1984; 246:F745-F756.
109. Hebert SC, Culpepper RM, Andreoli TE: NaCl transport in mouse medullary thick ascending limbs. I. Functional nephron heterogeneity and ADH-stimulated NaCl cotransport. *Am J Physiol* 1981; 241:F412–F431.
110. Hierholzer K, Wiederholt W, Holzgreve H, et al: Micropuncture study of renal transtubular concentration gradients of sodium and potassium in adrenalectomized rats. *Arch Ges Physiol* 1965; 285:193–210.
111. Higashihara E, DuBose TD Jr, Kokko JP: Direct examination of chloride transport across papillary collecting duct of the rat. *Am J Physiol* 1978; 235:F219–F226.
112. Hoffman N, Thees M, Kinne R: Phosphate transport by isolated renal brush border vesicles. *Pfluegers Arch* 1976; 362:147–156.
113. Hopfer U: Transport in isolated plasma membranes. *Am J Physiol* 1978; 234:F89–F96.
114. Howards SS, Davis BB, Knox FG, et al: Depression of fractional sodium reabsorption by the proximal tubule of the dog without sodium diuresis. *J Clin Invest* 1968; 47:1561–1572.
115. Imai M: Functional heterogeneity of the descending limbs of Henle's loop. II. Interspecies differences among rabbits, rats, and hamsters. *Pfluegers Arch* 1984; 402:393–401.
116. Imai M: Function of the thin ascending limb of Henle of rats and hamsters perfused in vitro. *Am J Physiol* 1977; 232:F201–F209.
117. Imai M, Hayashi M, Araki M: Functional heterogeneity of the decending limbs of Henle's loop. I. Internephron heterogeneity in the hamster kidney. *Pfluegers Arch* 1984; 402:385–392.
118. Ives HE, Yee VJ, Warnock DG: Asymetrc distribution of the Na^+/H^+ antiporter in renal proximal tubule epithelial cells. *J Biol Chem* 1983; 258:13513–13516.
119. Jacobson HR: Functional segmentation of the mammalian nephron. *Am J Physiol* 1981; 241:F203–F218.
120. Jacobson HR, Kokko JP: Intrinsic differences in various segments of the proximal convoluted tubule. *J Clin Invest* 1976; 57:818–825.
121. Jamison RL: Micropuncture study of superficial and juxtamedullary nephrons in the rat. *Am J Physiol* 1970; 218:46–55.
122. Jamison RL: Micropuncture study of segments of thin loop of Henle in the rat. *Am J Physiol* 1968; 215:236–242.
123. Jamison RL, Work J, Schafer JA: New pathways for potassium transport in the kidney. *Am J Physiol* 1982; 242:F297–F312.
124. Jard S: Vasopressin isoreceptors in mammals: Relation to cyclic AMP-dependent and cyclic AMP-independent transduction mechanisms. *Current Topics in Membranes and Transport*. New York, Academic Press, 1983, p 255–285.
125. Jewell PA, Verney EB: An experimental attempt to determine the site of the neurohypophyseal osmoreceptors in the dog. *Phil Trans* 1957; B240:197–324.
126. Johnston CI, Davis JO, Howards SS, Wright FS: Cross-circulation experiments on the mechanism of

the natriuresis during saline loading in the dog. *Circ Res* 1967; 20:1–10.

127. Kady NN, Nashat FS, Tappin JW, Wilcox CS: Glomerulotubular balance in the whole kidney of the anesthetized dog during infusion of hypertonic saline. *J Physiol (Lond)* 1974; 242:114–116P.

128. Kahn AM, Dolson GM, Hise MK, et al: Parathyroid hormone and dibutyryl cAMP inhibit Na^+/H^+ exchange in renal brush border vesicles. *Am J Physiol* 1985; 248:F212–F218.

129. Kaissling B: Structural aspects of adaptive changes in renal electrolyte excretion. *Am J Physiol* 1982; 243:F211–F226.

130. Kaissling B: Ultrastructural organization of the transition from the distal nephron to the collecting duct in the desert rodent Psammomys obesus. *Cell Tissue Res* 1980; 212:475–495.

131. Kaissling B, Kriz W: Variability of intercellular spaces between macula densa cells: A TEM study in rabbits and rats. *Kidney Int* 1982; 12:9–17.

132. Kaissling B, Koeppen BM, Wade JB, LeHir M: Effects of mineralocorticoids on the structure of intercalated cells. *Acta Anat* 1981; III:72A.

133. Kaissling B, Kriz W: Structural analysis of the rabbit kidney. *Adv Anat Embryol Cell Biol* 1979; 56:1–123.

134. Kaissling B, Peter S, Kriz W: The transition of the thick ascending limb of Henle's loop into the distal convoluted tubule in the nephron of the rat kidney. *Cell Tissue Res* 1977; 182:111–118.

135. Kamm O, Aldrich TB, Grote IW, et al: The active principles of the posterior lobe of the pituitary gland. I. The demonstration of the presence of two active principles. II. The separation of the two principles and their concentration in the form of potent solid preparations. *J Am Chem Soc* 1928; 50:573–601.

136. Katz AI: Renal Na-K-ATPase: Its role in tubular sodium and potassium transport. *Am J Physiol* 1982; 242:F207–F219.

137. Katz AI, Doucet A, Morel F: Na-K-ATPase activity along the rabbit, rat, and mouse nephron. *Am J Physiol* 1979; 237:F114–F120.

138. Katz B, Epstein FH: The role of sodium-potassium-activated adenosine triphosphatase in the reabsorption of sodium by the kidney. *J Clin Invest* 1967; 46:1999–2011.

139. Kaufman JS, Hamburger RJ: Passive potassium transport in the proximal convoluted tubule. *Am J Physiol* 1985; 248:F228–F232.

140. Kawashima H, Kurokawa K: Unique hormonal regulation of vitamin D metabolism in the mammalian kidney. *Mineral Electrolyte Metab* 1983; 9:227–235.

141. Kinne R, Schwartz JL: Isolated membrane vesicles in the evaluation of nature, localization and regulation of renal transport processes. *Kidney Int* 1978; 14:547–556.

142. Kinsella JL, Freiberg JM, Sacktor B: Glucocorticoid activation of Na^+/H^+ exchange in renal brush border vesicles: Kinetic effects. *Am J Physiol* 1985; 248:F233–F239.

143. Kinsella JL, Aronson PS: Determination of the coupling ratio for Na^+/H^+ exchange in renal microvillus membrane vesicles. *Biochim Biophys Acta* 1982; 689:161–164.

144. Kinsella JL, Aronson PS: Amiloride inhibition of the $Na^+ - H^+$ exchanger in renal microvillus membrane vesicles. *Am J Physiol* 1981; 241:F374–F379.

145. Kinsella JL, Aronson PS: Properties of the Na^+-H^+ exchanger in in renal microvillus membrane vesicles. *Am J Physiol* 1980; 238:F461-F469.

146. Kirk KL, Schafer JA, DiBona DR: Quantitative analysis of the structural events associated with antidiuretic hormone-induced volume reabsorption in the rabbit cortical collecting tubule. *J Membrane Biol* 1984; 79:65–74.

147. Knochel JP, Whale MG: The role of aldosterone in renal physiology. *Arch Intern Med* 1973; 131:876–884.

148. Koch KM, Aynedjian HS, Bank N: Effect of acute hypertension on sodium reabsorption by the proximal tubule. *J Clin Invest* 1968; 47:1696–1709.

149. Koeppen BM, Biagi BA, Giebisch G: Intracellular microelectrode characterization of the rabbit cortical collecting duct. *Am J Physiol* 1983; 244:F35–F47.

150. Koeppen BM, Helman SI: Acidification of luminal fluid by the rabbit cortical collecting tubule perfused in vitro. *Am J Physiol* 1982; 242:521–531.

151. Kokko JP: Membrane characteristics governing salt and water transport in the loop of Henle. *Fed Proc* 1974; 33:25–30.

152. Kokko JP: Proximal tubule potential difference dependence on glucose, HCO_3^- and amino acids. *J Clin Invest* 1973; 52:1362–1367.

153. Kon V, Ichikawa I: Effector loci for renal nerve control of cortical microcirculation. *Am J Physiol* 1983; 245:F545–F553.

154. Kone CK, Madson KM, Tisher CC: Ultrastructure of the thick ascending limb of Henle in the rat kidney. *Am J Anat* 1984; 171:217–226.

155. Koushanpour E: *Renal Physiology: Principles and Functions*, ed 1. Philadelphia, WB Saunders Co, 1976.

156. Koushanpour E, Tarica RR, Stevens WF: Mathematical simulation of normal nephron function in rat and man. *J Theor Biol* 1971; 31:177–214.

157. Kruck F: Influence of humoral factors on renal tubular sodium handling. *Nephron* 1969; 6:205–216.

158. Landwehr DM, Klose RM, Giebisch G: Renal tubular sodium and water reabsorption in the isotonic sodium chloride loaded rat. *Am J Physiol* 1967; 212:1327–1333.

159. Landwehr DM, Schnermann J, Klose RM, Giebisch G: Effect of reduction in filtration rate on renal tubular sodium and water reabsorption. *Am J Physiol* 1968; 215:687–695.

160. Lassiter WE, Mylle M, Gottschalk CW: Net transtubular movement of water and urea in saline diuresis. *Am J Physiol* 1964; 206:669–673.
161. Lassiter WE, Mylle M, Gottschalk CW: Micropuncture study of urea transport in rat renal medulla. *Am J Physiol* 1966; 210:965–970.
162. Lauson HD: Metabolism of antidiuretic hormones. *Am J Med* 1967; 42:713–744.
163. Lewy JE, Windhager EE: Peritubular control of proximal tubular fluid reabsorption in the rat kidney. *Am J Physiol* 1968; 214:943–954.
164. Lifschitz MD, Schrier RW, Edelman IS: Effect of actinomycin D on aldosterone-mediated changes in electrolyte excretion. *Am J Physiol* 1973; 224:376–380.
165. Lipton P, Edelman IS: Effects of aldosterone and vasopressin on electrolytes of toad bladder epithelial cells. *Am J Physiol* 1971; 221:733–741.
166. Lo C-S, Lo TN: Time course of the renal response to triiodothyronine in the rat. *Am J Physiol* 1979; 236:F9–F12.
167. Lockett MF: Effects of saline loading on the perfused cat kidney. *J Physiol (Lond)* 1966; 187:489–500.
168. Lowitz HD, Stumpe KO, Ochwadt B: Micropuncture study of the action of angiotensin II on tubular sodium and water reabsorption in the rat. *Nephron* 1969; 6:173–187.
169. Malnic G, De Mello Aires M: Micropuncture study of chloride, bicarbonate and sulfate transfer in proximal tubules of rat kidney. *Am J Physiol* 1970; 218:27–32.
170. Malnic G, Klose RM, Giebisch G: Micropuncture study of distal tubular potassium and sodium transport in the rat nephron. *Am J Physiol* 1966; 211:529–547.
171. Malnic G, Klose RM, Giebisch G: Micropuncture study of renal potassium excretion in the rat. *Am J Physiol* 1964; 206:674–686.
172. Malvin RL, Wilde WS, Sullivan LP: Localization of nephron transport by stop flow analysis. *Am J Physiol* 1958; 194:135–142.
173. Marsh DJ: Solute and water flows in thin limbs of Henle's loop in the hamster kidney. *Am J Physiol* 1970; 218:824–831.
174. Maunsbach AB: Ultrastructure of the proximal tubule, in Orloff J, Berliner RW (eds): *Handbook of Physiology, Section 8, Renal Physiology*. Washington DC, American Physiological Society, 1973, pp 31–79.
175. Mausbach AB: Observations of the segmentation of the proximal tubule in the rat kidney. *J Ultrastruct Res* 1966; 16:239–258.
176. McDonald SJ, De Wardener HE: The relationship between the renal arterial perfusion pressure and the increased sodium excreion which occurs during an infusion of saline. *Nephron* 1965; 2:1–14.
177. Morel F: Sites of hormone action in the mammalian nephron. *Am J Physiol* 1981; 240:F159–F164.
178. Morel F, Mylle M, Gottschalk CW: Tracer microinjection studies of effect of ADH on renal tubular diffusion of water. *Am J Physiol* 1965; 209:179–187.
179. Morgan T, Berliner RW: Permeability of the loop of Henle, vasa recta, and collecting duct to water, urea and sodium. *Am J Physiol* 1968; 215:108–115.
180. Morris DJ: The metabolism and mechanism of action of aldosterone. *Endocrinol Rev* 1981; 2:234–247.
181. Moses AM, Scheinman SJ, Schroeder ET: Antidiuretic and PGE_2 responses to AVP and dDAVP in subjects with central and nephrogenic diabetes insipidus. *Am J Physiol* 1985; 248:F354–F359.
182. Moss NG: Renal function and renal afferent and efferent nerve activity. *Am J Physiol* 1982; 243:F425–F433.
183. Muller J: *Regulation of Aldosterone Biosynthesis*. New York, Springer-Verlag, 1971.
184. Mulrow PJ, Forman BH: The tissue effects of mineralocorticoids. *Am J Med* 1972; 53:561–572.
185. Murer H, Hopfer U, Kinne R: Sodium-proton antiport in brush-border membrane vesicles isolated from rat small intestine and kidney. *Biochem J* 1976; 154:597–604.
186. Myers CE, Bulger RE, Tisher CC, Trump BF: Human renal ultrastructure. IV. Collecting duct of healthy individuals. *Lab Invest* 1966; 15:1921–1950.
187. Nashat FS, Tappin JW, WilcoxCS: The renal blood flow and the glomerular filtration rate of anesthetized dogs during acute changes in plasma sodium concentration. *J Physiol (Lond)* 1976; 256:731–745.
188. Norgaard JOR, Maunsbach AB: Use of isolated glomeruli in studies of renal physiology and pathology, in Leaf A, et al (eds): *Renal Pathophysiology—Recent Advances*. New York, Raven Press, 1980, pp 75–90.
189. O'Neil RG, Boulpaep EL: Effect of amiloride on the apical cell membrane cation channels of a sodium-absorbing, potassium-secreting renal epithelium. *J Membr Biol* 1979; 50:365–387.
190. O'Neil RG, Boulpaep EL: Ionic conductive properties and electrophysiology of the rabbit cortical collecting tubule. *Am J Physiol* 1982; 243:F81–F95.
191. Orloff J, Handler JS: The cellular mode of action of antidiuretic hormone. *Am J Med* 1964; 36:686–697.
192. Orloff J, Handler JS: The role of adenosine 3',5'-phosphate in the action of antidiuretic hormone. *Am J Med* 1967; 42:757–768.
193. Peter K: *Untersuchungen uber Bau und Entwicklung der Niere*. Jena, Gustav Fischer, 1909.
194. Petty KJ, Kokko JP, Marver D: Secondary effect of aldosterone on Na-K ATPase activity in the rabbit cortical collecting tubule. *J Clin Invest* 1981; 68:1514–1521.
195. Pollock DM, Banks RO: Effect of atrial extract on renal function in the rat. *Clin Sci* 1983; 65:47–55.
196. Pricam C, Humbert F, Perrelet A, Orci L: A freeze-etch study of the tight junction of the rat kidney tubules. *Lab Invest* 1974; 30:286–291.
197. Rector FC Jr: Sodium, bicarbonate, and chloride ab-

sorption by the proximal tubule. *Am J Physiol* 1983; 244:F461–F471.
198. Richards AN, Walker AM: Methods of collecting fluid from known regions of the renal tubules of amphibia and of perfusing the lumen of a single tubule. *Am J Physiol* 1937; 118:111–120.
199. Rocha AS, Kokko JP: Sodium chloride and water transport in the medullary thick ascending limb of Henle. Evidence for active chloride transport. *J Clin Invest* 1973; 52:612–623.
200. Rocha AS, Kudo LH: Water, urea, sodium, chloride, and potassium transport in the in vitro isolated perfused papillary collecting duct. *Kidney Int* 1982; 22:485–491.
201. Rodicio J, Herrera-Acosta J, Sellman JC, et al: Studies on glomerulotubular balance during aortic constriction, ureteral obstruction and venous occlusion in hydropenic and saline-loaded rats. *Nephron* 1969; 6:437–456.
202. Roemmelt JC, Sartorius OW, Pitts RF: Excretion and reabsorption of sodium and water in the adrenalectomized dogs. *Am J Physiol* 1949; 159:124–136.
203. Roesinger B, Schiller A, Taugner R: A freeze-fracture study of tight junctions in the pars convoluta and pars recta of the renal proximal tubule. *Cell Tissue Res* 1978; 186:121–133.
204. Sachs H: Biosynthesis and release of vasopressin. *Am J Med* 1967; 42:687–700.
205. Sansom SC, O'Neil RG: Mineralocorticoid regulation of apical cell membrane Na^+ and K^+ transport of the cortical collecting duct. *Am J Physiol* 1985; 248:F858–F868.
206. Sardet C, Pisam M, Maetz J: The surface epithelium of teleostean fish gills. Cellular and junctional adaptations of the chloride cell in relation to salt adaptation. *J Cell Biol* 1979; 80:96–117.
207. Schlatter E, Greger R, Weidtke C: Effect of "high ceiling" diuretics on active salt transport in the cortical thick ascending limb of Henle's loop of rabbit kidney: Correlation of chemical structure and inhibitory potency. *Pfluegers Arch* 1983; 396:210–217.
208. Schmidt U, Dubach UC: Activity of (Na^+-K^+)-stimulated adenosintriphosphatase in the rat nephron. *Pfluegers Arch* 1969; 306:219–226.
209. Schmidt U, Guder WG: Sites of enzyme activity along the nephron. *Kidney Int* 1976; 9:233–242.
210. Schrier RW, McDonald KM, Marshall RA, Lauler DP: Absence of natriuretic response to acute hypotonic intravascular volume expansin in dogs. *Clin Sci* 1968; 34:57–72.
211. Schwartz GJ, Al-Awquati Q: Carbon dioxide causes exocytosis of vesicles containing H^+ pumps in isolated perfused proximal and collecting tubules. *J Clin Invest* 1985; 75:1638–1644.
212. Schwartz GJ, Weinstein AM, Steele RE, et al: Carbon dioxide permeability of rabbit proximal convoluted tubules. *Am J Physiol* 1981; 240:F231–F244.

213. Schwartz MM, Venkatachalam MA: Structural differences in thin limbs of Henle: Physiological implications. *Kidney Int* 1974; 6:193–208.
214. Seely JF, Boulpaep EL: Electrical potentials across proximal and distal tubules of dog kidney. *Am J Physiol* 1971; 221:1084–1096.
215. Seifter J, Kinsella JL, Aronson PS: Mechanism of Cl transport in *Necturus* renal microvillus membrane vesicles. *Kidney Int* 1981; 19:257A.
216. Selkurt EE, Womack I, Dailey WN: Mechanism of natriuresis and diuresis during elevated renal arterial pressure. *Am J Physiol* 1965; 209:95–99.
217. Share L: Vasopressin, its bioassay and the physiological control of its release. *Am J Med* 1967; 42:701–712.
218. Share L, Kimura T, Matsui K, Shade RE, Crofton JT: Metabolism of vasopressin. *Fed Proc* 1985; 44:59–61.
219. Sharp GWG, Leaf A: Mechanisms of action of aldosterone. *Physiol Rev* 1966; 46:593–633.
220. Shipp JC, Hanenson IB, Windhager EE, et al: Single proximal tubules of Necturus kidney. Methods for micropuncture and microperfusion. *Am J Physiol* 1958; 195:563–569.
221. Siegenthaler WE, Peterson RE, Frimpter GW: The renal clearance of aldosterone and its major metabolites, in Baulier EE, Rubel P (eds): *Aldosterone, A Symposium*. Philadelphia, FA Davis Co, 1964, pp 51–72.
222. Simpson SA, Tait JF: Recent progress in methods of isolation, chemistry and physiology of aldosterone. *Recent Prog Horm Res* 1955; 11:183–210.
223. Sonnenberg H: Atrial natriuretic factor—A new hormone affecting kidney function. *Klin Wochenschr* 1985; 63:886–890.
224. Spring KR, Giebisch G: Kinetics of Na^+ transport in Necturus proximal tubule. *J Gen Physiol* 1971; 70:307–328.
225. Spring KR, Hope A: Size and shape of the lateral intercellular spaces in living epithelium. *Science* 1978; 200:54–58.
226. Stanton BA: Role of adrenal hormones in regulating distal nephron structure and ion transport. *Fed Proc* 1985; 44:2717–2722.
227. Stetson DL, Wade JB, Giebisch G: Morphologic alterations in the rat medullary collecting duct following potassium depletion. *Kidney Int* 1980; 17:45–56.
228. Stokes JB, Ingram MJ, Williams AD, Ingram D: Heterogeneity of the rabbit collecting tubule: Localization of mineralocorticoid hormone action to the cortical portion. *Kidney Int* 1981; 20:340–347.
229. Stokes JB: Effect of prostaglandin E_2 on chloride transport across rabbit thick ascending limb of Henle. Selective inhibition of the medullary portion. *J Clin Invest* 1979; 64:495–502.
230. Stokes JB, Tisher CC, Kokko JP: Structural-functional heterogeneity along the rabbit collecting tubule. *Kidney Int* 1978; 14:585–593.

231. Sutherland EW, Robison GA, Butcher RW: Some aspects of the biological role of adenosine 3',5'-monophosphate (cyclic AMP). *Circulation* 1968; 37:279–306.
232. Thibault G, Garcia R, Cantin M, Genest J: Atrial natriuretic factor. Characterization and partial purification. *Hypertension* 1983; 5 (suppl I):I-75-I-80.
233. Tisher CC: Anatomy of the kidney, in Brenner BM, Rector FC (eds): *The kidney.* vol I. Philadelphia, WB Saunders Co, 1976, pp 3–64.
234. Tisher CC, Bulger RE, Trump BF: Human renal ultrastructure. III. The distal tubule in healthy individuals. *Lab Invest* 1968; 18:655–668.
235. Tisher CC, Bulger RE, Trump BF: Human renal ultrastructure. I. Proximal tubule of healthy individuals. *Lab Invest* 1966; 15:1357–1394.
236. Teulon J, Anagnostopoulos T: Proximal cell K^+ activity: Technical problems and dependence on plasma K^+ concentration. *Am J Physiol* 1982; 243:F12–F18.
237. Tobian L, Coffee K, McCrea P: Evidence for a humoral factor of non-renal and non-adrenal origin which influences renal sodium excretion. *Trans Assoc Am Physicians* 1967; 80:200–206.
238. Torikai S, Kurokawa K: Effect of PGE_2 on vasopressin-dependent cell cAMP in isolated single nephron segments. *Am J Physiol* 1983; 245:F58-F66.
239. Ullrich KJ: Sugar, Amino acid, and Na^+ cotransport in the proximal tubule. *Ann Rev Physiol* 1979; 41:181–195.
240. Ussing HH: The distinction by means of tracers between active transport and diffusion. *Acta Physiol Scand* 1949; 19:43–56.
241. Ussing HH, Zerahn K: Active transport of sodium as the source of electric current in the sort-circuited isolated frog skin. *Acta Physiol Scand* 1951; 23:110–127.
242. Vander AJ, Malvin RL, Wilde WS, et al: Effects of adrenalectomy and aldosterone on proximal and distal tubular sodium reabsorption. *Proc Soc Exp Biol Med* 1958; 99:323–325.
243. Vandewalle A, Wirthensohn G, Heidrich HG, Gruder WG: Distribution of hexokinase and phosphoenolpyruvate carboxykinase along the rabbit nephron. *Am J Physiol* 1981; 240:F492–F500.
244. Velazquez H, Wright FS, Good DW: Luminal influences on potassium secretion: Chloride replacement with sulfate. *Am J Physiol* 1982; 242:F46–F55.
245. Verney EB: The antidiuretic hormone and the factors which determine its release. *Proc Roy Soc B* 1947; 135:25–106.
246. Wade JB, O'Neil RG, Pryor JL, Boulpaep EL: Modulation of cell membrane area in renal cortical collecting tubules by corticosteroid hormones. *J Cell Biol* 1979; 81:439–445.
247. Warnock DG, Eveloff J: NaCl entry mechanisms in the luminal membrane of the renal tubule. *Am J Physiol* 1982; 242:F561–F574.
248. Welling LW, Welling DJ: Shape of epithelial cells and intercellular channels in the rabbit proximal nephron. *Kidney Int* 1976; 9:385–394.
249. Welling LW, Welling DJ: Surface areas of brush border and lateral cell walls in the rabbit proximal nephron. *Kidney Int* 1975; 8:343–348.
250. Welling LW, Welling DJ, Hill JJ: Shape of cells and intercellular channels in rabbit thick ascending limb of Henle. *Kidney Int* 1978; 13:144–151.
251. Wiederholt M: Effect of actinomycin-D on renal Na and K transport. Proc of 2nd Ann Meeting Am Soc Nephrol, Washington DC, 1968.
252. Williams GH, Cain JP, Dluhy RG, Underwood RH: Studies of the control of plasma aldosterone concentration in normal man. I. Response to posture, acute and chronic volume depletion, and sodium loading. *J Clin Invest* 1972; 51:1731–1742.
253. Williams GH, Dluhy RG: Aldosterone biosynthesis: interrelationship of regulatory factors. *Am J Med* 1972; 53:595–605.
254. Windhager EE: Some aspects of proximal tubular salt reabsorption. *Fed Proc* 1974; 33:21–24.
255. Windhager EE, Boulpaep EL, Giebisch G: Electrophysiological studies on single nephron. In: Proc 3rd Intern Congr Nephrol, Washington DC, vol 1, Basel, S Karger, 1967, pp 35–47.
256. Windhager EE, Lewy JE, Spitzer A: Intrarenal control of proximal tubular reabsorption of sodium and water. *Nephron* 1969; 6:247–259.
257. Wirz H, Hargitay B, Kuhn W: Lokalisation des Konzentrierungsprozesses in der Niere durch direkte Kryoskopie. *Helv Physiol Pharmacol Acta* 1951; 9:196–207.
258. Wirz H, Dirix R: Urinary concentration and dilution, in Orloff J, Berliner RW (eds): *Handbook and Physiology, Section 8, Renal Physiology.* Washington DC, American Physiological Society, 1973, pp 415–430.
259. Woodhall PB, Tisher CC: Response of the distal tubule and cortical collecting duct to vasopressin in the rat. *J Clin Invest* 1973; 52:3095–3108.
260. Wong NLM, Quamme GA, Sutton RAL, Dirks JH: Effects of mannitol on water and electrolyte transport in the dog kidney. *J Lab Clin Med* 1979; 94:683–692.
261. Wright FS, Giebisch G: Renal potassium transport: Contributions of individual nephron segments and populations. *Am J Physiol* 1978; 235:F515–F527.
262. Wright FS, Knox FG, Howards SS, Berliner RW: Reduced sodium reabsorption by the proximal tubule of DOCA-escaped dogs. *Am J Physiol* 1969; 216:869–875.
263. Wuu TC, Crumm S, Saffran M: Amino acid sequence of porcine neurophysin-I. *J Biol Chem* 1971; 246:6043–6063.
264. Zabel M, Schiebler TH: Histochemical, autoradiographic and electron microscopic investigations of the renal proximal tubule of male and female rats after castration. *Histochemistry* 1980; 69:255–276.

10

Tubular Reabsorption and Secretion: Classification Based on Overall Clearance Measurements

In the preceding chapter we analyzed the sequential processing of the filtrate and the mechanisms of tubular transport of major electrolytes and water along the various segments of the nephron. That analysis was largely based on successful application of in-vivo micropuncture, in-vitro microperfusion, and electrophysiological techniques at the level of a single nephron. However, such techniques have been used to delineate the sequential processing of only a limited number of substances. To ascertain the renal handling of those solutes (mostly nonelectrolytes) for which micropuncture data are scarce or as yet not available, we must resort to the standard overall clearance technique. Therefore, in this chapter, we expand on the materials presented in Chapter 7 and consider in more detail the overall (as opposed to sequential) processing of the major nonelectroyte constituents of the filtrate, using the standard clearance measurements. From such information we can determine the net tubular transport of a substance by the kidney as well as classify the type of cellular transport process involved.

Insofar as the renal tubular processing of the filtrate is concerned, the solutes contained in the tubular fluid are transported by any one or a combination of three processes: bulk flow, simple passive diffusion, and carrier-mediated transport. The latter type of transport, however, is by and large the major mechanism for the net reabsorption or secretion of nearly all the important solutes present in the filtrate and plasma. It should be recalled from Chapter 4 that reabsorption and secretion refer to the direction of transport and not the underlying mechanisms. Thus, reabsorption refers to the transport of solutes from the tubular lumen into the peritubular capillary blood, and secretion refers to the transport of solutes from the peritubular capillary blood into the tubular lumen.

Renal Titration Curves: Classification of Tubular Transport

The capacity of the renal tubules for reabsorption and secretion varies, depending on the type of solute being transported. To simplify this presentation, we classify renal tubular transport into a passive process, which requires no expenditure of energy (e.g., urea transport) and an active process, which does require expenditure of energy (e.g., glucose and sodium transport). Note that the terms passive and active transport, as used here, imply that these processes represent the major contributing components in the overall transport of the given solute.

The renal handling of an actively transported solute may be further subdivided into two categories, depending on whether there is a maximum tubular transport capacity for that solute: (1) Those solutes for which there is a definite upper limit for unidirectional rate of transport, either reabsorptive or secretory in direction. Thus, as the quantity of solute delivered (filtered load) to the tubules for reabsorption is increased, the rate of reabsorption will increase only up to a certain maximal rate. Further increase in the filtered load presented to the tubules will not increase the rate of reabsorption. This limiting rate of reabsorption is called the *maximum tubular reabsorptive capacity*[41, 90] and is designated by the symbol $\dot{T}_m$. It denotes the maximum amount of a solute reabsorbed per minute, and its value varies with the substance involved. Similarly, there is also a *maximum tubular secretory capacity*, which defines the limiting rate at which the tubules can secrete specific substances. Renal reabsorption of glucose and secretion of *p*-aminohippurate (PAH) are examples of actively transported solutes exhibiting tubular transport maximum. (2) Those solutes for which there is no definite upper limit for unidirectional rate of transport, and hence no $\dot{T}_m$. Reabsorption of sodium along the nephron is a notable example. Although sodium reabsorption is not a $\dot{T}_m$-limited process, its net transepithelial reabsorption, as described in Chapter 9, is limited by the tubular flow rate, a factor influencing the cellular contact time which in part determines the rate of sodium diffusion from the tubular lumen into the renal tubule cell. Consequently, the reabsorption of sodium and like substances is called *gradient-time-limited* transport.

Whether a substance is transported by a $\dot{T}_m$-limited process or by a gradient-time-limited process can be determined by simultaneous measurements of its filtered load and urinary excretion rate at varying plasma concentrations. From these measurements we can determine the net rate of reabsorption or secretion of the substance and whether or not it is transported by a $\dot{T}_m$-limited process.

To illustrate the procedure, let us consider the renal transport of substance x with a permeability ratio of unity in a healthy subject. Using the techniques described in Chapter 7, we measure the renal clearance of substance x along with the clearance of inulin (C_{In}) as the concentration of x in the plasma ($[x]_p$) is progressively increased. Since inulin clearance is a measure of glomerular filtration rate (GFR), the filtered load of the substance for a given $[x]_p$ would be $C_{In} \cdot [x]_p$ or, more directly, GFR $\cdot [x]_p$. Assuming that in such an experiment GFR remains relatively constant, the filtered load of substance x should increase in direct proportion to the increasing plasma concentration of x. Since the urinary excretion rate of substance x (i.e., $\dot{V}_u \cdot [x]_u$), which can also be determined, is equal to the difference between the filtered load and the rate of tubular transport ($\dot{T}_x$) (see Equation 7.16), the value of the latter at a given plasma concentration would be

$$\dot{T}_x = GFR \cdot [x]_p - \dot{V}_u \cdot [x]_u \quad (10.1)$$

It should be recalled from Chapter 7 that if the substance x is reabsorbed, the sign of $\dot{T}_x$ would be positive, but if the substance x is secreted, the sign of $\dot{T}_x$ would be negative.

To determine whether the renal transport of x is by a $\dot{T}_m$-limited process or not, we simply construct simultaneous plots of the filtered load (GFR $\cdot [x]_p$), the urinary excretion rate ($\dot{V}_u \cdot [x]_u$), and their calculated difference, i. e., the transport rate ($\dot{T}_x$), against the increasing plasma concentration of x ($[x]_p$). Such a combined plot, like those shown in Figures 10.1 and 10.2 (upper panels), is known as the *renal titration curves* for that substance; it determines the plasma concentration of the substance at which the "membrane carrier" is fully saturated (see discussion under *Glucose*).

Now if the substance x is transported by a $\dot{T}_m$-limited process, we would expect the plot of $\dot{T}_x$ against $[x]_p$ to plateau at some value of $[x]_p$ and to remain constant thereafter. This is depicted by the theoretical reabsorption line parallel to the abscissa in Figure 10.1 (upper left graphs) for a substance transported by net reabsorption (e.g., glucose), and in Figure 10.2 (upper left graphs) for a substance transported by net secretion (e.g., PAH). Note that in both of these figures, the filtered load, the tubular transport, and the renal excretion rate are plotted against the filtered load and not the plasma concentration, as mentioned above. This is a more common way of plotting the titration curves. Note that plotting the filtered load against itself yields a 45 degree line through the origin, serving as a standard reference with which to compare the renal excretion rate.

On the other hand, if the renal transport of substance x is by a gradient-time-limited process, such a plateau should not occur within the physiological range of concentrations. This is depicted by the curvilinear line in Figure 10.1 (upper right graphs) for a substance transported by net reabsorption, and in Figure 10.2 (upper right graphs) for a substance transported by net secretion.

The lower graphs in Figures 10.1 and 10.2 depict the theoretical relationships between the fraction of the filtered load excreted by the kidneys ($\dot{V}_u \cdot [x]_u$/GFR $\cdot [x]_p$) as a function of the filtered load. Note that this fraction is another expression of clearance ratio (see Chapter 7) and is obtained by dividing Equation 10.1 by the filtered load and rearranging terms:

$$\frac{\dot{V}_u \cdot [x]_u}{GFR \cdot [x]_p} = 1 \pm \frac{\dot{T}_x}{GFR \cdot [x]_p} \quad (10.2)$$

In the case of a substance that is reabsorbed (Figure 10.1) by a $\dot{T}_m$-limited process (e.g., glucose), the clearance ratio increases sharply and asymptotically toward the unity line. This is because when $\dot{T}_x = \dot{T}_m$ (depicted by the vertical dashed line), the right-hand term in Equation 10.2 approaches zero as the filtered load is infinitely increased. In contrast, in the case of a substance not reabsorbed by a $\dot{T}_m$-limited process (e.g., sodium), the clearance ratio gradually approaches the unity line.

216   10. Tubular Reabsorption and Secretion: Classification Based on Overall Clearance Measurements

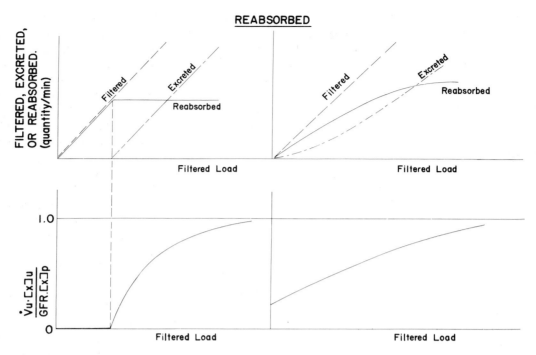

FIGURE 10.1. Theoretical relationships of renal titration curves (*upper graphs*) and clearance ratios (*lower graphs*) for a substance reabsorbed by a $\dot{T}_m$-limited process (*left panel*) and a substance reabsorbed by a gradient-time-limited process (*right panel*). Redrawn from Koch, The Kidney; in Ruch and Patton: Physiology and Biophysics, Philadelphia, W. B. Saunders Co., 1973. Reprinted by permission.

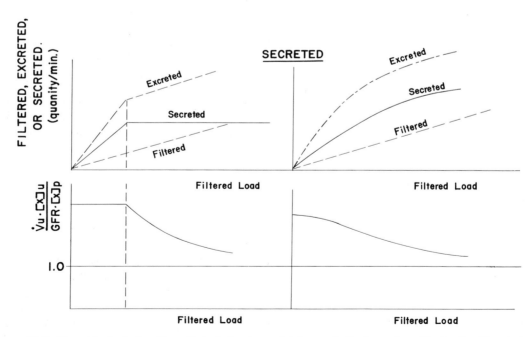

FIGURE 10.2. Theoretical relationships of renal titration curves (*upper graphs*) and clearance ratios (*lower graphs*) for a substance secreted by a $\dot{T}_m$-limited process (*left panel* and a substance secreted by a gradient-time-limited process (*right panel*). Redrawn from Koch, The Kidney; in Ruch and Patton: Physiology and Biophysics, Philadelphia, W. B. Saunders Co., 1973. Reprinted by permission.

In the case of a substance that is secreted (Figure 10.2) by a $\dot{T}_m$-limited process (e.g., PAH), the clearance ratio begins at a value above unity (because $\dot{T}_x$ is positive in Equation 10.2) and sharply and asymptotically approaches the unity line. In contrast, in the case of a substance not secreted by a $\dot{T}_m$-limited process (e.g., potassium), the clearance ratio gradually approaches the unity line.

Let us now inquire into the mechanisms and direction of net tubular transport of some important constituents of the filtrate as revealed from the application of the above concepts and titration studies. To facilitate presentation, we begin by considering those solutes that are transported by net reabsorption and then those that are transported by net secretion.

Tubular Reabsorption

In this section we consider the renal transport of D-glucose, phosphate, sulfate, amino acids, organic anions, uric acid, and proteins, most of which with a few exceptions are reabsorbed by a $\dot{T}_m$-limited process. Because renal transport of glucose has been more extensively studied, its reabsorptive characteristics will be presented in more detail, so that it may serve as the prototype for the renal transport of the other substances mentioned above.

D-Glucose

The classic clearance studies of Shannon and associates[86, 87] were the first in which the principles of titration curves were used to characterize the mechanism of renal transport of D-glucose in normal dogs. The conclusions and inferences drawn from these experiments have since become the framework for many of the subsequent studies designed to elucidate further the nature of the D-glucose transport system. Consequently, to facilitate understanding of these newer findings, we shall begin by briefly enumerating the highlights of their conclusions:

1. Since at normal plasma glucose concentration the renal clearance of glucose was found to be zero, they concluded that all the filtered glucose must have been reabsorbed.
2. Because reabsorption was virtually complete, they reasoned that glucose must have been reabsorbed from the tubular fluid into the peritubular blood against a concentration gradient by an active process.
3. As the plasma concentration of glucose was progressively increased, they found that the tubular reabsorptive mechanisms became gradually saturated, so that glucose began to appear in the urine.
4. As the plasma glucose concentration was further increased, they observed that the transport mechanism became fully saturated, and the tubular reabsorption reached its maximum limit ($\dot{T}_m$). Thereafter, $\dot{T}_m$ for glucose remained constant and unaffected by further increases in the plasma glucose concentration.
5. For a given animal, $\dot{T}_m$ for glucose was found to be stable during acute variations in the glomerular filtration rate, whether the changes were spontaneous or induced by obstruction of the abdominal aorta above the renal arteries or by hemorrhage.

On the basis of these findings, Shannon and co-workers suggested that the filtered glucose is reabsorbed by an active, carrier-mediated mechanism, the precise nature of which and the tubular site where glucose is reabsorbed remained to be determined. Micropuncture studies in rats have subsequently shown that the tubular site where virtually all the filtered glucose is reabsorbed is the proximal tubule.[31, 107]

That D-glucose is in fact reabsorbed by a $\dot{T}_m$-limited process was subsequently confirmed both in man[91] and in dog.[14] The human studies showed that in normal subjects, $\dot{T}_m$ for glucose was 375 ± 79.7 in 24 men and 303 ± 55.3 mg/min in 11 women tested. Both $\dot{T}_m$ values are corrected to 1.73 m^2 body surface area. Furthermore, the dog studies showed that $\dot{T}_m$ for glucose remained relatively constant in the face of modest changes in GFR, but was markedly decreased when GFR was greatly reduced.[21] This latter finding was attributed to a reduced overall function in those nephrons that are functioning and not to the existence of a glomerulotubular balance for glucose in any nephron.

Characteristics of D-Glucose Titration Curves

From the studies cited above as well as a number of others, a unifying pattern of the general characteristics of renal transport of D-glucose has emerged. This is depicted in Figure 10.3, which illustrates the renal titration curves for glucose in man. To begin with, note that as the plasma concentration of glucose, $[G]_p$, is progressively increased, the filtered load of glucose (GFR · $[G]_p$) increases linearly. Similarly, the tubular reabsorption of glucose ($\dot{T}_G$) (Equation 10.1) matches the filtered load and increases linearly with the plasma glucose concentration, so long as the latter remains below a value of 200 mg/dL. However, as the plasma concentration of glucose begins to exceed this value, a portion of the filtered glucose escapes tubular reabsorption and is therefore excreted in the urine (glucosuria). The plasma concentration at which glucose begins to appear in the urine (i.e., $[G]_p$ = 200 mg/dL in Figure 10.3) is known as the *renal threshold concentration* for glucose. Note that this concentration is not synonymous with the plasma concentration that completely saturates the transport mechanism (i.e., $[G]_p$ = 400 mg/dL in Figure 10.3). Note also that glucosuria occurs long before the transport mechanism is fully saturated.

As shown in Figure 10.3, the quantity of filtered glucose that escapes reabsorption, and hence is excreted in the urine, increases curvilinearly with the increasing plasma concentration until the latter exceeds a value of 400 mg/dL. Thereafter, the transport system becomes completely saturated and the quantity of filtered glucose reabsorbed will remain constant and independent of further increases in plasma glucose concentration. This constant reabsorption rate is called the tubular reabsorption maximum for glucose and is designated by the symbol $\dot{T}_{mG}$. At the maximal tubular reabsorption rate, the glucosuria will also become constant and a linear function of the plasma glucose concentration. Indeed, as the plasma glucose concentration exceeds 400 mg/dL, the slope of the excretion curve becomes parallel to the slope of the filtered curve.

As depicted in Figure 10.3, when the plasma glucose concentration varies between 200 and 400 mg/dL, both the amount of glucose reabsorbed and the amount of glucose excreted increase gradually, rather than abruptly, while the filtered load increases linearly. The rounded regions of both the reabsorbed and excreted curves, which graphically relate $\dot{T}_G$ to the filtered load of glucose, are known as the splay of the titration curves. Of the various solutes that are reabsorbed by a $\dot{T}_m$-limited process, glucose reabsorption shows a minimal degree of splay.[68]

The degree of splay in the glucose titration curves has been attributed to the kinetics of transepithelial transport of glucose. Let us briefly examine its effects on renal glucose transport.

Kinetics of Glucose Transport. As noted briefly in Chapter 8 and further detailed below, the transepithelial transport of glucose in the proximal tubule involves active sodium and glucose cotransport into the cell across the luminal membrane, followed by passive diffusive efflux across the basolateral membrane. It is believed that both cotransport influx and

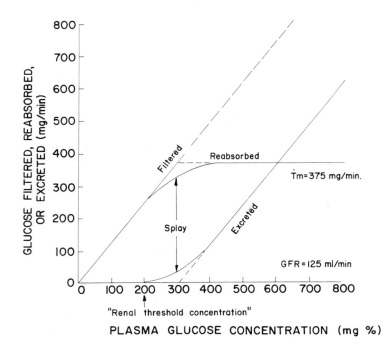

FIGURE 10.3. A typical titration curve for glucose in man. The solid lines for reabsorption and excretion rates represent the actual experimental data, while the dashed lines represent the theoretical and extrapolated lines. Modified and reproduced with permission from Pitts, R. F.: Physiology of the Kidney and Body Fluids, 3rd Edition. Copyright © 1974 by Yearbook Medical Publishers, Inc., Chicago.

diffusive efflux of glucose are mediated by specific membrane carriers. Thus, the kinetics of carrier-mediated glucose transport has been suggested as a possible explanation for the degree of splay observed in its renal titration curves. This explanation can best be understood if we assume that the proximal tubular reabsorption of glucose involves the combination of the glucose molecule (G) with a membrane carrier (C) in a reversible reaction described by the following equilibrium equation:

$$K = \frac{[G] \cdot [C]}{[GC]} \quad (10.3)$$

where [G] is the concentration of glucose in the tubular fluid, [C] is the concentration of the free membrane carrier, [GC] is the concentration of the glucose-carrier complex at the surface of the luminal membrane, and K is the dissociation constant for the GC-complex.

As described in Chapter 8, the greater the affinity of glucose for the carrier molecule, the greater will be [GC], and hence the smaller the value of K. Conversely, the smaller the affinity of glucose for the carrier, the larger will be the product [G] · [C], and the larger the value of K. As long as [C] is large relative to [G], essentially all of the filtered glucose will be reabsorbed, and the clearance of glucose will be zero. However, as the plasma concentration of glucose and hence [G] in the proximal tubule increase, some glucose molecules will escape reacting with the free carrier molecules. Since the carrier has a finite affinity for glucose (that is, the value of K must be finite), a "supersaturating" concentration or one much higher than the renal threshold concentration of glucose is required to fully saturate the transport system. Consequently, glucose will appear in the urine long before the transport system is fully saturated or $\dot{T}_{mG}$ is reached.

Clinically, the glucosuria seen in an otherwise healthy, young diabetic patient is due to an elevated plasma glucose concentration, since both GFR and T_m for glucose are within normal limits. However, in elderly patients with long-standing diabetes, glucosuria may be absent despite the high plasma glucose concentration. This is due to a reduced GFR caused by deposition of mucopolysaccharide-protein complex in the glomerular capillary, a condition known as intercapillary glomerulosclerosis.[72]

A number of titration studies have shown that the renal transport of D-glucose exhibits a high degree of structural specificity; that is, glucose reabsorption is subject to competitive inhibition (see Chapter 8) when other sugars or substances of similar molecular structure are present in the filtrate. Thus, the sugar molecule having the highest affinity (low K_m) for the carrier molecule will tend to displace the sugar molecule with lower affinity (high K_m).

The most widely used substance for studying the structural specificity of D-glucose transport has been phlorizin, a phenolic glucoside. Lotspeich and Woronkow[63] showed that intravenous infusion of small doses of phlorizin in dogs virtually blocked proximal reabsorption of D-glucose. This inhibition was found to be reversible when phlorizin infusion was stopped. These investigators suggested that the inhibitory effect of phlorizin on renal glucose reabsorption involves the competitive combination of phlorizin with the carrier of the D-glucose transport system in accordance with Michaelis-Menten saturation kinetics (see Chapter 8). These results have been substantiated in the cat[17] and in other mammalian species.[61] Recent studies,[45, 100] using membrane-derived isolated microvesicles, have shown that phlorizin competitively inhibits the sodium-dependent concentrative uptake of D-glucose by the brush border membrane. Phlorizin has no effect at the basolateral membrane, where D-glucose transport is sodium-independent.

On the basis of such studies it is now accepted that the same D-glucose carrier system is used to transport other sugars, such as xylose, fructose, and galactose. However, glucose appears to have the highest affinity for the carrier, compared with these other sugars, so that when present it is preferentially reabsorbed.

Factors Affecting $\dot{T}_m$ for D-Glucose

Earlier studies cited above have suggested that $\dot{T}_m$ for D-glucose remains stable and is independent of spontaneous or induced changes in the glomerular filtration rate. However, other studies have challenged this conclusion. Thus, Kruhoffer[55] found in the rabbit a reversible reduction in the rate of D-glucose reabsorption despite the presence of a large glucose load when the GFR was reduced by dehydration or hemorrhage. He suggested that these changes were due to intermittency in the GFR in the single nephron, rather than to any change in the rate of glucose reabsorption by individual proximal convoluted tubules. Subsequently, Van Liew and coworkers,[106] using clearance techniques in the rat, observed a linear relationship between the rate of glucose reabsorption and the spontaneous reduction in the GFR. They suggested that this relationship may be an expression of the existence of glomerulotubular balance for glucose in a single nephron similar to that which exists for sodium or fluid reabsorption (see Chapter 9). Similar results were obtained by Keyes and Swanson[47] who examined $\dot{T}_{mG}$/GFR ratio in the dog, as an index of glomerulotubular function, during spontaneous or induced changes in GFR. From these studies they suggested that the value of this ratio depends on (a) the number of functioning nephrons, and (b) the number of transport sites per nephron. The latter was assumed to vary directly (but not necessarily linearly) with the proximal tubular volume, a direct consequence of changes in the GFR. Accordingly, any factor that increases GFR would tend to increase the proximal tubular volume, thereby exposing more of the transport sites located on the luminal microvilli (brush border) to the tubular fluid and resulting in an increase in glucose reabsorption and hence $\dot{T}_{mG}$. Conversely, any factor that reduces the GFR should have the opposite effect. Thus, for a given GFR and within the physiological range, it is the combination of the number of functioning nephrons and the available transport sites that ultimately determine the maximum reabsorptive capacity of the kidney for D-glucose.

As was described in Chapter 9, any factor that tends to increase the GFR, such as expansion of the extracellular fluid volume (subsequent to intravenous fluid infusion, an experimental procedure employed in all renal titration studies), will depress fluid reabsorption, whereas any factor that reduces GFR will enhance fluid reabsorption in the proximal tubule. Since this segment of the nephron is also the site of glucose reabsorption, it follows that any factor that alters fluid reabsorption here may secondarily induce parallel changes in glucose reabsorption. Hence, the experimentally determined $\dot{T}_m$ for glucose may represent an "apparent" and not an absolute value for the tubular reabsorptive maximum, subject to body fluid homeostasis.

That alterations in sodium or fluid reabsorption in the proximal tubule, subsequent to spontaneous or reduced changes in GFR, do indeed induce parallel changes in glucose reabsorption has been well documented in both rats and dogs. For example, Robson and co-workers[78] found that the $\dot{T}_{mG}$/GFR ratio decreased in saline-loaded rats. This was attributed to an increase in GFR and a decrease in $\dot{T}_{mG}$ following saline diuresis. Since sodium reabsorption was also reduced subsequent to saline-loading (see also Chapter 9), they suggested that glucose reabsorption is in some manner coupled to sodium reabsorption in the proximal tubule.

In support of this concept, Kurtzman and associates[56] found that in the dog D-glucose reabsorption is highly sensitive to changes in both GFR and sodium reabsorption. Furthermore, subsequent studies by these and other investigators demonstrated that the effect of sodium reabsorption on solute reabsorption in the proximal tubule is not unique to D-glucose and represents a general phenomenon. Thus, alterations in sodium and water reabsorption in this nephron segment have a profound effect on reabsorption of other solutes besides D-glucose, such as phosphate (see discussion below) and bicarbonate (see Chapter 11).

In the case of D-glucose, the mechanism coupling its reabsorption to that of sodium has been clarified by direct micropuncture studies in the rat by Stolte and collaborators.[95] These investigators found that the net amount of D-glucose reabsorbed in the proximal tubule is directly influenced by the net amount of sodium and water reabsorbed by this nephron segment. Furthermore, they showed (a) that the rate of D-glucose reabsorption by the proximal tubule depends on both the concentration of glucose in the tubular fluid and the net rate of sodium and water reabsorption in this segment, and (b) that the glucose-sodium cotransport accounted for a large fraction of the total amount of D-glucose reabsorbed by this nephron segment.

Finally, recent studies have shown that $\dot{T}_m$ for D-glucose in the proximal tubule is modified by several factors: tubular transit time,[56] net reabsorption of sodium,[85] fluid reabsorption,[78] the number of receptor sites available on the surface of the brush-border membrane,[47] and back diffusion of glucose.[41]

Cellular Mechanism of D-Glucose Transport

The evidence presented thus far clearly suggests that the quantity of D-glucose normally filtered is almost completely reabsorbed in the proximal tubule by a sodium-dependent cotransport process. However, the precise cellular mechanisms involved in the transepithelial transport of D-glucose remains to be elucidated.

Our current concept of the cellular mechanism for the active sodium-dependent D-glucose cotransport across the proximal tubular cell is largely adapted from direct studies in the small intestine. In this tissue, D-glucose is actively reabsorbed from the lumen into the cells, where its concentration reaches a value higher than that in either the blood or the intestinal fluid.[26] In an apparent contradiction to this concept, however, Krane and Crane,[54] using normal D-glucose concentration in the bathing medium, failed to show cellular accumulation of of D-glucose by the kidney cortical slices. In contrast, in a later study, Kleinzeller and co-workers,[49] using a low concentration of D-glucose in the bathing medium, found significant cellular uptake of the sugar, with the intracellular concentration of D-glucose reaching a level higher than that in the extracellular fluid, thereby suggesting active transport of D-glucose at the luminal cell membrane. In a subsequent study, Tune and Burg,[101] who measured glucose concentration simultaneously in the tubular fluid, cells, and peritubular bath in isolated, perfused rabbit renal proximal tubules, have conclusively demonstrated that D-glucose is actively transported and that the site of the active transport is localized at the luminal cell membrane.

Although these and similar studies have contributed significantly to our understanding of the overall transepithelial transport of glucose in the kidney, they have not provided direct information on the cellular mechanism of the D-glucose transport system. Such information, however, has been obtained from recently developed techniques for isolation and biochemical characterization of the luminal cell membrane of the proximal tubule. An example is the comprehensive kinetic studies of the isolated luminal brush border of rabbit proximal tubule by Chesney and associates.[18] Using this preparation, these investigators clearly demonstrated that the initial step in the D-glucose transport involves rapid interaction of glucose with the luminal cell membrane, a process that is both saturable and dependent on the concentration of D-glucose. Furthermore, this initial binding of D-glucose with the membrane is reversible and requires presence of calcium and magnesium and is inhibited competitively by phlorizin. Moreover, analysis of the relationship between the concentration of D-glucose and the binding of the sugar revealed the existence of two distinct receptor sites for the binding of D-glucose with the brush-border membrane, one having a much greater binding affinity for D-glucose. Despite a number of differences in the kinetics characteristics of these two receptor sites, present evidence does not rule out the possibility that the two binding sites are mediated by a single membrane carrier protein whose affinity for binding is highly sensitive to the concentration of D-glucose.

Recent electrophysiological studies[32, 82] have provided unequivocal evidence that D-glucose transport across the luminal membrane initiates Na^+ flux in the same direction, with a coupling ratio of approximately 1:1. This finding strongly supports the concept that a "carrier" mechanism may be responsible for transluminal transport of D-glucose. The resulting coupled Na^+-glucose transport into the cell generates the lumen-negative transepithelial voltage described in Chapter 9. Thus, D-glucose transport is a *"rheogenic"* process, meaning that its transport generates an electric current carried by sodium ions.[83] However, this 1:1 stoicheometry does not apply to the net sodium transport along the entire length of the proximal tubule. Several studies[4, 46] have shown that the net sodium transport far exceeds the net D-glucose transport. This discrepancy is because the net sodium transport involves three distinct mechanisms (see Chapter 9): active transport (accounting for approximately 38% of the total), passive electrodiffusion, and solvent drag, which together account for the remaining 62% of the net transepithelial sodium transport. However, it has been shown that the active component of net sodium transport correlates with the active D-glucose cotransport, suggesting that $\dot{T}_m$ for D-glucose is determined by active sodium transport.[103]

On the basis of the various studies cited above, the following represents the probable steps involved in the cellular mechanisms of D-glucose transport across the mammalian proximal tubule cell. The first step in the transepithelial transport of D-glucose involves the active sodium-dependent cotransport of the sugar from the tubular lumen into the proximal cell, across the brush border of the luminal cell membrane. Once inside the cell, because the intracellular concentration of D-glucose is higher than that in the peritubular fluid, glucose will be transported across the basolateral cell membrane down its concentration gradient by passive diffusion. However, because of the limited passive permeability of this membrane to D-glucose, transport across the basolateral cell membrane may also involve facilitated diffusion. Since solutes are reabsorbed iso-osmotically along the proximal tubule, the transepithelial reabsorption of D-glucose is accompanied by an osmotically equivalent quantity of water. In short, D-glucose is reabsorbed as an isotonic solution.

Phosphate

Phosphate is the major constituent of the skeletal structures, which account for 80% of the total body stores of this substance. The remaining 20% is contained in the intracellular fluid and is associated with glycogen. Because of its prevalence in bone, the major function of phosphate concerns the metabolism of this tissue. However, phosphate also plays a vital role in carbohydrate metabolism and energy transformation and serves as an important blood buffer in the regulation of acid-base balance (see Chapter 11). Additionally, phosphate represents a significant constituent in a number of important compounds, such as phospholipids, phosphoproteins, nucleic acids, and nucleoproteins.

The daily requirement of phosphate is about 0.9 g for adults and somewhat higher for children (about 1.5 g) and pregnant women (about 2.5 g). The dietary phosphate is absorbed from the small intestine into the blood, where it exists in three fractions: (1) A lipid-phosphate fraction, with a concentration of 8 mg/dL, (2) an ester-phosphate fraction, with a concentration of 1 mg/dL, and (3) a completely ionized inorganic fraction, with a concentration of 3 mg/dL.[38] Since the plasma concentration of the inorganic phosphate fraction is the only one that is stabilized at 2.0 mM/L by renal reabsorption, the present discussion will be limited to this fraction.

Characteristics of Phosphate Titration Curves

Classic studies of Pitts and Alexander[75] in normal and acidotic dogs and those of Lambert and associates[57] and Anderson[2] in man represent the earliest attempt to characterize the renal transport of phosphate by the application of titration techniques. Their findings, confirmed by subsequent studies in rat,[30] dog,[1] monkey,[105] and man[13] have generally established that the renal handling of phosphate is similar in many respects to that of D-glucose, but with two notable exceptions:

1. Since the $\dot{T}_m$ for D-glucose reabsorption is set at such a high value (about 375 mg/min in man), it is normally never exceeded, indicating that the kidneys do not regulate the plasma concentration of D-glucose. In contrast, the $\dot{T}_m$ for phosphate reabsorption is set at such a low value (0.1 mM/min in man) that a slight change in its plasma concentration produces marked changes in urinary excretion of phosphate. This implies that the kidneys do regulate the plasma concentration of phosphate.

2. As discussed earlier, the $\dot{T}_m$ for D-glucose is relatively stable and is not influenced by ionic composition or hormone levels of the plasma. The $\dot{T}_m$ for phosphate, however, is variable and is markedly influenced by the ionic composition of plasma and the circulating levels of parathyroid and adrenocortical hormones.

Figure 10.4 presents the renal titration curves for phosphate in the dog, which graphically illustrates the salient features of the renal transport of phosphate observed in this and other mammalian species, including man. As can be seen from this figure, the value of $\dot{T}_m$ for phosphate in these studies averaged about 0.1 mM/min and was attained at a filtered load of about 0.125 mM/min. Furthermore, when GFR was altered by adjustment of dietary protein intake, a factor known to markedly influence GFR in dogs, $\dot{T}_m$ was not affected.

Like D-glucose, the reabsorption of phosphate is virtually complete at normal plasma concentration. But, as the plasma concentration exceeds the renal threshold, the reabsorptive mechanism becomes saturated, resulting in urinary excretion of phosphate (phosphaturia). Analogous to glucose, these findings suggest that the filtered phosphate is reabsorbed by an active, carrier-mediated process with $\dot{T}_m$-limited characteristics. Likewise, the observed splay in the titration curves has been attributed to the kinetics of phosphate transport, which have already been described.

A number of micropuncture studies in rats,[30, 97] dogs,[1] and monkeys[105] have established that like glucose, the bulk of filtered phosphate is reabsorbed in the proximal tubule. However, in the rat it has been shown that the loops of Henle and the "distal tubule" and collecting duct also have a limited capacity to reabsorb phosphate.[70] Furthermore, recollection micropuncture and whole kidney clearance studies by Puschett and associates[71] showed that the proximal reabsorption of phosphate, like glucose, is coupled to that of sodium under a variety of experimental conditions, including the expansion of extracellular fluid volume, isotonic saline diuresis, and aortic constriction.

Factors Affecting $\dot{T}_m$ for Phosphate

A number of studies designed to elucidate the cellular mechanism of renal reabsorption of phosphate have revealed that several factors alter the $\dot{T}_m$ for phos-

FIGURE 10.4. Renal reabsorption and excretion of phosphate as functions of the filtered load in the normal dog. (Redrawn from Pitts and Alexander.[75])

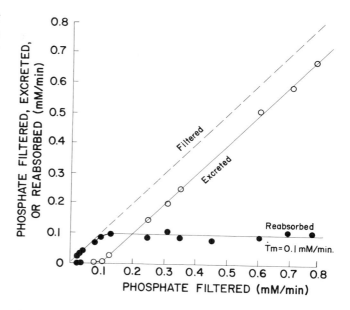

phate and hence its renal transport. These findings are summarized in Figure 10.5. Pitts and Alexander[75] showed that in the dog infusion of glucose sufficient to saturate the transport mechanism reduced the $\dot{T}_m$, whereas infusion of phlorizin increased the $\dot{T}_m$ for phosphate. From these observations they suggested that the cellular reabsorption of phosphate and glucose probably share a common mechanism.

By far the most important factors affecting renal reabsorption of phosphate are the circulating levels of the PTH and adrenocortical hormone. Prolonged administration of PTH markedly reduced $\dot{T}_m$ for phosphate,[13] thereby depressing tubular reabsorption and increasing urinary excretion of phosphate. Micropuncture studies in the dog[1,50] subsequently showed that the induced phosphaturia is due to the PTH inhibition of phosphate reabsorption in both the proximal and "distal" tubules. Thus, administration of PTH increased the tubular fluid to ultrafiltrate (TF/UF) phosphate concentration ratio toward unity, whereas parathyroidectomy decreased this ratio considerably below unity compared with normal (Figure 10.5).

A number of studies reviewed by Talmadge and Belanger[99] have provided convincing evidence that the renal action of PTH is mediated by cAMP. The probable mechanism of such an action was outlined briefly in Chapter 8. Accordingly, PTH present in the blood stimulates the adenylate cyclase located on the outer surface of the peritubular membrane, thereby increasing the intracellular production of

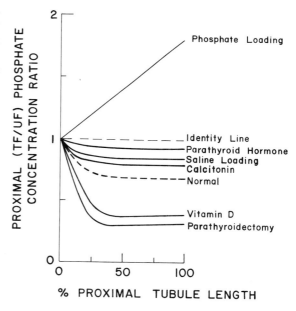

FIGURE 10.5. The proximal tubular fluid-to-ultrafiltrate (TF/UF) phosphate concentration ratios as functions of percent of proximal tubule length under a variety of experimental conditions. The *light dashed line* is the identity line to which all the other lines should be compared. (Redrawn from Knox et al.[51])

cAMP. The latter, by an as yet unknown mechanism, alters the permeability of the luminal cell membrane to phosphate, thereby decreasing its cellular uptake and reabsorption.

Like PTH, the administration of excessive amounts of the adrenocortical hormone cortisone reduces the $\dot{T}_m$ for phosphate, thereby promoting phosphaturia. This effect may represent the underlying mechanism for phosphaturia observed in patients with osteomalacia of hyperadrenocorticism.[72]

Since the discovery of thyrocalcitonin (TCT), the hypocalcemic-hypophosphatemic principle of the thyroid gland by Hirsch and co-workers,[42] its possible role in the renal transport of calcium and phosphate has been extensively studied. It has been found that TCT, besides stimulating the uptake of calcium and phosphate by the bone and thereby lowering the concentration of these ions in the blood, also stimulates urinary excretion of phosphate (increasing TF/UF ratio in Figure 10.5) in both normal and parathyroidectomized subjects.[89] Since TCT causes hypocalcemia by the mechanism just mentioned, and because plasma concentration of calcium indirectly influences plasma phosphate concentration (normally concentrations of calcium and phosphate in the blood are reciprocally related), it has been suggested that the renal action of TCT may be a secondary phenomenon.

Another factor that influences renal reabsorption of phosphate is vitamin D. Although its effect on urinary excretion of phosphate may reflect induced changes in the intestinal absorption of calcium and phosphate, parathyroid hormone, and bone metabolism, vitamin D has been shown to have a direct stimulating effect on renal tubular reabsorption of phosphate. This effect is qualitatively similar to that of parathyroidectomy, that is, vitamin D increases renal reabsorption of phosphate, as evidenced by a marked decrease in the proximal TF/UF phosphate concentration ratio shown in Figure 10.5.[51] It should be pointed out, however, that vitamin D has a dual action: it stimulates the renal reabsorption of phosphate while it simultaneously increases the plasma concentration of calcium. The latter effect, in turn, decreases the blood level of PTH, which in turn increases renal reabsorption of phosphate.

Both calcium and phosphate loading have been found to have a direct effect on renal tubular reabsorption of phosphate. Whereas hypercalcemia stimulates phosphate reabsorption, hyperphosphatemia depresses phosphate reabsorption. Indeed, consistent with the view that phosphate reabsorption is by an active, carrier-mediated process, the progressive increase in plasma phosphate concentration tends to saturate the transport mechanism and leads to phosphaturia. This has been confirmed by micropuncture studies by Strickler and associates[97] who found a linear increase in the proximal TF/UF concentration ratio with hyperphosphatemia (Figure 10.5).

Like D-glucose, tubular reabsorption of phosphate has been shown to be closely coupled to the reabsorption of sodium and water in the proximal tubule. Thus, any factor, such as saline loading, which depresses sodium reabsorption in this nephron segment, will decrease phosphate reabsorption, as evidenced by an increase in TF/UF concentration ratio. The mechanisms underlying the phosphate-sodium cotransport are presumably similar to those mentioned earlier for glucose.

Finally, studies in man have demonstrated a definite circadian rhythm for the urinary excretion of phosphate.[93] Phosphate excretion was found to be minimal during the morning, becoming maximal during the afternoon and evening. No such study has been done in the dog or in any other mammalian species. In man, changes in the urinary excretion rate of phosphate were found to parallel those in plasma concentration. Furthermore, these changes followed parallel changes in the blood level of the parathyroid hormone. It has also been suggested that the normal variability of the $\dot{T}_m$ for phosphate may be due to daily fluctuations in the rate of secretion of the parathyroid hormone.[3] Despite this evidence, the exact nature of this circadian rhythm remains obscure.

Sulphate

Plasma concentration of inorganic sulfate is maintained within normal limits of about 2.0 mM/L by a balance between the rate of its production from sulfur-containing amino acids and the rate of its excretion by the kidney. Renal titration studies in most mammalian species, including man,[69] similar to that depicted in Figure 10.6, have shown that the urinary excretion of sulfate is normally far less than the filtered load, suggesting that the dominant mode of renal tubular transport of sulfate is by net reabsorption.

These same studies have further shown that sulfate is reabsorbed by a $\dot{T}_m$-limited process, exhibiting very little splay. Furthermore, they suggest that sulfate is reabsorbed by a mechanism similar to that for phosphate; that is, sulfate is reabsorbed by an active,

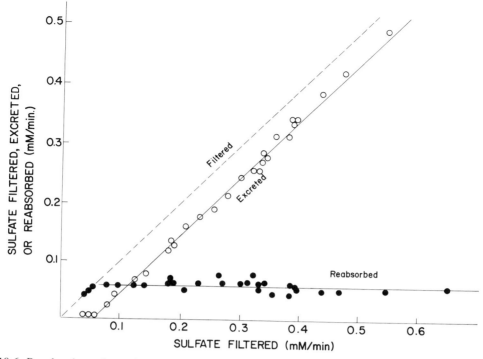

FIGURE 10.6. Renal reabsorption and excretion of sulfate as functions of the filtered load in the normal dog. (Redrawn from Lotspeich.[62])

sodium-dependent, carrier-mediated process. This has been confirmed by recent studies using brush border membrane vesicles.[33, 64]

Administration of glucose, phosphate, and amino acids, all of which are reabsorbed in the proximal tubule, has been found to inhibit sulfate reabsorption in this nephron segment. This finding suggests that the sulfate must also be reabsorbed in the proximal tubule, a conclusion that has been confirmed by stop-flow studies in the dog.[40]

The $\dot{T}_m$ for sulfate, as measured by titration studies, was found to be of the same order of magnitude as that for phosphate. Like glucose, $\dot{T}_m$ for sulfate was found to be sensitive to changes in GFR.[8] Furthermore, $\dot{T}_m$ for sulfate could be depressed by infusion of glucose, sufficient to saturate the transport mechanism, and by infusion of hypertonic saline. As described in Chapter 9, the latter effect is due to the presence of an excessive amount of chloride, a readily reabsorbed anion, compared with the poorly reabsorbed sulfate anion. These studies suggest that sulfate is probably reabsorbed by the same cellular transport processes and carrier system that transport glucose and phosphate.

Amino Acids

Plasma concentration of amino acids is maintained within normal limits of 2.5 to 3.5 mM/L by a dynamic balance between the rate of dietary intake and its metabolic conversion, as well as the rate of renal reabsorption of the quantity normally filtered. In addition, the kidneys participate in the transamination process (conversion of one amino acid to another) as well as the formation and excretion of ammonia to conserve major body electrolytes, such as sodium and potassium. This latter process constitutes an important mechanism in the renal regulation of acid-base balance and is discussed in Chapter 11.

Normally, 99% of the filtered amino acids is reabsorbed mainly in the proximal tubule.[82] However, under pathological conditions, basolateral uptake of amino acids may lead to their secretion and eventual excretion in the urine.

Kinetic studies have shown that the transepithelial transport of amino acids involves an active carrier-mediated process via the transcellular pathway and a small back-diffusion via the paracellular pathway (see Chapter 8). These two opposing fluxes become equal at the end of the late proximal tubule, where a steady-state concentration is established. The active reabsorption involves a $\dot{T}_m$-limited process, with the maximum affinity for the carrier being dependent on the specific amino acid being transported.

Characteristics of Amino Acid Titration Curves

As early as 1943, Pitts[74] demonstrated that in the dog amino acids are reabsorbed by a $\dot{T}_m$-limited process. In a subsequent study, when the effects of a number of amino acids on creatine (a compound structurally similar to glycine) reabsorption were compared, Pitts[73] observed a competition for transport between structurally similar compounds. From these studies, Pitts concluded that there must be a single transport mechanism for the renal reabsorption of all amino acids. However, more recent evidence has clearly shown that there are at least three or more independent transport systems for the renal reabsorption of amino acids. For example, studies by Beyer and co-workers[11,12] and Wright and associates[112] have shown that some amino acids (e.g., lysine and arginine) are reabsorbed by a $\dot{T}_m$-limited process, while others (e.g., histidine and methionine) are reabsorbed without exhibiting $\dot{T}_m$-limited characteristics. Subsequent studies by Webber[108,109] have confirmed earlier findings and have further demonstrated the existence of separate but sometimes overlapping pathways for the renal transport of neutral, basic, and acidic amino acids. Furthermore, stop-flow studies[15] have shown that most amino acids are reabsorbed in the proximal tubule. Also, reabsorption is active, since it occurs against a concentration gradient, as evidenced by a TF/P amino acid concentration ratio significantly below 1.0 in the proximal tubule.

Figures 10.7 and 10.8 illustrate the renal titration curves in the dog for two representative amino acids (glycine and lysine). These titration curves demonstrate many of the transport characteristics common to most of the amino acids that have been studied. Note that the titration curves for glycine exhibit considerable splay, whereas those for lysine show no splay. Also, the measured $\dot{T}_m$ for glycine is some 15-fold greater than that for lysine.

A number of studies have explored the possibility that factors other than the already mentioned glomerular-tubular balance may be responsible for the observed variations in the splay of the titration curves. A review of the available evidence[114] supports the concept that the shape of the titration curve is largely determined by the equilibrium constant of the transport system. Thus, as the equilibrium constant K (Equation 10.3) approaches infinity (i.e., the Michaelis-Menten constant K_m approaches zero), the splay in the titration curve approaches zero. The low K_m value for glucose transport determined by micropuncture techniques and the high K_m value for amino acid transport obtained from studies of kidney cortex slices provide further corroborative evidence for this concept.

Several studies reviewed by Young and Freedman[114] have clearly demonstrated that the renal transport of amino acids is influenced by the presence of a number of substances, such as sodium and glucose, in the glomerular filtrate. For example, it has been shown that the uptake of amino acids by kidney cortex slices is diminished in a low-sodium medium and that the inhibition is complete in a sodium-free medium. More direct evidence for the underlying mechanism has come from studies of the effect of sodium on amino acid transport in isolated proximal tubules and brush-border preparations. According to these studies, reabsorption of amino acids in the proximal tubule involves at least two sequential steps—namely, binding to the brush-border membrane and subsequent active transport into the cell. The binding step is found to be less sodium-dependent than the active transport step.

The influence of glucose on amino acid transport is best exemplified by *Fanconi's syndrome*, a disease characterized by glucosuria associated with aminoaciduria. The underlying defect is believed to be due to a disruption of some common step in the reabsorptive pathway for glucose and amino acid. A number of clearance studies lend credence to the validity of this hypothesis. Thus, it has been found that intravenous infusion of glucose reduces renal reabsorption of amino acid. Similar inhibition is obtained by infusion of phlorizin, a competitive inhibitor of glucose transport, as well as some amino acids, such as lysine, glycine, and alanine. These findings are consistent with the above concept and further suggest that the competitive interaction between glucose and amino acid transport may be mediated via a sodium-dependent ATPase system that provides energy for transport.

Recent studies[104] have confirmed that the amino acid reabsorption in the proximal tubule is clearly sodium-dependent. Furthermore, the results of studies with brush-border vesicles,[39] coupled with in-vivo[81] and in-vitro[53] electrical potential measure-

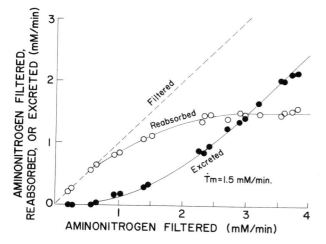

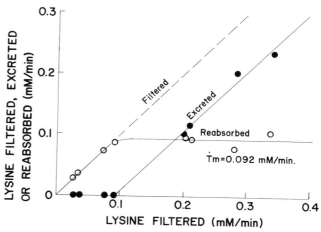

FIGURE 10.7. Reabsorption and excretion of aminonitrogen as functions of the filtered load in dogs in which glycine was infused to increase plasma concentration of aminonitrogen. (Redrawn from Pitts.[74])

FIGURE 10.8. Reabsorption and excretion of lysine as functions of filtered load in the dog. (Redrawn from Wright et al.[112])

ments, have shown that the luminal uptake of the amino acids occurs by a rheogenic, sodium-amino acid cotransport mechanism driven by the transluminal amino acid concentration and the sodium electrochemical gradients. The amino acids thus accumulated inside the proximal tubule cell leave across the basolateral membrane via a specific carrier by facilitated diffusion. Taken together, these studies suggest that, with the exception of acidic amino acids, the basic and neutral amino acids are reabsorbed via an active, sodium-dependent, carrier-mediated mechanism across the luminal membrane. Amino acids leave the cell across the basolateral membrane by facilitated diffusion. In the case of acidic amino acids, the sodium-dependent luminal uptake is enhanced by a cell-to-lumen potassium concentration gradient.[71, 84]

In summary, the available evidence indicates that amino acids are reabsorbed by an active, sodium-dependent, carrier-mediated transport process with Michaelis-Menten type kinetics in the proximal tubule similar to that described for D-glucose. But unlike the latter, there appear to be at least four stereospecific and separate transport systems for the reabsorption of the filtered amino acids. Furthermore, except for aspartate, the naturally occurring L-isomers show a much greater affinity for the carrier system than the D-isomers. Thus, one mechanism transports the *neutral amino acids*, namely, alanine, cysteine, histidine, isoleucine, leucine, methionine, phenylalanine, serine, threonine, tryptophan, tyrosine, and valine. Another mechanism transports the *basic amino acids*, namely, arginine, ornithine, and lysine. The third mechanism transports the *acidic amino acids*, namely, aspartic and glutamic acids. The fourth mechanism transports the *iminoglycine amino acids*, namely, proline, and hydroxyproline. Moreover, each transport mechanism is capable of reabsorbing more than one amino acid, but is influenced by competitive inhibition due to both structural and stereoisomeric configurations. Finally, because $\dot{T}_m$ for most amino acids studied is set very high, all the amino acids normally filtered are reabsorbed and none is excreted. Thus, like D-glucose, it appears that the kidney does not regulate the plasma concentration of amino acids.

Organic Anions

The kidney normally excretes a number of organic anions. The magnitude of their urinary excretion rate is markedly influenced by disturbances in body acid-base balance,[24, 34] being increased in metabolic alkalosis and decreased in metabolic acidosis (see Chapter 11 for definition of these terms). Cooke and associates[24] have suggested that organic aciduria represents a mechanism to regulate plasma chloride concentration in conditions of acid-base imbalance. Thus, in metabolic alkalosis the filtered chloride is reabsorbed while sodium is excreted in combination with organic anions. In contrast, in metabolic acidosis the filtered sodium is excreted in combination with chloride, instead of organic anions.

We shall now briefly describe the renal handling of some of the important organic anions found in the urine.

Citrate

Citrate is the initial substrate of the tricarboxylic acid (TCA) cycle. It is both the source of energy for the kidney and the most abundant organic acid in the urine. Its urinary excretion helps to solubilize calcium, thereby reducing the possibility of calcium phosphate stone formation.

In humans and other mammalian species, plasma concentrations of citrate range between 0.05 and 0.3 mM.[88] At normal plasma pH of 7.41, most of the plasma citrate is in the form of triply charged citrate ions and circulates unbound to any large molecules. Therefore, citrate is freely filtered at the glomerulus. Stop-flow studies have shown that citrate is reabsorbed predominantly in the proximal tubule.[22,35] The fraction of filtered citrate excreted varied in different species. In man, approximately 10% to 35% of filtered citrate is excreted.[43] However, a much smaller fraction is excreted in the rat,[29] and in the dog.[6]

Because high concentration of citrate is toxic to the heart, conventional titration analysis of renal handling of citrate has been attempted only in few studies. Thus, within the limits of citrate concentration tolerated by the body, animal studies have shown that citrate titration curves exhibit considerable splay, suggesting that a $\dot{T}_m$ for citrate reabsorption may actually exist. This conclusion is in part confirmed by studies of Grollman and co-workers[34] who found that infusion of some intermediate components of the TCA cycle—such as oxaloacetate, succinate, fumarate, and malate—blocks tubular reabsorption of citrate. In these studies the citrate clearance became equal to GFR and never exceeded it, thus suggesting that these intermediate components of the TCA cycle competitively inhibit citrate reabsorption by the renal tubules.

The major factor affecting citrate excretion is the disturbance in systemic acid-base balance. As early as 1931 (see review by Simpson[88]), metabolic alkalosis was found to increase citrate excretion in humans, a finding subsequently confirmed in other mammalian species. Recent evidence suggests that the effects of systemic acid-base disturbances are mediated via alterations in the pH gradient across the inner mitochondrial membrane.[88] Thus, under normal acid-base balance there is a continuous influx of citrate from both the luminal and the basolateral surfaces of the proximal convoluted cells. The luminal influx of citrate is mediated by a carrier in the brush-border membrane and presumably by another carrier in the basolateral membrane. The influx of citrate by these two routes raises the cytoplasmic concentration of citrate above that in the plasma. Cytoplasmic citrate is transported into mitochondria by the tricarboxylate carrier. In the mitochondria, the citrate undergoes oxidative metabolism in the TCA cycle to form carbon dioxide, water, and adenosine triphosphate (ATP). Normally, the rate of citrate metabolism exceeds the rate of citrate reabsorption, causing citrate uptake from the peritubular blood. However, some filtered citrate escapes reabsorption and is excreted in the urine. The magnitude of citrate excretion varies in different mammalian species. The rate of citrate transport by the tricarboxylate carrier across the mitochondrial membrane depends on the concentrations of citrate in the cytoplasm and in the mitochondrial matrix and the pH gradient across the inner mitochondrial membrane. The greater these gradients, the higher would be the rate of mitochondrial uptake of citrate. During metabolic acidosis, the mitochondrial pH gradient increases, leading to stimulation of the tricaboxylate carrier in the inner mitochondrial membrane. As a result, citrate enters the mitochondrial matrix causing the cytoplasmic citrate concentration to fall, thereby favoring enhanced citrate reabsorption and reduced excretion. In this manner, metabolic acidosis decreases citrate excretion. In contrast, during metabolic alkalosis, the opposite sequence of events takes place. Thus, there will be a decrease in mitochondrial pH gradient, inhibition of tricarboxylate carrier, reduced mitochondrial uptake of citrate, elevation of cytoplasmic citrate concentration, reduced reabsorption, and finally enhanced citrate excretion.

Besides the effects of acid-base disturbances, renal transport of citrate is influenced by several other factors. These include organic acids, metabolic inhibitors, and vitamin D. Several studies have shown that infusion of citric acid cycle compounds, such as malate, succinate, and fumarate, blocks citrate reabsorption, thereby increasing its urinary excretion.[6] The effect of infusion of these compounds on citrate metabolism is somewhat similar to that of metabolic alkalosis. Likewise, administration of some inhibitors of citric acid cycle, such as malonate, which inhibits succinic dehydrogenase, causes an increase in citrate excretion.[88] Finally, it has been shown that administration of vitamin D to normal rats increases citrate excretion.[7] Subsequent studies[27] showed that this effect was due to vitamin D inhibition of mitochondrial metabolism in the kidney.

Alpha-Ketoglutarate

This substance is an important component of the TCA cycle and hence of the energy supply system of all the cells, including those of the renal tubules. Its plasma concentration is normally quite low, about 0.1 μM/mL, despite the fact that all the filtered α-ketoglutarate is reabsorbed. Its tubular reabsorption is active, since the cellular uptake from the tubular fluid occurs against the concentration gradient. Furthermore, tubular transport is $\dot{T}_m$-limited, with the reabsorptive mechanism being saturated at a plasma concentration more than 20 times normal. The reabsorptive process exhibits relatively little splay. Because the renal threshold of α-ketoglutarate is considerably higher than its plasma concentration, the kidneys do not regulate the plasma concentration of this substance. The reabsorptive mechanism, therefore, serves to prevent the urinary loss of this important intermediary metabolite.

Besides being transported by the tubules, α-ketoglutarate is used by the renal tubule cells as a source of metabolic energy. It has been estimated that the kidneys use about 60% of the intravenously administered α-ketoglutarate, the remaining amount being used by the liver. This suggests that α-ketoglutarate is transported across both luminal and basolateral renal cell membranes against the electrochemical gradient by a $\dot{T}_m$-limited process.[98] Furthermore, these investigators observed that the $\dot{T}_m$ for α-ketoglutarate is increased in acute respiratory and metabolic acidosis and is decreased in acute respiratory and metabolic alkalosis. These changes in the $\dot{T}_m$ and the resulting effects on the tubular reabsorption of α-ketoglutarate were attributed to the induced changes in the intracellular pH of the renal tubule cells.

Acetoacetate

Normally, all the filtered acetoacetate is reabsorbed actively by a $\dot{T}_m$-limited process. However, in starvation and in uncontrolled diabetes mellitus, the urinary excretion rate rises in proportion to the increased plasma concentration.

Beta-Hydroxybutyrate

This substance is also reabsorbed by an active, $\dot{T}_m$-limited process. Its titration curves exhibit marked splay, with the renal threshold concentration being about 20 mg/dL in man.

Lactate

Lactate is normally completely reabsorbed. Whether it is reabsorbed by a $\dot{T}_m$-limited process is currently unresolved. However, in man when plasma lactate concentration exceeds the renal threshold concentration of about 60 mg/dL, its urinary excretion rate varies in proportion to increasing plasma concentration, suggesting a $\dot{T}_m$-limited reabsorptive process.[72]

Uric Acid

Uric acid is the end product of purine metabolism. Approximately 70% of uric acid formed is eliminated by the kidney, and the remainder is degraded by the intestinal flora.[92] Normally, the plasma concentration of uric acid is maintained between 4 to 6 mg/dL by a balance of three processes: (a) the rate of synthesis, (b) the rate of glomerular filtration, and (c) the rate of tubular transport.[36] Plasma uric acid concentration varies with sex and age.[67] It is lower in children of both sexes (3 to 4 mg/dL). At puberty, plasma uric acid concentration in men show a further elevation of 1 to 2 mg/dL. In women, plasma uric acid concentration approaches adult men after menopause.

In man a plasma uric acid concentration above 7 mg/dL is associated with increased risk for gout. This is a metabolic disease characterized by painful inflammation of joints caused by deposition of crystals of monosodium citrate. Although only a small number of patients with hyperuricemia ever develop gout, all patients with gout have hyperuricemia at some stage in their clinical course.

In man the bulk of the excreted urinary nitrogen is in the form of urea and ammonia, with excretion of uric acid accounting for only 5% of the total. Despite this, there is a great deal of interest in the mechanism of renal transport of uric acid, the principal reason being the observed elevation of plasma uric acid concentration (hyperuricemia) in patients with gout. The hyperuricemia is generally attributed to either an increase in the rate of uric acid synthesis, or a decrease in the rate of its renal excretion, or possibly a combination of the two.[113]

A number of studies in most mammalian species, including man, have shown that the tubular transport of uric acid is bidirectional in the proximal tubule, with reabsorption and secretion occurring within the same tubular cell.[44,110] In the dog, however, stop-flow studies have shown that uric acid is only reabsorbed in the proximal tubule, but is both reabsorbed and secreted in the "distal tubule." Micropuncture studies in rat[79, 111] have revealed that uric acid is reabsorbed throughout the length of the proximal tubule. Uric acid secretion is also detected in the same region. The mechanism of uric acid secretion appears to be similar to that of PAH (see below). In both human and rat, volume expansion enhances[28] and volume contraction reduces[94] the fractional excretion of uric acid. Both effects are exerted in the proximal tubule.[80]

Titration studies by Berliner and associates[9] have shown that uric acid is reabsorbed actively by a $\dot{T}_m$-limited process, with a maximum transport rate of about 15 mg/min per 1.73 m² body surface area in man. However, the maximum transport capacity normally is not exceeded, but can be saturated at plasma uric acid concentrations exceeding 15 to 20 mg/dL. Uric acid titration curves exhibit considerable splay, suggesting that the extent of splay, rather than the magnitude of $\dot{T}_m$, may govern the urinary excretion rate of uric acid. Normally, more than 90% of the filtered uric acid is reabsorbed in the proximal tubule. Hence, tubular secretion must largely account for the bulk of uric acid excreted in the urine. In addition, because uric acid is lipid insoluble, nonionic passive diffusion (see Chapter 11) plays no role in its tubular transport and, therefore, its urinary excretion. Consequently, changes in urinary pH that may occur in acid-base disturbances have no effect on its urinary excretion.[69]

Clinically, the hyperuricemia of gout may result from abnormal functioning of any one or a combination of those factors responsible for maintaining the plasma uric acid concentration at normal levels. Thus, hyperuricemia may result from (a) an increase in the rate of synthesis of uric acid, (b) a reduction in the rate of glomerular filtration, or (c) an increase in the tubular reabsorption or a decrease in the tubular secretion. In man, the pyrazinamide (PZA) suppression test has been used extensively to localize the primary lesion causing hyperuricemia and to assess the relative effect of tubular secretion.[37] The premise for using this test has been that PZA (an antituberculosis drug) selectively blocks the tubular secretion of uric acid. Therefore, the quantity of uric acid excreted in the urine after PZA test must represent the quantity that escaped tubular reabsorption. Furthermore, the difference between the quantity of uric acid excreted before and after the PZA test represents the amount of uric acid that is secreted by the tubules.

However, in a recent review of laboratory and clinical experiences with PZA, Holmes and associates[44] concluded that the inhibitory effect of this substance on tubular secretion is not as selective and complete as had been assumed. Therefore, further clinical use of PZA suppression test must be reevaluated.

Proteins

As mentioned in Chapter 5, the ultrafiltrate is virtually protein-free. However, stop-flow studies in dogs showed that albumin, hemoglobin, and other low molecular weight proteins are filtered at the glomerulus and are subsequently reabsorbed in the proximal tubule.[60] Recent studies by Cortney and associates,[25] in which labeled proteins were injected into the rat proximal and "distal" tubules, further revealed that the injected proteins are largely reabsorbed in the proximal tubule, but a smaller quantity is also reabsorbed in the "distal tubule."

Low molecular weight proteins (LMWP) include all circulating proteohormones and peptides, such as insulin, glucagon, parathyroid hormone, growth hormone, and angiotensin II, as well as a variety of secretory and tissue proteins, such as immunoprotein fragments (Bence-Jones proteins). Evidence indicates that LMWP are freely filtered and subsequently reabsorbed by proximal tubular epithelia and eventually hydrolyzed within the renal tubule cells into their constituent amino acids. As a result, unlike the large proteins, such as albumin, which have a plasma half-life of days, LMWP have a plasma half-life that is measured in hours or less.[66]

Although the glomerular filtrate contains only a low concentration of protein, the large volume of the filtrate formed per day contains several grams of protein. In man, about 0.5% of the total plasma albumin is filtered, yielding a filtrate albumin concentration of less than 30 mg/dL, compared with 4.0 g/dL in the plasma. Normally, virtually all the filtered albumin is reabsorbed in the proximal tubule by an active, $\dot{T}_m$-limited process. In man, the $\dot{T}_m$ for albumin is estimated to be about 30 mg/dL, with the reabsorptive mechanism being saturated at a plasma concentration of about 6 to 7 g/dL.[72]

Physiological studies using stop-flow,[59] micropuncture,[58] and microperfusion[25] techniques have shown that the proximal tubule is the major site of protein reabsorption in the kidney. Morphological studies[20, 23, 65, 96] have proposed the following possible endocytic mechanism for reabsorption of the filtered proteins in the proximal tubule. According to this scheme, the filtered proteins are *adsorbed* to the endocytic sites at the luminal brush-border membrane and then separated in endosomes. The latter migrate into the cell interior, where they fuse with lysosomes. Protein digestion takes place within endolysosomes. The amino acids, thus formed, leave the endolysosomes and cross the basolateral membrane to enter the peritubular blood. Recent studies[16] suggest that small linear peptides, such as angiotensin II and bradykinin, are transported by a different mechanism. They are hydrolyzed by brush border enzymes at the luminal membrane. The liberated amino acids are then transported by the amino acids transport systems.

Taken together, these studies suggest that the reabsorption of proteins may be achieved by two different mechanisms: Proteins or peptides are sequestered by brush border proteases and the amino acids are subsequently reabsorbed by specific amino acids transport systems, or all the proteins are taken up by endocytosis. In this case, the proteins (as has been shown experimentally by tracers) are trapped within tubular invaginations of the apical membrane situated between the bases of the microvilli, which then bud off as small vesicles into the cytoplasm. Here, they fuse to larger vesicles and vacuoles and with primary lysosomes to secondary lysosomes (phagosomes), in which the digestion to amino acids takes place. Undigested residues within lysosomes are released by exocytosis from the luminal membrane (Figure 10.9).

In summary, the amount of filtered protein is directly proportional to GFR, the plasma concentration, and the permeability ratio (see Chapter 7). Low molecular weight proteins are filtered at the glomerulus in significant amounts. Since urine is normally protein-free, it follows that all the filtered proteins must be reabsorbed in the renal tubules, primarily in the proximal tubule. Since tubular reabsorption of proteins is not a saturable process over a wide range of filtered loads, it follows that the filtration process is the rate-limiting step in the overall clearance of low molecular weight proteins from the circulation. Thus, filtered load of LMWP provides the means by which the kidney regulates the quantity and turnover rate of LMWP in the plasma.

The filtered protein is reabsorbed by an adsorptive endocytic mechanism, characterized by a high capacity and a relatively low affinity. Furthermore, reabsorption is a selective process, depending on the charge, size, and stereospecificity of the endocytic sites for the particular protein. The reabsorbed proteins are subsequently catabolized to amino acids

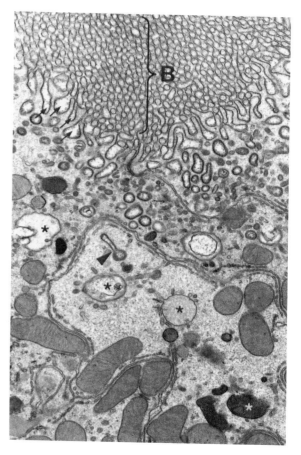

FIGURE 10.9. An electron micrograph of the apical portion of a rat proximal tubule cell, where (B) indicates the cross-section of brush border. The tubular invaginations between the bases of microvilli (*arrows*) bud off as small vesicles, which then fuse with primary lysosomes (*arrowhead*) and each other to form the various kinds of phagosomes (*).

within the endolysosomes. The rate of hydrolysis depends on the nature of the protein. It varies from minutes for proteohormones to many hours or even days for some low molecular weight proteins. Finally, the amino acids generated by lysosomal hydrolysis return to the blood by permeation across the basolateral membrane. Smaller linear peptides, such as angiotensin and bradykinin, are hydrolyzed at the luminal surface by the bursh border proteases. The liberated amino acids are transported by the various amino acid transport systems described above.

We conclude this section with Table 10.1, which summarizes the salient characteristics of the renal handling of several important constituents of the filtrate that have already been discussed.

TABLE 10.1. Summary of renal transport of some important constituents of the filtrate.

Substance	Mean clearance in physiological conditions	Permeability ratio	Tubular reabsorption	Tubular secretion
1. Magnesium	6 mL/min	50%	Proximal tubule; active	Possible
2. Calcium	1–3 mL/min	65%	Proximal tubule; active	No
3. Phosphate	3–10 mL/min	100%	Proximal tubule; active, $\dot{T}_m$-limited; dependent on calcium.	Theoretically possible but never shown
4. Chloride	0.2–2 mL/min	100%	Proximal tubule; passive, follows sodium reabsorption; active in the thick ascending limb of Henle's loop	Unknown
5. Sulfate	10 mL/min	50%–100%	Both proximal and "distal" tubules; active with $\dot{T}_m$ about the same as that for phosphate	Unknown
6. Amino acids	Varies with nature of amino acid; but less than 5 mL/min	Not known, probably 100%	Proximal tubule; $\dot{T}_m$-limited	Unknown
7. Endogenous creatinine	Grossly equal to glomerular clearance	100%	Apparently none	None or slight (dog)
8. Exogenous creatinine	130% of inulin clearance	100%	Apparently none	Certain (man)
9. Uric acid	7–10 mL/min	Probably 100%	Both proximal and "distal" tubules; active, $\dot{T}_m$-limited	Occurs in "distal tubule"

Tubular Secretion

With the exception of potassium (see Chapter 9), hydrogen ions, and ammonia (see Chapter 11), most of the substances normally secreted by the renal tubules are either weak acids or bases. These substances fall into one or more of the following classes: (1) They are not normal constituents of the plasma, such as drugs (e.g., penicillin and salicylate, a metabolic product of aspirin); (2) they are not metabolized by the body and are secreted virtually unchanged into the urine [e.g., PAH, phenol red, and iodopyracet (Diodrast), a radiopaque contrast material]; and (3) they are slowly but incompletely metabolized (e.g., thiamine or vitamin B_1).

Because PAH has been widely used to measure the renal plasma flow (see Chapter 7) and is secreted by an active, $\dot{T}_m$-limited process, its renal transport is the only one described here. Its tubular transport, however, should serve as the prototype for the renal handling of the other substances secreted by the tubules.

p-Aminohippurate

As mentioned in Chapter 7, PAH, which is a weak organic acid, is secreted into the proximal tubule by a $\dot{T}_m$-limited process. Its tubular transport exemplifies the renal handling of other similar substances, such as phenol red and iodopyracet.

Although PAH is mostly secreted by the proximal tubule,[5, 102] Cho and Cafurny[19] have demonstrated that a smaller quantity of PAH is reabsorbed in this segment. Thus, there appears to be a bidirectional transport mechanism for PAH in the kidney. Although this small bidirectional transport of PAH does not appreciably influence the clearance of PAH, it certainly has a profound effect on the magnitude of $\dot{T}_m$ for this substance.

Studies with isolated rat basolateral vesicles[10, 48] have shown that PAH is transported across the basolateral membrane by an active, carrier-mediated mechanism that could be inhibited competitively. Other studies[76] have shown that the rate of PAH entry was dependent on the transbasolateral sodium concentration gradient, a factor favoring the basolateral uptake of the PAH anion. This finding suggests that PAH and sodium are cotransported across the basolateral membrane, with the extracellular-to-cytoplasmic sodium concentration gradient serving as the driving force.

Taken together, the above studies suggest that PAH is secreted only in the second half of the prox-

Factors increasing clearance	Factors decreasing clearance	Organs involved in elimination from body
Increased plasma magnesium; aldosterone	Low plasma magnesium	⅓ Renal excretion ⅔ GI excretion
Plasma calcium; vitamin D; parathyroid hormone; presence of anions forming nonionizable salts of calcium, citrate, sulfate, etc.	Lowered plasma calcium; thiazide diuretics	¼ Renal excretion ¾ GI excretion
Diurnal variation, between noon and 8 PM; parathyroid hormone, acidosis; hypercalcemia	Diurnal variation from midnight to midday; alkalosis; hypocalcemia; anterior pituitary growth hormone	Renal excretion is predominant but not exclusive
Same as for sodium clearance; competition with bicarbonate; saline diuresist;	Same as for sodium clearance; competition with bicarbonate; carbonic anhydrase inhibitors	Renal excretion is almost exclusive in the absence of vomiting
Unknown	Unknown	Almost exclusively by kidney
Tubular competition between groups of amino acids	Unknown	Minimal physiological amino aciduria
Anything that increases glomerular filtration	Anything that lowers glomerular filtration	Exclusively by kidney
Anything that increases glomerular filtration	Anything that lowers glomerular filtration	Exclusively by kidney
Tubular inhibitors (e.g., probenecid) in high dose	Tubular inhibitors (probenecid in low dose); and above all thiazide diuretics in normal therapeutic dose	Exclusively by kidney; diminished in hyperuricemia

imal tubule by the following mechanisms. First PAH enters into the proximal cell across the basolateral membrane by an active, sodium-dependent, carrier-mediated process. The subsequent intracellular accumulation of PAH provides the driving force for its secretion across the luminal membrane into the

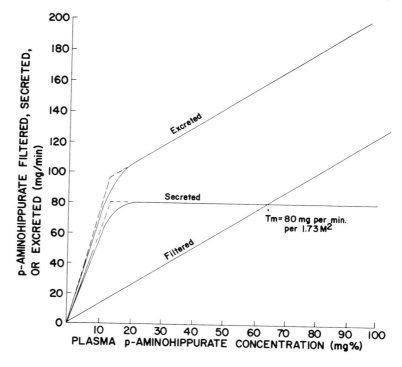

FIGURE 10.10. Secretion and excretion of PAH as a function of plasma concentration in man. Modified and reproduced with permission from Pitts, R. F.: *Physiology of the Kidney and Body Fluids,* 3rd Edition. Copyright © 1974 by Yearbook Medical Publishers, Inc., Chicago.

proximal tubule lumen along its concentration gradient by facilitated diffusion. The luminal transport is independent of transluminal sodium and potassium concentration gradients. The luminal carrier for PAH transport is believed to be responsible for urate uptake from lumen into the proximal tubule cell. In short, the transepithelial transport of PAH appears to be active and occurring against a concentration gradient, since as mentioned in Chapter 9, there is no significant electrical potential gradient across the proximal cell membrane.

Figure 10.10 presents the renal titration curves for PAH in man. Note that like reabsorption of D-glucose, secretion of PAH and like substances involves active transport, characterized by a $\dot{T}_m$-limited process. There is, however, considerable splay in the titration curves for the same reasons as those given for glucose reabsorption. In man, $\dot{T}_m$ values for PAH, iodopyracet, and phenol red average 80, 57, and 36 mg/min per 1.73 m² body surface area, respectively.[72]

Intravenous infusion of PAH or iodopyracet decreases tubular secretion of phenol red, suggesting that these substances are secreted by a common carrier system. However, it appears that PAH and iodopyracet have greater affinity for the carrier, compared with phenol red, with competition for the carrier and eventual tubular secretion obeying the same kinetic laws that were outlined in Chapter 8.

Problems

10.1. Calculate the quantity of glucose filtered and the quantity reabsorbed by the tubules under the following conditions:
   Plasma glucose concentration = 500 mg/dL
   Urinary excretion rate = 275 mg/min
   Inulin clearance = 130 mL/min
   Permeability ratio = 1.0

10.2. Calculate the quantity of iodopyracet (Diodrast) expressed as iodine, which is filtered, and the quantity secreted by the tubules under the following conditions:
   Plasma iodopyracet concentration = 6 mg/dL
   Urinary excretion rate = 42 mg/min
   Inulin clearance = 130 mL/min
   Permeability ratio = 0.73

10.3. Complete Table 10.a below by calculating the following:
   Quantity of glucose filtered, mg/min
   Quantity of glucose reabsorbed, mg/min
   Total quantity of glucose brought to the kidney, mg/min
   Plasma clearance of glucose, mL/min
   Extraction ratio for glucose

   (Assume GFR = 130 mL/min; renal plasma flow (RPF) = 700 mL/min; and glucose permeability ratio = 1.0.)

10.4. Construct a graph with plasma glucose concentrations in mg/dL as abscissa (0 to 1,000 mg/dL) and the quantity of glucose filtered in mg/min as ordinate (0 to 1,500 mg/min). Plot the data from Table 10.a.

10.5. On the same graph plot the following data from Table 10.a against plasma glucose concentration:
   a. Glucose excreted, mg/min
   b. Glucose reabsorbed, mg/min
   c. Total glucose brought to kidney, mg/min

10.6. At what plasma concentration does glucose first appear in the urine? _____

10.7. What is the value of $\dot{T}_m$ for glucose? _____

10.8. Fill in the following table by reading the appropriate values from your graph:

	Plasma Glucose Concentration (mg/dL)			
	200	400	600	900
a. Glucose filtered, mg/min				
b. Glucose reabsorbed, mg/min				
c. Glucose excreted, mg/min				
d. % of filtered glucose excreted				

   (Note: calculated by dividing glucose excreted by glucose filtered and multiplying by 100.)

10.9. How does the % of the filtered glucose that is excreted vary as the plasma glucose level increases? _____

10.10. If 100% of the filtered glucose were excreted, what would the value be for:
   a. Glucose clearance? _____
   b. Glucose extraction ratio? _____

10.11. Complete Table 10.b below by making the appropriate calculations. (Assume GFR = 130 mL/min; RPF = 700 mL/min; and PAH permeability ratio = 0.83.)

10.12. Construct a graph similar to that for glucose and plot the following ordinates against the plasma PAH concentration as abscissa (ordinate range, 0 to 200 mg/min; abscissa range, 0 to 100 mg/dL):
   a. PAH filtered, mg/min
   b. PAH secreted by tubules, mg/min

TABLE 10.a

Plasma glucose (mg/dL)	Urinary excretion rate (mg/min)	Glucose filtered (mg/min)	Glucose reabsorbed (mg/min)	Total glucose brought to Kidney (mg/min)	Plasma glucose clearance (mL/min)	Glucose extraction ratio
a. 50	0					
b. 100	0					
c. 288	0					
d. 500	275					
e. 700	535					
f. 1,000	925					

    c. PAH excreted in urine, mg/min
    d. Total PAH brought to kidney, mg/min

10.13. At what plasma level does tubular secretion level off? _____

10.14. What is the value of $\dot{T}_m$ for PAH? _____

10.15. From your graph, fill in the following table:

    Plasma PAH concentration (mg/dL)
    4 8 15 30 50 70

    a. PAH filtered, mg/min
    b. PAH secreted by tubules, mg/min
    c. PAH excreted in urine, mg/min
    d. $\dfrac{\text{PAH filtered}}{\text{PAH secreted by tubules}}$

10.16. How does the ratio of filtered PAH to secreted PAH vary as the plasma PAH level increases? _____

10.17. If the quantity of PAH secreted by the tubules became insignificantly small compared with the quantity filtered, what would the value be for:
    a. PAH clearance? _____
    b. PAH extraction ratio? _____

10.18. Construct a graph as follows:
    a. As the left hand ordinate scale: Plasma clearance in mL/min. Range 0 to 700 mL/min. Scale: 25 mm = 100 mL/min.
    b. As the right hand ordinate scale: Extraction ratio. Range 0 to 1.0. Scale: 25 mm = 0.14.
    c. As the abscissae: Three separate scales, arranged one below the other, representing plasma concentrations in mg/dL as follows:
      (1) Upper scale: Plasma inulin concentration range: 0 to 500 mg/dL. Scale: 25 mm = 50 mg/dL.
      (2) Middle scale: Plasma glucose concentration range: 0 to 1,000 mg/dL. Scale: 25 mm = 100 mg/dL.
      (3) Lower scale: Plasma PAH concentration range: 0 to 100 mg/dL. Scale: 25 mm = 10 mg/dL.

10.19. On your graph plot the following clearances against the corresponding plasma concentrations:
    a. Inulin clearance from Table 7.c, Chapter 7.
    b. Glucose clearance from Table 10.a, above.
    c. PAH clearance from Table 10.b, below.

10.20. At an infinitely high plasma glucose level, which of the following would be true? (Check correct answer.)
    a. Glucose clearance = GFR _____
    b. Glucose clearance > GFR _____

TABLE 10.b

Plasma PAH (mg/dL)	Urinary Excretion Rate (mg/min)	PAH Filtered (mg/min)	PAH Secreted by Tubules (mg/min)	Total PAH brought to Kidney (mg/min)	Plasma PAH clearance (mL/min)	PAH extraction ratio
a. 2	14					
b. 6	42					
c. 13	91					
d. 20	99.5					
e. 40	120.5					
f. 60	142.5					

    c. Glucose clearance < GFR _____

10.21. At an infinitely high plasma PAH level, which of the following would hold? (Check correct answer.)
    a. PAH clearance = GFR _____
    b. PAH clearance > GFR _____
    c. PAH clearance < GFR _____

10.22. Above what plasma PAH concentration does PAH clearance fall below the renal plasma flow?

References

1. Agus ZS, Puschett JB, Senesky D, Goldberg M: Mode of action of cyclic adenosine 3',5'-monophosphate on renal tubular phosphate reabsorption in the dog. *J Clin Invest* 1971; 50:617–626.
2. Anderson J: A method for estimating Tm for phosphate in man. *J Physiol (Lond)* 1955; 130:268–277.
3. Aurbach GD, Potts JT Jr: Parathyroid hormone. *Am J Med* 1967; 42:1–8.
4. von Baeyer H, Haeberle DA, van Liew JB, Hare D: Glomerular tubular balance of renal D-glucose transport during hyperglycemia. Clearance and micropuncture studies on its characteristics at saturated transport conditions. *Pfluegers Arch* 1980; 384:39–47.
5. Baines AD, Gottschalk CW, Lassiter WE: Microinjection study of p-aminohippurate excretion by rat kidneys. *Am J Physiol* 1968; 214:703–709.
6. Baruch SB, Burich RL, Eun CE, King VF: Renal metabolism of citrate. *Med Clin North Am* 1975; 59:569–582.
7. Bellin SA, Herting DC, Cramer JW, et al: The effect of vitamin D on urinary citrate in relation to calcium, phosphorus and urinary pH. *Arch Biochem Biophys* 1954; 50:18–24.
8. Berglund F, Lotspeich WD: Renal tubular reabsorption of inorganic sulfate in the dog, as affected by glomerular filtration rate and sodium chloride. *Am J Physiol* 1956; 185:533–538.
9. Berliner RW, Hilton JG Jr, Yr TF, Kennedy TJ Jr: The renal mechanism for urate excretion in man. *J Clin Invest* 1950; 29:396–401.
10. Berner W, Kinne R: Transport of p-aminohippuric acid by plasma membrane vesicles isolated from rat kidney cortex. *Pfluegers Arch* 1976; 361:269–277.
11. Beyer KH, Wright LD, Russo HF, et al: The renal clearance of essential amino acids: tryptophane, leucine, isoleucine and valine. *Am J Physiol* 1946; 146:330–335.
12. Beyer KH, Wright LD, Skeggs HR, et al: Renal clearance of essential amino acids: their competition for reabsorption by the renal tubules. *Am J Physiol* 1947; 151:202–210.
13. Bijvoet OLM: Relation of plasma phosphate concentration to renal tubular reabsorption of phosphate. *Clin Sci* 1969; 37:23–36.
14. Bradley SE, Laragh IH, Wheeler HO, et al: Correlation of structure and function in the handling of glucose by nephrons of the canine kidney. *J Clin Invest* 1961; 40:1113–1131.
15. Brown JL, Samiy AH, Pitts RF: Localization of aminonitrogen reabsorption in the nephron of the dog. *Am J Physiol* 1961; 200:370–372.
16. Carone FA, Peterson DR: Hydrolysis and transport of small peptides by the proximal tubule. *Am J Physiol* 1980; 238:F151–F158.
17. Chan SS, Lotspeich WD: Comparative effects of phlorizin and phloretin on glucose transport in the cat kidney. *Am J Physiol* 1962; 203:975–979.
18. Chesney RW, Sacktor B, Rowen R: The binding of D-glucose to the isolated luminal membrane of the renal proximal tubule. *J Biol Chem* 1973; 218:2182–2191.
19. Cho KC, Cafruny EJ: Renal tubular reabsorption of p-aminohippuric acid (PAH) in the dog. *J Pharmacol Exp Ther* 1970; 173:1–12.
20. Christensen EI, Maunsbach AB: Intralysosomal digestion of lysozyme in renal proximal tubular cells. *Kidney Int* 1974; 6:396–407.
21. Coello JB, Bradley SE: Function of the nephron population during hemorrhagic hypotension in the dog with special reference to the effects of osmotic diuresis. *J Clin Invest* 1964; 43:386–400.
22. Cohen RD, Prout RES: The origin of urinary citrate. *Clin Sci* 1964; 26:237–245.
23. Cojocel C, Maita K, Baumann K, Hook JB: Renal processing of low molecular weight proteins. *Pfluegers Arch* 1984; 401:333–339.
24. Cooke RE, Segar WE, Reed C, et al: The role of potassium in the prevention of alkalosis. *Am J Med* 1954; 17:180–195.
25. Cortney MA, Sawin LL, Weiss DD: Renal tubular protein absorption in the rat. *J Clin Invest* 1970; 49:1–4.
26. Crane R: Absoprtion of sugars, in Code CF (ed): *Handbook of Physiology, Section 6, Alimentary Canal*. Washington DC, American Physiology Society, 1968, pp 1323–1351.
27. DeLuca HF, Steenbock H: An in vitro effect of vitamin D on citrate oxidation by kidney mitochondria. *Science* 1957; 126:258.
28. Diamond H, Meisel A: Influence of volume expansion, serum sodium, and fractional excretion of sodium on urate excretion. *Pfluegers Arch* 1975; 356:47–57.
29. Franklin R, Costello LC, Stacey R, Stephens R: Calcitonin effects on plasma and urinary citrate level in rats. *Am J Physiol* 1973; 225:1178–1180.
30. Frick A: Reabsorption of inorganic phosphate in the rat. I. Saturation of transport mechanism. II. Suppression of fractional phosphate reabsorption due to expansion of extracellular fluid volume. *Arch Ges Physiol* 1968; 304:351–364.

31. Frohnert P, Hohmann B, Zwiebel R, Baumann K: Free flow micropuncture studies of glucose transport in the rat nephron. *Arch Ges Physiol* 1970; 315:66–85.
32. Fromter E: Electrophysiological analysis of rat renal sugar and amino acid transport. I. Basic phenomena. *Pfluegers Arch* 1982; 393:179–189.
33. Grinstein S, Turner BJ, Silverman M, Rothstein A: Inorganic anion transport in kidney and intestinal brush border and basolateral membranes. *Am J Physiol* 1980; 238:F452–F460.
34. Grollman AP, Harrison HC, Harrison HE: The renal excretion of citrate. *J Clin Invest* 1961; 40:1290–1296.
35. Grollman AP, Walker WG, Harrison HC, Harrison HE: Site of reabsorption of citrate and calcium in the renal tubule of the dog. *Am J Physiol* 1963; 205:697–701.
36. Gutman AB, Yu TF: A three-component system for regulation of renal excretion of uric acid in man. *Trans Assoc Am Physicians* 1961; 74:353–365.
37. Gutman AB, Yu TF, Berger L: Renal function in gout. III. Estimation of tubular secretion and reabsorption of uric acid by use of pyrazinamide. *Am J Med* 1969; 47:575–592.
38. Hall PF: *The Functions of the Endocrine Glands*. Philadelphia, W B Saunders Co, 1959.
39. Hammerman MR, Sacktor B: Na^+-dependent transport of glycine in renal brush border membrane vesicles. *Biochim Biophys Acta* 1982; 686:189–96.
40. Hierholzer K, Cade R, Gurd R, et al: Stop-flow analysis of renal reabsorption of sulfate in the dog. *Am J Physiol* 1960; 198:833–837.
41. Higgins IT, Meinders AE: Quantitative relationship of renal glucose and sodium reabsorption during ECF expansion. *Am J Physiol* 1975; 229:66–71.
42. Hirsch PF, Voelkel EA, Munson PL: Thyrocalcitonin: Hypocalcemic hypophosphatemic principle of the thyroid gland. *Science* 1964; 146:412–413.
43. Hodgkinson A: Citric acid excretion in normal adults and in patients with renal calculus. *Clin Sci* 1962; 23:203–212.
44. Holmes EW, Kelley WN, Wyngaarden JB: The kidney and uric acid excretion in man. *Kidney Int* 1972; 2:115–118.
45. Hopfer U: Transport in isolated plasma membranes. *Am J Physiol* 1978; 234:F89–F96.
46. Kawamura J, Mazumdar DC, Lubowitz H: Effect of albumin infusion on renal glucose reabsorption in the rat. *Am J Physiol* 1977; 232:F286–F290.
47. Keyes JL, Swanson RE: Dependence of glucose Tm on GFR and tubular volume in the dog kidney. *Am J Physiol* 1971; 221:1–7.
48. Kinsella JL, Holohan PD, Pessah NI, Ross CR: Transport of organic ions in renal cortical luminal and antiluminal membrane vesicles. *J Pharmacol Exp Ther* 1979; 209:443–450.
49. Kleinzeller A, Kolinska J, Benes I: Transport of glucose and galactose in kidney cortex cells. *Biochem J* 1967; 104:843–851.
50. Knox FG, Osswald H, Marchand GR, et al: Phosphate transport along the nephron. *Am J Physiol* 1977; 233:F261–F266.
51. Knox FG, Schneider EG, Willis LR, et al: Site and control of phosphate reabsorption by the kidney. *Kidney Int* 1973; 3:347–353.
52. Koch A: The Kidney, in Ruch TC, Patton HD (eds): *Physiology and Biophysics*, ed 20. Philadelphia, WB Saunders Co, 1973, pp 844–872.
53. Kokko JP: Proximal tubule potential difference: Dependence on glucose, HCO_3^- and amino acids. *J Clin Invest* 1973; 52:1362–1367.
54. Krane S, Crane R: The accumulation of D-galactose against a concentration gradient by slices of rabbit kidney cortex. *J Biol Chem* 1959; 234:211–216.
55. Kruhoffer P: *Studies on Water and Electrolyte Excretion and Glomerular Activity in the Mammalian Kidney*. Copenhagen, Rosenkilde and Bagger, 1950, pp 76–85.
56. Kurtzman NA, White MG, Rogers PW, Flynn JJ III: Relationship of sodium reabsorption and glomerular filtration rate to renal glucose reabsorption. *J Clin Invest* 1972; 51:127–133.
57. Lambert PP, Van Kessel E, Leplat C: Etude sur l'elimination des phosphates inorganiques chez l'homme. *Acta Med Scand* 1947; 128:386–410.
58. Landwehr DM, Carvalho JS, Oken DE: Micropuncture studies of the filtration and absorption of albumin by nephrotic rats. *Kidney Int* 1977; 11:9–17.
59. Lathem W, Davis BB: Renal tubular reabsorption of protein: Demonstration and localization of egg-albumin and β-lactoglobulin reabsorption in the dog. *Am J Physiol* 1960; 199:644–648.
60. Lathem W, Davis BB, Zweig PH, Dew R: The demonstration and localization of renal tubular reabsorption of hemoglobin by stop flow analysis: *J Clin Invest* 1960; 39:840–845.
61. Lotspeich WD: Phlorizin and the Cellular Transport of Glucose. *Harvey Lectures, Ser. 56*. New York, Academic Press, 1961, pp 63–91.
62. Lotspeich WD: Renal tubular reabsorption of inorganic sulfate in the normal dog. *Am J Physiol* 1947; 151:311–318.
63. Lotspeich WD, Woronkow S: Some quantitative studies on phlorizin inhibition of glucose transport in the kidney. *Am J Physiol* 1958; 195:331–336.
64. Lucke H, Stange G, Murer H: Sulphate-ion/sodium-ion co-transport by brush-border membrane vesicles isolated from rat kidney cortex. *Biochem J* 1979; 182:223–229.
65. Maack T, Mackensie DDS, Kinter WB: Intracellular pathways of renal absorption of lysozyme. *Am J Physiol* 1971; 221:1609–1616.
66. Maack T, Johnson V, Kau ST, et al: Renal filtration,

transport, and metabolism of low-molecular weight proteins: A review. *Kidney Int* 1979; 16:251–270.
67. Mikkelsen WM, Dodge HJ, Valkenburg H: The distribution of serum uric acid values in a population unselected as to gout or hyperuricemia: Tecumsch, Michigan, 1959–1960. *Am J Med* 1965; 39:242–251.
68. Mudge GH: Clinical patterns of tubular dysfunction. *Am J Med* 1958; 24:785–804.
69. Mudge GH, Berndt WO, Valtin H: Tubular transport of urea, glucose, phosphate, uric acid, sulfate, and thiosulfate, in Orloff J, Berliner JW (eds): *Handbook of Physiology, Section 8, Renal Physiology*. Washington DC, American Physiological Society, 1973, pp 587–652.
70. Murayama Y, Morel F, LeGrimellec C: Phosphate, calcium, and magnesium transfers in proximal tubules and loops of Henle, as measured by single nephron microperfusion experiments in the rat. *Arch Ges Physiol* 1972; 333:1–16.
71. Murer H, Leopolder A, Kinne R, Burckhardt G: Recent observations on the proximal tubular transport of acidic and basic amino acids by rat renal proximal tubular brush border vesicles. *Int J Biochem* 1980; 12:223–228.
72. Pitts RF: *Physiology of the Kidney and Body Fluids*, ed 3. Chicago, Year Book Medical Publishers 1974.
73. Pitts RF: A comparison of the renal reabsorptive processes for several amino acids. *Am J Physiol* 1944; 140:535–547.
74. Pitts RF: A renal reabsorptive mechanism in the dog common to glycine and creatine. *Am J Physiol* 1943; 140:156-168.
75. Pitts RF, Alexander RS: The renal absorptive mechanism for inorganic phosphate in normal and acidotic dogs. *Am J Physiol* 1944; 142:648–662.
76. Podevin RA, Boumendil-Podevin EF, Priol C: Concentrative PAH transport by rabbit kidney slices in the absence of metabolic energy. *Am J Physiol* 1978; 235:F278–F285.
77. Puschett JB, Agus ZS, Senesky D, Goldberg M: Effects of saline loading and aortic obstruction on proximal phosphate transport. *Am J Physiol* 1972; 223:851–857.
78. Robson AM, Srivastava PL, Bricker NS: The influence of saline loading on renal glucose reabsorption in the rat. *J Clin Invest* 1968; 47:329–335.
79. Roch-Ramel F, Diezi-Chomety F, De Rougemont D, et al: Renal excretion of uric acid in the rat: A micropuncture and microperfusion study. *Am J Physiol* 1976; 230:768–776.
80. Roch-Ramel F, Granges F, Roth L, et al: Renal handling of urate by nondiuretic and diuretic rats. HPLC-amperometric determination of urate concentrations. *Renal Physiol* 1980; 2:122–129.
81. Samarija I, Fromter E: Electrophysiological analysis of rat renal sugar and amino acid transport. V. Acidic amino acids. *Pfluegers Arch* 1982; 393:215–221.
82. Samarzija I, Hinton BT, Fromter E: Electrophysiological analysis of rat renal sugar and amino acid transport. II. Dependence of various transport parameters and inhibitors. *Pfluegers Arch* 1982; 393:190–197.
83. Schafer JA, Andreoli TE: Rheogenic and passive Na$^+$ absorption by the proximal nephron. *Ann Rev Physiol* 1979; 41:211–227.
84. Schneider EG, Sacktor B: Sodium gradient-dependent L-glutamate transport in renal brush border membrane vesicles. Effect of an intra vesicular>extravesicular potassium gradient. *J Biol Chem* 1980; 255:7645–7649.
85. Schultze RG, Berger H: The influence of GFR and saline expansion on Tm$_G$ of the dog kidney. *Kidney Int* 1973; 3:291–297.
86. Shannon JA, Farber S, Troast L: The measurement of glucose T$_m$ in the normal dog. *Am J Physiol* 1941; 133:752-761.
87. Shannon JA, Fisher S: The renal tubular reabsorption of glucose in the normal dog. *Am J Physiol* 1938; 122:765–774.
88. Simpson DP: Citrate excretion: A window on renal metabolism. *Am J Physiol* 1983; 244:F223–F234.
89. Singer FR, Woodhouse NJY, Parkinson DK, Joplin GF: Some acute effects of administered porcine calcitonin in man. *Clin Sci* 1969; 37:181–190.
90. Smith HW: *The Kidney, Structure and Function in Health and Disease*. New York, Oxford University Press, 1955.
91. Smith HW, Goldring W, Chasis H, et al: The application of saturation methods to the study of glomerular and tubular function in the human kidney. *J. Mt Sinai Hosp NY* 1943; 10:59–108.
92. Sorensen LB: Extrarenal disposal of uric acid, in Kelley WN, Weiner IM (eds): *Uric Acid, Handbook of Experimental Pharmacology, vol 51*, Berlin, Springer-Verlag, 1978, pp 325–336.
93. Stanbury SW: Some aspects of disordered renal tubular function. *Adv Intern Med* 1958; 9:231–282.
94. Steele TH, Oppenheimer S: Factors affecting urate excretion following diuretic administration in man. *Am J Med* 1969; 47:564–574.
95. Stolte H, Hare D, Boylan JW: D-glucose and fluid reabsorption in proximal surface tubule of the rat kidney. *Arch Ges Physiol* 1972; 334:193–206.
96. Straus W: Occurrence of phagosomes and phagolysosomes on different segments of the nephron in relation to the reabsorption, transport, digestion, and extrusion of intravenously injected horseradish peroxidase. *J Cell Biol* 1964; 21:295–308.
97. Strickler JC, Thompson DD, Klose RM, Giebisch G: Micropuncture study in inorganic phosphate excretion in the rat. *J Clin Invest* 1965; 43:1596–1607.
98. Sulamita B, Pitts RF: Renal handling of α-ketoglutarate by the dog. *Am J Physiol* 1964; 207:483–494.
99. Talmadge RV, Belanger LF (eds): *Parathyroid Hor-*

mone and Thyrocalcitonin (Calcitonin). New York, Excerpta Medica Foundation, 1968.
100. Thierry J, Poujeol P, Ripoche P: Interaction between Na$^+$-dependent uptake of D-glucose, phosphate, L-alanine in rat renal brush border membrane vesicles. *Biochim Biophys Acta* 1981; 647:203–210.
101. Tune BM, Burg MB: Glucose transport by proximal renal tubules. *Am J Physiol* 1971; 221:580–585.
102. Tune BM, Burg MB, Patlak CS: Characteristics of *p*-aminohippurate transport in proximal renal tubules. *Am J Physiol* 1969; 217:1057–1063.
103. Ullrich KJ: Sugar, Amino acid, and Na$^+$ cotransport in the proximal tubule. *Ann Rev Physiol* 1979; 41:181–196.
104. Ullrich KJ, Rumrich G, Kloss S: Sodium dependence of the amino acid transport in the proximal convolution of the rat kidney. *Pfluegers Arch* 1974; 351:49–60.
105. Vander AJ, Cafruny EJ: Stop flow analysis of renal function in the monkey. *Am J Physiol* 1962; 202:1105–1108.
106. Van Liew JB, Deetjen P, Boylan JW: Glucose reabsorption in the rat kidney—dependence on glomerular filtration. *Arch Ges Physiol* 1967; 295:232–244.
107. Walker A, Bott P, Oliver J, MacDowell M: The collection and analysis of fluid from single nephrons of the mammalian kidney. *Am J Physiol* 1941; 134:580–595.
108. Webber WA: Characteristics of acidic amino acid transport in mammalian kidney. *Can J Biochem Physiol* 1963; 41:131–137.
109. Webber WA: Interactions of neutral and acidic amino acids in renal tubular transport. *Am J Physiol* 1962; 202:577–583.
110. Weiner IM: Urate transport in the nephron. *Am J Physiol* 1979; 237:F85–F92.
111. Weinman EJ, Sansom SC, Steplock DA, et al: Secretion of urate in the proximal convoluted tubule of the rat. *Am J Physiol* 1980; 239:F383–F387.
112. Wright LD, Russo HF, Skeggs HR, et al: The renal clearance of essential amino acids: arginine, histidine, lysine and methionine. *Am J Physiol* 1947; 149:130–134.
113. Wyngaarden JB: On the dual pathogenesis of hyperuricemia in primary gout. *Arthritis Rheum* 1960; 3:414–420.
114. Young JA, Freedman BS: Renal tubular transport of amino acids. *Clin Chem* 1971; 17:245–266.

11

Regulation of Acid-Base Balance

In a healthy man, the arterial plasma concentration of hydrogen ions, $[H^+]_p$, expressed in pH units (pH = $-\log [H^+]$) is alkaline and ranges from 7.35 to 7.45, but may range from 7.0 to 7.8 in pathological states.

The alkalinity of blood is continuously threatened by endogenouus influx of acids produced from the metabolism of the normal average diet. The acids produced are in two forms: (1) *Volatile acid* in the form of carbon dioxide produced (about 200 mM/kg of body weight) from the oxidative metabolism of the foodstuffs, which on hydration yields the weak carbonic acid (H_2CO_3):

$$CO_2 + H_2O \rightleftarrows H_2CO_3 \rightleftarrows H^+ + HCO_3^- \quad (11.1)$$

The carbon dioxide thus produced is eliminated exclusively by the pulmonary system. (2) *Nonvolatile* or *fixed acid* produced from catabolism of the proteins in the diet. This yields an additional 1 mM of acid per kilogram of body weight in adults and slightly higher amounts in infants and young children. For individuals on vegetarian diet, the nonvolatile component is mainly alkali.

Besides the diet, the net influx of nonvolatile acid may increase in certain physiological and pathological conditions. Examples include lactic acid production in exercise and acetoacetic acid and β-hydroxybutyric acid production during uncontrolled diabetes mellitus.

If the nonvolatile acid is allowed to remain in the body, it clearly poses a serious threat to all normal life processes. Therefore, to prevent such a disaster, the nonvolatile acid must be neutralized and then excreted in suitable forms by the renal system.

It is clear that the normal alkalinity of the body fluids must be defended against the continuous daily influx of acid. This task of maintaining the blood pH within limits compatible with life is accomplished by the integrated actions of three important systems: (1) The *blood buffer* system, whose response is quick but can only make partial adjustment; (2) the *pulmonary system*, whose response is similarly quick but can make partial correction; and (3) the *renal system*, whose response is slow but completely corrects deviations in blood pH.

To better understand the mechanisms of renal regulation of blood pH, we begin this chapter with an overview of the functions of the blood buffers and the pulmonary system.

Physiological Chemistry of Blood Buffers

Regulation of normal blood pH involves both *passive* (or chemical) buffering, such as occurs in vitro, and *active* (or physiological) buffering, such as occurs in vivo, in which pulmonary and renal systems manipulate the plasma concentrations of the components of some passive blood buffers.

A chemical buffer consists of a mixture of a weak acid (or alkali) and its salt (Equation 11.1) and has the property of resisting changes in pH that would otherwise result from adding acid or alkali. This property may be defined quantitatively by plotting the amount of acid or alkali added against the pH. Such a plot yields a sigmoidal curve and is called the titration curve of the buffer.

Another way of defining this property is to apply the mass-action law to such a buffer mixture, yielding the well-known Henderson-Hasselbalch equilibrium equation:

$$pH = pK + \log \frac{[salt]}{[acid]} \quad (11.2)$$

where [salt] and [acid] are the concentrations of salt and acid components of the buffer mixture, pH and pK are the negative logs of the $[H^+]$ and the dissociation constant (K) of the weak acid, respectively. Accordingly, the strength of the buffer, i.e., its *buffering capacity*, depends on (a) the concentration of the buffer (the sum of the acid and salt concentrations) and (b) the nearness of the pH to the pK of the buffer. At equal salt and acid concentrations, the ratio term in Equation 11.2 becomes unity and its log equal to zero, in which case the pH becomes equal to the pK. At this pH, the buffering capacity is maximal, but because of the sigmoidal nature of the titration curve, two thirds of this capacity is lost by moving the pH one unit to either side of the pK.

Whole blood is a *heterogeneous* mixture consisting of plasma and cell phases. Its major buffers are inorganic phosphate, bicarbonate, and plasma proteins, all residing in the plasma phase, and the hemoglobin, residing in the red cell phase.

Insofar as *passive buffering* is concerned, the inorganic phosphate with a plasma concentration of 2 mM/L and a pK of 6.8 is very poor. The bicarbonate buffer, however, with a plasma concentration of 25 mM/L and a pK of 6.1, is a much stronger passive buffer. In contrast, hemoglobin and plasma proteins are by far the strongest passive buffers of the blood. These polyvalent buffers provide a total concentration of some 50 mM/L (there are normally 15.8 g/dL of hemoglobin and 3.5 g/dL of plasma proteins in the blood) and their pK's are so distributed as to yield a uniform maximal buffering capacity over the full pathological pH range.

When it comes to *active buffering*, however, the important blood buffers are inorganic phosphate and bicarbonate. The reason is that selective renal excretion of the inorganic phosphate provides a means of adjusting its plasma concentration by the renal system. The bicarbonate buffer (Equation 11.1) is even more important, because the concentration of its volatile acid component (CO_2) is manipulated by the pulmonary system, while the concentration of its salt component (HCO_3^-) is manipulated by the renal system.

Despite these differences in buffering capacity, all blood buffers exist in equilibrium with the same plasma $[H^+]$. Therefore, the ratio term in Equation 11.2 for one buffer determines the ratio term for all other buffers. This phenomenon is called the *isohydric principle* and may be expressed as

$$[H^+] = \frac{[HA]_1}{[A^-]_1} K_1 = \frac{[HA]_2}{[A^-]_2} K_2 \ldots \quad (11.3)$$

where [HA] and $[A^-]$ are the acid and salt concentrations of the buffer pair, K is its dissociation constant, and subscripts 1 and 2 refer to the buffer species.

Because of the isohydric principle, a change in the buffering capacity of the entire blood buffer system is reflected by a change in the buffering capacity of one buffer pair. Hence, to understand the response of the blood buffer system to acid-base disturbances, it is sufficient to examine the behavior of only one of the blood buffer pairs. For reasons mentioned earlier, of the four major blood buffers, the *bicarbonate buffer* serves as the best candidate. Therefore, disturbances in acid-base balance can be ascertained from studying the behavior of the bicarbonate buffer and the regulation of its two components by the pulmonary and renal systems.

Carbon Dioxide Absorption Curve of True Plasma

The pulmonary and renal systems monitor the pCO_2 and $[HCO_3^-]$ of the true plasma, not the separated plasma. *True plasma* is plasma that includes the buffering effect of the all-important Hb buffer. *Separated plasma* is plasma that excludes this effect. In acid-base regulation, the important differences between the two types of plasma are in the behavior of their carbon dioxide absorption curves. For true plasma, this curve is obtained as follows. Whole blood is exposed at 37 °C to gas mixtures containing various known tensions of carbon dioxide and sufficiently high O_2 tension to completely saturate the hemoglobin with oxygen. After equilibration, the blood is centrifuged anaerobically and the total concentration of carbon dioxide in the plasma and the volume of carbon dioxide in the gas phase are determined. The latter is converted to pCO_2 and concentration units and is then subtracted from the total carbon dioxide in the plasma, to yield $[HCO_3^-]$ in the plasma. The plot of $[HCO_3^-]$ thus obtained at various carbon dioxide tensions against the pCO_2 of the gas mixture to which the blood was exposed yields the carbon dioxide absorption curve for the true plasma shown in Figure 11.1. In contrast, if we had separated the plasma first and then exposed it to various carbon

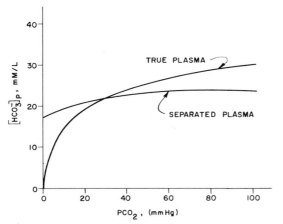

FIGURE 11.1. Carbon dioxide absorption curve for true and separated plasma. (From Koushanpour.[15])

dioxide tensions, we would get the other curve shown in Figure 11.1. Note that the carbon dioxide absorption curve for the separated plasma is much flatter and does not go through the origin, because the buffering effect of hemoglobin is now lacking. Note also that the steeper the carbon dioxide absorption curve, the more closely it follows the iso-pH line (see Figure 11.4 and Equation 11.7), and hence the better is the buffering capacity.

It is clear that to accurately describe the behavior of the whole blood buffers in acid-base disturbances from the bicarbonate buffer system, we must use the carbon dioxide absorption curve of the true plasma.

The necessity of dealing with the bicarbonate buffer system in true plasma has an important advantage, for the Henderson-Hasselbalch equation can be applied to the homogeneous plasma phase or the homogeneous red cell phase, but not to the heterogeneous whole blood. Applying the Henderson-Hasselbalch equation to the bicarbonate buffer of the true plasma yields

$$\text{pH} = 6.1 + \log \frac{[HCO_3^-]_p}{[H_2CO_3]_p} \quad (11.4)$$

where the subscript refers to the plasma. Since in the plasma the equilibrium for the formation of H_2CO_3 from carbon dioxide and water (Equation 11.1) is far to the left, then the $[H_2CO_3]_p$ term in Equation 11.4 is proportional to the pCO_2 and is defined by Henry's law of gas solubility:

$$[H_2CO_3]_p = \alpha \, pCO_2 \quad (11.5)$$

where α is the CO_2 solubility coefficient and has a

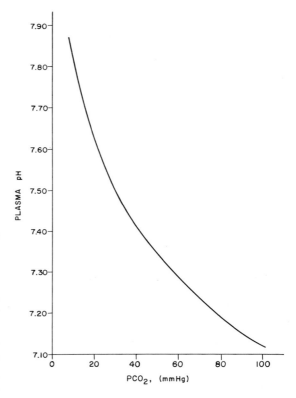

FIGURE 11.2. Relationship between plasma pH and pCO_2. (From Koushanpour.[15])

value of 0.03 mM/L per mm Hg. Substituting Equation 11.5 into 11.4 yields

$$\text{pH} = 6.1 + \log \frac{[HCO_3^-]_p}{0.03 \, pCO_2} \quad (11.6)$$

Note that the $[HCO_3^-]_p$ term is equal to the difference between the total carbon dioxide content of the plasma and that which is physically dissolved (0.03 pCO_2).

Mathematically, as defined by Equation 11.6, and physiologically, pH is a dependent variable; its changes are determined by variations in pCO_2, which is the primary independent variable (direct input), and $[HCO_3^-]_p$, which is the secondary independent variable (indirect input). With this in mind, there are only three ways of graphically relating the three variables in this equation.

Figure 11.2 shows the plot of pH as a function of pCO_2. The data used to construct this and the other two plots (Figures 11.3 and 11.4) are listed in Table 11.1. Note that as pCO_2 rises, the pH falls in clearly curvilinear fashion. Despite the fact that this plot

TABLE 11.1. Carbon dioxide absorption of normal oxygenated true plasma.

pCO_2 (mm Hg)	$[HCO_3^-]_p$ (mM/L)	pH
10	15.2	7.82
20	19.4	7.62
30	22.0	7.50
40	24.0	7.41
50	25.6	7.36
60	26.9	7.29
70	28.0	7.24
80	28.9	7.19
90	29.8	7.15
100	30.4	7.12

Data from Henderson.[13]

represents the direct input-output relationship for the bicarbonate buffer system, it is not used.

Another way of relating these variables is by plotting $[HCO_3^-]_p$ as a function of pH. This plot, called the *bicarbonate-pH diagram*, was first introduced by Davenport.[8] As shown in Figure 11.3, it is linear, and the line is called the normal buffer line. The curvilinear line shown in this figure is called the $pCO_2 = 40$ mm Hg isobar, meaning that the carbon dioxide tension is the same anywhere along the curve. The data for plotting the pCO_2 isobar is obtained by first solving the Henderson-Hasselbalch equation for $[HCO_3^-]_p$:

$$[HCO_3^-]_p = 0.03 \, pCO_2 \times 10^{(pH - 6.1)} \quad (11.7)$$

and then solving the equation for a given value of pCO_2 (such as $pCO_2 = 40$ mm Hg) and different values for pH ranging from 7.0 to 7.8. Note that for a $pCO_2 < 40$ mm Hg, the pCO_2 line will be shifted to the right, whereas for a $pCO_2 > 40$ mm Hg, it will be shifted to the left.

A third way of plotting these variables is to plot $[HCO_3^-]_p$ as a function of pCO_2. Such a plot, first introduced by Clark,[6] is called the *carbon dioxide absorption curve*. It is denoted as the "normal respiratory pathway" in Figure 11.4. The pH corresponding to any given pair of values for $[HCO_3^-]_p$ and pCO_2, as defined by Equation 11.6, is depicted by an iso-pH line radiating from the origin. Thus, the iso-pH = 7.41 line is defined by a line going from the origin through the normal point (N) defined here by $[HCO_3^-]_p = 25$ mM/L and $pCO_2 = 40$ mm Hg. Note that when pH > 7.41, the iso-pH line will lie to the left, whereas for pH < 7.41 the iso-pH line will lie to the right of the normal iso-pH = 7.41 line.

The most striking feature of the carbon dioxide absorption plot is that it allows for complete visualization of the three variables of the plasma bicarbonate buffer system in one operation. Consequently, in this chapter we shall use this plot to discuss acid-base disturbances rather than the bicarbonate-pH diagram, although the latter is equally useful.

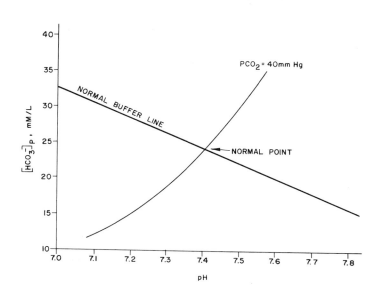

FIGURE 11.3. The bicarbonate-pH diagram of true plasma, with $pCO_2 = 40$ mm Hg isobar. (Redrawn from Davenport.[8])

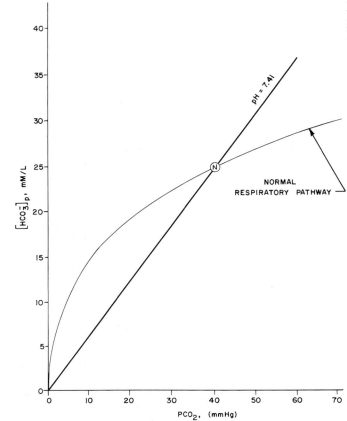

FIGURE 11.4. The bicarbonate-pCO$_2$ plot of true plasma, with radiating iso-pH = 7.41 line. (From Koushanpour.[15])

A closer examination of the carbon dioxide absorption curve reveals that as pCO$_2$ increases, such as may occur in hypoventilation, [HCO$_3^-$]$_p$ also increases, while the pH decreases. The blood reactions involved are shown in Figure 11.5. The rise in [HCO$_3^-$]$_p$ is due to neutralization of most of the added carbonic acid by the polyvalent proteinate, Pr$^-$ (plasma protein and oxyhemoglobin), whose concentration decreases, but leaves the sum of the concentrations of bicarbonate and proteinate buffers unchanged. Furthermore, as the added carbon dioxide reduces the pH, the labile cations, B$^+$ (mainly Na$^+$ and K$^+$), formerly associated with the proteinate anions, are released to become associated with the bicarbonate ions. The fall in pH is buffered, however, because the protein acids (HPr) that are formed ionize much less than the added carbonic acid. A precisely analogous but opposite sequence of reac-

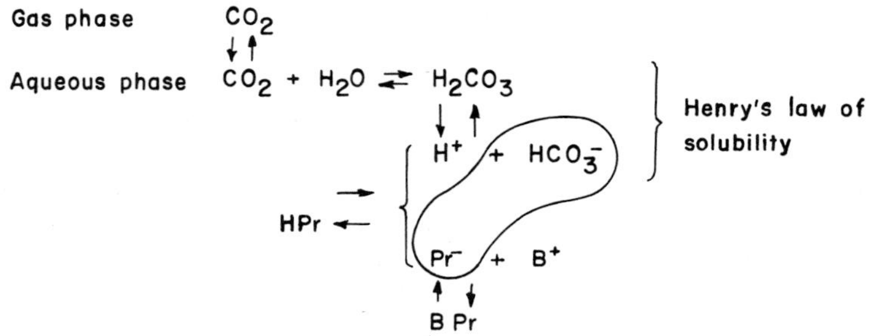

FIGURE 11.5. Reactions illustrating solubility of carbon dioxide in blood. (Redrawn from Muntwyler.[22])

tions occurs in hyperventilation as the blood moves to the left of the normal point along the normal respiratory pathway.

Standard Bicarbonate Content, $[HCO_3^-]_{40}$

Influx of fixed acids and alkalis into the blood causes disturbances in acid-base balance. How can this best be detected and measured? Obviously, determination of $[HCO_3^-]_p$ is not satisfactory because it is influenced as much by changes in pCO_2 as by influx of acid or alkali. The best way is to determine the level of the carbon dioxide absorption curve by measuring $[HCO_3^-]_p$ at some fixed, or standard, pCO_2. The standard pCO_2 usually used is 40 mm Hg, thereby eliminating the pulmonary compensation for acid-base disturbance. In practice, the whole blood sample is equilibrated at a pCO_2 of 40 mm Hg and a pO_2 of 150 mm Hg (to assure saturation of hemoglobin with O_2) at a temperature of 37 °C. The plasma bicarbonate, measured by subtracting dissolved CO_2 (= 0.03 pCO_2) from total carbon dioxide content, is called the *standard bicarbonate content* of true plasma[2] and is symbolized as $[HCO_3^-]_{40}$. This is also a measure of the carbon dioxide content of the blood[34] or of the body stores of alkali—or, as conventionally called, the *alkaline reserve* for neutralizing fixed acids. Thus, a decrease in $[HCO_3^-]_{40}$ indicates an increase in the influx of fixed acids, whereas an increase in $[HCO_3^-]_{40}$ indicates an increase in the influx of fixed alkali. These changes are reflected in downward and upward displacement of the carbon dioxide absorption curve, respectively. Just as the arterial pCO_2 is regulated by the pulmonary system, the $[HCO_3^-]_{40}$ is regulated by the renal system.

Clinically, the standard bicarbonate is not directly measured, but its deviation from normal, called *base excess* (BE) or *base deficit* (BD), is routinely approximated from measurement of arterial pCO_2 and pH and total hemoglobin. The validity of this estimation depends largely on the care with which the arterial blood is drawn and how quickly it is analyzed.[31,32] The analyses of pCO_2 and pH are usually carried out on a blood gas analyzer, and of hemoglobin on a co-oximeter. These values are then used to calculate, for example, base excess (or deficit) using the following equation.[33]

$$\text{Base Excess} = (1 - 0.0143\,\text{Hb})\{([HCO_3^-]_p - (9.5 + 1.63\,\text{Hb})(7.4 - pH_p)) - 24\} \quad (11.8)$$

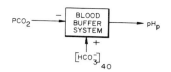

FIGURE 11.6. Scheme illustrating the elements of the blood buffer system. (From Koushanpour.[15])

where 0.0143 is the number of milliliters of oxygen per gram of hemoglobin in 100 mL of blood, Hb is the number of grams of hemoglobin in 100 mL of blood, $(9.5 + 1.63\,\text{Hb})$ is the correction factor for absorption of CO_2 by Hb, and 7.4 and 24 are the assumed normal values for pH_p and $[HCO_3^-]_p$, which have already been defined. Note that the value of $[HCO_3^-]_p$ used in Equation 11.8 is calculated from pCO_2 and pH using the rearranged Henderson-Hasselbalch equation (Equation 11.7).

Elements of Acid-Base Balance

The regulation of acid-base balance involves two physiological regulators, the pulmonary and renal systems, which share one component in common—the blood buffer system (Figure 11.6). The respiratory regulator controls arterial pCO_2, as the direct input to this component, whereas the renal regulator controls $[HCO_3^-]_{40}$, as the indirect input to the same component. Before detailing the mechanisms whereby these two regulators stabilize the true plasma pH, the output of the blood buffer system, we need to understand some basic terminology,[36] which is used to describe acid-base disturbances.

Acidosis and alkalosis are states characterized by patterns of abnormalities in pCO_2, $[HCO_3^-]_p$, $[HCO_3^-]_{40}$, and pH_p. The first two are always disturbed, but the last two sometimes are and sometimes are not. It is, therefore, well to remember that acidemia (low "plasma" pH), such as may occur in hypercapnia (high pCO_2), and alkalemia (high "plasma" pH), such as may occur in hypocapnia (low pCO_2), refer only to one element in the patterns and are not synonymous with acidosis and alkalosis.

Every disturbance in acid-base balance has an initial cause, on the basis of which they are classified as *respiratory* or *metabolic*. Respiratory disturbances result from some initial forcing of the respiratory regulator that alters pCO_2, the direct input to the blood buffer system. Metabolic disturbances result from some initial forcing of the renal regulator (gain of fixed acid or alkali by the body) that alters $[HCO_3^-]_{40}$, the indirect input to the blood buffer system. Thus, the block diagram in Figure 11.6 neatly distinguishes the two classic causes of acid-base disturbances. In this block diagram, a minus sign near pCO_2 is a reminder that a step increase in this quantity causes a decrease in pH. Likewise, a plus sign near $[HCO_3^-]_{40}$ is a reminder that a step increase in this quantity causes an increase in pH. Further details of these relationships is discussed later.

Disturbances in acid-base balance may be uncompensated (steady-state error in pH) or compensated to various degrees (some or no steady-state error in pH). The respiratory regulator compensates for metabolic disturbances, and the renal regulator compensates for respiratory disturbances. Each regulator will attempt to compensate for the effects of initial causes impinging on its own system, but because of proportional control (see Appendix A for further details and definitions), a steady-state error in pH will result. (It should be noted that this within-compensation, though it is tacitly recognized, is not called compensation.) The error in pH will then elicit a second compensation from the other regulator. Compensations for disturbances in acid-base balance refer only to this cross-compensation from the other regulator and not to the within-regulator compensation.

Referring to Figure 11.4, uncompensated respiratory disturbances are portrayed by a displacement along the normal carbon dioxide absorption curve on the "normal respiratory pathway." Respiratory acidosis causes a displacement to the right, whereas respiratory alkalosis causes a displacement to the left, with no change in $[HCO_3^-]_{40}$. On the other hand, metabolic disturbances are portrayed by a shift in the carbon dioxide absorption curve, determined by a change in $[HCO_3^-]_{40}$. Thus, a fall in $[HCO_3^-]_{40}$, as in metabolic acidosis, causes a downward shift, whereas a rise in $[HCO_3^-]_{40}$, as in metabolic alkalosis, causes an upward shift in the normal carbon dioxide absorption curve.

Let us now examine closely how pulmonary and renal regulators compensate for disturbances in acid-base balance.

An Overview of the Pulmonary pH Regulator

Figure 11.7 shows an orienting functional diagram of the elements of the pulmonary pH regulator. For ease of presentation, it is divided into three components: the respiratory centers, the lungs and thorax, and the pulmonary gas exchanger. Briefly, the respiratory centers, located in the medulla oblongata and pons, monitor (designated by a small *box* on some *arrows*) at least six signals (shown by the *arrows* impinging on the left). As before, the *plus* or *minus* signs on any *arrow* indicate the directional effect of that forcing on the output of the appropriate box. The algebraic sum of these signals (neural integration) produces a net change in motor nerve impulses issued to the diaphragm and the thoracic muscles which control pulmonary ventilation ($\dot{V}_e$). As shown, the frequency of motor nerve impulses is a positive function of the sensitivity of the respiratory centers (K_c) to the given input signals. Likewise, the magnitude of pulmonary ventilation for a given motor nerve impulse is a negative function of the mechanical impedance (Z) of the lungs and thorax. Because ventilation is a cyclic process, the mechanical impedance is a function of airflow resistance (R), breathing frequency (f), and lung compliance (C):

$$Z = \sqrt{R^2 + \frac{1}{(\pi f C)^2}} \qquad (11.9)$$

The third component of the pulmonary pH regulator is the pulmonary gas exchanger. For clarity of presentation, it is divided into two phases: an O_2 exchanger and a CO_2 exchanger. The input to both exchangers is the ventilation equivalent ($\dot{V}_eE$), defined as the ratio of the pulmonary ventilation to the metabolic rate (MR). The outputs of the pulmonary gas exchanger are the partial pressures of O_2 and CO_2 in the arterial blood (p_aO_2, and p_aCO_2, respectively). The partial pressures of O_2 and CO_2 in the inspired air (p_iO_2 and p_iCO_2, respectively) represent the indirect inputs or properties of the exchanger. Note that for each component, the indirect input (property) determines its gain (= output/input) (see Appendix A for further details and definitions).

The other two systems shown in Figure 11.7, though not an integral part of the pulmonary pH regulator, are included for the sake of completeness and subsequent integration of this regulator with the renal pH regulator.

The lower left-hand box represents the metabolic system (muscle, etc.) with metabolic rate and lactate as its outputs, and exercise as its input. The triangle on the *arrow* labeled "lactate" indicates that its production must exceed a critical threshold level before it can effectively alter $[HCO_3^-]_{40}$. The final element

An Overview of the Pulmonary pH Regulator

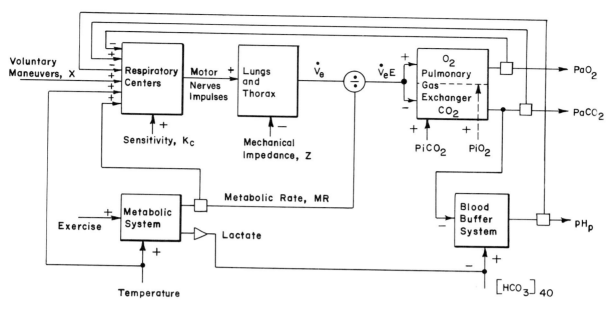

FIGURE 11.7. Components of the pulmonary pH regulator. (From Koushanpour.[15])

is the blood buffer system, shown by the box in the lower right-hand side, which we have already described.

Forcings and Responses of the Pulmonary pH Regulator

The steady-state responses of the pulmonary pH regulator to some selected forcings are summarized in Table 11.2. For each forcing, the deviation from normal is indicated by (+) for increase, (−) for decrease, and (0) for no change. Guided by the block diagram in Figure 11.7, verify the entries in this table by first considering the effect of each forcing at the level of the appropriate component and then follow the effect through the various feedback loops throughout the entire system. To illustrate, let us see how the pulmonary pH regulator cross-compensates for metabolic acidosis and alkalosis, the subject of present concern.

Metabolic Acidosis

Suppose a patient developed metabolic acidosis (as in diabetes mellitus or renal insufficiency) sufficient to lower his $[HCO_3^-]_{40}$ to 19 mM/L. If his ventilation remained unchanged, his arterial pCO_2 will remain constant, in which case the decrease in $[HCO_3^-]_{40}$ acting at the blood buffer system will lower the pH to 7.28. Consequently, as shown in Figure 11.8, the patient moves down the $pCO_2 = 40$ mm Hg isobar line to point A. But a fall in pH, acting through its feedback loop, stimulates ventilation via activation of the peripheral and central chemoreceptors, so that both $\dot{V}_e$ and $\dot{V}_eE$ rise. The increase in $\dot{V}_eE$ (induced by hyperventilation), acting at the exchanger, elevates p_aO_2 and lowers p_aCO_2. The latter effect causes the patient to move to the left of point A toward point B along a new carbon dioxide absorption curve, thereby raising the pH back toward normal. This induced respiratory alkalosis constitutes the cross-compensation of the pulmonary pH regulator for metabolic acidosis.

How far left does the patient move along the new carbon dioxide absorption curve? Obviously not so far as complete pH compensation, for the lowered p_aCO_2, acting through its feedback loop, partially inhibits pH stimulation of ventilaton. Thus, the patient's blood pattern settles down somewhere short of complete pH compensation, say at point B. Because the respiratory response to the metabolic acidosis is immediate, the patient's blood pattern does not follow the "dog-leg" pathway N → A → B, in separate steps. Actually, it follows the continuous pathway N → B shown on the "normal metabolic pathway."

Metabolic Alkalosis

Exactly analogous, but opposite, changes occur if we induce a metabolic alkalosis sufficient to raise

TABLE 11.2. Steady-state responses of the pulmonary regulator to some selected forcings.

Disturbance Forcings	Maximum ventilatory gain (L/min)	Responses as deviation from normal						
		MR	Motor nerve impulses	$\dot{V}_e$	$\dot{V}_eE$	Arterial pO$_2$	Arterial pCO$_2$	Arterial pH
Normal resting man		0.272		6.85	2.5	95	40	7.41
CO$_2$ inhalation: Increased level of P$_I$CO$_2$	60	0	+	+	+	+	+	−
Altitude anoxia: Decreased level of P$_I$O$_2$	15	0	+	+	+	−	−	+
Metabolic acidosis: Decreased level of [HCO$_3^-$]$_{40}$	40	0	+	+	+	+	−	−
Metabolic aklalosis: Increased level of [HCO$_3^-$]$_{40}$	−	0	−	−	−	−	+	+
Pulmonary impedance: Increased Z	−	0	+	−	−	−	+	−
Respiratory depression: Decreased K$_c$	−	0	−	−	−	−	+	−
Voluntary hyperventilation: Increased X	170	0	+	+	+	+	−	+
Fever: Increased temperature	20	+	+	+ +	+	+	−	+
Moderate exercise: Increased MR	60	+	+	+	0	0	0	0
Severe exercise: Increased MR, Decreased level of [HCO$_3^-$]$_{40}$	120	+ +	+ + +	+ + +	+	+	−	−

From Koushanpour.[15]
MR = metabolic rate.

[HCO$_3^-$]$_{40}$ to 30 mM/L (point C, Figure 11.8). The patient hypoventilates (respiratory acidosis) and moves to the right of point C toward point D along a new carbon dioxide absorption curve, thereby lowering the pH toward normal. However, as before, the compensation is incomplete, and the patient's blood pattern settles down at point D. Again, the actual pathway that his blood follows is from point N to point D on the normal metabolic pathway, which represents the partially cross-compensated metabolic pathway.

It is clear that the pulmonary regulator provides an immediate, though incomplete, pH compensation in metabolic disturbances. The complete pH compensation is provided by adjustment of [HCO3$^-$]$_{40}$ by the renal pH regulator, which is described next.

An Overview of the Renal pH Regulator

Disturbances in blood pH are prevented by continuous adjustment of the two inputs to the blood buffer system by the pulmonary and renal regulators. As described in the previous section, the pulmonary regulator manipulates pCO$_2$, which is one of the inputs determining pH (Figure 11.6). The other input is the standard bicarbonate concentration, [HCO$_3^-$]$_{40}$, which is manipulated by the renal pH regulator. In this section we shall examine the operation of this regulator, component by component, as well as the intrarenal mechanisms mediating its response to acid-base disturbances.

The Standard Bicarbonate Pool

Standard bicarbonate, [HCO$_3^-$]$_{40}$, was defined earlier as the bicarbonate concentration measured when blood is equilibrated in vitro with a pCO$_2$ of 40 mm Hg. Normally, its value is 25 mM/L.

Standard bicarbonate pool is determined by the influx or efflux of bicarbonate. Regardless of the route, two kinds of bicarbonate flux may be distinguished. Controlled fluxes consist of the renal excretion of acid or alkali, symbolized by $\dot{E}$. Acid excretion leads to accumulation of alkali, which will be neutralized by retained carbon dioxide, producing an influx of bicarbonate into the body pool. Excretion of alkali leads to acid accumulation, which displaces carbon dioxide from bicarbonate, resulting in an ef-

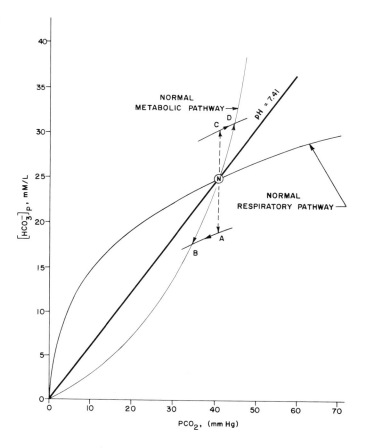

FIGURE 11.8. Pathways involved in pulmonary cross-compensation for metabolic disturbances. (From Koushanpour.[15])

FIGURE 11.9. Factors determining the net flux of bicarbonate $H\dot{C}O_3^-$ into the blood. (From Koushanpour.[15])

flux of bicarbonate from the body pool. The other kind of flux is a fortuitous, or accidental, flux, symbolized by $\dot{F}$, consisting of gains or losses of fixed acid or alkali by injection or via the GI tract or skin. The net bicarbonate flux ($H\dot{C}O_3^-$) is then the algebraic sum of these two fluxes, as illustrated in Figure 11.9.

The standard bicarbonate pool is the primary component in the renal pH regulator. Its functional input is the bicarbonate flux ($H\dot{C}O_3^-$), and its output is the standard bicarbonate content, $[HCO_3^-]_{40}$. An important property is time delay, emphasizing the slow renal adjustment of blood bicarbonate content. A second property is the fluid volume of the whole pool, which converts an increment of $H\dot{C}O_3^-$ to a concentration, but we shall leave this as a constant and ignore it. But note that in the steady state of this pool (when $[HCO_3^-]_{40}$ is unchanging), the input ($H\dot{C}O_3^-$), which is the algebraic sum of $\dot{F}$ and $\dot{E}$, must be zero! For the first time in this book we have encountered a component whose steady-state output can be determined only by following its changes during a transient. Its behavior is therefore governed by an integral equation (see Appendix A for further details and definitions), but fortunately, it is easier to verbalize than symbolize it, as illustrated schematically in Figure 11.10.

If HCO_3^- influx exceeds HCO_3^- efflux during the transient, the steady-state $[HCO_3^-]_{40}$ will have increased above its initial value, and vice versa. Note also that since $H\dot{C}O_3^-$ may be small compared with the large body pool, the transient tends to be long, even days or weeks. In the block diagram, the property "time delay" is a reminder of this long transient and of the fact that we accumulate $H\dot{C}O_3^-$ over time to obtain the steady-state output.

FIGURE 11.10. Factors determining standard bicarbonate concentration, $[HCO_3^-]_{40}$, in the blood. (From Koushanpour.[15])

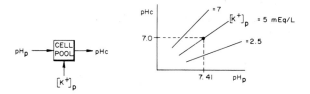

11.11. Factors affecting intracellular pH. (From Koushanpour.[15])

The Cellular Pool

We have just described the body standard bicarbonate pool as though it were homogeneous. For most situations this simplified description is adequate, for the extracellular and intracellular compartments of the whole pool behave alike. In certain situations, however, the two compartments behave oppositely, so that the plasma $[HCO_3^-]_{40}$ may rise, while that of body cells falls, and vice versa. The result is a *dissociated* metabolic disturbance in acid-base balance; an intracellular acidosis and extracellular alkalosis, or vice versa, may occur at the same time.

The key element in this process appears to consist of opposite transfers of H^+ and K^+ ions across the cell membrane between the extracellular and intracellular fluid compartments. If potassium salts are injected intravenously to produce a hyperkalemia, the K^+ ions enter body cells and H^+ ions leave them; this leaves an intracellular metabolic alkalosis and produces extracellular metabolic acidosis. Conversely, if body K^+ is depleted to produce hypokalemia, K^+ ions leave body cells and H^+ ions enter them; this leaves an extracellular alkalosis and produces an intracellular acidosis.[11]

We shall represent this dissociated behavior by introducing another component, called the cellular pool. Its input is pH_p, normally 7.41. Its output is the pH of body cells (pH_c); although not easily measurable, it is lower than that of plasma and we assign to it an arbitrary normal value of 7.0. Its property is the plasma potassium concentration, $[K^+]_p$, normally about 5 mEq/L. For lack of precise information we relate these variables in the simplest possible form, as illustrated in Figure 11.11. So long as $[K^+]_p$ is normal, the cellular and plasma pH simply rise and fall together. But, as noted in Chapter 9, a change in $[K^+]_p$ produces an inverse change in pH_p. Thus, an increase in $[K^+]_p$ will raise pH_c and lower pH_p, whereas a decrease in $[K^+]_p$ will produce the opposite effect. The possible mechanisms involved will be discussed later.

The Kidneys

The kidneys have the ability to selectively excrete acid or alkali in the urine, measured not in terms of urine pH, but as milliequivalents per day, which we have already symbolized as $\dot{E}$. To eliminate alkali, the kidneys excrete Na_2HPO_4 and $NaHCO_3$, both measured by titrating the urine with hydrochloric acid back to pH 7.41. To eliminate acid, they excrete NaH_2PO_4 and synthesize NH_3 (largely from glutamine) for excretion as NH_4Cl; the acid phosphate is measured by back titration with NaOH, but the NH_4^+ is determined separately and added.

The average diet has a somewhat acid residue, so that $\dot{E}$ is normally about 50 mEq/day. In disturbances of acid-base balance, however, $\dot{E}$ may be as low as -300 or as high as $+500$ mEq/day. (Note: A negative $\dot{E}$ means excretion of alkali in urine.) The sign and magnitude of $\dot{E}$ appear to be determined by the pH within the renal tubular cells, which we can only assume behaves like pH_c of the cells in general. Again, for lack of precise data, we assume a simple, negative, linear relationship for this behavior, as illustrated in Figure 11.12. Note that we have set the normal pH_c at 7.0 but specify a pH_c of 7.1 as the level at which the kidneys excrete a neutral urine ($\dot{E} = 0$). This allows the normal excretion of about 50 mEq of acid per day. The kidney property (K_k) determines the rate of acid or alkali excretion for a given pH error and is thus the controller sensitivity (see Appendix A for further details and definitions). This property may be reduced by renal disease, as

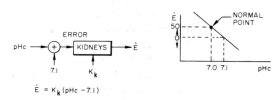

FIGURE 11.12. Factors determining net renal acid excretion. (From Koushanpour.[15])

in renal insufficiency, resulting in the accumulation of fixed acid, to yield one kind of metabolic acidosis.

Forcings and Responses of the Renal pH Regulator

Synthesis of the foregoing components of the renal pH regulator yields the block diagram shown in Figure 11.13. Note that one component, the blood buffer system, is shared by both the pulmonary and renal regulators, and that its output (pH_p) is a loop variable for both pH regulators. In studying the renal regulator in isolation, we can make pCO_2 an arbitrary constant, although we know it is a loop variable in the pulmonary pH regulator (Figure 11.7). Each regulator will cross-compensate for the other. Just as the pulmonary pH regulator cross-compensated for metabolic disturbances in acid-base balance, we shall find that the renal pH regulator also cross-compensates for respiratory disturbances in acid-base balance.

Excluding cross-compensation, all the disturbance forcings that initially impinge on the renal pH regulator produce metabolic disturbances in acid-base balance: (a) changes in fortuitous fluxes ($\dot{F}$) produce either metabolic acidosis or alkalosis, (b) reductions in K_k produce a metabolic acidosis, and (c) changes in $[K^+]_p$ produce dissociated metabolic disturbances. Since these forcings are so important, let us describe them more systematically:

Causes of Metabolic Alkalosis (Elevated $[HCO_3^-]_{40}$)

1. Gain of base from ingestion or injection of alkali, such as alkaline salts for treatment of peptic ulcer.
2. Loss of acid from excessive vomiting of gastric hydrochloric acid. If this persists, K^+ ion depletion may result.

Causes of Metabolic Acidosis (Reduced $[HCO_3^-]_{40}$)

1. Gain of acid from ingestion of acid salts (NH_4Cl, for example), or production of lactic acid in exercise, or of ketoacids in fasting or uncontrolled diabetes mellitus, or loss of renal ability to excrete acid (renal insufficiency).
2. Loss of base in alkaline body fluids, such as plasma lost from burned skin surface, or severe diarrhea. If the latter persists, K^+ ion depletion may be severe.

Causes of Dissociated Metabolic Disturbances

1. Hyperkalemia from ingestion or injection of K^+ salts produces extracellular acidosis and intracellular alkalosis.
2. Hypokalemia from K^+ ion loss (GI fluids contain higher $[K^+]$ than plasma), produces extracellular alkalosis and intracellular acidosis.

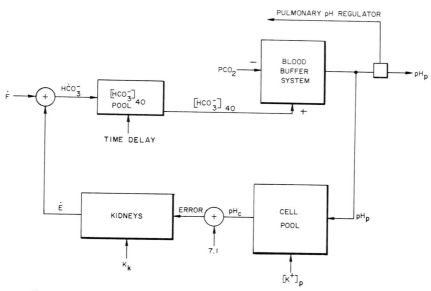

FIGURE 11.13. Synthesis of the renal pH regulator. (From Koushanpour.[15])

Still excluding cross-compensation, the renal regulator operates by proportional control, but with a new twist. A change in any of the three properties ($\dot{F}$, K_k, or $[K^+]_p$) will induce a rate of error development in pH; the error in pH will induce a compensatory rate of error correction. Since both are rates, the control is proportional and exhibits steady-state error. The new twist is that with no control ($\dot{E}$ is not adjustable), the pH error will grow endlessly, with no steady-state, but a fatal outcome.

Let us illustrate with the example of ingesting NH_4Cl to produce a metabolic acidosis; the NH_3 is converted to urea by the gut and liver, leaving an equivalent of hydrochloric acid. One dose of NH_4Cl amounts to a pulse forcing, which will disturb the system, but since the forcing does not persist, the pH will recover to its original normal value. To examine the steady-state response of NH_4Cl, we must use a step forcing, consisting of, say, three doses a day for as long as necessary to attain the steady state.

Guided by the block diagram, we start with a negative value for $\dot{F}$, the fortuitous influx of acid. The negative sign indicates that the fortuitous acid influx tends to reduce $[HCO_3^-]_p$ and hence $[HCO_3^-]_{40}$. This initiates a progressive fall in $[HCO_3^-]_{40}$, then pH_p, then pH_c. The latter initiates a progressive rise in renal excretion of acid, $+\dot{E}$, as a compensatory response. All these changes progress in magnitude until acid excretion matches acid ingestion; at this time the steady state is reached, though it may take a week or more. Now we find that $[HCO_3^-]_{40}$, pH_p, and pH_c are all low, as steady-state errors, for it is the pH error that maintains the compensatory renal excretion.

Why did it take so long to reach the steady state? Not because of the kidney, for it responds promptly to changes in its own pH. The lag is due to the large $[HCO_3^-]_{40}$ pool, acting as a large alkaline reserve; the fortuitous influx of acid, even though continuous, was small compared with the size of the pool. What would have happened in the absence of control, i.e., with the feedback loop opened and $\dot{E}$ unadjustable? The persisting influx of acid would produce an endlessly falling $[HCO_3^-]_{40}$ and pH to zero, or death. Renal regulation is clearly a vital function.

Precisely analogous but opposite changes occur in metabolic alkalosis. To ensure your understanding, trace through the block diagram in the same manner and analyze the renal response to such a disturbance.

Changes in $[K^+]_p$ produce dissociated metabolic acid-base disturbances. Guided by the block diagram, note that a rise in $[K^+]_p$, as by injection of K^+, will raise pH_c, lower acid excretion, or produce alkali excretion, thereby making $\dot{E}$ negative. Since $\dot{F}$, normally negative, is unchanged, their algebraic sum will be increasingly negative, causing $[HCO_3^-]_{40}$ and then pH_p to fall. In the steady state, there will be an extracellular metabolic acidosis, and an intracellular metabolic alkalosis.

Renal Cross-Compensation

We have just seen that all the forcings that impinge initially on the renal regulator induce metabolic disturbances in acid-base balance, despite within-compensation by the renal regulator. As noted earlier, in acid-base balance, this within-compensation is tacitly recognized, but is not called compensation; the latter term is reserved for what we here call cross-compensation between the two pH regulators. We learned that the pulmonary pH-regulator cross compensates for metabolic disturbances arising in the renal pH-regulator. The abnormal pH_p induces a feedback adjustment of ventilation which corrects the pH_p, but only partially because of proportional control. This is called partial compensation (actually, partial cross-compensation). Furthermore, we saw that in metabolic disturbances, even with cross-compensation added to within-compensation, the correction of pH was never complete; there was only partial compensation. In fact, if the respiratory cross-compensation were somehow complete, disaster would result. The restoration of a normal pH would completely prevent renal within-compensation, and this would induce endlessly progressive and therefore fatal changes.

Let us now examine the reverse situation. All the forcings that impinge initially on the pulmonary regulator induce respiratory disturbances in acid-base balance, despite compensation within the pulmonary regulator. Again, this within-compensation is tacitly recognized, though it is not called compensation, but the cross-compensation provided by the renal regulator is called compensation. The abnormal pH_p induces a feedback adjustment of renal excretion which gradually corrects pH_p. Although the kidney starts the altered excretion immediately, it is slow in altering $[HCO_3^-]_{40}$ and pH_p. Accordingly, the respiratory disturbance may remain for some hours without detectable cross-compensation in the blood, although it is detectable in the urine. During this interval it is said to be uncompensated. Then, as time passes, the respiratory disturbance becomes partially compensated and eventually completely compensated, leaving no steady-state error. The reason is that in response to pCO_2 forcing, the renal regulator operates by integral control. Let us see why.

Referring to the block diagram of the renal pH-regulator (Figure 11.13), a respiratory disturbance consists of an abnormal pCO_2, elevated in acidosis, for example, with a low pH_p. The latter will lower pH_c, which will cause the kidneys to excrete extra acid ($\dot{E}$ becomes more positive). Since $\dot{F}$ is unaltered and normally negative, $H\dot{C}O_3^-$ becomes positive and $[HCO_3^-]_{40}$ begins to rise slowly, thereby slowly correcting pH_p. We might say the kidneys are inducing a metabolic alkalosis to counteract the respiratory acidosis. The key feature is that as long as pH_p is low, $[HCO_3^-]_{40}$ and pH_p will continue to rise, and this process cannot stop until pH_p is completely corrected. At that time, $\dot{E}$ will have returned to normal and become equal and opposite to $\dot{F}$. This is integral control, because *the rate of error correction is proportional to the error.* Improvement must continue as long as any error remains.

We learned that complete respiratory cross-compensation for a metabolic disturbance would be disastrous, but the reverse is not true. Complete renal cross-compensation for a respiratory disturbance has beneficial effects. Not only is the pH restored, but other errors may undergo further correction. For example, restoration of pH in the respiratory alkalosis that occurs during residence at high altitudes removes the respiratory inhibition otherwise produced by alkalemia. As a result, ventilation undergoes an additional compensatory increase, further correcting the low pO_2. The pCO_2 will be further depressed, but this alone (without alkalemia) is very well tolerated.

Mechanisms of Renal pH Regulation

Having just described the elements of renal regulation of acid-base balance, we are now in a position to examine the mechanisms involved in such a regulation.

As described in Chapter 4, the primary function of the kidney is to conserve the major anions and cations of the body fluids, a consequence of which is acid-base regulation. To maintain the total quantities and concentrations of the major electrolytes within normal limits, the kidneys must (a) stabilize the standard bicarbonate pool by obligatory and controlled reabsorption of the filtered bicarbonate and (b) excrete a daily load of 50 to 100 mEq of fixed (nonvolatile) acids produced by metabolism. The fixed acids are excreted as titratable acids and ammonium salts. The capacity of the kidneys to perform these two functions is directly related to their ability to stabilize the intracellular and extracellular $[H^+]$ at normal levels. Let us now examine the intrarenal mechanisms involved for each of these functions.

Obligatory Bicarbonate Reabsorption in the Proximal Tubule

As depicted in Figure 9.10 (see Chapter 9), normally 80% to 90% of the filtered bicarbonate is reabsorbed along the proximal tubule. As decribed in Chapter 9, the key element of the reabsorption process, first proposed by Pitts and Alexander,[25] is the electroneutral *exchange of hydrogen ions for sodium ions* across the luminal border of the proximal cell membrane. As shown in Figure 11.14, Na^+ moves across the luminal membrane down its electrochemical gradient (*dashed arrow*) into the cell. To maintain electrical neutrality, H^+ is secreted actively (*circle* at the luminal membrane) from the cell into the tubular lumen.[1, 10] Once inside the tubular lumen, H^+ combines with the filtered HCO_3^- to form H_2CO_3. The carbonic acid thus formed is readily dissociated into carbon dioxide and water by the catalytic action of the enzyme *carbonic anhydrase*, located in the brush border of the luminal membrane.[9] The presence of membrane-bound carbonic anhydrase at this site serves to catalyze dehydration of carbonic acid formed in the lumen. Thus, the membrane-bound carbonic anhydrase acts to reduce the steady-state concentration of carbonic acid that might otherwise limit H^+ secretion, thereby enabling the proximal tubule to maximally reabsorb bicarbonate. The carbon dioxide thus formed in the lumen readily diffuses into the cell along its partial pressure gradient, and equilibrates with the carbon dioxide in the blood.

Because tubular H^+ is buffered by HCO_3^-, the entry of Na^+ into the cell must be electrically balanced by an anion. This is accomplished by either (a) passive diffusion of Cl^- down its electrochemical gradient, or (b) regeneration of HCO_3^- from cellular hydration of carbon dioxide to form carbonic acid in the presence of cytoplasmic carbonic anhydrase, with hydroxyl ion as its substrate.[20] The result of either reaction leads to a net reabsorption of $NaHCO_3$ or sodium chloride into the capillary blood. Note that Na^+ is reabsorbed by an active process from the

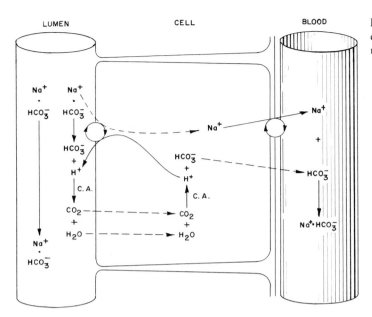

FIGURE 11.14. Mechanism of obligatory bicarbonate reabsorption in the proximal tubule. (From Koushanpour.[15])

cell into the blood (see Chapter 9 for mechanism). This is depicted by the *solid arrow* and the *circle* on the basolateral side of the renal cell in Figure 11.14.

Similarly K^+ is reabsorbed as $KHCO_3$ or potassium chloride into the capillary blood. However, it should be remembered (see Chapter 9) that in contrast to Na^+, the transluminal reabsorption of K^+ is active while transbasolateral transport is passive.

Controlled Bicarbonate Reabsorption in the Distal Tubule and Collecting Duct

The remaining 10% to 20% of the filtered bicarbonate is reabsorbed mainly in the "distal tubule" and collecting duct by a mechanism that involves the *exchange of sodium ions for potassium or hydrogen ions*. The essence of this exchange mechanism, first proposed by Berliner and associates,[4] is illustrated in Figure 11.15. As shown, these authors postulated a "competitively shared" secretory pathway for hydrogen and potassium in exchange for sodium. This is depicted by the *circle* on the luminal side of the renal cell. Furthermore, they assumed that the intracellular $[H^+]$ and $[K^+]$ were the limiting factors in determining the competition between the net rate of active and passive secretion of H^+ and K^+, respectively, at the luminal pump in exchange for Na^+. As mentioned earlier, the intracellular $[H^+]$ and $[K^+]$ are, in part, determined by the plasma concentrations of these ions.

Recent histochemical studies[9] have revealed a high degree of carbonic anhydrase activity in the intercalated cells in all mammalian nephrons. However, as the collecting ducts descend into the outer medulla the principal cells show an increase in carbonic anhydrase activity. There is no evidence of carbonic anhydrase activity in the papillary collecting ducts.

The absence of carbonic anhydrase activity in the distal convoluted tubule suggests that this nephron segment plays a minor role in bicarbonate reabsorption and acid-base regulation.[10] Available evidence strongly suggests that the collecting duct is a major site for bicarbonate reabsorption and urinary acidification. McKinney and Burg[21] showed that acetazolamide (a carbonic anhydrase inhibitor) completely inhibited bicarbonate reabsorption in the in-vitro perfused rabbit cortical collecting duct. This finding indicates the presence of an hydrogen ion-secreting mechanism in the cortical collecting duct similar to that described for the proximal tubule.

Except for the Na^+-H^+ and Na^+-K^+ exchange pump at the luminal border of the collecting duct cell membrane, the reabsorptive pathways for bicarbonate are analogous to those described for the proximal tubule. Thus, as shown in Figure 11.15, the secreted H^+ ion combines with the filtered HCO_3^- to form H_2CO_3. The latter dissociates into carbon dioxide and water. But unlike the proximal tubule, there is no luminal membrane-bound carbonic anhydrase involved in the dissociation reaction.[5,9] The

FIGURE 11.15. Mechanism of controlled bicarbonate reabsorption in the collecting duct. (From Koushanpour.[15])

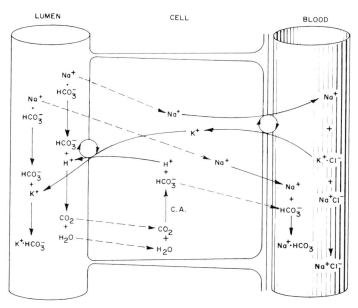

carbon dioxide thus generated in the lumen diffuses into the cell and equilibrates with carbon dioxide in the blood. In the cell, the generated carbon dioxide is hydrated under the influence of cytoplasmic carbonic anhydrase to yield HCO_3^- and H^+. The latter enters the secretory pathways, while the former is reabsorbed along with Na^+ into the blood. Also, as shown, the exchange of potassium for sodium leads to eventual reabsorption of 1 mole of sodium chloride for every mole of $KHCO_3$ excreted. As mentioned in Chapter 9, the mechanism of Na^+-K^+ exchange on the basolateral side is by an active process and is mediated by the Na^+-K^+ stimulated adenosine triphosphatase (ATPase) system (depicted by the *circle* in Figure 11.15). Present evidence (see Chapter 9) indicates that on the luminal side the Na^+-K^+ exchange occurs mainly in the principal cells and is influenced by blood levels of aldosterone, while potassium reabsorption is coupled to hydrogen ion secretion in the intercalated cells in the collecting duct, though this is not shown in Figure 11.15.

The Na^+-H^+ ion exchange mechanism is specialized to achieve different goals in the proximal tubule and collecting duct. In the proximal tubule, a large quantity of hydrogen ions is exchanged for sodium ions against a low cell-to-lumen hydrogen ion concentration gradient. The net result is the reabsorption of nearly three fourths of the filtered $NaHCO_3$. In the collecting duct, a small quantity of hydrogen ions is exchanged for sodium ions against a high lumen-to-cell hydrogen ion concentration gradient. The net result is the reabsorption of the $NaHCO_3$, which escaped reabsorption in the proximal tubule.

Factors Regulating Bicarbonate Reabsorption

Both obligatory and controlled bicarbonate reabsorption are influenced by at least five factors: (a) the "apparent" tubular reabsorptive capacity for bicarbonate, (b) the arterial pCO_2, (c) the plasma concentration of potassium, (d) the plasma concentration of chloride, and (e) the plasma levels of the adrenocortical hormones (glucocorticoids and mineralocorticoids). Let us examine the effects of each and their significance.

The "Apparent" Tubular Reabsorptive Capacity for Bicarbonate

Several earlier studies, using the previously described (see Chapter 10) overall clearance methods and renal titration curves to characterize bicarbonate transport, demonstrated that the normal kidney handles bicarbonate transport as though there were a maximum tubular transport rate for this substance (Figure 11.16). As shown, bicarbonate reabsorption increases linearly with increasing plasma concentration ($[HCO_3^-]_p$) until the latter reaches a value of about 25 mM/L, at which point the filtered load of the bicarbonate begins to exceed the tubular reabsorptive capacity. Thereafter, further increases in $[HCO_3^-]_p$ result in no further increase in tubular reab-

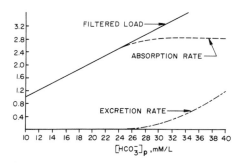

FIGURE 11.16. Renal reabsorption and excretion of bicarbonate as a function of the plasma concentration. (Redrawn from Pitts et al.[26] Reproduced from *The Journal of Clinical Investigation,* 1949, 28:35–44 by copyright permission of The American Society for Clinical Investigation.)

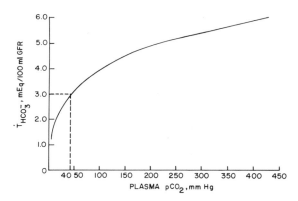

FIGURE 11.17. Relationship between net bicarbonate reabsorption, $\dot{T}_{HCO_3^-}$, and arterial pCO_2 in the dog. The curve approximates the experimental data obtained by Rector, et al.[29] Reproduced from *The Journal of Clincial Investigation,* 1960, 39:1706–1721 by copyright permission of The American Society for Clinical Investigation.

sorption, suggesting the existence of a tubular transport maximum ($\dot{T}_m$) for bicarbonate.

Subsequent studies, however, have questioned the existence of a "true" $\dot{T}_m$ for bicarbonate reabsorption. For example, Purkerson and associates[28] found that in the rat the so-called maximum tubular transport capacity was highly sensitive to changes in glomerular filtration rate (GFR) induced by expansion of the extracellular fluid volume, a procedure usually employed in $\dot{T}_m$ determinations. Since expansion of extracellular volume is known to depress proximal sodium reabsorption and to reset the glomerular-tubular balance (Chapter 9), the possibility that the so-called "normal" patterns of titration curves for bicarbonate reabsorption (Figure 11.16) may be an experimental artifact cannot be excluded. Similar conclusions have been reached by Slatopolsky and co-workers[35] and Kurtzman,[16] who studied bicarbonate reabsorption in man and dog, respectively. Kurtzman[16] found that expansion of the extracellular volume depressed bicarbanate reabsorption and that this depression was related not to changes in GFR or $[HCO_3^-]_p$, but rather to the increase in the fractional sodium excretion. Although expansion of the extracellular volume decreased plasma potassium concentration, the depression of bicarbonate reabsorption was not related to the decrease in plasma potassium.

These findings strongly suggest that the net bicarbonate reabsorption is a function of GFR and hence of the degree of extracellular fluid (ECF) volume expansion: the reabsorption rate, and hence $\dot{T}$, being high when ECF volume expansion is minimal, and low when ECF volume expansion is maximal,

suggesting an *"apparent"* and not a *"true"* $\dot{T}_m$ for renal transport of bicarbonate.

Bennett and associates[3] reinvestigated the renal titration curves for bicarbonate and more specifically the effect of variations in GFR on its tubular reabsorption. In both hydropenic and ECF volume-expanded dogs, they found a close functional relationship between the absolute rate of bicarbonate reabsorption and GFR at any plasma bicarbonate concentration. This tends to support the existence of an "apparent" but not a "true" $\dot{T}_m$ for tubular reabsorption of bicarbonate. They further suggested that the apparent reciprocal relationship between bicarbonate $\dot{T}_m$ and ECF volume expansion observed previously and cited above might be due to an imperfect glomerulo-tubular balance.

The Arterial pCO_2

A number of studies[7, 29, 30] have demonstrated that the apparent $\dot{T}_m$ for the reabsorption of bicarbonate is very sensitive to changes in the arterial pCO_2. Thus, an increase in the arterial pCO_2 increases the apparent $\dot{T}_m$ and, by the mechanisms depicted in Figure 11.14, increases the cellular regeneration of HCO_3^- and H^+. The resulting elevation of cellular $[H^+]$ increases the availability of H^+ for the Na^+-H^+ exchange process, thereby further increasing the cellular regeneration of HCO_3^- and its subsequent reabsorption into the blood. This relationship is illustrated in Figure 11.17. On the basis of these findings, Rector and associates[29] have proposed that bicarbonate reabsorption may be mediated by two separate processes. One process has an apparent $\dot{T}_m$ for HCO_3^-, which is dependent on the arterial pCO_2 and independent of carbonic anhydrase activity. A second process is dependent on carbonic anhydrase and independent of the arterial pCO_2. Although these two processes can account for most of the bicarbonate reabsorbed by obligatory and controlled mechanisms, they lack direct experimental verification.

The effect of arterial pCO_2 on net bicarbonate reabsorption represents the mechanism whereby the kidney cross-compensates for respiratory disturbances in acid-base balance. Thus, in *respiratory acidosis*, the rise in pCO_2 induces the kidneys, via the mechanism just described, to produce a metabolic alkalosis. Because of the integral control characterizing the renal cross-compensation, the kidneys will slowly increase $[HCO_3^-]_{40}$ by increasing HCO_3^- reabsorption as well as increasing acid excretion, thereby returning the blood back toward normal. Similarly, in *respiratory alkalosis*, the fall in pCO_2 decreases HCO_3^- reabsorption, thereby reducing $[HCO_3^-]_{40}$. Once again, because of the integral control, the kidneys slowly increase HCO_3^- excretion, thereby returning the blood pH back toward normal.

The Plasma Concentration of Potassium

A number of studies have shown that oral ingestion of potassium chloride (hyperkalemia) results in increased excretion of potassium bicarbonate in urine, and despite the development of the extracellular metabolic acidosis, the urine remains alkaline. This results in the so-called dissociated or "paradoxical" metabolic acid-base disturbance. Furthermore, it has been shown that administration of potassium salts produces a marked depression of HCO_3^- reabsorption, even when the filtered load of HCO_3^- is sufficient to saturate the reabsorptive mechanism.[11] This observed inverse relationship between the rate of HCO_3^- reabsorption and plasma potassium concentration (Figure 11.18) is consistent with the view mentioned earlier that there appears to be a competition between K^+ and H^+ secretion in the collecting duct and that the magnitude of ion secretion is related to the reciprocal of their concentrations in the renal tubular cells.

Changes in plasma potassium concentration, which produce dissociated metabolic acid-base disturbances, exert their effects on bicarbonate reabsorption in the collecting duct. As described in Chapter 9 and shown in Figure 11.15, hyperkalemia, induced by

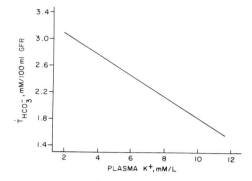

FIGURE 11.18. The relationship between net renal bicarbonate reabsorption, $\dot{T}_{HCO_3^-}$, and plasma potassium concentration in the dog. The line approximates the experimental data obtained by Fuller, et al.[11]

injection of potassium salt, produces three effects: (1) It enhances the transport of K^+ into the renal cells in exchange for cellular H^+ to maintain electrical neutrality. (2) This increased plasma hydrogen ion concentration is partly buffered by the plasma bicarbonate, resulting in the formation of water and carbon dioxide. The latter is eliminated by the lungs. Thus, hyperkalemia results in extracellular metabolic acidosis. (3) In the renal cells the increased cellular potassium concentration favors the exchange of sodium with potassium; the latter is excreted as potassium bicarbonate. The increased filtered load of potassium and the associated anions (HCO_3^-) further augment renal excretion of $KHCO_3$. The result is the formation of alkaline urine. In short, potassium chloride loading results in extracellular metabolic acidosis accompanied by excretion of alkaline urine.

Of more clinical importance and interest is the reverse of the above condition, namely, hypokalemic metabolic alkalosis. The depletion of body stores of potassium due to diarrhea, vomiting, or chronic hypersecretion of adrenocortical hormones increases cellular hydrogen ion concentration at the expense of the plasma hydrogen ion concentration. The net result is the condition of extracellular hypokalemic metabolic alkalosis accompanied by excretion of acid urine.

In mild K^+ depletion, the contraction of the extracellular volume may be the underlying cause of alkalosis. Bicarbonate reabsorption can be decreased and hence alkalosis controlled by merely administering isotonic saline solution without supplementing K^+.

Micropuncture studies in rats, subjected to a variety of respiratory and metabolic acid-base disturbances, by Malnic and associates[18, 19] have futher elucidated the roles of extracellular alkalosis and acidosis on K^+ and H^+ secretion in the distal nephron (see Appendix C for nomenclature). These investigators found that K^+ secretion was enhanced in both respiratory and metabolic alkalosis, despite hypokalemia, while K^+ secretion was depressed in acidosis, despite hyperkalemia. Furthermore, the increase in K^+ secretion was associated with a rise in intratubular pH and hence a fall in intracellular pH. These findings, in part, confirm the K^+-H^+ competitive secretory concept and provide direct evidence that the intracellular H^+ ions determine the degree of K^+ secretion, while the intratubular bicarbonate load is the key factor determining the rate of H^+ secretion in the collecting duct.

The Plasma Concentration of Chloride

Like potassium, there is an inverse relationship between the plasma chloride concentration and the rate at which the filtered bicarbonate is reabsorbed. At present, the mechanism of this reciprocal relationship is not fully understood. One explanation may be the existence of the well-known *chloride-shift* in response to changes in HCO_3^- to maintain the electrical neutrality between the plasma and cell compartments. As a consequence of this reciprocal relationship, as plasma $[HCO_3^-]$ rises, $[Cl^-]$ in the plasma falls. This inverse relationship serves to maintain the sum of the plasma concentrations of bicarbonate and chloride approximately constant.

The Plasma Levels of the Adrenocortical Hormones

The rate of bicarbonate reabsorption in the collecting duct is markedly influenced by changes in the secretion rate and blood levels of the adrenocortical hormones—glucocorticoids (cortisones) and mineralocorticoids (aldosterone). Chronic hypersecretion of the adrenocortical hormones, as in Cushing's syndrome, and hyperaldosteronism, as in Conn's syndrome, both produce hypokalemia and hence dissociated extracellular metabolic alkalosis. The alkalosis is the result of the elevated plasma bicarbonate concentration caused by increased renal bicarbonate reabsorption (Figure 11.18) and acid excretion. Conversely, chronic hyposecretion of adrenocortical hormones or adrenal insufficiency, such as may occur in Addison's disease, results in hyperkalemia and hence extracellular metabolic acidosis, consequent to reduced renal bicarbonate reabsorption. In both of these conditions, changes in bicarbonate reabsorption are secondary to hypokalemia and hyperkalemia produced by hypersecretion and hyposecretion of the adrenocortical hormones, respectively. The net effect is the development of a dissociated metabolic acid-base disturbance.

Controlled Excretion of Titratable Acids and Ammonia

Reabsorption of Na^+ and K^+ along with HCO_3^- and Cl^- in the proximal tubule results in no net gain of H^+ by the tubular fluid. This is because the H^+ secreted is buffered mostly by the bicarbonate in the tubular fluid, thereby resulting in minimal acidification of urine by the end of this segment. The major site of acidification of urine is the collecting duct. To understand the mechanism of acidification of urine in this segment, it should be recalled that in addition to HCO_3^- and Cl^-, the glomerular filtrate contains phosphate and sulfate, as well as organic acid anions.

Depletion of the standard bicarbonate pool results from the neutralization of strong acids, such as sulfuric and phosphoric acids, which are formed from the metabolism of dietary proteins and phospholipids. The neutralization of these acids results in the formation of carbon dioxide, which is eliminated by the lungs, and the neutral salts (Na_2SO_4 and Na_2HPO_4), which are transported to the kidneys:

$$H_2SO_4 + 2NaHCO_3 \rightarrow Na_2SO_4 + 2H_2O + 2CO_2 \quad (11.10)$$

$$H_3PO_4 + 2NaHCO_3 \rightarrow Na_2HPO_4 + 2H_2O + 2CO_2 \quad (11.11)$$

The kidneys replenish the standard bicarbonate pool by excreting these neutral salts in two ways: (1) transformation into acid salts, by exchanging sodium ions with *ammonium ions*, and (2) conversion into free *titratable acids*, which can be excreted at the prevailing urinary pH. What determines by which method the kidney excretes the neutral salt is the difference between the pK_a of the salt and the hydrogen ion concentration of the tubular urine, or pH_u. The greater the ionization constant (K_a) for a given neutral salt, the smaller would be the pK_a. Thus, the larger the difference between pH_u and pK_a ($pH_u - pK_a$), the larger would be the quantity of the ionized neutral salt to be excreted. These salts are excreted exclusively as ammonium salts. Conversely, the smaller the ionization constant of the neutral salt, the larger would be the pK_a. Thus, the smaller the $(pH_u - pK_a)$, the smaller the quantity of the ionized neutral salts to be excreted. These salts can be excreted as titratable acids. For example, the sulfate salt, having a large ionization constant (low pK_a), is excreted as ammonium salt:

$$Na_2SO_4 + 2H_2CO_3 + 2NH_3$$
$$\rightarrow (NH_4)_2SO_4 + 2NaHCO_3$$
$$\text{(excreted)} \quad \text{(reabsorbed)} \quad (11.12)$$

while the phosphate salt, having a smaller ionization constant (high pK_a) is excreted as titratable acid:

$$Na_2HPO_4 + H_2CO_3 \rightarrow NaH_2PO_4 + NaHCO_3$$
$$\text{(excreted)} \quad \text{(reabsorbed)} \quad (11.13)$$

Therefore, the sum of the urinary excretion of titratable acid and ammonia is a measure of the total renal replacement of the standard bicarbonate pool.

Because chloride and sulfate anions are completely ionized at normal blood pH, they cannot be excreted as hydrochloric acid or H_2SO_4. This is because the presence of a small quantity of these acids in the urine, if not buffered, lowers the pH of tubular urine to a value of 1.0, which is far below the minimum pH of 4.5 generated by the collecting duct cells. Therefore, these anions must be excreted in combination with a substance that makes the tubular urine less acid. The kidney excretes chloride and sulfate anions in combination with ammonium ions (NH_4^+), which are synthesized by the cells of the proximal tubules and transported by a specialized mechanism (see below) to the collecting duct where it is secreted into the urine.

The normal kidney excretes almost twice as much acid combined with ammonia than it excretes titratable acid. However, as shown in Table 11.3, the diseased kidney excretes a lesser amount of acid combined with ammonia.

Let us now examine the mechanisms of formation of titratable acids and ammonium salts and their excretion in urine.

Controlled Excretion of Titratable Acid

Formation and excretion of titratable acid is one process by which the kidney conserves Na^+ and replenishes the body bicarbonate pool and thereby acidifies the urine. The probable mechanism for this process is illustrated in Figure 11.19. The key element in the process is the quantity of filtered phosphate buffers. In the blood and hence the glomerular filtrate, the phosphate exists as both Na_2HPO_4 and

TABLE 11.3. Urinary excretion of acid in man in health and disease.

Condition	mEq of Acid excreted/day	Ratio of ammonia/titratable acid
Normal:		
Acid combined with ammonia	30–50	
Titratable acid	10–30	1–2.5
Diabetic ketosis:		
Acid combined with ammonia	300–500	
Titratable acid	75–250	1–2.5
Chronic renal insufficiency:		
Acid combined with ammonia	0.5–15	
Titratable acid	2.0–20	0.2–1.5

From Pitts.[25]

NaH_2PO_4, with a concentration ratio ($[Na_2HPO_4]/[NaH_2PO_4]$) of 4:1. As shown, the exchange of hydrogen ions (produced from cellular hydration of carbon dioxide for sodium ions converts disodium phosphate (Na_2HPO_4) to monosodium phosphate (NaH_2PO_4), which is excreted as titratable acid in urine. This Na^+-H^+ exchange results in reabsorption of three of the nine Na^+ ions filtered, thereby altering the ($[Na_2HPO_4]/[NaH_2PO_4]$) ratio to 1:4 in the excreted urine. As a consequence of this exchange, sodium and bicarbonate are returned to the peritubular capillary blood, thereby replenishing the body stores of bicarbonate.

The rate of formation of titratable acid and its subsequent excretion are influenced by at least four factors: (1) the amount of buffer anions filtered and remaining in the tubular lumen; (2) the pK_a of these buffers; (3) the concentration of hydrogen ions in the tubular fluid; and (4) the plasma hydrogen ion concentration. Thus, the greater the filtered load of buffer anions and the closer the pK_a to pH_u, the greater is the quantity of the titratable acid excreted. Furthermore, the greater the hydrogen ion concentration in the blood (the lower the pH_p), as in acidosis, and hence that in the ultrafiltrate, the greater would be the rate of Na^+-H^+ exchange and ultimately the greater the reabsorption of $NaHCO_3$ into the blood.

Controlled Excretion of Ammonium Salt

If the excretion of titratable acid was the only mechanism for renal acidification of urine, the magnitude of acid excretion would be limited by the amount of filterd phosphate buffers. That the kidney can excrete

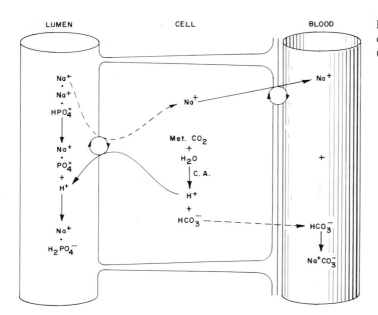

FIGURE 11.19. Mechanism of controlled excretion of titratable acid in the collecting duct. (From Koushanpour.[15])

considerably more acid than can be accounted for by the amount of excreted titratable acid, even though the urinary pH normally would not fall below 4.5, is indicative that another process exists by which the kidney acidifies urine. This process is the formation of ammonia for the excretion of highly ionized neutral salts as ammonium salts, which in the process conserves sodium and replenishes the body bicarbonate pool.

The ability of the kidney to excrete titratable acid is limited by the amount of filtered buffers (e. g., phosphate buffers) that reach the collecting duct and by the ability of the kidney to lower the pH of the urine. Therefore, the net acid excretion by the kidney is largely controlled by the factors that regulate urinary excretion of ammonia.

Several in-vivo micropuncture and in-vitro microperfusion studies (see review by Good and Knepper[12]) have clearly established that renal synthesis of ammonia is the principal source of ammonia excreted in the urine. These studies have further shown that the major portion of renal ammonia is synthesized in the proximal tubule and that this nephron segment is also the principal site for adaptive changes in ammonia synthesis in response to systemic acid-base disturbances. A comprehensive analysis of the available in-vivo and in-vitro data[12] has led to the formulation of the following steps for the renal transport of ammonia and its excretion in the urine. (1) The ammonia is synthesized in the proximal tubule and preferentially secreted into the lumen by a mechanism that includes nonionic diffusion (see below). (2) The ammonia secreted by the proximal tubule is then transported into Henle's loop, where a large fraction is reabsorbed as NH_4^+ from the thick ascending limb. (3) Ammonia reabsorption in this segment is driven indirectly by active sodium chloride cotransport, which leads to accumulation of ammonia in the renal medullary intersitim. And (4) accumulation of ammonia in the medullary interstitium enhances the driving force for diffusion of NH_3 into the collecting duct. Presence of H^+ ions in the lumen of the collecting duct leads to the preferential trapping of ammonia and formation of NH_4^+.

In addition to the ammonia transferred from the proximal tubule, microperfusion studies[14] have shown that the collecting duct is capable of synthesizing a small amount of ammonia and has the capacity to increase its ammonia production in response to chronic metabolic acidosis. Figure 11.20 illustrates the salient features of ammonia synthesis and secretion in the collecting duct. As shown, the key element is the formation of ammonia within the renal cell mainly from glutamine extracted from the peritubular blood. The deamidation and deamination of each mole of glutamine, in the presence of the enzyme glutaminase, yields two moles of un-ionized ammonia (NH_3). In the tubular cells 99% of the ammonia thus produced exists as ammonium ions (NH_4^+) and only 1% as free NH_3. The un-ionized NH_3 is lipid-soluble

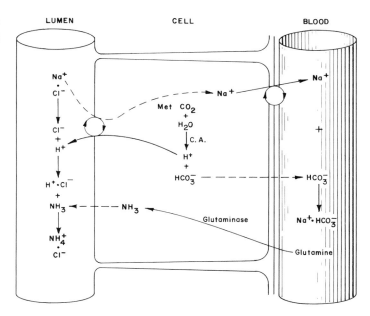

FIGURE 11.20. Mechanism of controlled excretion of ammonium ions in the collecting duct. (From Koushanpour.[15])

and can diffuse freely across the renal cell membranes into the blood or lumen. In contrast, the ionized NH_4^+ is water-soluble and can diffuse much less readily from the cell into the lumen or peritubular blood. But as rapidly as one molecule of NH_3 diffuses out of the cell, it is immediately replaced by another molecule of NH_3 from dissociation of H^+ from NH_4^+:

$$NH_4^+ \rightarrow NH_3 + H^+ \quad (11.14)$$

The H^+ ions thus produced become part of the cellular H^+ pool and therefore available for secretion into the lumen.

According to Bronstedt formulation, NH_4^+ is considered an acid because it yields an H^+ in alkaline solution, whereas NH_3 is considered a base because it accepts H^+ in acid solution. Therefore, as soon as the un-ionized NH_3 enters the lumen, it binds with H^+ and becomes trapped within the lumen as relatively nondiffusible NH_4^+. In this manner, cellular ammonia is secreted into the lumen by passive *nonionic diffusion* without the expenditure of energy to maintain cell-to-urine pH gradient. This asymmetric diffusibility of NH_3 and NH_4^+ and the high pK of the NH_3/NH_4^+ buffer system (pK = 9.2) are, therefore, well-suited for Na^+-H^+ exchange with a minimum lowering of urinary pH.

The nonionic diffusion of ammonia just described exemplifies a common biological phenomenon in which the lipid-soluble component of a buffer pair (e.g., NH_3) can readily diffuse across the cell membrane, while its water-soluble component (e.g., NH_4^+) cannot. Such a process has potential clinical importance in promoting renal excretion of weak acids and bases. A notable example is its application in the treatment of phenobarbital poisoning. The treatment involves intravenous infusion of mannitol and $NaHCO_3$ solutions. Infusion of the poorly reabsorbable mannitol will induce osmotic diuresis (see Chapter 9 for mechanism), thereby reducing tubular reabsorption of phenobarbital. Infusion of $NaHCO_3$ will alkalinize the glomerular filtrate, thereby further reducing the passive reabsorption of the phenobarbital, a weak acid, by the nonionic diffusion process.

The rate of renal ammonia production and its urinary excretion is influenced by at least three factors[24, 27]: (1) the degree of urinary acidity, (2) the degree of acidosis, and (3) the relative rates of flow of tubular fluid and peritubular capillary blood.

1. The effect of urinary pH on the net ammonia excretion ($\dot{E}_{NH_3}$) (production and secretion) is illustrated in Figure 11.21. As shown, during both normal acid-base state and metabolic acidosis, there is an inverse relationship between $\dot{E}_{NH_3}$ and pH_u. Thus, the lower the urinary pH (the more acid the urine), the greater is the net ammonia excretion.

2. Figure 11.21 also illustrates the effect of severity of acidosis on the rate of net ammonia excretion. It shows that for any given urinary pH, the rate of net ammonia excretion is higher during acidosis

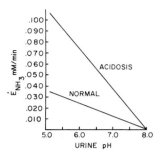

FIGURE 11.21. The relation between ammonia excretion, $\dot{E}_{NH_3}$ and urine pH in a single animal under conditions of normal acid-base state and chronic acidosis. Urine pH was varied by intravenous infusion of $NaHCO_3$. (Redrawn from Pitts.[23])

than during normal acid-base state. The increase in $\dot{E}_{NH_3}$ during acidosis is most probably due to adaptive increase in renal production and secretion of ammonia and to a lesser extent to increased nonionic NH_3 diffusion and its luminal trapping. Thus, Lotspeich[17] found that in NH_4Cl acidosis, a condition known to increase ammonia excretion, renal extraction of glutamine from peritubular blood was increased in rat, dog, and man. Furthermore, in rat he found a parallel adaptive increase in the synthesis of the enzyme glutaminase. The increase in ammonia excretion during NH_4Cl acidosis was found to be mediated by NH_4Cl-stimulation of DNA-dependent RNA synthesis by the kidney. Despite these and other extensive studies, the mechanism of adaptive increase in renal ammonia excretion in acidosis is at present not fully understood.

3. Besides urinary pH, the rate of diffusion of cellular NH_3 into the lumen is determned by the relative rates of flow of tubular fluid and peritubular blood. If the tubular fluid had the same pH as the blood, the peritubular blood, by virtue of its greater flow rate, would provide a favorable sink for diffusion of cellular NH_3. However, the slow tubular flow rate is more than offset by its lower pH, as compared with blood, for it makes it possible for 75% of the NH_3 produced within the renal cell to be diffused into the lumen and only 25% to be carried off by the blood. Of course, the situation can be reversed by any condition that causes the pH of the tubular fluid to approach or exceed the pH of the blood.

We conclude this section with a synthesis of the foregoing components of the intrarenal mechanisms of pH regulation, yielding the block diagram shown in Figure 11.22. Examination of this diagram reveals three important features of the renal pH regulator: (1) The intracellular $[H^+]$ is the main factor determining renal excretion of acid or alkali. (2) All variations in renal excretion of acid or alkali translate into variations in HCO_3^- ions entering renal blood. (3) Excretion of HCO_3^- may deplete body K^+.

Table 11.4 summarizes the responses of the renal pH-regulator to some selected forcings already mentioned. It is clear that the *renal cell pH* is the key factor determining the renal excretion of acids or alkalis.

The Dual pH-Regulator: Disturbances in Acid-Base Balance and Their Diagnoses

From the foregoing analysis it is clear that the ultimate regulation of acid-base balance entails coordinated, integrated function of the blood buffers and the pulmonary and renal systems. Having studied the response characteristics of each system in isolation, we are now in a position to consider their combined response to acid-base disturbances, much the same way they operate in vivo in the body. For this purpose, we use the system's functionaal block diagram to analyze the mechanism of within-and cross-compensation, and the $[HCO_3^-]_p$-pCO_2-pH plot to diagnose the disturbances in acid-base balance and as an aid in following the recovery and response to treatment.

Figure 11.23 presents the functional diagram of the dual pH-regulator. Note that the blood buffer system is shared by both the pulmonary and renal pH-regulators. To analyze the response of the dual pH-regulator to any acid-base disturbances, it is most helpful to proceed in two steps. First, analyze the within-compensation of the regulator to the forcing impinging on it. Then, if the forcing persists, proceed to analyze the cross-compensation from the other regulator. In such an analysis, you should recall that initial forcings for a respiratory disturbance in acid-base balance always impinge on the pulmonary system, which induces a within-compensated disturbance in pCO_2 and pH_p. If this disturbance persists, the real pH-regulator cross-compensates by adjusting $[HCO_3^-]_{40}$, slowly but eventually completely. In contrast, the initial forcings for a metabolic disturbance in acid-base balance always impinge on the renal

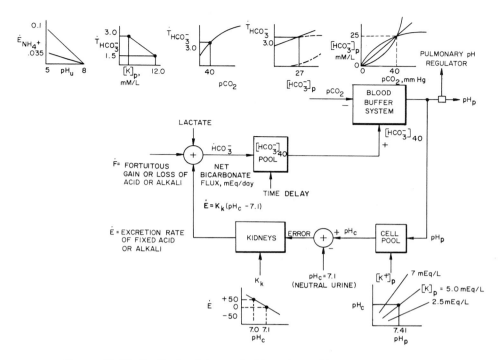

FIGURE 11.22. Synthesis of the renal pH-regulator. (From Koushanpour.[15])

system, which induces a within-compensated disturbance in [HCO$_3^-$]$_{40}$ and hence pH$_p$. This disturbance induces immediate but partial cross-compensation from the pulmonary pH-regulator.

To diagnose disturbances in acid-base balance and to follow the recovery and response to treatment, we use the [HCO$_3^-$]$_p$-pCO$_2$-pH plot, similar to that shown in Figures 11.8 and 11.24. For this purpose, we need to elaborate further on some of the main features of such a plot germane to the present discussion.

The *normal respiratory pathway* is the familiar in-vitro carbon dioxide absorption curve of true plasma. It identifies the blood response to changes in pCO$_2$ when [HCO$_3^-$]$_{40}$ is constant at 25 mM/L. Note that there is a whole family of carbon dioxide absorption curves, one for each level of [HCO$_3^-$]$_{40}$ other than 25 mM/L. In Figure 11.24, we have plotted two such curves, labeled high-gain and low-gain respiratory pathways. The data used to plot these, as well as the normal curve, are given in Table 11.5.

It should be apparent that at any moment a patient's blood pattern may lie on one of these curves. For each case, if we wish to know the value of [HCO$_3^-$]$_{40}$, we simply move along the patient's respiratory pathway to where the curve intersects with the iso-pCO$_2$ = 40 mm Hg line. Graphically, this procedure is

TABLE 11.4. Responses of the renal pH regulator to some selected forcings.

Forcing	Condition	Plasma pH	Renal Cell pH	Urine
1. NH$_4$Cl ingestion	Metabolic acidosis	−	−	Aciduria
2. CO$_2$ inhalation (acute)	Respiratory acidosis	−	−	Aciduria
3. K$^+$ depletion	Hypokalemic alkalosis	+	−	Aciduria
4. NaHCO$_3$ ingestion	Metabolic alkalosis	+	+	Alkaluria
5. Voluntary hyperventilation	Respiratory acidosis	+	+	Alkaluria
6. K$^+$ injection	Hyperkalemic acidosis	−	+	Alkaluria

(From Koushanpour.[15])

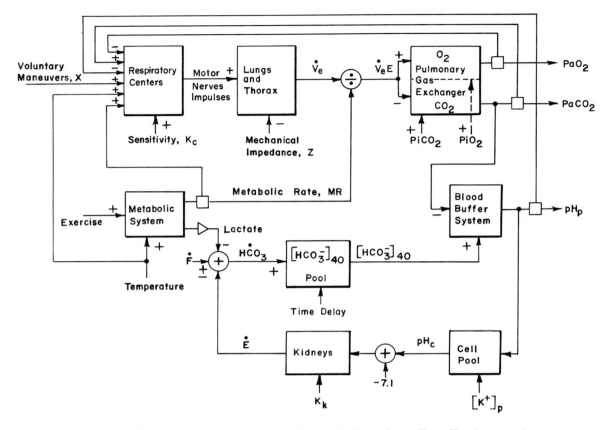

FIGURE 11.23. A functional block diagram of the dual pH-regulator. (From Koushanpour.[15])

analogous to exposing the patient's blood sample to $pCO_2 = 40$ mm Hg and determining its $[HCO_3^-]$. Moreover, if we wish to know what changes occur in the patient's blood pattern when the pulmonary regulator raises (hypoventilation) or lowers (hyperventilation) pCO_2 without renal cross-compensation (i.e., $[HCO_3^-]_{40}$ is constant), we simply move to the right (as for example from point N to point 4 in Figure 11.24) or to the left (from point N to point 1), respectively, along the patient's respiratory pathway.

The normal respiratory pathway, for which $[HCO_3^-]_{40} = 25$ mM/L, represents the *completely uncompensated* (no renal cross-compensation) respiratory pathway. On the other hand, the radiating iso-pH = 7.41 line represents the *completely cross-compensated* respiratory pathway.

The other two curves, labeled high-gain and low-gain respiratory pathways, are followed during pulmonary cross-compensation for metabolic alkalosis and acidosis. The term "gain" here refers to the level of $[HCO_3^-]_{40}$: high-gain refers to elevated $[HCO_3^-]_{40}$, as in metabolic alkalosis, and low-gain refers to reduced $[HCO_3^-]_{40}$, as in metabolic acidosis. Thus, in metabolic acidosis, in which $[HCO_3^-]_{40}$ is reduced, the pulmonary cross-compensation causes the blood to move along a low-gain respiratory pathway toward the iso-pH = 7.41 line. As shown in Figure 11.8, the combined effects of the metabolic acidosis and the induced respiratory alkalosis (partial pulmonary cross-compensation) are that the blood moves down along the curve labeled "normal metabolic pathway," such as that depicted by the curve connecting the normal point N to point 3 in Figure 11.24. Precisely analogous but opposite changes occur in metabolic alkalosis, so that the blood moves up along the normal metabolic pathway, such as that depicted by the curve connecting point N to point 6. Note that because changes in pH and pCO_2 have opposite effects on the respiratory center (Figure 11.23) in both metabolic acidosis and alkalosis, there will only be a partial cross-compensation by the pulmoary regulator.

The *normal metabolic pathway* just described

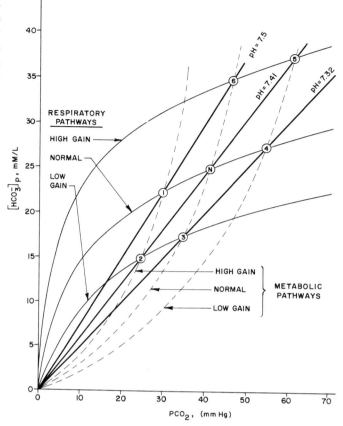

FIGURE 11.24. Respiratory and metabolic pathways drawn for data listed in Table 11.5. The three radiating iso-pH lines were obtained by drawing a straight line from the origin to these points: $pCO_2 = 48$ and $[HCO_3^-]_p = 37$; $pCO_2 = 59$ and $[HCO_3^-]_p = 37$; and $pCO_2 = 70$ and $[HCO_3^-]_p = 36$, respectively. (From Koushanpour.[15])

identifies the blood response to changes in $[HCO_3^-]_{40}$ when the gain for the pulmnary cross-compensation is constant. The term "gain" here refers to the state of pulmonary ventilation. Like the respiratory pathways, there is a whole family of metabolic pathways, depending on the *ventilatory state or gain* of the pulmonary cross-compensation. In Figure 11.24 we have plotted two such curves, labeled high-gain and low-gain metabolic pathways, respectively.

The curve labeled the high-gain (or hyperventilatory) metabolic pathway is followed during renal cross-compensation for respiratory alkalosis, which is initiated by some respiratory stimulation other than changes in pH or pCO_2. Examples include high altitude, anoxia, fever, and hysterical hyperventilation, all of which increase ventilation, yielding a lower pCO_2 for a given $[HCO_3^-]_{40}$. Thus, in chronic respiratory alkalosis, such as may occur during temporary residence at high altitudes (point 1, Figure 11.24), the renal regulator will slowly but eventually completely cross-compensate for the alkalosis by in- ducing a metabolic acidosis, thereby returning the pH back toward normal. In Figure 11–24, this is depicted by moving the blood from point 1 to point 2, along the high-gain metabolic pathway.

The curve labeled the low-gain (or hypoventilatory) metabolic pathway is followed during renal cross-compensation for respiratory acidosis, which is initiated by some respiratory inhibition other than changes in pH or pCO_2. Examples include emphysema, asthma, chronic partial obstruction of airway, and depression of the respiratory centers, all of which reduce ventilation yielding a high pCO_2 for a given $[HCO_3^-]_{40}$. Thus, in chronic respiratory acidosis, such as may occur in pulmonary obstruction (point 4 in Figure 11.24), the renal regulator will slowly but eventually completely cross-compensate for the acidosis by inducing a metabolic alkalosis, thereby returning the pH back toward normal. In Figure 11.24 this is depicted by moving the blood from point 4 to point 5, along the low-gain metabolic pathway. The only exception is the respiratory acidosis induced

TABLE 11.5. Bicarbonate concentration and pCO_2 values for true plasma used to plot respiratory and metabolic pathways in Figure 11.24.

	Respiratory pathways				Metabolic pathways		
	$[HCO_3^-]_p$ mM/L				$[HCO_3^-]_p$ mM/L		
pCO_2 (mm Hg)	Normal $[HCO_3^-]_{40}$	Low $[HCO_3^-]_{40}$	High $[HCO_3^-]_{40}$	pCO_2 (mm Hg)	Normal gain	High gain	Low gain
0	0	0	0	0	0	0	0
2.5	7.5	3.5	12.5	5.0	1.5	2.0	1.0
5.0	10.0	6.0	17.5	10.0	3.0	4.0	2.0
10.0	15.0	9.0	22.5	15.0	5.0	7.5	3.0
15.0	17.5	11.5	26.5	20.0	7.5	10.0	5.0
20.0	19.0	13.5	28.5	25.0	10.0	15.5	7.0
25.0	21.0	15.0	30.0	30.0	14.0	23.5	9.0
30.0	22.5	16.5	31.5	35.0	17.5	36.5	12.0
35.0	23.5	17.5	32.5	40.0	25.0		15.0
40.0	25.0	18.0	34.0	45.0	35.0		18.5
45.0	26.0	19.5	35.0	50.0			22.5
50.0	27.0	20.5	36.0	55.0			30.0
55.0	28.0	21.0	37.0	60.0			37.5
60.0	28.5	21.5	38.0				
65.0	29.5	22.0	38.5				
70.0	30.0	22.5	39.0				

From Koushanpour.[15]

by carbon dioxide inhalation, which causes hyperventilation and hypercapnia. Although the pulmonary gain appears to be increased (hyperventilation), the renal cross-compensation follows the low-gain (hypoventilatory) pathway.

Thus, any patient's blood at any moment must lie on one of these families of metabolic pathways. If we wish to know the gain of the pulmonary cross-compensation, we simply note whether the patient's blood pattern lies to the left or to the right of the "normal metabolic pathway." Furthermore, if we wish to know what changes occur in the patient's blood pattern when the renal regulator raises or lowers $[HCO_3^-]_{40}$, we simply follow the normal metabolic pathway up or down, respectively. It should be remembered that in Figure 11.24, the radiating iso-pH line of 7.41 represents also the *completely cross-compensated* metabolic pathway, while the iso-pCO_2 = 40 mm Hg line represents no cross-compensation. Since respiratory cross-compensation for metabolic disturbances is immediate but only partial, the actual pathway folowed is along neither the iso-pH = 7.41 line nor th iso-pCO_2 = 40 mm Hg line, but somewhere in between, namely the "normal metabolic pathway." Thus, the normal metabolic pathway represents the *partially cross-compensated* metabolic pathway by the pulmonary regulator with a normal gain.

Using the functional diagram of the dual pH-regulator shown in Figure 11.23 and the respiratory and metabolic pathways shown in Figure 11.24, it is now easy to follow the development, compensation, recovery, or response to treatment in acid-base disturbances. To facilitate the use of the graph in Figure 11.24 in conjunction with the problems at the end of this chapter, the graph is divided into a number of areas demarcated by intersection of the various pathways, each defining a type of disturbance in acid-base balance. For any point plotted on this graph, such as those numbered 1 through 6, we can read off five pieces of information: $[HCO_3^-]_p$, pCO_2, pH, $[HCO_3^-]_{40}$, and the pulmonary gain. From this information we can readily diagnose the disturbances in acid-base balance as acidosis or alkalosis, respiratory or metabolic or both, uncompensated, and partially or completely cross-compensated. Note that if both respiratory and metabolic acidosis (or alkalosis) are present, neither can or will be cross-compensated. Furthermore, since the graph in Figure 11.24 pertains only to plasma, it cannot at the same time display the cellular behavior in *dissociated* metabolic acid-base disturbances.

Now to help you discover for yourself the simplicity of diagnosing disturbances in acid-base balance and to master the materials presented, we end this chapter with a number of problems. To work out these problems, use the $[HCO_3^-]_p$-pCO_2-pH graph (Figure 11.24) in conjunction with the block diagram of the dual pH-regulator (Figure 11.23). Remember that whenever the block diagram dictates a change in $[HCO_3^-]_{40}$, follow a metabolic pathway, and whenever it dictates a change in pCO_2, follow a respiratory pathway.

Problems

11.1. If a subject on the normal respiratory pathway has a pCO_2 of 40 mm Hg, (a) what is the pH? _____. (b) $[HCO_3^-]_p$? _____. (c) $[HCO_3^-]_{40}$? _____.

11.2. If he breathes CO_2 for 45 minutes to elevate his pCO_2 to 53.5 mm Hg, (a) what is the pH? _____. (b) $[HCO_3^-]_p$? _____. (c) $[HCO_3^-]_{40}$? _____. (d) Is this an acidemia or alkalemia? _____. (e) Acidosis or alkalosis? _____. (f) Metabolic or respiratory? _____. (g) Compensated or uncompensated? _____. (h) A displacement on the normal respiratory pathway, or shift to a new one? _____.

11.3. If he voluntarily hyperventilates for 45 minutes to lower his pCO_2 to 29 mm Hg, (a) what is the pH? _____. (b) $[HCO_3^-]_p$? _____. (c) $[HCO_3^-]_{40}$? _____. (d) Is this an acidemia or alkalemia? _____. (e) Acidosis or alkalosis? _____. (f) Metabolic or respiratory? _____. (g) Compensated or uncompensated? _____. (h) Displacement on the normal respiratory pathway, or shift to a new one? _____.

11.4. In the above examples, (a) do pH and pCO_2 vary together or oppositely? _____. (b) Do $[HCO_3^-]_p$ and pCO_2 vary together or oppositely? _____. (c) Can $[HCO_3^-]_p$ change without change in $[HCO_3^-]_{40}$? _____.

11.5. (a) What is the $[HCO_3^-]_{40}$ for the higher curve? _____. (b) The lower? _____. (c) What kind of primary disturbance would shift a person to one of these curves? _____. (d) Compensation for what kind of disturbance would shift a person to one of these curves? _____.

11.6. A normal person develops a metabolic acidosis sufficient to reduce his $[HCO_3^-]_{40}$ to 19 mM/L. (a) If his respiratory regulator made no cross-compensation, what pathways would he follow? _____. What would his pCO_2 be? _____. $[HCO_3^-]_p$? _____. pH? _____. (b) If now his respiratory regulator is allowed to respond, what pathway would he follow? _____. What would his pCO_2 become? _____. $[HCO_3^-]_p$? _____. pH? _____. (c) Would he normally follow this "dog-leg" pathway? _____. Why? _____. (d) What pathway will he actually follow? _____.

11.7. A normal person develops a metabolic alkalosis sufficient to raise his $[HCO_3^-]_{40}$ to 34 mM/L. (a) If his respiratory regulator made no cross-compensation, what pathway would he follow? _____. What would his pCO_2 be? _____. $[HCO_3^-]_p$? _____. pH? _____. (b) If now his respiratory regulator is allowed to respond, what pathway would he follow? _____. What would his pCO_2 become? _____. $[HCO_3^-]_p$? _____. pH? _____. (c) Would he normally follow this "dog-leg" pathway? _____. Why? _____. (d) What pathway will he actually follow? _____.

11.8. A normal person rides swiftly on a train from sea level to an altitude that induces hyperventilation sufficient to lower his pCO_2 to 29 mm Hg. (a) What pathway did he follow? _____. (b) Identify the disturbances in acid-base balance (i.e., respiratory or metabolic, acidosis or alkalosis, uncompensated, partially compensated, or fully compensated). _____.

11.9. This person now remains at altitude for the next month. (a) Does the initial cause persist? _____. (b) Will compensation occur? _____. (c) What regulator will compensate? _____. (d) Is it fast or slow? _____. (e) Will the pH be corrected to normal? _____. (f) What input to the blood buffer system will be adjusted? _____. (g) What pathway will be followed? _____. (h) What will the steady-state pH be? _____. $[HCO_3^-]_p$? _____. $[HCO_3^-]_{40}$? _____. pCO_2? _____. (i) Identify the disturbance in acid-base balance. _____. (j) Is he hyperventilating more, or less, than when he arrived? _____. (k) Is his pO_2 correction better, or worse, than when he arrived? _____.

11.10. This person now returns swiftly by train to sea level. (a) Does this correct the initial cause? _____. (b) What is his $[HCO_3^-]_{40}$? _____. (c) Will his hyperventilation disappear, or merely moderate? _____. (d) What pathway did he follow? _____. (e) What is his pCO_2? _____. pH? _____. $[HCO_3^-]_p$? _____. (f) Identify this disturbance in acid-base balance. _____.

11.11. This person now remains at sea level for the next month. (a) What will his $[HCO_3^-]_{40}$ do now? _____. (b) What pathway will he follow?

_____. (c) What will his pH be? _____. [HCO$_3^-$]$_p$? _____. pCO$_2$? _____. (d) Was this fast or slow? _____. (e) Have we gone full circle? _____.

11.12. To make sure you are getting the idea, trace through in the same four steps the case of a person who has a sudden, persistent respiratory depression, which later is suddenly corrected.

11.13. (a) In what disturbances of acid-base balance may the pH be normal? _____. (b) in what disturbance may the pCO$_2$ be low? _____. (c) In what disturbances may the [HCO$_3^-$]$_{40}$ be low? _____. (d) In what disturbances may the renal excretion of acid be high? _____. (e) Can you fully diagnose the type of disturbance from any one of these? _____.

11.14. Now that you have become familiar with the graph, it can be simplified to three lines (Figure 11.8), each running through the normal point; the normal pH line, the normal respiratory pathway, and the normal metabolic pathway. For any point plotted on such a graph you can tell whether pCO$_2$ is high or low, pH is high or low, and [HCO$_3^-$]$_{40}$ is high or low. From this information you can diagnose any single disturbance and any combination of disturbances in acid-base balance.

11.15. Suppose a patient with a low [HCO$_3^-$]$_{40}$ was diagnosed as having metabolic acidosis, when it was really a compensated respiratory alkalosis; the patient was therefore given NaHCO$_3$ by mouth. (a) What pathway will be followed? _____. (b) What will happen to the pH? _____. (c) Does this help or hinder? _____.

11.16. Suppose a patient with a low pCO$_2$ was diagnosed as having respiratory alkalosis, when it was really a metabolic acidosis; the patient was therefore given CO$_2$ to breathe. (a) What pathway will be followed? _____. (b) What will happen to the pH? _____. (c) Does this help or hinder? _____.

11.17. The table below gives the blood gas findings for a normal person and for eight patients with acid-base disturbances (none breathing CO$_2$).

Patients	pCO$_2$ (mm Hg)	[HCO$_3^-$]$_p$ (mM/L)	[HCO$_3^-$]$_{40}$ (mM/L)	pH
Normal	40	25.0	25.0	7.41
1	45	35.0	33.0	7.50
2	60	37.5	33.0	7.41
3	54	27.5	25.0	7.32
4	47	20.0	18.5	7.25
5	35	17.5	18.5	7.32
6	25	15.0	18.5	7.41
7	29	22.0	25.0	7.50
8	33	32.0	33.0	7.60

Using the data in this table in conjunction with the plot shown in Figure 11.24, answer the following questions:

(1) Which patient has:
   a. A fully compensated respiratory alkalosis? _____
   b. A fully compensated respiratory acidosis? _____
   c. A partially compensated metabolic alkalosis? _____
   d. A partially compensated metabolic acidosis? _____
   e. An uncompensated respiratory alkalosis? _____
   f. An uncompensated respiratory acidosis? _____
   g. A combined respiratory and metabolic acidosis? _____
   h. A combined respiratory and metabolic alkalosis? _____

(2) Which patient(s), if any, would you expect to see excreting acid _____ or alkali _____ in his urine, relative to normal?

(3) If patients 1 and 5 in the table had a dissociated acid-base disturbance, which one would you expect to be excreting acid _____ or alkali _____ in his urine?

(4) Which patient(s)'s blood pattern(s) lie(s) on a high-gain _____ or low-gain _____ metabolic pathway?

(5) Which patient(s)'s blood pattern(s) lie(s) on a high-gain _____ or low-gain _____ respiratory pathway?

(6) Which patient(s), if any, have acidemia _____ or alkalemia _____?

(7) Which patient(s), if any, have elevated [HCO$_3^-$]$_{40}$? _____

(8) Which patient(s), if any, have reduced [HCO$_3^-$]$_{40}$? _____

(9) Starting with the normal point (N) in Figure 11.24 and using the numbers on this graph, indicate the sequential changes in blood pattern that occur in (1) sudden respiratory obstruction _____, (2) lasting a month _____, (3) sudden removal of the obstruction _____, and (4) a month later _____.

(10) Which patient(s), if any, would benefit if given NH$_4$Cl? _____

(11) Which patient(s), if any, would benefit if given NaHCO$_3$? _____

(12) Which patient(s), if any, represent a person in the apneic stage of asphyxia? _____

References

1. Aronson PS: Mechanisms of active H^+ secretion in the proximal tubule. *Am J Physiol* 1983; 245:F647–F659.
2. Astrup P, Siggaard-Andersen O, Jorgensen K, Engel K: The acid-base metabolism—a new approach. *Lancet* 1960; 1:1035–1039.
3. Bennett CM, Springberg PD, Falkinburg NR: Glomerular-tubular balance for bicarbonate in the dog. *Am J Physiol* 1975; 228:98–106.
4. Berliner RW, Kennedy TJ Jr, Orloff J: Factors affecting the transport of potassium and hydrogen ions by the renal tubule. *Arch Intern Pharmacodyn* 1954; 97: 299–312.
5. Clapp JR, Watson JF, Berliner RW: Osmolality, bicarbonate concentration and water reabsorption in proximal tubule of the dog nephron. *Am J Physiol* 1963; 205:273–280.
6. Clark WM: *Topics in Physical Chemistry*. Baltimore, William & Wilkins Co, 1948.
7. Cogan MG, Alpern RJ: Regulation of proximal bicarbonate reabsorption. *Am J Physiol* 1984; 247:F387–F395.
8. Davenport HW: *The ABC of Acid-Base Chemistry*, ed 4. Chicago, University of Chicago Press, 1958.
9. Dobyan DC, Bulger RE: Renal carbonic anhydrase. *Am J Physiol* 1982; 311:F311–F324.
10. DuBose TJ Jr: Application of the disequilibrium pH method to investigate the mechanism of urinary acidification. *Am J Physiol* 1983; 245:F535–F544.
11. Fuller GR, MacLeod MB, Pitts RF: Influence of administration of potassium salts on the renal tubular reabsorption of bicarbonate. *Am J Physiol* 1955; 182:111–118.
12. Good DW, Knepper MA: Ammonia transport in the mammalian kidney. *Am J Physiol* 1985; 248:F459–F471.
13. Henderson LJ: *Blood: A Study in General Physiology*. New Haven, Conn, Yale University Press, 1928.
14. Knepper MA, Good DW, Burg MB: Mechanism of ammonia secretion by cortical collecting ducts of rabbits. *Am J Physiol* 1984; 247:F729–F738.
15. Koushanpour E: *Renal Physiology: Principles and Functions*, ed 1. Philadelphia, WB Saunders Co, 1976.
16. Kurtzman NA: Regulation of renal bicarbonate reabsorption by extracellular volume. *J Clin Invest* 1970; 49:586–595.
17. Lotspeich WD: Metabolic aspects of acid-base change. *Science* 1967; 155:1066–1075.
18. Malnic G, De Mello Aires M, Giebisch G: Potassium transport across renal distal tubules during acid-base disturbances. *Am J Physiol* 1971; 221:1192–1208.
19. Malnic G, De Mello Aires M, Giebisch G: Micropuncture study of renal tubular hydrogen ion transport in the rat. *Am J Physiol* 1972; 222:147–158.
20. Maren TH: Carbonic anhydrase: chemistry, physiology and inhibition. *Physiol Rev* 1967; 47:595–781.
21. McKinney TD, Burg MB: Bicarbonate absorption by cortical collecting tubules in vitro. *Am J Physiol* 1978; 234:F141–F145.
22. Muntwyler E: *Water and Electrolyte Metabolism and Acid-Base Balance*. St Louis, CV Mosby Co, 1968.
23. Pitts RF: The renal excretion of acid. *Fed Proc* 1948; 7:418–426.
24. Pitts RF: The renal regulation of acid base balance with special reference to the mechanism for acidifying the urine. I and II. *Science* 1945; 102:49–54, 81-85.
25. Pitts RF, Alexander RS: The nature of the tubular mechanism for acidifying the urine. *Am J Physiol* 1945; 144:239–254.
26. Pitts RF, Ayer JL, Schiess WL: The renal regulation of acid base balance in man: III. The reabsorption and excretion of bicarbonate. *J Clin Invest* 1949; 28:35–44.
27. Pitts RF: The role of ammonia production and excretion in regulation of acid-base balance. *N Engl J Med* 1971; 284 (1):32–38.
28. Purkerson ML, Lubowitz H, White RW, Bricker NS: On the influence of extracellular fluid volume expansion on bicarbonate reabsorption in the rat. *J Clin Invest* 1969; 48:1754–1760.
29. Rector FC Jr, Seldin DW, Roberts AD Jr, Smith JS: The role of plasma CO_2 tension and carbonic anhydrase activity in the renal reabsorption of bicarbonate. *J Clin Invest* 1960; 39:1706–1721.
30. Rector FC Jr: Acidification of the urine, in Orloff J, Berliner RW (eds): *Handbook of Physiology, Section 8, Renal Physiology*. Washington DC, American Physiological Society, 1973, pp 431–454.
31. Severinghaus JW, Stupfel M, Bradley AF: Accuracy of blood pH and pCO_2 determinations. *J Appl Physiol* 1956a; 9:189–196.
32. Severinghaus JW, Stupfel M, Bradley AF: Variations of serum carbonic acid pK' with pH and temperature. *J Appl Physiol* 1956b; 9:197–200.
33. Siggaard-Anderson O: *The Acid-Base Status of the Blood*, ed 2. Baltimore, William & Wilkins Co, 1964.
34. Singer RB, Hastings AB: Improved clinical method for estimation of acid-base balance of human blood. *Medicine* 1948; 27:223–242.
35. Slatopolsky E, Hoffsten P, Purkerson M, Bricker NS: On the influence of extracellular fluid volume expansion and of uremia on bicarbonate reabsorption in man. *J Clin Invest* 1970; 49:988–998.
36. Winters RW: Terminology of acid-base disorders. *Ann NY Acad Sci* 1966; 133:211–247.

12

Mechanism of Concentration and Dilution of Urine

In Chapter 9, while discussing the sequential processing of the filtrate along Henle's loop, we alluded to the role of this nephron segment in concentrating and diluting the urine. In this chapter we consider the principle of countercurrent multiplication and the evidence for its application to the kidney. We then examine the mechanism of concentration and dilution of urine, the measurement of the ability of the kidney to concentrate urine, and finally the action of some selective diuretics and their potential therapeutic effects. We begin with a comprehensive presentation of the structural organization of the renal medulla.

Architectural Organization of the Renal Medulla

To understand the overall structural organization of the renal medulla it is helpful to begin with some functional considerations. Phylogenetically, the renal medulla has been developed in mammals (and a similar structure in birds) for the purpose of conserving water. Hence, mammals are able to excrete superfluous salts and other waste products in little water, i.e., to produce a urine whose overall osmotic concentration surpasses plasma concentration. The primary function of the renal medulla is to concentrate urine above plasma levels. Most lower vertebrates and invertebrates are able to produce a dilute urine. Phylogenetically, the ability to excrete water in excess of salt (with respect to the concentration of body fluids) is much older than the opposite function. Mammalian kidneys have both abilities, namely, to dilute and to concentrate the urine. Both functions, as will become clear later, are intimately related to each other: *The ability to concentrate the urine depends on the ability to dilute it.*

There seem to be no major differences among mammals in their diluting capacity. In general, the urine can be diluted down to levels of 60 to 50 mOsm/L H_2O.[2,36] Some species are able to dilute the urine this much, even though they probably never need to during their lifetime. In contrast, there are great differences among mammals with respect to their ability to concentrate urine.[75] Species living in an aquatic environment can only moderately concentrate their urine, up to some 600 mOsm/L H_2O (i.e., roughly twice the concentration of the systemic plasma). On the other hand, there are desert species that can produce urine having an osmolality of up to 6,000 mOsm/L H_2O (i.e., 20 times the plasma concentration). Human kidney is considered to be a "poor concentrator," achieving a maximal urine concentration of about 1,500 mOsmol/L H_2O. These functional differences correlate with the overall structural elaboration of the renal medulla. All species with a high urine-concentrating ability have a well-developed renal medulla; in aquatic species the renal medulla is less complexly organized.

It has long been believed that species with the highest urine-concentrating ability have only long loops. This has turned out to be incorrect. All highly concentrating species, like rat, mouse, golden hamster, and desert rodents, have a greater number of short loops than of long loops. Only two species, the cat and dog, are known to have 100% long loops; in comparison with other species, their urine-concentrating ability is average. On the other hand, aquatic species with a low urine-concentrating ability, like the mountain beaver, have only short loops and consequently only an outer medulla. Hence, to achieve the highest urine concentrations, both types of loops are necessary. As we shall see later, with only short loops the maximal possible urine concentration is only about twice the plasma concentration.

There are several biological problems arising from the requirement to produce a fluid with such an excessive osmotic concentration. (1) This fluid has to be insulated from body fluids of the usual low concentration. (2) Since a biological mechanism for active water transport has never been developed (and would not be very economical), salt has to be primarily transported, and water will follow by osmosis. (3) It is hard to imagine a salt-transporting epithelium that would be able to tolerate a gradient of from 300 to 2,900 mOsm/L H_2O across the epithelium, and simultaneously be able to increase this difference by active transport, thereby lowering the concentration from 300 to 290 mOsm/L H_2O on one side of the epithelium and increasing the concentration of the fluid on the other side from 2,900 to 3,000 mOsm/L H_2O. Active salt transport can only be possible against moderate concentration gradients. (4) The tissue that serves this concentrating function has to be nourished; hence the blood supply has to be adapted to the high osmotic concentrations.

The renal medulla fulfills all these requirements and also occupies an insulated position. Toward the cortex, the insulation is effected by the outermost region of the medulla, namely, the outer stripe of the outer medulla. At all other sides, the urine within the renal pelvis is the insulator. To guarantee moderate concentration gradients, the structure that is responsible for the creation of the high osmolality is shaped like a hairpin loop (Henle's loop), which allows for a gradual increase in the concentration of the tubular fluid in the descending limb toward the bend and a gradual decrease again in the ascending limb. Since transports responsible for the creation of this corticomedullary gradient are all effected in a transverse plane (with respect to the loop) the gradient at any level remains small. The blood supply, too, is effected by loop-shaped vessels that allow the blood to gradually increase in osmolality when entering and gradually decrease in osmolality when flowing out.

The individual components of the renal medulla have been presented in Chapters 4 and 9. To understand the urinary-concentrating mechanism, we must first examine in greater detail the architectural organization of the renal medulla, i.e., the way the loops, the collecting ducts, and the vessels are built together to form a renal medulla.

The three regions of the medulla, the outer stripe, the inner stripe, and the inner medulla, are presented separately. They contain different populations of nephron segments. The vascular bundles are present in each of these zones; they may be considered as central axes around which the tubules are arranged in a certain pattern. We, therefore, always start our description with the bundle axis. This allows us to follow the structures in a longitudinal direction from region to region toward the papillary tip (Figures 12.1 and 12.2). A prior knowledge of the architectural organization of the cortex, presented in Chapter 4 and in Figures 12.1 and 12.2, is necessary to a full understanding of the ensuing discussion.

Outer Stripe of the Outer Medulla

The outer stripe contains straight parts of the proximal tubule, straight parts of the distal tubule (thick ascending limbs), and collecting ducts. The vascular bundles develop in the outer stripe. With respect to the cortex, one may interpret their position as to continue the axes of interlobular arteries down into the medulla.[62] The loop limbs of juxtamedullary nephrons are grouped immediately around the vascular bundles. In continuation of the medullary rays of the cortex, the loop limbs of superficial and midcortical nephrons travel at some distance from the bundle together with the collecting ducts[50] (Figures 12.1 and 12.2).

Cross sections through the outer stripe exhibit a dominance of the proximal tubules, which occupy almost 70% of the total tissue volume in this region. This is because proximal tubules are much thicker in diameter than straight distal tubules and the straight proximal tubules of juxtamedullary nephrons are not really straight (as their name indicates) but descend tortuously through the outer stripe (Figure 12.3).

As outlined in Chapter 4, the tubules of the outer stripe are supplied by a very specific capillary plexus. True capillaries, derived from direct branches of efferent arterioles, appear to be few. The dominating capillary vessels are the ascending vasa recta that contact the tubules like capillaries. Thus, the outer stripe tubules are mainly supplied by venous blood coming up from deeper parts of the medulla. This organization is virtually a countercurrent arrangement between ascending vasa recta and descending proximal tubules[48] and may function as an ultimate trap for solutes, preventing their loss from the renal medulla into the venous outflow (Figure 12.6c). In this context another peculiarity is noteworthy: Compared with all other regions of the kidney, the interstitial spaces in the outer stripe are very scarcely developed. This might serve to minimize the diffusional loss of salts from the medulla to the cortex by way of the interstitial space. Taken together, the countercurrent arrangement between vessels and tubules and the scarcity of an interstitium suggest that the outer stripe, in addition to other functions, serves to insulate the hypertonic medulla from the cortex.

Inner Stripe of the Outer Medulla

Considerable interspecies differences exist with respect to the architectural organization of this region. Two types of an inner stripe may be distinguished.[38] In most species, including rabbit, cat, dog, and man, vascular bundles of the *simple* type are present, which only are established by descending and ascending vasa recta arranged in a checkerboard pattern. Hence, these bundles represent a classic countercurrent arrangement, a "rete mirabile," consisting of ascending and descending vessels. The tubules are arranged around these bundles in a pattern similar to that in the outer stripe. The loop limbs of juxtamedullary nephrons, i.e., generally the long loops, are grouped immediately around the bundles, whereas those loops derived from superficial nephrons, i.e., generally the short loops, together with the collecting ducts lie most distant from the bundles. Altogether, these tubules are supplied by the dense capillary plexus derived from descending vasa recta (Figures 12.1, 12.2, and 12.4).

In several rodent species with a high urinary concentrating ability (e.g., rat, mouse, and desert rodent) the architectural pattern of the inner stripe is different. The vascular bundles are of the so-called *complex* type, which differs from the simple type by the fact that the descending thin limbs of short loops descend within the vascular bundles. All other tubules keep their position similar to that found in the simple type of an inner stripe. In the complex bundles a countercurrent arrangement of three tubes is established: One type of ascending tube, the ascending vasa recta, is closely packed together with two types of descending tubes, the descending vasa recta and the descending thin limbs of short loops (Figure 12.5).

Both types of vascular bundles provide the possibility of a countercurrent exchange between the descending and ascending vasa recta serving the inner medulla (Figure 12.6). Since the ascending vasa recta from the inner stripe, for the major part, do not ascend within the bundles, it must be emphasized

FIGURE 12.1. Architectural organization of the kidney as revealed by four successive cross-sections: (a) cortex, (b) outer stripe, (c) inner stripe, and (d) inner medulla.

(a) Cross-section through the cortex shows that the parenchyma consists of the cortical labyrinth (*shaded area*) and the medullary rays (the two *white oval areas*). The labyrinth contains the interlobular blood vessels, glomeruli, and the arcades. For simplicity of presentation, the convoluted tubules, which are the major components of the labyrinth, are not shown. The medullary rays contain the collecting duct and the straight parts of the proximal and distal tubules (thick ascending limbs). In the lower left, the typical grouping of the collecting ducts within the medullary ray (*white oval*) is emphasized.

(b) Cross-section through the outer stripe shows two vascular bundles that replace the interlobular vessel axis of the cortex. The continuation of the medullary rays of the cortex is surrounded by *hatched oval lines*. Within the oval area, the collecting ducts and the loop limbs of the superficial and midcortical nephrons are found. The loop limbs of the juxtamedullary nephrons are situated immediately around the vascular bundles.

(c) Cross-section through the inner stripe shows the fully developed vascular bundles. Like the outer stripe, the loop limbs of the juxtamedullary nephrons are situated near the bundles, and those from the superficial and midcortical nephrons together with the collecting ducts are located distant from the bundles. Note the heterogeneity of the thin limbs: Those of the juxtamedullary nephrons lie near the bundles and are thicker in diameter, whereas those of the superficial nephrons lie distant from the bundles.

(d) Cross-section through the inner medulla. The area defined by the *dashed rectangle* corresponds to the entire area shown in section (c). This marked reduction in size is because short loops and many vasa recta have already turned back in the inner stripe. Note the grouping of the collecting ducts, which reflect the usual arrangement seen in the medullary rays of the cortex, and the absence of the thick ascending limbs. Thin limbs (both descending and ascending) are associated with the vasa recta or collecting ducts.

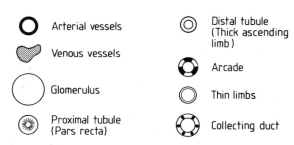

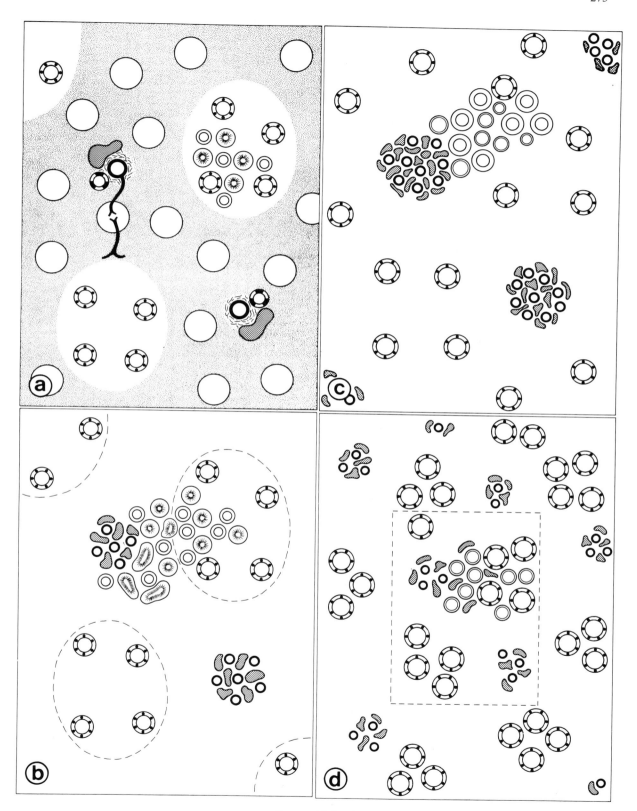

FIGURE 12.1a–d

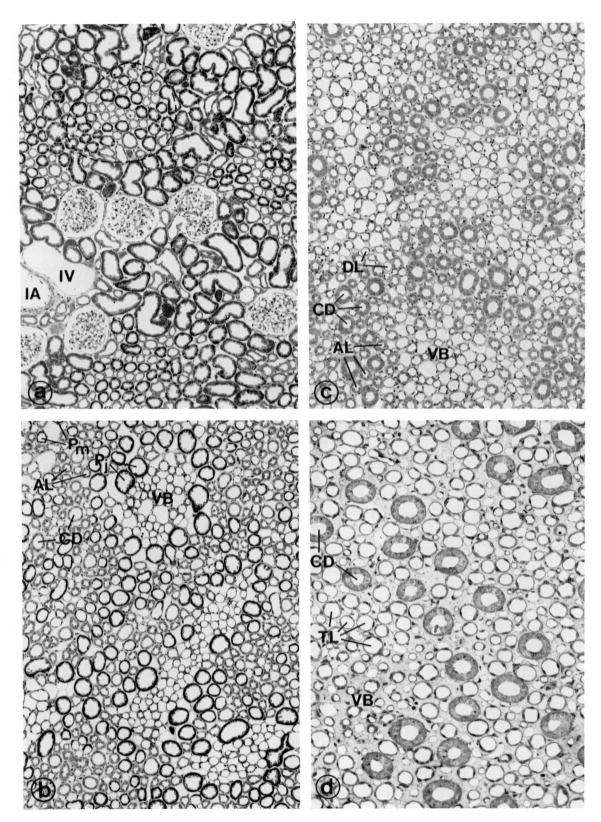

FIGURE 12.2a–d

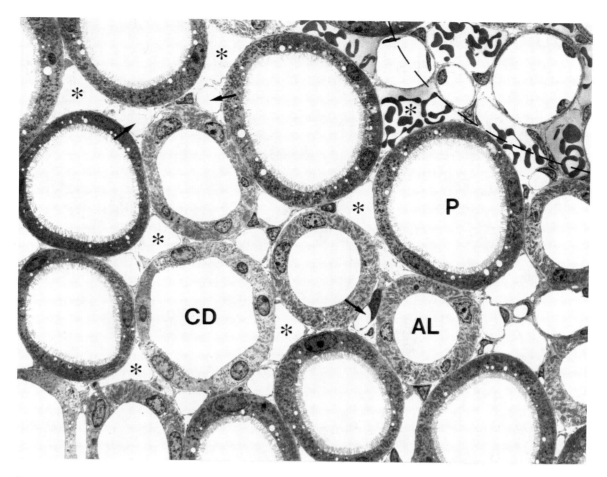

FIGURE 12.3. Electron micrograph of a cross-section through the outer stripe of the rabbit kidney. The corner of a vascular bundle is marked by a *dashed line*. Straight proximal tubules (P), thick ascending limbs (AL), and collecting ducts (CD) are shown. Note the cross-sections through several large capillary vessels marked by astericks (*) which correspond to the ascending vasa recta; the smaller capillary profiles marked by *arrows* presumably belong to the real capillaries supplied by the efferent arterioles.

◁ FIGURE 12.2. Light micrographs of a series of cross-sections through the (a) cortex, (b) outer stripe, (c) inner stripe, and (d) inner medulla of the human kidney.

(a) Demonstrates the two portions of the cortical parenchyma, the cortical labyrinth and the medullary rays (one is marked by a *dashed line*). The labyrinth contains the interlobular arteries (IA) and veins (IV), the glomeruli and the convoluted tubular segments. The medullary rays contain the straight tubular segments.

(b) In the outer stripe the tubules are longitudinally arranged around the vascular bundles (VB). The straight proximal tubules of juxtamedullary nephrons (P_j) are situated at the periphery of the vascular bundles and are larger in diameter than those of superficial and midcortical nephrons (P_m) lying distant from the bundles. AL = thick ascending limb and CD = collecting duct.

(c) In the inner stripe the vascular bundles (VB) are fully developed. The collecting ducts (CD) are arranged in distant rings around the bundles. DL = descending thin limb and AL = thick ascending limb.

(d) In the inner medulla the vascular bundles (VB) are small and cannot clearly be separated from the surrounding thin descending and ascending limbs. TL = thin limbs and CD = collecting duct.

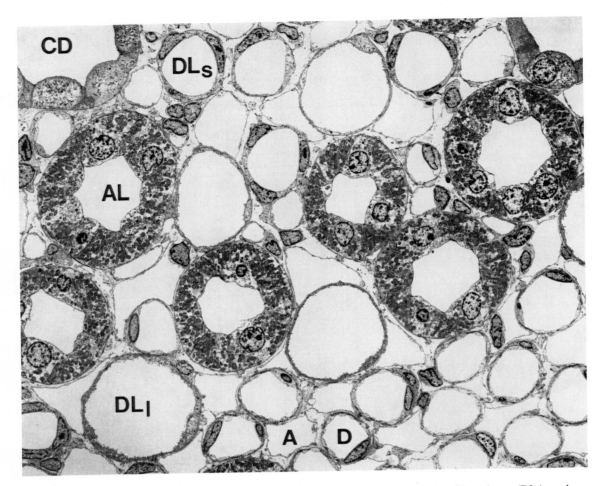

FIGURE 12.4. Electron micrograph of a cross-section through the inner stripe of the rabbit kidney. The vascular bundle, a portion of which is shown in the right lower corner, contains the descending (D) and ascending (A) vasa recta. The descending thin limbs of long loops (DL_l) are larger in diameter than those of short loops (DL_s). AL = thick ascending limb and CD = collecting duct.

that the vascular bundles serve as a countercurrent trap that is predominantly located in the inner stripe of the outer medulla, but mainly concerns the blood flow to and from the inner medulla. In addition to this possibility of a countercurrent exchange mechanism between descending and ascending vasa recta, both types of vascular bundles provide a route by which solutes leaving the inner medulla with the venous blood may be shifted to the inner stripe of the outer medulla.[47] In the simple type of bundles a countercurrent exchange of solutes between the ascending vasa recta from the inner medulla and the descending vasa recta leaving the bundles for the inner stripe will carry the solutes to the inner stripe capillaries and offer them to the inner stripe tubules.

In the complex type of bundles, the countercurrent arrangement between the ascending vasa recta from the inner medulla and the descending thin limbs of short loops is dominant and offers a more direct possibility that solutes from the inner medulla are shifted to short loops in the inner stripe of the outer medulla (see later: "Recycling of Urea").

Inner Medulla

The inner medulla is very differently developed among various species. However, all species with a high urinary-concentrating ability have a well-developed inner medulla.

The inner medulla contains thin descending and

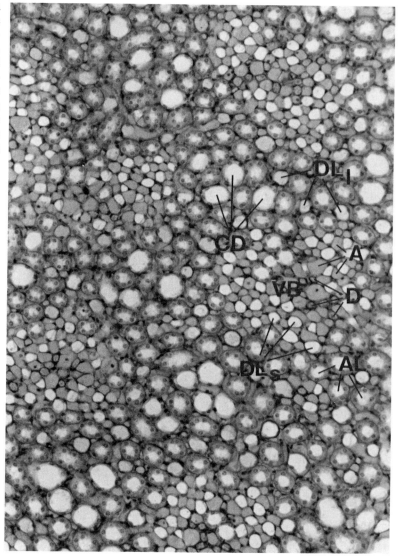

FIGURE 12.5. Light micrograph of a cross-section through the inner stripe of a mouse kidney showing the complex type of vascular bundles (VB). In addition to descending (D) and ascending (A) vasa recta, these bundles contain the descending thin limbs of short loops (DL_s). The descending thin limbs of long loops (DL_l) are thicker in diameter and lie outside the bundles. AL = thick ascending limb and CD = collecting duct.

thin ascending limbs of the long loops and the collecting ducts. A clear architectural pattern within the inner medulla is less apparent than in the outer medulla (Figures 12.1, 12.2, and 12.7). The vascular bundles are already drastically decreased in number of vasa recta when entering the inner medulla. Toward the papilla they are further diminished and finally they are fully absent. As far as bundles are discernible, the collecting ducts are generally situated at some distance from them. At the very beginning of the inner medulla, the collecting ducts are still arranged in groups reflecting their grouping within the medullary rays of the cortex. Joining of collecting ducts first occurs among the ducts of one group. Thin limbs of Henle, whether descending or ascending, may be associated with both collecting ducts and vasa recta. Hence, constant histo-topographical relationships between certain structures or special separations of others do not seem to be of major importance for the function of the inner medulla.

The interstitium is well developed in the inner medulla, increasing in volume toward the depth of the papilla. It contains a unique cell type, the so-called lipid-laden interstitial cell.[10] These cells are characterized by their many lipid droplets. In addition, they are arranged in a characteristic pattern: Like the rungs of a ladder they are transversely interposed between the longitudinal running tubules

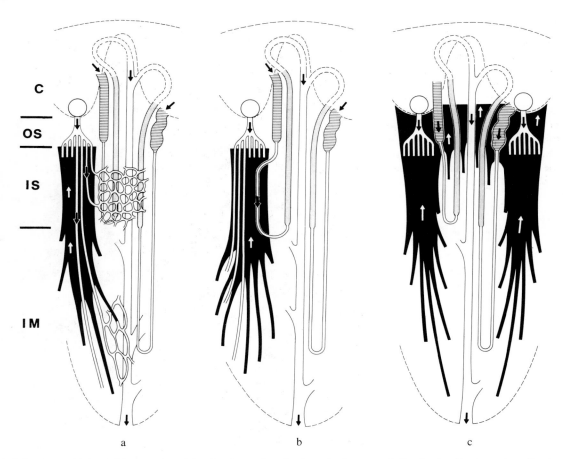

FIGURE 12.6. Possible connections established by the medullary circulation. In each of the three panels a short loop, a long loop, and a colleting duct are shown. The straight parts of proximal tubules are shown as *hatched*, the thin limbs and collecting ducts are shown in *white*, and the thick ascending limbs are shown as *dotted*. Descending vasa recta (DVR) (derived from efferent arterioles of juxtamedullary glomeruli) and capillaries are shown as *white*; for simplicity of presentation only a few DVR are shown in full length. Ascending vasa recta (AVR), drawn en bloc, are shown as *black*. C = cortex, OS = outer stripe, IS = inner stripe, and IM = inner medulla. The *left panel* (a) shows the simple type of medulla, where vascular bundles represent a countercurrent arrangement between DVR and AVR. Countercurrent exchange of some substance (e.g., urea) from ascending vasa recta to descending vasa recta will trap the substance in the inner medulla. As AVR from IM are also arranged along DVR destined for IS, there is a possibility that same substance (e.g., urea, originating from IM) will also be transferred to the IS capillary plexus. Entry into tubules of IS (descending thin limbs of short loops!) will open a recycling possibility back to the inner medulla via the distal tubule/collecting duct route. Reentry into the ascending vasa recta within the inner medulla will restart the process. The *middle panel* (b) shows the complex type of medulla, where in addition to DVR, descending thin limbs of short loops descend within bundles in a countercurrent arrangement with AVR from IM. In addition to countercurrent trapping between AVR and DVR, a recycling route via short loops of Henle is seen. Countercurrent exchange from AVR to descending limbs of short loops will return some substance to IM via the normal nephron and the collecting duct route. The *right panel* (c) shows the relationships between AVR and tubules in OS (valid for both simple and complex type of medulla). The total venous blood from the medulla traverses the outer stripe in wide tortuous channels contacting tubules of OS like capillaries. Whether this arrangement serves as an ultimate trap for medullary solutes and/or is part of a recycling route for substances coming up from the medulla and secreted into the proximal tubules is unknown. (Modified from Kriz.[47]) *For color art see front matter*.

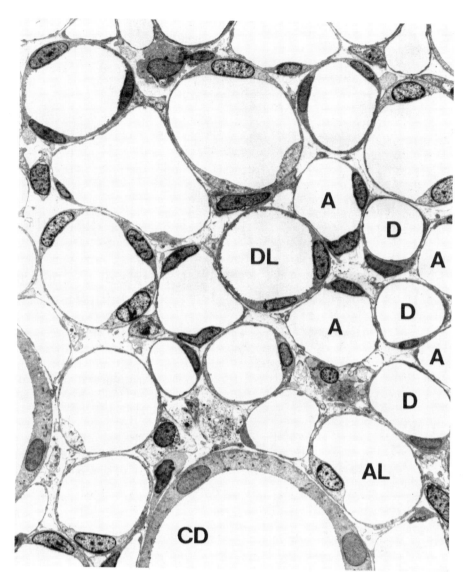

FIGURE 12.7. Electron micrograph of a cross-section through the inner medulla of a rat kidney. Descending (D) and ascending (A) vasa recta are more or less intermingled with descending (DL) and ascending (AL) thin limbs. CD = collecting duct.

and vessels (Figure 12.8). These cells are thought to serve two specific functions. First, they synthesize large amounts of prostaglandins. Whether their lipid droplets are related to this function is not known. Second, these cells appear to cut the interstitium into individual compartments that are successively arranged from the base to the tip of the inner medulla. Therefore, a function as obstacles to interstitial diffusion in the axial direction may be supposed.

The inner medulla has a very peculiar shape. It is characteristic for the inner medulla to taper from a broad base to a thin papilla (or crest). The mass of the inner medulla is therefore unevenly distributed along the longitudinal axis. A reconstruction study in the rat has shown that the inner medulla is shaped like a mushroom, consisting of a broad hat and a thin stalk (the papilla). Calculations in the model have shown that the first half of the inner medulla comprises roughly 80% of the total inner medullary volume and consequently only 20% is left for the papillary half.[38] This outer shape perfectly reflects what happens with the structures within the inner

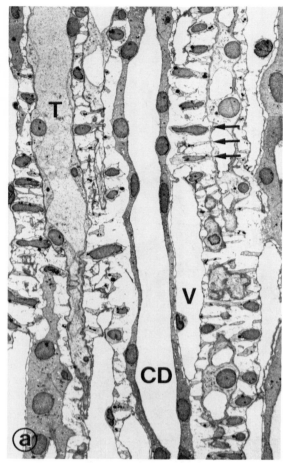
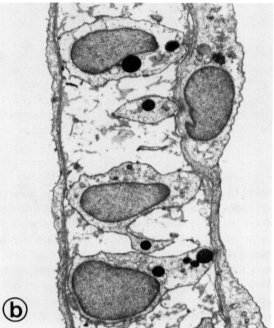

medulla: Loops of Henle, collecting ducts, and vasa recta all decrease rapidly in number from the base to the tip of the papilla. It has been calculated that in the rat, out of an estimated number of 10,000 long loops entering the inner medulla at its base, only about 1,500 reach the papillary half of the inner medulla and only a few reach the tip of the papilla.[38] Thus, the highest osmolalities are finally generated only by a very small proportion of long loops within an extremely small volume fraction of the medulla.

Let us now see how these anatomical features of Henle's loop and its associated blood supply contribute to concentration and dilution of urine as perceived by the countercurrent multiplication hypothesis described next.

Principle of Countercurrent Multiplication

In 1951 Hargitay and Kuhn[33] proposed a theory for the mechanism of concentration and dilution of urine based on the premise that Henle's loop, by virtue of its anatomical position within the kidney, acts as a countercurrent multiplier system. According to this theory, if the epithelial cells of the ascending limb of Henle's loop were able to establish a small osmolar concentration difference between the fluid contents of the ascending and descending limbs at each level along the length of the loop, this small transverse gradient could be multiplied into a significant longitudinal gradient along the length of the loop by the countercurrent (opposite in direction) flow of tubular fluid in the two limbs.

Applying this concept to the kidney, one would expect that the isotonic fluid issuing from the proximal tubule should become progressively concentrated as it traverses along the descending limb of the loop, attaining maximum osmolality in the loop bend. Then, as the fluid travels up the ascending limb, it should become progressively hypotonic as it enters the early distal tubule. The osmotic gradient thus established between the two limbs along the length of the loop could result from either (1) net

◁ FIGURE 12.8. Light (a) and electron (b) micrographs showing the lipid-laden interstitial cells of the inner medulla. Note the characteristic arrangement comparable to the rungs of a ladder (*arrows*). In (b) the characteristic lipid droplets (*black*) are seen. T = thin limb, V = vas rectum, and CD = collecting duct.

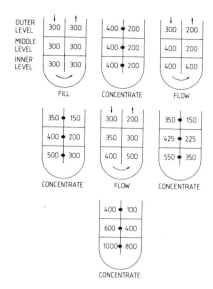

FIGURE 12.9. A model depicting the operation of the countercurrent multiplication in Henle's loop.

of 300 mOsm/L, which is iso-osmotic with the plasma. The loop is subdivided into three levels, the cells of which are capable of active solute transport (mainly sodium chloride) from the ascending to descending limbs. They are designated outer, middle, and inner levels. The first step in the operation of the countercurrent multiplication is the establishment of the transverse gradient. This is accomplished by the active transport of enough sodium chloride from the ascending limb to the descending limb to create a maximum transverse gradient of about 200 mOsm/L. Note that since the concentration of the fluid in the descending and ascending limbs increase in parallel toward the loop bend, the uphill transport of salt necessary to generate the transverse gradient remains small at any level. The next step involves the flow of the contents of the compartment in the outer level out of the ascending limb into the distal convoluted tubule, while an equal volume of fluid having a concentration of 300 mOsm/L enters from the proximal tubule into the descending limb. The net effect is the separation of salt and water; salt is trapped within the medulla and water is carried up into the cortex. The subsequent steps are the repeat of the previous ones. That is, each flow step is followed by a concentrate step until the final steady-state concentrate condition shown at the bottom of Figure 12.9 is reached. For the operation of this system within the renal medulla, it is necessary that equilibration of the loop fluid with the surrounding interstitium occur at every level, resulting in a similar longitudinal osmotic gradient throughout the medullary interstitium.

From this model, it is evident that the magnitude of the longitudinal gradient depends on (a) the magnitude of the transverse gradient, (b) the length of Henle's loop, and (c) the tubular flow rate. Thus, in the steady state, the mechanism of urine concentration depends on the operation of two processes: (1) The countercurrent multiplication of the continuously generated concentration difference between the two limbs of Henle's loop, which establishes an osmolar concentration gradient within the medullary interstitium from the corticomedullary border to the tip of the papilla. (2) The osmotic equilibration of the adjacent collecting duct fluid with the surrounding hypertonic medullary interstitium. As we shall see later, the medullary vasa recta complements the operation of both processes and enhances their efficiency by removing water in excess of solutes so as to preserve the hypertonicity generated within the medullary interstitium.

transfer of water from the descending limb to the ascending limb (an improbable biological mechanism), or (2) net solute transport in the opposite direction. In either case, the initial transverse gradient at each level along the loop, which could be quite small, is multiplied into a significant gradient along the longitudinal medullary axis of Henle's loop. The net result would be the development of a high osmolar concentration gradient along the meullary interstitium to which the collecting tubules would be exposed. The mechanism of formation of hypertonic urine is perceived to be the consequence of the osmotic equilibration of the collecting duct fluid with the surrounding hypertonic medullary interstitium. The osmolality of the excreted urine would thus depend on the magnitude of the generated longitudinal gradient, which in turn is governed by (a) the magnitude of the transverse gradient (the so-called "single effect"), (b) the length of Henle's loop, (c) the solute and water permeability of the epithelia of the descending and ascending limbs, and (d) the flow rate in Henle's loop, the collecting duct, and the vasa recta, as well as interactions between these structures.

To better visualize the operation of the countercurrent multiplication system, let us consider the model shown in Figure 12.9. Initially, at time zero (upper left corner) the descending and ascending limbs are shown to be filled with fluid containing 150 mM/L of sodium chloride, for a total osmolality

Principle of Countercurrent Exchange

A countercurrent solute exchanger in its simplest form may be represented by two countercurrent tubes attached to each other. Fluid flowing in the two tubes may belong to different circulations (as for instance is the case in the placenta, where solute transfer between the fetal and maternal blood is enhanced by countercurrent exchange) or, as in the renal medulla, may belong to the same circulation when both tubes form a hairpin loop.[73] We will only consider this latter case. In such a system there is an inflowing limb and an outflowing limb and some fluid circulating through the loop. The separating membrane along and between both loop limbs has to be highly permeable to the solutes involved to allow diffusive equilibration between both loop limbs (Figure 12.10).

In the renal medulla, the blood vessels are arranged in a countercurrent fashion and act as a countercurrent solute exchanger. The descending vasa recta represent the inflowing tube, and the ascending vasa recta represent the outflowing tube. They are packed together in a checkerboard pattern within the vascular bundles. The entire complex arrangement represents the solute exchanger system. Descending and ascending vasa recta are not directly connected to each other by a bend; the connection is effected by the capillaries.

A countercurrent exchanger can only be of functional significance if an axial gradient has already been created and is maintained by some different mechanism. In the kidney the axial osmotic gradient along the cortico-papillary axis is created and maintained by countercurrent multiplication within Henle's loop. To create a concentration difference of any kind, some sort of an active, energy-consuming process is necessary. Within the kidney this active process is represented by the active salt reabsorption from the ascending limbs of Henle's loops.

Since the renal medulla has to be nourished, some blood has to flow through the medulla. Such an additional flow of fluid through an hypertonic compartment would quickly wash out the solutes and abolish the gradient. To minimize the washing out effect, a countercurrent exchanger is built in. Within the medullary capillaries the blood reaches osmotic equilibration with the surrounding hypertonic interstitium. When the blood is flowing out in the ascending vasa recta, the countercurrent arrangement of descending and ascending vasa recta within the

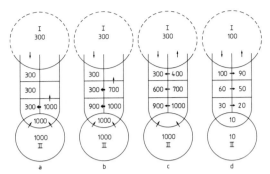

FIGURE 12.10. This figure visualizes the operation of a countercurrent exchanger involving the same circulation. A hairpin loop connects compartment I (equivalent to the renal cortex and systemic circulation) to compartment II (equivalent to the papilla of the renal medulla). Schematics (a), (b), and (c) demonstrate the trapping effect. In compartment I there is an overall osmolality of 300 mOsm/L H_2O compared with an osmolality of 1,000 mOsm/L H_2O in compartment II, which has been built up and is maintained by the countercurrent multiplication process (Figure 12.9). Blood circulating through compartment II will equilibrate by solute entry with the high osmolalities at this site; thus, the concentration of the outflowing blood will be raised. Therefore, at any level of the loop, solutes will diffuse from the outflowing tube into the inflowing tube, and, consequently, the inflowing blood will gradually increase in concentration and the outflowing blood will gradually decrease almost down to systemic concentrations. Note that in contrast to a countercurrent multiplier system operating by active solute transport (Figure 12.9), the concentration of the outflowing fluid in a countercurrent exchanger is always slightly higher than the concentration of the inflowing fluid, reflecting that osmotic equilibration in a flowing system is never perfect. Schematic (d) shows the excluding effect of a countercurrent exchange system. Oxygen, whose tension is always higher in the arterial blood entering the system than in the venous blood leaving the system, will diffuse at any level of the loop from the inflowing to the outflowing tube, resulting in very low oxygen tensions at the loop bend in compartment II. Hence, oxygen is passively excluded from entering the system.

vascular bundles allows solutes to diffuse from the ascending vasa recta into the descending vasa recta. Consequently, the outflowing fluid will successively decrease in concentration, whereas the concentration of the inflowing fluid will successively increase. Since full equilibration in a flowing system can not be reached by passive diffusion, the osmotic concentration of the blood leaving the medulla by the ascending vasa recta will still be somewhat higher than the concentration of the blood entering the descending vasa recta. Hence, even by means of a countercurrent exchanger, the washout of solutes can not be fully prevented. A countercurrent exchanger can neither create a concentration difference along the axis of the system nor maintain those differences; it can only minimize the dissipation of these differences.[38]

A countercurrent exchanger may also act in the opposite way. Think of some substance that always has a higher concentration in the inflowing tube than in the outflowing tube and to which the separating membrane between both tubes is permeable. In this case the substance will be excluded from entering the system (Fig. 12.10d). In the kidney this is the case with oxygen, which is in higher concentration in the arterial blood of the inflowing tube than in the venous blood of the outflowing tube. Therefore, at any level of the countercurrent exchange system, oxygen will diffuse from the inflowing tube to the outflowing tube, resulting in very low oxygen concentrations at the loop bend. From an overall point of view, countercurrent exchange systems of the hairpin type (involving the same circulation) have a barrier function.[49] Depending on the direction of the axial gradient for some solute, countercurrent exchange will trap some solute within the system or will exclude it from entering the system. In contrast, countercurrent exchange between two different circulations (e.g., placenta) will produce an equilibrating effect.

Experimental Evidence for Existence of the Countercurrent Multiplication System in Henle's Loop

The original experimental verification of the countercurrent multiplication hypothesis was provided by Wirz and colleagues.[90] These investigators found that when in situ frozen sections of the kidney of hydropenic (thirsty) rats were thawed under microscope, the cortical tissue fluid melted at the same temperature as the arterial blood. This implied that the fluids in the tubules, which are located in the cortex, were iso-osmotic with the blood. However, the fluid contained in the tissues from the outer and inner medulla melted at progressively lower and lower temperatures, compared with the arterial blood, indicating that the fluids in the medullary structures, namely, Henle's loop, collecting duct, and vasa recta, became

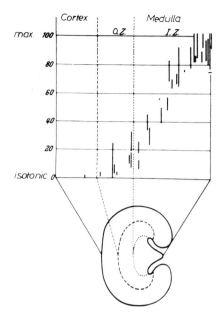

FIGURE 12.11. Osmolality of tissue slices from the cortex, the outer zone (OZ) and the inner zone (IZ) of the medulla of the kidney of a hydropenic rat. Ordinate scale is the measured tonicity expressed as a percentage of maximum. (From Wirz et al.[90])

increasingly hyperosmotic relative to the blood. Figure 12.11 graphically presents a summary of their findings. It shows that as we proceed from the cortex toward the papilla, in the longitudinal direction (in the figure from left to right across the kidney), the tissue fluid becomes progressively hyperosmotic, while in a given transverse level (in the figure along the vertical direction within a zone) the fluid osmolality remains the same. The earlier observations of Wirz and associates[90] have subsequently been confirmed in several studies.

Gottschalk and Mylle[32] were the first to make a systematic micropuncture study of the various accessible segments of the nephron in both rat and hamster, thereby providing the most impressive evidence in support of the countercurrent multiplication in Henle's loop and its role in concentration and dilution of urine. They found that in rats the fluids collected from the proximal tubule remained iso-osmotic (fluid-to-plasma or F/P osmolality ratios equal to unity), regardless of whether the ureteral urine was hypo-osmotic or hyperosmotic. Furthermore, they observed that in hydropenic rats forming highly concentrated urine, the fluid collected from the early distal convoluted tubule was consistently hypo-os-

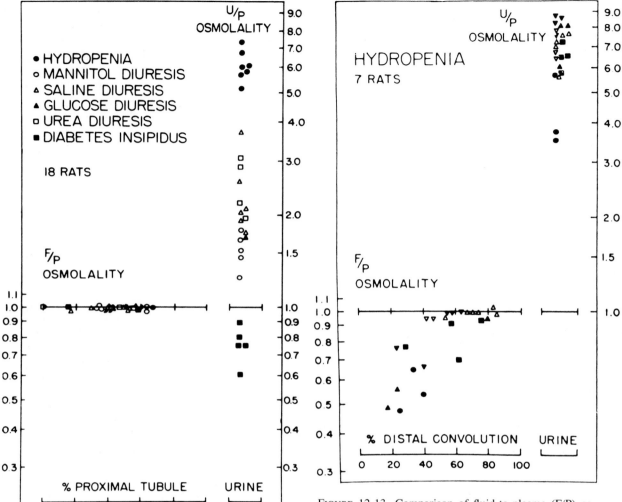

FIGURE 12.12. Comparison of fluid-to-plasma (F/P) osmolality ratios of fluid collected from the proximal convoluted tubule with the urine-to-plasma (U/P) osmolality ratios in rats under different diuretic conditions. (From Gottschalk and Mylle[32])

FIGURE 12.13. Comparison of fluid-to-plasma (F/P) osmolality ratios of fluid collected from the distal convolution with the urine-to-plasma (U/P) osmolality ratios during hydropenia. Different symbols refer to different rats. Note that in accordance with recent findings, the term distal convolution used in this figure denotes a heterogeneous segment consisting of the distal convoluted tubule, the connecting tubule, and part of the cortical collecting duct. (From Gottschalk and Mylle.[32])

motic (F/P osmolality ratios less than unity) and became iso-osmotic in the later segments of the renal tubule at the cortical surface, but never hyperosmotic. These findings, which are summarized in Figures 12.12 and 12.13, clearly demonstrate that the formation of hyperosmotic urine is a consequence of the collecting duct function, thus providing compelling experimental evidence in support of the countercurrent multiplication theory.

As mentioned earlier, the observed hypotonicity of the fluid entering the early distal convoluted tubule can be the result of either two processes: (1) solute transport from the ascending limb to the descending limb, or (2) water transport in the opposite direction.

To differentiate between these two processes, Gottschalk and Mylle[32] induced osmotic diuresis by infusion of either a nonreabsorbable solute, such as mannitol, or sodium chloride, and determined the F/

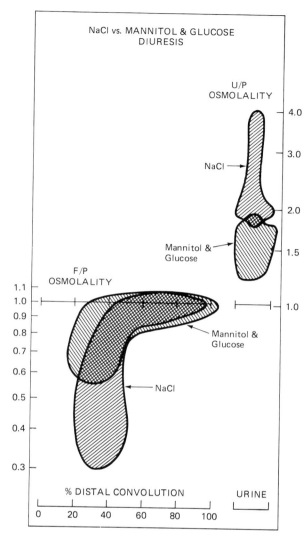

FIGURE 12.14. Comparison of fluid-to-plasma (F/P) osmolality ratios of fluid collected from the distal convolution and urine-to-plasma (U/P) osmolality ratios during hypertonic sodium chloride diuresis with those during hypertonic mannitol and glucose diureses. Note that in accordance with recent findings, the term distal convolution used in this figure denotes a heterogeneous segment consisting of the distal convoluted tubule, the connecting tubule, and part of the cortical collecting duct. (From Gottschalk and Mylle.[32])

P osmolality ratios in the fluid collected from the early distal convoluted tubule. As shown in Figure 12.14, when diuresis was induced by intravenous infusion of a 25% solution of mannitol, urine flow increased up to 80 times that of the hydropenic state, and the F/P osmolality ratios in the early distal convoluted tubule were low but not less than 0.6. Similar results were obtained during glucose diuresis, a solute that is not reabsorbed in Henle's loop. However, when diuresis was induced by intravenous infusion of a 5% sodium chloride solution, urine flow rate increased to the same level as with mannitol and glucose, but the F/P osmolality ratios in the early distal convoluted tubule were as low as 0.3. During both sodium chloride diuresis and mannitol or glucose diuresis, the fluid collected from the later segment of the renal tubule at the cortical surface (presumably the early portion of the cortical collecting duct) was iso-osmotic with the plasma.

The increase in the degree of hypotonicity of the fluid in early distal convoluted tubule during sodium chloride diuresis, compared with mannitol or glucose diuresis, was taken as a strong indication that this observed hypotonicity is due to hyperosmotic reabsorption of sodium chloride from the ascending limb of Henle's loop and not the addition of water. Thus, they suggested that it is this hyperosmotic solute reabsorption that is responsible for the development of the transverse gradient, which when multiplied leads to the establishment of a significant longitudinal osmolality gradient in the surrounding medullary interstitium.

A "First Approach" Model of the Operation of the Countercurrent Multiplication System in Henle's Loop

On the basis of these micropuncture studies and other evidence cited above, Gottschalk and Mylle[32] proposed a working model for the countercurrent multiplication hypothesis as it applies to the kidney. The salient features of their model are depicted in Figure 12.15.

This model envisions that sodium and chloride ions are reabsorbed by an active mechanism from the relatively water-impermeable ascending limb of Henle's loop into the surrounding medullary interstitium until a limiting gradient of 200 mOsm/kg of water has been established between the medullary interstitium and the fluid in the ascending limb. This *"single effect"* of osmolar concentration gradient is then multiplied as the fluid in the descending limb comes into osmotic equilibrium with the surrounding hypertonic medullary interstitium by diffusion of water out of and sodium chloride into the descending

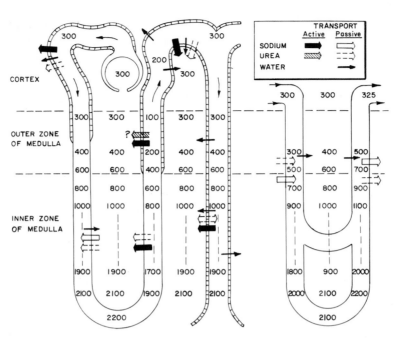

FIGURE 12.15. Simplified model for the operation of the countercurrent multiplication system in a nephron with a long loop and in the vasa recta. The numbers represent hypothetical osmolality values. No quantitative significance is to be attached to the number of arrows, and only net movements are indicated. As is the case with the vascular loops, not all the loops of Henle reach the tip of the papilla, and hence the fluid in them does not become as concentrated as that of the final urine, but only as concentrated as the medullary interstitial fluid at the same level. (Redrawn from Gottschalk and Mylle.[32])

limb. The overall effect would be that an increasing osmolar concentration gradient is established in the longitudinal direction within the interstitium, extending from the corticomedullary border to the tip of the papilla, as was found by Wirz and associates.[90] The osmolality of the urine issuing from the collecting duct was assumed to be the consequence of passive osmotic equilibration of the fluid in this tubule segment as it traverses through the hypertonic medullary interstitium. Ullrich and Jarausch[87] found that the hypertonicity of the medullary tissues was due largely to increased concentration of sodium chloride and urea in this region. Furthermore, consistent with the micropuncture studies summarized in Figure 12.12, Gottschalk and Mylle[32] suggested that in the presence of antidiuretic hormone (ADH), the epithelium of the collecting duct is maximally water-permeable, thereby ensuring the osmotic equilibration of the fluid in the collecting duct with the surrounding hypertonic medullary interstitium. Moreover, they postulated that the degree to which the urine in the collecting duct is concentrated depends on the relative rates of tubular flow along Henle's loop and the collecting duct. As described in Chapter 9, a slow rate of tubular flow tends to increase the tubular transit time, thereby prolonging the contact time between the luminal fluid and the tubular epithelium. This would favor a maximum attainment of osmotic equilibration of tubular fluid with the surrounding hypertonic medullary interstitium and hence a greater concentration of urine.

As illustrated in Figure 12.15 and envisioned by this model, the vasa recta play an important role in the maintenance of the longitudinal hypertonicity of the medullary interstitium with which the collecting duct fluid comes into osmotic equilibrium. Furthermore, renal handling of urea within the medulla contributes significantly to the efficiency of the urinary concentration by Henle's loop. Because of their importance, the roles of both vasa recta and urea in the operation of the countercurrent multiplication process is discussed later in separate sections of this chapter.

As predicted by the countercurrent hypothesis, Gottschalk found that in hydropenic rats, in which plasma ADH concentration is high, fluid collected from Henle's loop and the vasa recta near the tip of the papilla are essentially in osmotic equilibrium with urine collected from the adjacent collecting duct.[31] This is illustrated in Figure 12.16, which also shows that in animals with diabetes insipidus the collecting duct fluid was hypo-osmotic relative to the fluid collected from adjacent Henle's loop and vasa recta. However, the hypertonicity of the latter fluids was much less compared with that of the hydropenic rats. Furthermore, a comparison of the osmolality of the fluid in the early distal convoluted tubule between intact hydropenic rats and rats with diabetes insipidus during sodium chloride diuresis showed a marked reduction in the F/P ratios in this segment, with eventual decrease in U/P osmolality ratios. These results, reproduced in Figure 12.17, provide unequivocal evidence for the effect of ADH on the water permeability of the epithelia of the cortical collecting duct and the role of this hormone in concentration and dilution of urine.

Modern Concepts of Urinary Concentrating Mechanisms

As has been shown in the preceding section, the fundamental predictions of the countercurrent multiplication hypothesis have been largely confirmed. However, there remain several facts that cannot be explained by the straight-forward model of medullary countercurrent multiplication as presented in the preceding section. First of all, the structural organization of the renal medulla deviates in many respects from the simple model of a countercurrent multiplier system. There are three structurally different parts of the medulla, and these should be expected to have different functions. In addition, there are short and long loops of Henle, with structurally different segments in the descending and ascending limbs, which also should be expected to differ in function. Most discrepant to the proposed function is the histological simplicity of the thin ascending limb, whose epithelium does not appear to possess the capability of transcellular active salt transport and, in functional studies, has never been shown to have these properties. Moreover, there is a medullary vascular system, consisting of descending and ascending vasa recta connected by capillaries, whose role within the concentrating process is not explained by Kuhn's original countercurrent hypothesis. Functionally, the fact that urea administration to protein-deprived rats increases the total urine concentration, including the non-urea solutes, which has been known since 1934[28] and repeatedly confirmed subsequently, can not be explained by the countercurrent hypothesis. Finally, it has been shown that the reabsorbed water from

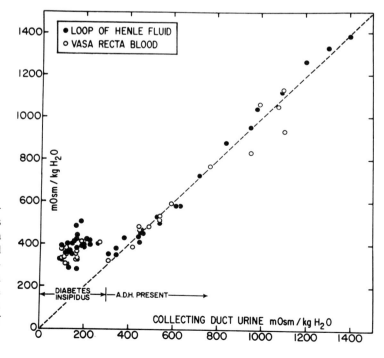

FIGURE 12.16. Relation between the osmolality of collecting duct urine and fluids collected from Henle's loop and vasa recta blood in various normal desert rodents and in hamsters with experimental diabetes insipidus. Soime of the values obtained in the presence of antidiuretic hormone (ADH) are from hamsters with diabetes insipidus following the administration of exogenous ADH. (From Gottschalk.[31])

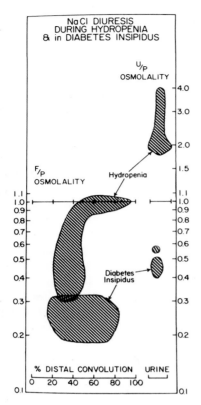

FIGURE 12.17. Comparison of fluid-to-plasma (F/P) osmolality ratios of fluid collected from the distal convolution and urine-to-plasma (U/P) osmolality ratios during sodium chloride diuresis in normal hydropenic rats and in rats with experimental diabetes insipidus. Note that in accordance with recent findings, the term distal convolution used in this figure denotes a heterogenous segment consisting of the distal convoluted tubule, the connecting tubule, and part of the cortical collecting duct. (From Gottschalk.[31])

the collecting ducts does not reenter the loops of Henle as required by the original model but leaves the medulla by the ascending vasa recta (see later).

These discrepancies have stimulated several alternative hypotheses. Before presenting these alternative hypotheses, it seems appropriate to briefly discuss what is known about the structural and functional properties of the individual tubular segments involved in the concentrating process.

Descending Thin Limb

As presented in Chapter 9, the descending thin limb is not a homogeneous segment as regards its epithelial organization. The descending thin limbs of short loops are made up of a different type of epithelium than those of long loops. In the descending thin limbs of long loops, two segments, an upper and a lower segment, must be distinguished. Moreover, the upper segment is subject to considerable interspecies differences. In the rabbit kidney, the descending thin limbs of long loops have a much simpler epithelial organization than in most other species. In microperfusion experiments of descending thin limbs of the rabbit a very high hydraulic conductance combined with a low permeability to salts and urea has been found.[44, 61] These findings do not seem to hold completely true for other species. Very recently, the structural interspecies differences in the long descending thin limbs have been shown to be paralleled by functional differences. In rats and hamsters, the upper part of long descending thin limbs are highly permeable to cations, in addition to a high water permeability.[35, 85] In these studies, for the first time, functional differences between short and long descending limbs (upper parts) have been uncovered: short descending thin limbs, like those of long loops, have a high hydraulic conductance but differ from them in the low cation permeability and relatively high permeability for urea (for more details see Chapter 9).

A major functional question concerning the role of descending thin limbs in the concentrating process is whether osmotic equilibration of descending limb fluid occurs by water extraction, solute addition (salt, urea), or a combination of both. From the first in-vitro studies by Kokko,[44] it has been calculated that equilibration of descending thin limb fluid occurs 96% by water extraction and only 4% by solute entry. These calculations have provided the basis for the development of the so-called passive model of urine concentration (see below). However, these calculations obviously do not hold true for most other species. Earlier studies on the descending thin limbs of long loops in the rat either in the excised papilla[64] or in-vivo in the exposed papilla[67] have concluded that solute entery into the descending thin limbs accounts for a much greater (up to 50%) fraction of the equilibration process along the descending thin limbs. Recent findings by Imai and his co-workers[35, 85] also favor a combination of solute entry and water extraction as being responsible for osmotic equilibration in the descending thin limbs in rats and hamsters. Studies in a desert rodent (*Psammomys obesus*)[18, 19] have shown that in this particular species the rise in osmolality in descending thin limb fluid occurs 85% by solute addition and only 15% by water extraction. These interspecies differences must be kept in mind for the subsequent discussion of proposed models.

Ascending Thin Limb

Ascending thin limbs are only found in long loops. Their epithelial organization is homogeneous along their entire course and appears to be similarly structured among various species. This corroborates with most of the functional studies performed in-vitro and in-vivo in rat, hamster, and rabbit (see Table 9.2 in Chapter 9), which all tend to show that (1) there is no active salt transport in this segment, (2) the epithelium is highly permeable to sodium and chloride (highest permeability to these ions found anywhere in the renal tubule), and (3) the epithelium has a very low osmotic water permeability. Consequently, equilibration of tubular fluid flowing up the ascending thin limbs occurs mainly by diffusive exit of sodium chloride. These findings are consistent with the hypothesis of a passive urinary concentration mechanism in the inner medulla (see below).

Thick Ascending Limb

The thick ascending limb of Henle's loop performs the key function in the urinary concentrating process: it separates salt from water. It actively reabsorbs sodium chloride (for the mechanism, see Chapter 9) from the tubular fluid, and since this epithelium is virtually completely impermeable to osmotic flow of water, the tubular fluid emerging from the end of this segment in the cortex is very dilute. Thus, the salt is trapped in the medulla and the water is carried away into the cortex. The major solute left in the tubular fluid is urea, even though a proportion of the urea is expected to diffuse out of the cortical part of the thick ascending limb into the interstitium.[42] This hypotonicity of tubular fluid at the end of the thick ascending limb is achieved regardless of whether the kidney produces a dilute or a concentrated urine. In case the kidney produces a dilute urine, the hypotonicity of the tubular fluid at the end of the thick ascending limb is maintained throughout the distal convoluted tubule, the connecting tubule, and the collecting ducts, even if the overall composition of this fluid with regard to its fractional content of electrolytes, urea, and H^+ ions is largely changed by distal tubules and collecting ducts. Thus, the diluting capacity of the kidney is established by the function of the thick ascending limb and can be enhanced by solute reabsorption from the collecting ducts (see below).

The thick ascending limb consists of two parts, a medullary part (MAL) and a cortical part (CAL). There are several differences between the two parts. As already described in Chapter 9, although the basic salt reabsorption mechanism is the same in both segments, the medullary thick ascending limb can reabsorb sodium chloride at a faster rate than the cortical segment, but cannot produce as great a transepithelial sodium chloride concentration difference. Thus, under normal conditions most of the sodium chloride is already reabsorbed in the medullary part and comparably little is left for the cortical segment to establish the final dilution. Consequently, most of the salts reabsorbed by this segment are added to the medullary interstitium and are trapped within the medulla, thereby enhancing the concentrating mechanism. The cortical part of the thick ascending limb is located within quite another environment and salts reabsorbed from this segment will dissipate into the systemic circulation. ADH stimulates salt reabsorption from the medullary thick ascending limb (see Chapter 9), and thereby, obviously, the fraction of salt reabsorbed from the medullary thick ascending limb is increased relative to the fraction reabsorbed from the cortical thick ascending limb.

At the end of the cortical thick ascending limb, most of the filtered sodium chloride has been reabsorbed. The macula densa, which is located within the terminal portion of the cortical thick ascending limb, appears to control the diluting function of the thick ascending limb. As has been pointed out by Knepper and Burg,[43] the sodium delivery to the distal convoluted tubule is obviously buffered against rapid short-term changes by various mechanisms, including glomerulo-tubular balance, flow-dependent reabsorption in Henle's loop and, finally, by the macula densa feedback mechanism.

Collecting Duct

The collecting ducts are the only tubular segment that is present in all kidney zones. The collecting ducts begin in the medullary rays of the cortex (or may have a short portion already developed in the cortical labyrinth) and accept the tubular fluid from several nephrons. They then descend through the outer medulla, joining successively within the inner medulla to open finally as large papillary ducts on the tip of the papilla into the renal pelvis.

The epithelial organization, as was described in detail in Chapter 9, is not uniform along the course of the collecting ducts. There are major difficulties in correlating the structural heterogeneity with differences in function of the various parts of the collecting ducts since the individual transport properties of principal cells and intercalated cells are not known. In the context of this discussion two points have to be considered in more detail, namely, *urea and water permeability*. The cortical and outer medullary collecting ducts have a very low permeability to urea, whereas the inner medullary collecting ducts are highly permeable to urea.[68] So far, no structural component is known to account for this difference, which suggests that some unknown properties of the luminal membrane must be responsible. It is necessary to mention here that the structures of the luminal membrane (pores, channels, etc) affecting the urea permeability are different from those responsible for the water transfer across the luminal membrane of collecting ducts. The urea permeability does not change with ADH (although a small permeability increase in the papillary collecting duct with ADH has been reported[68]) as does the water permeability. When considering the distal nephron, it should be noted that urea impermeability starts already in the thick ascending limb. However, recent evidence[42] suggests an exception to this pattern in the cortical thick ascending limb which has been found to have sufficiently high urea permeability to allow urea to exit this segment. Otherwise, the impermeabilty to urea includes the distal convoluted tubule, and the connecting tubule, and terminates in the outer medullary collecting ducts. The restriction of high urea permeability to the inner medullary collecting ducts is considered to limit the volume into which the reabsorbed urea is distributed and thus maximize urea concentration at the tip of the papilla for a given supply of urea.[43] The difference in urea permeability between the inner medullary collecting duct and the remainder of the collecting duct system represents the main basis for the passive model of urine concentration.

The water permeability of the collecting ducts along their entire length is low in the absence of ADH, but is dramatically increased when ADH is present. Water reabsorption from the water-permeable collecting ducts, beginning within the cortex and extending down to the papillary tip, is responsible for the final concentration of urine. Tubular fluid that has been diluted in the thick ascending limb flows through additional water impermeable segments (the distal convoluted tubule and the connecting tubule) before it reaches the cortical collecting duct. In antidiuresis, when ADH is present, the cortical collecting ducts are the site of rapid water reabsorption and are where equilibration of tubular fluid with the surrounding cortical interstitium (plasma concentration) is achieved. Thus, the net amount of water that has been separated from its salt in the medulla by the function of the thick ascending limb is already reabsorbed from the collecting ducts in the cortex and is carried away into the systemic circulation. The tubular fluid, which in antidiuresis enters the medulla from the cortex within the collecting ducts, is already isotonic. The subsequent concentration of urine in the medulla above plasma level is only possible if this requirement has been fulfilled.[80]

Water extraction from medullary collecting ducts finally concentrates the urine. The amount of water reabsorption required to raise tubule fluid osmolality to plasma concentration in the cortical collecting duct is greater than the additional amount required to concentrate the urine to any level above plasma osmolality within the renal medulla. The mechanism is the same as in the cortex: water extraction by osmosis through the ADH-dependent water permeable epithelium of the medullary collecting ducts. Overall, water extraction in the medulla is effected by the salt that has been trapped in the medulla owing to the function of the thick ascending limb; no other source of energy is so far known. The question, therefore, which has to be answered, is: How does the accumulated salt together with other solutes (urea) lead to a gradual increase in osmotic gradient, reaching its maximum at the tip of the papilla? The following is an attempt to answer this question.

Proposed Mechanisms

The various concepts advanced to explain the urinary concentrating mechanism all agree that the mechanism in the inner medulla must be different from that in the outer medulla. In the outer medulla (more precisely, in the inner stripe of the outer medulla), a basic mechanism operates to concentrate the interstitial fluid up to about twice the plasma concentration (i.e., 600–700 mOsm/L H_2O) at the border between the inner and outer medulla. The resulting concentration gradient induces osmotic water reabsorption from the collecting duct, thereby concentrating the urine to comparable levels. This concentration process in the outer medulla is basic in the sense that the process in the inner medulla, by which any additional concentration of the urine is achieved, is fully dependent on the process in the outer medulla. This agrees with comparative anatomical findings showing that the most consistent part of the renal medulla, found in any species, is an inner stripe. Species, like the mountain beaver, whose renal medulla lacks an inner medulla, but has an inner stripe, are able to concentrate the urine to only about twice the plasma concentration.

This basic mechanism in the outer medulla qualitatively equals the "Gottschalk and Mylle model," which was presented in the preceding section and which is based on Kuhn's proposal of a countercurrent multiplier driven by active solute transport out of the ascending tube.[51] Hypertonic salt reabsorption at any level along the thick ascending limb in the outer medulla is the driving force ("single concentrating effect"). Whether equilibration with descending limb fluid occurs by solute addition or water extraction or a mixture of both is still a controversial question. Differences between short and long loops as well as species differences must be anticipated. Regardless of which type of equilibration with descending limb fluid occurs, there is no doubt that this system can work and produce the increase in the interstitial concentration of the above mentioned magnitude of about 600 to 700 mOsm/L H_2O. Osmotic water reabsorption from the collecting duct into the interstitium raises the concentration of the collecting duct urine to corresponding levels. The net reabsorbed water is carried away by ascending vasa recta.

The major unresolved problems concern the concentrating mechanism in the inner medulla, where an active component, a single concentrating effect, has never been found. Therefore, the question that we raised at the beginning of this section may be phrased more precisely as follows: *How is a portion of the salt, which has been reabsorbed from the thick ascending limbs in the outer medulla, transferred into the inner medulla, and accumulated there together with urea to produce an osmotic gradient that reaches its maximum at the tip of the papilla?*

The most fully developed hypothesis proposed to answer this question is the so-called passive mechanism of urine concentration in the inner medulla, which was independently, but almost simultaneously, proposed by Stephenson[82] and by Kokko and Rector.[45] These models are based on the idea published by Kuhn and Ryffel[52] as early as 1942, that countercurrent multiplication is possible by the mixing of different solutes. The principle of this mechanism is depicted in Figure 12.18.

In the "Stephenson/Kokko and Rector model" urea mixes with sodium chloride in the inner medullary interstitium, creating at any level of the inner medulla

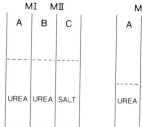

FIGURE 12.18. This figure illustrates the principle of countercurrent multiplication. The three compartments, A, B, and C, are separated by membranes of different permeabilities. MI is water permeable and solute impermeable (salt and urea), MII is water and urea impermeable, but permeable to salts. Initially, compartments A and B are filled with pure urea solutions and C with a pure salt solution, all of equal osmolality. Salt will diffuse from C to B raising B's osmolality; this will again induce osmotic water withdrawal from A. At equilibrium, different amounts of fluid will fill the three compartments: the volume will be unchanged in C, increased in B, and decreased in A. The osmolality will be decreased in C and increased in A and B to the same level. (Modified from Stephenson.[82])

a total osmolality higher than that in the descending limbs as well as in the collecting ducts. According to the permeability characteristics of both tubular segments, water is extracted from their lumens, thereby concentrating the descending limb fluid as well as the urine within the collecting ducts. However, where do the urea and the sodium chloride come from? According to this model, urea is delivered by the collecting ducts and the salt is delivered by the thin ascending limbs of Henle's loop.

Let us first consider the delivery of urea, which involves the following steps (Figure 12.19): (1) Hyperosmotic reabsorption of sodium chloride from the thick ascending limb in the outer medulla, which leaves behind urea as the main solute in the dilute tubular fluid. (2) Osmotic water extraction from the cortical and outer medullary collecting ducts. Since both collecting duct segments are relatively urea impermeable, the tubular fluid within the collecting ducts upon entering the inner medulla will have very high urea concentrations that surpass the interstitial concentration of this solute at this medullary level. (3) Since within the inner medulla the collecting ducts become very permeable to urea, urea will diffuse out of the collecting ducts into the interstitium, thereby raising the total osmolality of the interstitium.

The delivery of salt is most easily understood if

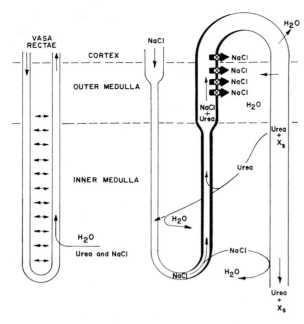

FIGURE 12.19. Countercurrent multiplication system without active transport in the inner medulla. Thick lines on the ascending limb indicates water-impermeability and X_s designates a nonreabsorbable solute. (Redrawn from Kokko and Rector.[45])

we consider an inner medulla with an already existing solute gradient. In this case, ascending limbs will carry up a fluid more concentrated in sodium chloride and in total osmolality than the surrounding interstitium at any level of the medulla. Since the ascending thin limb epithelium is highly ion permeable, sodium chloride will diffuse out into the interstitium. By mixing with the urea from the collecting ducts, the interstitial osmolality will reach sufficiently high values to induce water reabsorption from the descending thin limbs and collecting ducts. Water reabsorption from the *descending thin limbs* will concentrate the tubular fluid successively in the descending limb. After turning back, this concentrated, salt-rich fluid will successively reach less concentrated levels; consequently, salt will diffuse out of ascending thin limbs along their entire length. Water reabsorption from the *collecting ducts* will concentrate the collecting duct fluid. Since urea is a major solute in this fluid, urea will be successively concentrated toward deeper levels of the medulla. Compared with interstitial urea concentration, urea concentration within the collecting duct at any level

of the medulla will be slightly higher, allowing for diffusional exit of urea and thus urea addition to the interstitium at any inner medullary level.

To sum up, at any level of the inner medulla, mixing of urea from the collecting ducts and sodium chloride from the thin ascending limbs by diffusional entry into the interstitium will elevate total interstitial osmolality above that of descending limb fluid and above that of the collecting duct fluid (Figure 12.18). Water withdrawal from both will ensure the concentration of descending limb fluid as well as the concentration of the final urine. The net reabsorbed water will be carried away from the interstitium by ascending vasa recta (see later). It has to be mentioned that the model requires also a small proportion of urea uptake by the ascending (and/or descending) thin limbs to ensure urea recycling (see later).

As has become clear, this passive mechanism depends on very specific permeability properties of the tubular segments involved. The membrane properties found in the respective tubules of the rabbit appear to meet these requirements. Simulation studies by Stephenson and co-workers[83] have shown that even under these somewhat idealized conditions, only a modest concentration of urea in the inner medulla can result from the passive model. Equilibration of descending thin limb fluid has to be effected almost exclusively by water extraction; any major solute addition to the descending thin limbs in the inner medulla will further decrease the overall concentrating capacity. The permeability properties of the descending thin limbs which have been found in rats and hamsters, as in *Psammomys* (see above), are difficult to reconcile with such a passive mechanism. Among other objections against this model (see Jamison and Kriz[38]), a computer simulation of this mechanism for the rat kidney based on experimentally found tubular transport parameters failed to demonstrate the ability to increase salt concentrations within the inner medulla.[63]

For these reasons it seems likely that the passive model only partially explains the concentrating mechanism in the inner medulla. The Stephenson/Kokko and Rector models have two attractive features (both of which are discussed in more detail in the next section): First, it accounts for the complex manner by which urea is transported within the kidney, a fact that could not be explained if urea was only a waste product that had to be excreted. Second, these models stress the importance of short loops. The short loops are thought to provide urea to the collecting ducts in order to reach the necessary high urea concentration in the beginning of the inner medullary collecting ducts.[80]

Other approaches to explain the concentrating mechanism in the inner medulla focus on the descending thin limbs of long loops in the outer medulla.[11,47] One proposed explanation is the addition of solutes to the descending thin limbs of long loops within the outer medulla, which would elevate the tubular fluid concentration of descending limbs entering the inner medulla above the interstitial concentration. Bonventre and Lechene[11] showed that even a small increase (30 mOsm/L H_2O) in concentration of the descending thin limb fluid above that in the interstitium at the junction of the inner and outer medulla would allow for a mechanism in the inner medulla (similar to the passive model) to operate under less strict requirements on the permeability properties of the thin limbs in the inner medulla and to produce reasonable urine concentrations (2,200 mOsm/L H_2O in the calculation by Bonventre and Lechene). Also, Moore and Marsh,[63] through extensive simulation studies, concluded that a significantly higher sodium chloride concentration in the descending thin limb fluid than that prevailing in the interstitium at the inner-outer-medullary junction would provide the osmotic energy to improve considerably the efficiency of the passive mechanism.

From a structural viewpoint, active solute secretion into the descending thin limbs of long loops within the outer medulla seems reasonable. In most species the upper parts of long descending thin limbs lying in the outer stripe are equipped with a specifically differentiated epithelium with intensive cellular interdigitation, shallow junctions, and amplification of the basolateral cell membrane (see above). Remarkably similar features have been found to pertain to salt secreting epithelia in birds and fish.[25,72] Moreover, a membrane-bound sodium-potassium-adenosine triphosphatase (Na-K-ATPase) activity of this epithelium has been demonstrated.[26] Recently, the first data from microperfusion studies of this segment have been published,[35] which show that the epithelim is highly permeable to cations.

In summary, the key function of the urinary-concentrating mechanism is the hypertonic sodium chloride reabsorption by the thick ascending limbs in the outer medulla. Part of the osmotic energy created by this process is carried down into the inner medulla (1) by urea via the "distal tubule"-collecting duct route and (2) possibly by salt addition to the descending thin limbs in the outer medulla. Mixing of salt (derived from the loop) and urea (derived from the collecting ducts) within the inner medulla will create at any level of the inner medulla the necessary high interstitial osmolality to ensure water withdrawal from the collecting ducts and thus the concentration of the final urine.

The same basic mechanism, i.e., hypertonic sodium chloride reabsorption by the thick ascending limb, enables the dilution of the urine. In the absence of ADH, the hypotonic fluid that leaves the thick ascending limb basicly represents the final urine. Even if considerable alterations in the composition of individual solutes in this fluid occur along the distal convoluted tubule and collecting duct, it will not necessarily change the overall osmolality. An additional diluting effect, however, may come from active hypertonic salt reabsorption along the collecting ducts.[39,78]

Role of Urea in the Concentrating Mechanism

Urea is a waste product that has to be excreted by the kidney. Contrasting to this necessity is the fact that the urea clearance is always less than glomerular filtration rate (GFR), indicating that only about 50% of the filtered urea is finally excreted in the urine, and the remaining fraction is reabsorbed along the tubule. As has become clear from many studies, this reabsorbed fraction of urea within the kidney is necessary to achieve the maximum urinary osmolality.[69,74] Animals fed a protein-deficient diet have impaired urinary-concentrating ability. Administration of urea to protein-deficient animals increases total urine osmolality, including an increase of non-urea solutes.[12]

The passive model of urine concentration provides an explanation for this finding. According to this model, urea is the vehicle to transport osmotic energy (produced by salt reabsorption from the thick ascending limbs) via the normal nephron route down into the inner medulla. This mechanism implies that urea is necessary within the kidney in large amounts and is therefore retained in the kidney by several intrarenal recycling routes. To understand these routes, let us first examine the sequential processing of urea along the various nephron segments and collecting duct.

Micropuncture studies in hamsters,[55] rats,[88] and dogs[14] have shown that the tubular fluid-to-plasma concentration ratio for urea $(TF/P)_u$ was about one half that for inulin by the time the filtrate had reached the end of the proximal convoluted tubule, suggesting that about 50% of the filtered urea is reabsorbed along this nephron segment. Moreover, the tubular fluid-to-plasma (TF/P) urea-to-inulin concentration ratio approached a value of unity by the time the filtrate had entered the early distal convoluted tubule, suggesting that somehow an amount of urea equal to that reabsorbed in the proximal tubule was added to the tubular fluid as the filtrate traversed Henle's loop. Since the fluid collected from the loop bend had a urea-to-inulin (TF/P) concentration ratio considerably higher than unity, urea must have been added to the tubular fluid along the descending limb of Henle's loop, and a proportion must again have escaped from the thick ascending limb (probably the cortical thick ascending limb[42]). Furthermore, by the time the filtrate reaches the middle of the inner medullary collecting duct, the urea load is reduced by about 35%.[56] This decrease in urea load in the late portion of the collecting duct implies that urea is reabsorbed from this segment, presumably by passive diffusion into the medullary interstitium, an idea first suggested by Berliner and associates[8] and confirmed by several subsequent micropuncture studies.[91]

The results of the various experiments cited above are summarized in Table 12.1 for the rat, and their extrapolations to the human kidney are given in Table 12.2. These data show that normally in man about 50% of the filtered urea is reabsorbed along the nephron, while the remainder is excreted into the urine. Furthermore, comparing the percentage of filtered urea remaining in the fluid collected from different segments of the nephron, in Tables 12.1 and 12.2, reveals that urea concentrations in the fluids collected from the loop bend and the early distal convoluted tubules are much higher than can be accounted for by the urea that had escaped proximal reabsorption, suggesting that urea is effectively recirculated within the renal medulla. This recirculation of urea involves its reabsorption from the inner medullary collecting ducts and its subsequent diffusion/secretion into descending limbs (pars recta or descending thin limbs) and thin ascending limbs of Henle's loop (Figure 12.20).

Recycling of Urea

Recycling of urea returns to the inner medulla urea that had been carried out by the ascending limb fluid and ascending vasa recta plasma. The following sections present four possible pathways for urea recycling within the renal medulla.[43, 47]

1. Recycling Through Ascending Limbs, "Distal Tubules", and Collecting Ducts

As envisioned by the original model for passive urine concentration, a small proportion of the inner medullary interstitial urea enters the descending and ascending thin limbs.[81] This urea is then carried through the medullary and cortical (where some of the urea is reabsorbed; see pathway 3) thick ascending limb, distal convoluted tubule, and connecting tubule into the cortical and subsequently outer medullary collecting ducts, where it is concentrated subsequent to osmotic water withdrawal that occurs in both collecting duct segments during antidiuresis. On entering the inner medulla, the highly concentrated urea leaves the urea permeable inner medullary collecting ducts by passive diffusion into the interstitium. This last step closes the cycle.

TABLE 12.1. Relative flow rates and concentrations of water, sodium, urea, and potassium in different nephron segments of rat kidney.

Nephron segments	Percentage remaining*				Percentage of total osmolality		
	Water	Sodium	Urea	Potassium	Sodium	Urea	Potassium
Glomerulus	100.0	100.0	100.0	100.0	46.5	2.3	1.5
End of proximal tubules	23.0	23.0	44.7	16.4	46.5	4.4	1.0
Henle's loop (bend)	17.0	47.1	605.3	86.9	33.3	20.8	1.9
Early distal tubules	17.0	9.5	133.0	3.7	35.6	24.4	0.4
End of distal tubules	6.0	2.9	77.1	26.7	23.3	30.0	6.7
End of collecting tubules	0.24	0.1	26.3	11.9	4.3	55.0	15.7

*For the method of calculation see Chapter 9. (From Koushanpour et al.[46])

TABLE 12.2. Relative flow rates and concentrations of water, sodium, urea, and potassium in different nephron segments of human kidney.

Nephron segments	Percentage remaining*				Percentage of total osmolality		
	Water	Sodium	Urea	Potassium	Sodium	Urea	Potassium
Glomerulus	100.0	100.0	100.0	100.0	46.5	1.6	1.6
End of proximal tubules	20.0	20.0	50.0	16.0	46.5	4.0	1.3
Henle's loop (bend)	16.0	33.3	576.0	57.6	33.3	20.0	2.0
Early distal tubules	16.0	9.4	165.0	3.8	37.0	22.5	0.5
End of distal tubules	4.8	3.3	76.8	27.0	28.6	22.9	8.1
End of collecting tubules	0.8	0.7	48.0	9.6	14.4	33.3	6.6

*For the method of calculation see Chapter 9. (From Koushanpour et al.[46])

2. Recycling Through Ascending Vasa Recta, Short Loops of Henle, "Distal Tubules", and Collecting Ducts

Urea enters the capillaries and ascending vasa recta in the inner medulla. In the *complex type of vascular bundle* (Figure 12.6b), urea is expected to be transported by countercurrent exchange to the descending thin limbs of short loops, which are located within the bundles and which have a reasonably high permeability for urea.[35] The urea is then carried around short loops, through distal tubules and collecting ducts, and reenters the inner medulla by diffusion from the urea-permeable inner medullary collecting ducts. In the *simple type of vascular bundles*, an alternative pathway provides a comparable urea recycling via short loops similar to that in the complex vascular bundles (Figure 12.6a). In the inner medulla, urea enters the ascending vasa recta and is then transferred (within the vascular bundles of the inner stripe) by countercurrent diffusion to the descending vasa recta that supply the inner stripe capillaries. From these capillaries urea can be taken up by the thin descending limb of short loops, and from there it follws the same pathway outlined above for the complex vascular bundles.

3. Recycling Through Ascending Vasa Recta, Straight Parts of Proximal Tubules, and Descending Limbs

As in pathway 2, urea enters the ascending vasa recta in the inner medulla. Urea that escapes the countercurrent exchange with descending vasa recta or

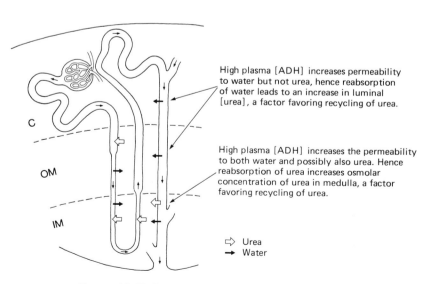

FIGURE 12.20. Recycing of urea within the kidney.

descending thin limbs (pathway 2) will be carried up into the outer stripe and basal parts of the medullary rays of the cortex. At this site the vasa recta are intimately related to the straight proximal tubules (Figure 12.6c). Active[41] or passive[42] secretion of urea into straight proximal tubules of short or long loops seems possible. If long loops are involved, the urea will directly reach the inner medulla within the descending thin limbs and will be concentrated by water reabsorption. Even if this urea never reaches the inner medullary interstitium (thus the cycle is not fully closed), this fraction of urea—as has been pointed out by Knepper and Burg[43]—nevertheless contributes to maintaining a high interstitial urea concentration by concentrating urea in the lumens of the loops of Henle and, therefore, reduces the driving force for loss of urea from the interstitium into the loops. If short loops are involved, urea may cycle around the short loops according to pathway 2.

4. Recycling Around the Loops of Henle[42]

As in pathway 1, urea enters descending and ascending thin limbs in the inner medulla and is carried up within thick ascending limbs. Within the outer stripe of the outer medulla and within the medullary rays of the cortex, the thick ascending limbs are relatively permeable to urea. A proportion of urea will dissipate by diffusion and will raise urea concentration in the interstitium of the outer stripe and medullary rays. Then, urea may be secreted actively or passively into the straight parts of proximal tubules, which represent the descending limb in this region. Cycles via short or via long loops are completed as in pathway 3.

Each of these pathways may contrbute to accumulation and maintenance of high urea concentration in the inner medullary interstitium. In each pathway, urea is concentrated by water reabsorption from either the descending limb in the outer medulla (in addition to a possible active secretion into the pars recta of proximal tubule, which represents the descending limb in the outer stripe of the outer medulla) or from collecting ducts in the cortex and outer medulla. Therefore, *the addition of urea to the inner medullary interstitium by these pathways includes a real concentrating effect in the inner medulla. This is in contrast to the simple countercurrent exchange of urea from ascending to descending vasa recta (Figure 12.6). Return of urea to the inner medulla by this mechanism does not include a concentrating effect and therefore only helps to minimize the dissipation of urea from the inner medulla* (see principle of countercurrent exchange).

Role of Medullary Blood Flow in the Concentrating Process

Compared with the high cortical blood flow, the medullary blood flow rate is low (see Chapter 7). Compared, however, with other organs, the medullary blood flow is in the usual range, about equal to that in the brain.[4] There are, however, major difficulties in measuring the medullary blood flow. None of the currently applied techniques is widely accepted. For a discussion of this topic, which is beyond the scope of this book, the reader is referred to a recent paper by Zimmerhackl and co-workers.[94]

The medullary blood circulation serves several functions: (1) It participates in the maintenance of the medullary solute gradient by countercurrent exchange, and it minimizes the dissipation of the solute gradient. (2) It functionally connects or separates the different renal zones, as, for instance, is obvious in the urea recycling pathways. (3) It removes from the medulla the net reabsorbed water by which the urine has been concentrated. (4) Last but not least, it supplies the renal medulla with nutrients and oxygen and removes wastes.

Countercurrent Exchange

The osmolality of blood flowing down the descending vasa recta equilibrates gradually with the increasing osmolality toward the tip of the papilla.[32] Compared with descending limb fluid equilibration with interstitial concentration, the descending vasa recta plasma lags behind in equilibration by some 10%.[37] Full equilibration is not reached before the blood has entered the capillaries belonging to the respective medullary level.[71] Ascending vasa recta plasma, in contrast, has not been found to be statistically different in osmolality (even if slightly higher) from the corresponding interstitial osmolality. Thus, equilibration of the outflowing blood with the gradually decreasing osmolalities toward the corticomedullary border occurs in a quicker and more perfect way than in the inflowing blood. This agrees with the finding that blood flow velocity in the descending vasa recta is higher than in the ascending vasa recta,[34] providing less time for equilibration and, secondly, that descending vasa recta have a much thicker wall that is probably less permeable than the fenestrated endothelium of the ascending vasa recta.

Equilibration in osmolality in the descending vasa recta occurs by a combination of solute entry and water extraction. Calculations by Jamison and Kriz[38] show that the major proportion (80% to 90%) is accounted for by solute entry, mainly urea and sodium chloride. The reverse must occur in the ascending vasa recta. Thus, *countercurrent equilibration between descending and ascending vasa recta occurs primarily by solute exchange*. The vasa recta must be considered as a countercurrent solute exchange system trapping urea and salt within the medulla. Structurally, this solute exchanger is represented by the vascular bundles.

In addition, about 15% to 20% of water is extracted from descending vasa recta plasma by the time the descending vasa recta reach the papilla. This water extraction accounts for the remaining part (10% to 20%) in osmolality equilibration, and, in addition, it *increases plasma protein concentration*, i.e., it increases the oncotic pressure of the plasma. This effect is essential since it provides the driving force by which the net water reabsorbed from the collecting duct is taken up by capillaries and ascending vasa recta and is carried out from the medulla into the systemic circulation (see below). It should be mentioned that vasa recta plasma increases in osmolality as it travels toward the papilla and decreases again as it travels toward the cortex. Likewise, erythrocytes increase and decrease their intracellular concentrations, respectively.[59]

The efficiency of a countercurrent exchanger is dependent on the velocity of the fluid circulating through the system.[73] There is an optimal velocity that provides enough time for full osmotic equilibration. With regard to the medulla, the higher the medullary blood flow, the less time will be available for osmotic equilibration and the greater will be the rate of solute washout from the medulla. Hence, for countercurrent exchange function the medullary blood flow has to have an optimal velocity; otherwise the medullary longitudinal solute gradient will be decreased.

How the medullary blood flow is regulated is not well understood. Previously it was believed that the medullary blood flow is not autoregulated, thus increasing with increasing systemic blood pressure. As described in Chapter 6, there is now good evidence that this is not true. First, the macula densa feedback regulation of blood flow and filtration rate of juxtamedullary glomeruli appears to be even more sensitive than this mechanism in the cortical glomeruli.[66] Moreover, recent evidence, based on in-vivo blood flow studies on the exposed papilla, indicate that the medullary blood flow is actually autoregulated.[17] The autoregulated pressure range, however, was found to be narrower than in the cortex. In the medulla the autoregulated pressure range was between 80 to 120 mm Hg, whereas in the cortex it was between 80 to 140 mm Hg.

There is general agreement that in water diuresis, medullary blood flow (or at least plasma flow) is considerably higher than in antidiuresis.[86] It is argued that the increased blood flow in water diuresis contributes, by a washout effect, to the diminished medullary solute gradient found in the medulla in animals undergoing water diuresis. It has been suggested that ADH, by a direct effect on the juxtamedullary afferent and/or efferent arterioles and descending vasa recta, slows down medullary blood flow in antidiuresis.[86] Some evidence for a direct effect of ADH on the medullary blood flow has recently been found.[93] Furthermore, medullary prostaglandins (PGE$_2$) increase medullary blood flow.[57,84] The mechanism by which this occurs is not known. Since medullary prostaglandins oppose the ADH effect at two other sites (water permeability of the collecting ducts and active sodium chloride reabsorption by the thick ascending limbs), it is tentatively speculated that possibly ADH and PGE$_2$ act at the same site with opposing effects on the medullary blood flow (see Chapter 9).

Functional Coupling or Separating

When considering the anatomical arrangement of the medullary vessels (Figure 4.3), it is obvious that blood flow to the inner stripe and that to the inner medulla are virtually separated from each other. Descending vasa recta serving the inner medulla traverse the inner stripe within the vascular bundles. Therefore, blood flowing to the inner medulla has not previously been exposed to tubules of the inner or outer stripe. All ascending vasa recta originating from the inner medulla pass through the inner stripe bundles. Accordingly, blood that has perfused tubules of the inner medulla does not then perfuse tubules of the inner stripe. On the other hand, the blood in total, which has perfused either the tubules of the inner medulla or the inner stripe, will afterwards perfuse the tubules of the outer stripe. It has been suggested that this arrangement functions as an ultimate trap to prevent solute loss from the medulla by some countercurrent mechanism between the ascending vasa recta and the descending tubules (P3 segments) in the outer stripe. Moreover, the urea recycling pathways 2 and 3 (described earlier) involve the medullary circulation. Within these routes the medullary vessels couple the function of different nephron segments, which are located in different medullary regions. Consequently, the medullary circulation is directly involved in the function of the medulla.

Water Removal

In Kuhn's original countercurrent multiplication hypothesis of urinary concentration, water reabsorbed from the collecting ducts reenters the loops of Henle to be transported out of the system. There is now overwhelming evidence that the net water reabsorbed from the collecting ducts in the medulla is taken up and carried out by the medullary circulation.[38] As has been mentioned above, water extraction from the descending vasa recta results in a rise of the oncotic pressure in the descending vasa recta plasma, reaching the highest possible value in the capillaries corresponding to a certain medullary level. At the same site, within the capillaries or beginning parts of ascending vasa recta, a decrease in hydraulic pressure can be expected. Comparing descending and ascending vasa recta, a pressure drop of about 1.5 mm Hg has been reported.[70] Although small, the algebraic sum of oncotic pressure increase (roughly 8 mm Hg at the tip of the papilla) and hydraulic pressure decrease is sufficient to explain the water uptake by medullary capillaries and ascending vasa recta. Therefore, net water uptake by the medullary capillaries and ascending vasa recta is obviously effected by the same combination of hydraulic and oncotic pressures as elsewhere in the body.[71] There are, however, no lymphatics in the medulla, and since water uptake by medullary tubules seems very improbable, any water entry into the medullary interstitium has to be balanced by the renal blood flow.

Nutrition

Several studies have shown that tissue oxygen tension within the medulla is much lower than in the cortex and, moreover, that there exists a corticomedullary gradient of oxygen starting with a pO_2 of about 70 to 80 mm Hg at the corticomedullary border and decreasing down to very low values (around 10 mm Hg or less) in the tip of the papilla.[4,5,13] As has been outlined above (principle of countercurrent exchange), a countercurrent exchanger establishes a barrier and may not only trap solutes within a system, but may also exclude substances from entrance into the system. This appears to be the case with oxygen. As everywhere in the body, oxygen concentrations are higher in the inflowing blood than in the outflowing blood. Countercurrent exchange of oxygen from the arterial blood in the descending vasa recta to the venous blood in the ascending vasa recta within the bundles short-circuits oxygen at any level of the medulla back to the corticomedullary border, resulting in the lowest pO_2 values within the papilla.

As outlined previously in this book, blood flow through the inner medulla is to some extent independent from that through the inner stripe. Recall that the descending vasa recta destined to supply the inner medulla and those destined to supply the inner stripe are already separately established within the outer stripe by branching from the efferent arterioles. Also, the blood drainage is separated. The blood from the inner medulla returns to the outer stripe by ascending vasa recta coming up within the bundles, whereas the blood from the inner stripe returns to the outer stripe mainly by vasa recta that ascend independently from the vascular bundles. This division of the medullary blood flow appears to correlate with different amounts of oxygen consumption: low in the inner medulla and high in the inner stripe. Blood flow through the inner stripe obviously maintains reasonably high pO_2 tensions to nourish the thick ascending limbs which are known to have a high metabolic rate.[16] Blood flow to the inner medulla is ineffective with respect to oxygen supply. Tubular segments within the inner medulla have obviously adapted to low oxygen tissue tensions. The inner medulla does not contain thick ascending limb or tubules with an high oxygen consumption. Possibly, as noted by Knepper and Burg,[43] oxygen supply to the inner medulla is too low for rapid active sodium chloride transport. One may speculate that this was the phylogenetic factor to develop an inner medulla with an obviously quite different concentrating mechanism than that in the outer medulla where the mechanism is based on O_2-consumptive, and active salt transport out of the thick ascending limbs.

Measurement of the Concentrating and Diluting Ability of the Kidney

From the foregoing discussion it is clear that the ability of the kidney to concentrate and dilute urine depends on the functions of Henle's loop and the collecting duct. We learned that the fluid emerging from the proximal tubule remains iso-osmotic, regardless of the osmolality of the excreted urine. However, as the filtrate traverses the descending limb, it becomes progressively hyperosmotic as a consequence of the countercurrent multiplication process following the hyperosmotic reabsorption of sodium chloride from the water-impermeable thick ascending limb of Henle's loop. The reabsorption of sodium chloride, without the osmotic reabsorption of water, in the ascending limb leaves behind an amount of water free of solutes, thereby making the fluid remaining in the tubule hypotonic. For this reason, this water, which is not reabsorbed, is referred to as the *solute-free* water, and the ascending limb is known as the *diluting segment*.

As was pointed out earlier, the degree of hypotonicity of the fluid in the ascending limb determines the magnitude of the transverse gradient (single effect), a factor that, in turn, governs the magnitude of the longitudinal gradient developed in the medullary interstitium. It is the latter osmolar concentration gradient that eventually determines the degree of concentration of fluid emerging from the collecting duct and hence the osmolality of the excreted urine. The latter is a consequence of the ability of the collecting duct to reabsorb water in excess of solute, a process greatly enhanced by ADH, and thereby to adjust its osmolality to that of the surrounding hypertonic medullary interstitium. For this reason, the collecting duct is referred to as the *concentrating segment*.

Thus, in the final analysis, the ability of the kidney to concentrate and dilute urine depends on the functions of the thick ascending limb of Henle's loop and the collecting duct. The former determines the volume of *solute-free* water excreted in the urine, whereas the latter determines the amount of osmotically active solutes or the *osmolar concentration* of the excreted urine. For this reason, it is of great clinical interest and potential diagnostic value to determine the relative amounts of water and osmotically active solutes that are not reabsorbed (and are therefore excreted in the urine).

Although in animal studies it is possible to assess the functions of the thick ascending limb and collecting duct directly (see Chapter 9), techniques for such a direct observation in man are not yet available. Therefore, in clinical practice a quantitative estimate of the functions of these tubular segments and hence of the concentrating and diluting ability of the kidney is obtained by measuring the osmolar concentration of urine relative to that of plasma as well as the volume of excreted urine, using the standard clearance techniques described in Chapter 7. The rationale for applying the clearance method to determining the concentrating and diluting ability of the kidney is as follows.

Micropuncture studies cited earlier have shown that almost two thirds of the filtrate is reabsorbed iso-osmotically along the proximal tubule, while most of the remaining portion is reabsorbed in Henle's loop, the "distal tubule," and the collecting duct, where the osmolality of the tubular fluid undergoes considerable modification. Thus, any deviation in the osmolality of the excreted urine from isotonicity must reflect relative reabsorptive changes in water and solutes along these latter tubular segments. Hence, excretion of a concentrated urine implies net reabsorption of water in excess of solute, whereas excretion of a dilute urine implies just the opposite. Accordingly, to obtain a quantitative estimate of the concentrating and diluting ability of the kidney, one needs simply to measure the renal clearance of the osmotically active solutes present in the blood and compare it with the renal clearance of solute-free water. Let us now examine the operational definitions of these clearances and their physiological meanings.[89]

Osmolar clearance, symbolized by C_{Osm}, is defined as the volume of plasma containing the same amount of osmotically active solutes that was present at the same osmolar concentration in the urine. Stated another way, it describes the rate at which osmotically active solutes are removed from the plasma along with a volume of water sufficient to contain the solutes in an iso-osmotic solution. Any surplus of water exceeding this volume in the urine is therefore considered solute-free water. Thus, the total volume of urine (V) may be considered to be the sum of two *virtual* volumes:

$$\text{Urine volume (V)} = C_{Osm} + C_{H_2O} \quad (12.1)$$

where C_{H_2O} is the volume that contains surplus water as solute-free water.

The osmolar clearance is calculated by dividing the total amount of osmotically active solutes excreted in urine per minute ($U_{Osm} \cdot V$) by the plasma osmolality (P_{Osm}):

$$C_{Osm} \text{ (mL/min)} = \frac{U_{Osm} \text{ (mOsm/L)} \cdot V \text{ (mL/min)}}{P_{Osm} \text{ (mOsm/L)}} \quad (12.2)$$

Because both U_{Osm} and P_{Osm} are measured by cryoscopical method—that is, by freezing-point depression—their values include the contribution of both electrolytes and nonelectrolytes. In a fasting but otherwise normal individual, osmolar clearance varies between 2 to 3 mL/min and is relatively independent of urine flow.

Solute-free water clearance is defined as the difference between the volume of urine excreted per minute (V) and the osmolar clearance (C_{Osm}) calculated from Equation 12.2:

Solute-free water clearance (mL/min) =
$$V \text{ (mL/min)} - C_{Osm} \text{ (mL/min)} \quad (12.3)$$

If the urine is neither concentrated nor dilute, that is, if it is iso-osmotic with the plasma ($U_{Osm} = P_{Osm}$), it follows from the definition of osmolar clearance (Equation 12.2) that $V = C_{Osm}$ in Equation 12.3, and the solute-free water clearance would be zero. However, if the urine is concentrated, that is, if U_{Osm} is much greater than P_{Osm}, and V is very small, as would be the case during dehydration, C_{Osm} would be greater than V, and their difference would be *negative*. This negative solute-free water clearance is designated by the symbol $T^c_{H_2O}$, and it means that there is a net reabsorption of water. That is why the excreted urine is *hypertonic*. The superscript "c" in the symbol $T^c_{H_2O}$ signifies that the net reabsorption of water occurs in the concentrating segment of the nephron, namely, the collecting duct. On the other hand, if the urine is dilute, that is, if U_{Osm} is much smaller than P_{Osm}, and V is relatively large, as would be the case during water ingestion or diuresis, C_{Osm} would be less than V, and their difference would be *positive*. This positive solute-free water clearance is designated by the symbol C_{H_2O}, and it means that water is not reabsorbed and hence is excreted in the urine. That is why the urine is *hypotonic*. In short, a negative solute-free water clearance means net reabsorption of water in excess of solutes and hence excretion of an hypertonic urine. In contrast, a positive solute-free water clearance means net excretion of water in excess of solutes and hence formation of an hypotonic urine. Thus, $T^c_{H_2O}$ is equal to $-(C_{H_2O})$.

To further illustrate the physiological meanings of osmolar clearance as well as the negative and positive solute-free water clearances, let us consider the following two cases:

1. In a normal man, restriction of fluid intake results in elaboration of urine some four to five times more concentrated than the plasma. In such a circumstance, urine flow may reach a value as low as 0.5 mL/min. Assuming a plasma osmolality of 300 mOsm/L, U_{Osm} would have a value of 5 × 300 mOsm/L = 1,500 mOsm/L, and the osmolar clearance would be

$$C_{Osm} \text{ (mL/min)} = \frac{1{,}500 \text{ (mOsm/L)} \times 0.5 \text{ (mL/min)}}{300 \text{ (mOsm/L)}} = 2.5 \text{ mL/min} \quad (12.4)$$

Because urine is hypertonic, there must have been a net reabsorption of water, and hence a negative solute-free water clearance ($T^c_{H_2O}$):

$$T^c_{H_2O} \text{ (mL/min)} = V - C_{Osm}$$
$$= 0.5 \text{ (mL/min)} - 2.5 \text{ (mL/min)}$$
$$= -2.0 \text{ mL/min} \quad (12.5)$$

This negative solute-free water clearance means that the osmotically active solutes that were present in 2.5 mL of blood (C_{Osm} = 2.5 mL/min) were excreted in 0.5 mL of hypertonic urine, resulting in a net reabsorption of 2.0 mL of solute-free water back into the blood. In this manner the kidneys have conserved water to dilute the body fluids, thereby decreasing the plasma osmolality toward normal.

2. Ingestion of a liter or more of water by a normally hydrated man will increase urine flow to as much as 20 mL/min, while at the same time it may decrease urine osmolality to 0.1 that of the plasma. Again, assuming a plasma osmolality of 300 mOsm/L, U_{Osm} = 0.1 × 300 mOsm/L = 30 mOsm/L, and

$$C_{Osm} \text{ (mL/min)} = \frac{30 \text{ (mOsm/L)} \times 20 \text{ (mL/min)}}{300 \text{ (mOsm/L)}} = 2.0 \text{ mL/min} \quad (12.6)$$

Because urine is hypotonic, there must have been relatively less water reabsorption, and hence a positive solute-free water clearance (C_{H_2O}):

$$C_{H_2O} \text{ (mL/min)} = V - C_{Osm}$$
$$= 20 \text{ (mL/min)} - 2.0 \text{ (mL/min)}$$
$$= 18.0 \text{ mL/min} \quad (12.7)$$

This positive solute-free water clearance means that the osmotically active solutes that were present in 2.0 mL of blood (C_{Osm}) = 2.0 mL/min) were excreted in 20 mL of hypotonic urine, resulting in a net loss of 18.0 mL of solute-free water from the body. In this manner the kidneys have excreted water to concentrate body fluids, thereby increasing the plasma osmolality toward normal.

Thus, by manipulating the reabsorption of water and osmotically active solutes, the kidneys adjust the volume and osmolality of the body fluids. The outcome of this adjustment is reflected in the volume and osmolality of the excreted urine as measured by osmolar and solute-free water clearances. However, as pointed out in Chapter 1, the kidneys appear to be more effective in protecting the body against dilution than concentration. The kidneys cannot replace the excessive fluid lost from the body, and therefore must be aided by direct fluid replacement.

Physiology of Diuresis: Diuretics and Their Actions

Diuresis and Its Classification

Diuresis is defined as an increase in urine volume flow (V), usually in excess of 2 mL/min. This increase in urine flow could result from interference with two normal tubular functions: (1) Reduced osmotic reabsorption of water, leading to increased solute-free water clearance (C_{H_2O}) and hence *water diuresis*; and (2) reduced solute reabsorption, primarily sodium ions with associated anions, leading to increased osmolar clearance (C_{Osm}) and hence *solute or osmotic diuresis*. In both cases there is an increase in urine volume flow, which is the basis for diuresis. Therefore, we may classify diuretic states as either water or osmotic diuresis. Let us now briefly examine the conditions that produce these diuretic states.

Water Diuresis

Normally, water diuresis results from excessive fluid intake. The resulting hemodilution reduces the blood concentration of ADH and hence diminishes its effectiveness in promoting osmotic reabsorption of water in the collecting duct.

Clinically, there are two pathological conditions that alter the normal effect of ADH and hence lead to water diuresis. One is the *central* diabetes insipidus, a condition characterized by subnormal synthesis and release of ADH, caused by some hypothalamic lesions. The second is the *nephrogenic* diabetes insipidus, a condition characterized by the loss of responsiveness of the collecting duct epithelium to ADH.

Osmotic Diuresis

As discussed in Chapter 9, any freely filterable, water-soluble, nonreabsorbable, or poorly reabsorbable solute increases urine flow by retarding osmotic reabsorption of water and hence sodium reabsorption in selective segments of the nephron. Such a substance is called an osmotic diuretic and the ensuing diuresis is called osmotic diuresis. In this type of diuresis, there is a parallel increase in both osmolar and solute-free water clearances.

Osmotic diuresis may also be induced by intravenous infusion of any solution that tends to increase blood volume and systemic arterial blood pressure and hence glomerular filtration rate (GFR) and medullary blood flow. As pointed out in Chapter 9, intravenous infusion of a large quantity of isotonic saline inhibits sodium reabsorption in the proximal tubule in both rat[54] and dog,[21] leading to increased urinary excretion of sodium and water. This natriuresis, resulting from reduced tubular reabsorption of sodium, occurs even in the presence of aldosterone and reduced GFR.[20] That the inhibition of the proximal sodium reabsorption may be a consequence of potentiation of a natriuretic hormone during isotonic saline infusion has been inferred from cross-circulation experiments.[40] Thus, cross-circulation of blood from a saline-loaded donor dog to a normal recipient dog produced significant natriuresis, even when sodium load was decreased by aortic constriction above the renal arteries. Despite extensive research, the existence of such a hormone lacks direct endocrinological proof at present.

The most widely used osmotic diuretic agent is mannitol, a nonreabsorbable solute that is freely permeable at the glomerulus and is neither reabsorbed nor secreted. Urea, which is freely filtered but is partially reabsorbed by flow-dependent passive diffusion along the nephron, is a close second. Intravenous infusion of these solutes has a dual effect: (1) They cause an expansion of the extracellular volume, which, through its cardiovascular effect, leads to an increase in GFR and medullary blood flow. Both of these factors contribute to enhanced natriuresis. (2) Presence of these solutes in the tubular filtrate will obligate water, retarding its osmotic reabsorption, a factor markedly limiting diffusive entry of sodium across the luminal membrane into the renal tubule cell. The result is a reduction in the transepithelial sodium transport, a factor contributing to enhanced natriuresis.

Recollection micropuncture studies in dogs have shown that the major inhibitory effect of mannitol on sodium reabsorption occurs along the ascending limb (the diluting segment) of Henle's loop, with a somewhat lesser inhibition of sodium reabsorption in the proximal tubule.[76] Thus, during mannitol diuresis, the proximal inhibition of sodium reabsorption, though comparatively small, results in an increased delivery of filtrate to the distally located nephron segments. In Henle's loop, the resulting increased tubular flow rate decreases sodium reabsorption in the ascending limb, thereby reducing the transverse gradient and hence the medullary longitudinal osmolar gradient. The latter effect reduces the extent of osmotic equilibration of the fluid in the collecting duct with the surrounding medullary interstitium, a factor accounting for enhanced natriuresis during mannitol diuresis. Another factor contributing to this natriresis is the increased medullary blood flow during the mannitol diuresis. Thus, the increase in both the medullary blood flow and the tubular flow rates contributes to solute washout from the medulla, thereby reducing the medullary concentration gradient, which leads to excretion of a relatively dilute urine.

Figure 12.12 shows the comparative diuretic effects of some commonly used osmotic diuretic agents. It is evident that the final osmolality of the excreted urine depends on several factors: (1) Whether the osmotically active solute present in the ultrafiltrate is reabsorbed or not. Thus, infusion of mannitol, which is not reabsorbed, causes excretion of a more dilute urine than infusion of sodium chloride, which is reabsorbed throughout the nephron. (2) The extent to which an osmotically active solute is reabsorbed determines the degree of induced osmotic diuresis. Thus, infusion of glucose, which is reabsorbed by a Tm-limited process only in the proximal tubule, results in a greater diuresis (and hence a more dilute urine) than urea infusion. The latter is reabsorbed to varying degrees throughout the nephron. (3) Whether the osmotically active solute infused dissociates in solution or not (and hence increases the number of particles in solution) is another determinant of the degree of induced diuresis. Thus, infusion of sodium sulfate (Na_2SO_4) results in a greater diuresis (and hence a more dilute urine) than sodium chloride infusion. This is because Na_2SO_4 yields three osmotically active particles in solution, whereas NaCl yields only two. Moreover, the sulfate ion is a poorly reabsorbable solute compared with the chloride ion. (4) Whether the subject is in a state of water imbalance or not. Thus, infusion of any osmotic diuretic agent causes a greater diuresis in a well-hydrated subject (positive water balance) than in a hydropenic subject (negative water balance). (5) Whether ADH or aldosterone levels in the blood are within normal limits or not, and whether the subject is receiving exogenous doses of these hormones. Thus, infusion of any osmotic diuretic agent to a patient with diabetes insipidus causes excretion of a more dilute urine (greater diuresis) compared with a normal individual.

Diuretics and Their Classification

Diuretics are a group of pharmacological agents that cause an increase in urine volume flow. However, the major purpose of diuretic therapy is not to increase urine volume flow, but rather to rid the body of excess sodium and water, to reduce the extracellular volume, and hence to eliminate edema and associated hypertension.

The commonly used diuretics may be classified in terms of (a) chemical structures, (b) sites of action along the nephron, or (c) whether they interfere with solute (primarily sodium chloride) or water reabsorption along the nephron. Most of the commonly used diuretics exert their effects primarily by interfering with sodium chloride reabsorption along one or more segments of the nephron. Their effects on water reabsorption occur secondarily, owing to the osmotic effect of unreabsorbed sodium and its associated anions. Consequently, most of the diuretics may be classified as osmotic diuretics, which, with few exceptions, cause a parallel increase in both osmolar and solute-free water clearances. Therefore, from a functional point of view, we will classify the various diuretic drugs as osmotic diuretics according to their sites of action along the nephron.

Methods Used to Assess Sites of Action of Diuretics

Our current knowledge of the sites and mechanisms of action of the commonly used diuretics has been derived from the application of clearance, stop-flow, micropuncture, and microperfusion techniques to the study of renal function.

As described in Chapter 7, measuring the renal clearance of a substance provides a means of assessing the integral function of the whole kidney. Application of this technique to assess renal effects of diuretics has yielded three kinds of information: (1) The magnitude of diuretic-induced natriuresis, expressed as the fraction of filtered sodium excreted ($U_{Na} \cdot V/GFR \cdot P_{Na}$); (2) diuretic-induced changes in urinary electrolyte composition; and (3) evaluation of diuretic-induced changes in concentrating and diluting mechanisms from changes in solute-free water clearance (C_{H_2O}). Of these, the latter information has provided considerable insight into the possible sites of diuretic action. Thus, it has been shown that any diuretic agent that interferes with sodium chloride reabsorption in the proximal tubule will tend to increase C_{H_2O}, whereas any diuretic agent that interferes with sodium chloride reabsorption in the distal nephron will tend to reduce C_{H_2O}.[77]

Stop-flow technique, as described previously in Chapter 9, though a useful tool, has several shortcomings. Nevertheless, its application has provided "gross" segmental localization of the diuretic effects on tubular transport along the nephron.

By far the most direct information about the sites and mechanisms of diuretic action has come from those studies in which micropuncture and microperfusion techniques were employed. Without going into specifics of these techniques (see Chapter 9 for details), we shall now describe the sites and modes of action of some commonly used diuretics along the nephron (Figure 12.21) as revealed by these studies.

Sites and Modes of Action of Specific Diuretics Along the Nephron

Proximal Tubule

In the proximal tubule the filtered sodium is reabsorbed as sodium chloride and as $NaHCO_3$. As described in Chapters 9 and 11, in both cases the transepithelial reabsorption of sodium results from its passive entry into and its active extrusion out of the tubule cell (Figure 11.14). Reabsorption of filtered chloride ion occurs passively along the electrochemical gradient established by sodium movement. However, as illustrated in Figure 11.14, reabsorption of filtered HCO_3^- ion occurs by a more complicated mechanism. It involves *carbonic anhydrase* catalysis of cellular hydration of carbon dioxide and the subsequent cellular generation of H^+ and HCO_3^- ions. The H^+ ion thus generated is exchanged for Na^+ ion at the luminal cell membrane (a process promoting carbon dioxide formation and its cellular entry) while the generated HCO_3^- ion is reabsorbed by passive diffusion into the blood. Thus, the key to HCO_3^- ion reabsorption is the carbonic anhydrase-catalyzed generation of H^+ ion. Accordingly, diuretic agents that exert their effects in the proximal tubule may be classified as those that directly interfere with sodium reabsorption by inhibiting carbonic anhydrase-catalyzed H^+ ion generation and its exchange for Na^+ ion. The latter has become the most popular type of diuretic for inhibiting sodium reabsorption in the proximal tubule and hence inducing natriuresis.

Carbonic Anhydrase Inhibitors. Acetazolamide

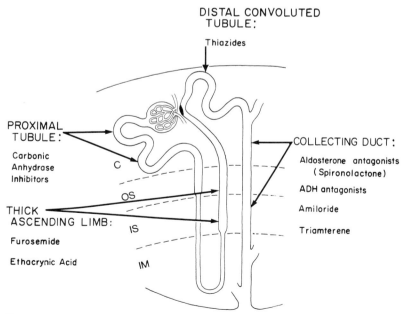

FIGURE 12.21. Sites of action of the commonly used diuretics along the nephron.

(Diamox) is a sulfonamide and a weak inhibitor of carbonic anhydrase activity. For a detailed discussion of the chemistry and pharmacology of this class of drugs, the interested reader is referred to an excellent review by Maren.[60]

The major effect of acetazolamide is the inhibition of Na^+-H^+ exchange at the luminal membrane of the proximal tubule (Figure 12.21), where the major fraction of sodium bicarbonate is reabsorbed. This drug inhibits the intracellular hydration of carbon dioxide, a primary source of cellular H^+ ion. Consequently, it inhibits sodium bicarbonate reabsorption, thereby increasing its urinary excretion and making the urine alkaline. Similarly, it inhibits sodium bicarbonate reabsorption in the "distal tubule," a factor contributing to increased luminal negativity and enhanced potassium secretion (see Chapter 9 for details).

Acute administration of acetazolamide leads to production of an alkaline urine due to marked increase in urinary excretion of sodium, potassium, and bicarbonate (Table 12.3). Additionally, it causes 10% to 30% reduction in GFR and inconsistent changes in renal plasma flow (RPF). Thus, the hemodynamic effect of the drug must be taken into consideration when evaluating its renal effect.

Chronic administration of acetazolamide leads to the development of mild to moderate hyperchloremic, metabolic acidosis, and eventually results in the loss of diuretic effectiveness and the return of urinary electrolyte composition to predrug pattern. This latter response is called the *refractory state* and its mechanism is not known. For this reason, acetazolamide is considered to be both a weak carbonic anhydrase inhibitor and a weak diuretic. The major clinical use of acetazolamide is for nondiuretic indications, such as the treatment of chronic glaucoma.

Another weak carbonic anhydrase inhibitor is chlorothiazide (Diuril), a heterocyclic sulfonamide. It is an orally effective diuretic that inhibits sodium chloride reabsorption mainly in the "distal tubule" (see below). It is widely used in clinical treatment of edema and associated hypertension.

Henle's Loop

As described previously in Chapter 9, the major site of sodium chloride reabsorption in Henle's loop is the thick ascending limb (the diluting segment). Thus, any agent that interferes with sodium chloride reabsorption in this nephron segment will tend to reduce the transverse gradient, which through the operation of the countercurrent multiplication mechanism will eventually lead to excretion of a dilute urine. For this reason the diuretics that inhibit sodium chloride reabsorption in the thick ascending limb are by far the most potent natriuretics (Table 12.3), since they inhibit the renal concentrating and diluting mechanism. There are two diuretics that fall into this category: ethacrynic acid and furosemide.

Ethacrynic Acid and Furosemide. Although ethacrynic acid (a derivative of phenoxyacetic acid) and furosemide (a sulfamylbenzene derivative of anthranilic acid) are chemically dissimilar, they produce similar diuretic effects in the kidney. Numerous micropuncture studies in rats,[23,65] dogs,[6,15,22] and monkeys[7] have clearly shown that the major site of renal action of both of these diuretic drugs is the thick ascending limb of Henle's loop. Both drugs

TABLE 12.3. Changes in urinary composition and relative natriuretic potency of some commonly used diuretics.

Diuretics	Sites of action	Effects on Excretion				% Maximal fraction of filtered sodium excreted
		Na	K	Cl	HCO₃	
Ethacrynic acid	TAL	+	+	+	0	25
Furosemide	TAL	+	+	+	0	25
Chlorothiazide	PT, DT	+	+	+	+*	8
Acetazolamide	PT, DT	+	+	+*	+	5
Spironolactone	CD	+	−	+*	+	3
Amiloride	DT, CD	+	−	+	+	3
Triamterene	CD	+	−	+*	+	3

(*) Means that the effect is not consistently seen, (+) means increase, (−) means decrease, and (0) means no change. PT = proximal tubule; TAL = thick ascending limb of Henle's loop; DT = "distal tubule"; CD = collecting duct. (Modified from Goldberg.[30])

inhibit sodium chloride reabsorption in the thick ascending limb, thereby increasing the concentration of these ions in the tubular fluid and eventually their urinary excretion (Table 12.3). The resulting increase in the osmolality of the ascending limb fluid reduces the transverse gradient, which through the operation of the countercurrent multiplication mechanism would lead to a reduction in the medullary longitudinal osmolar concentration gradient and hence production of a dilute urine. Additionally, both diuretics produce marked kaliuresis (Table 12.3), secondary to enhanced "distal" tubular flow rate (see Chapter 9 for mechanism).

Unlike other diuretics, furosemide and ethacrynic acids have no effect on GFR. Both diuretics inhibit chloride flux into the macula densa cells, thereby abolishing the tubuloglomerular feedback and preventing the expected decrease in GFR secondary to enhanced tubular flow rate.[92] In contrast, both furosemide and ethacrynic acid affect renal hemodynamics. Thus, systemic administration of both diuretics causes renal vasodilation and an increase in renal blood flow (RBF). The increase in RBF is primarily in the inner cortical and medullary blood flows.[79] Although the precise mechanism mediating the renal vasodilation is not known, the increase in the medullary blood flow is important for the diuresis induced by these agents. The increase in the medullary blood flow leads to washout of the medullary hypertonicity, thereby decreasing water reabsorption in the collecting duct and the eventual excretion of a dilute urine.

Distal Tubule and Collecting Duct

As noted in Chapter 9, the "distal tubule" is a heterogeneous segment with a capacity for reabsorption of sodium and water as well as secretion of hydrogen and potassium ions. Micropuncture studies[53] have shown that like acetazolamide, chlorothiazide is a weak carbonic anhydrase inhibitor that blocks Na^+/H^+ exchange in the luminal membrane of the proximal convoluted tubule, and hence sodium reabsorption in this segment. Chlorothiazide also inhibits sodium and chloride reabsorption in the distal convoluted tubule by a mechanism other than inhibition of Na^+/H^+ exchange in this segment.[9,60]

Acute administration of this drug to both dog and man[9] leads to marked increase in urinary excretion of sodium, potassium, chloride, and bicarbonate (Table 12.3). However, in contrast to acetazolamide, chronic administration of chlorothiazide leads to development of hypochloremic, hypokalemic, metabolic alkalosis. Moreover, the chlorouretic potency of this drug is impaired by neither metabolic acidosis nor metabolic alkalosis (in contrast to acetazolamide). In addition, administration of chlorothiazide to dogs[24] and man[3] has no significant effect on the urinary concentrating mechanism. These findings suggest that besides inhibiting carbonic anhydrase activity, chlorothiazide must also interfere with chloride reabsorption in the early distal convoluted tubule, where urine is hypotonic. The molecular mechanism mediating this effect is, however, not known.

In the collecting duct, reabsorption of sodium occurs by two separate mechanisms; one is aldosterone-dependent and the other is not. Thus, inhibition of sodium reabsorption in the collecting duct may be accomplished by two types of diuretics; those that interfere with the action of aldosterone and those that do not. Spironolactone is an example of the first category, and triamterene and amiloride are examples of the second category. The feature common to all three is that they inhibit potassium secretion and indirectly induce natriuresis.

Spironolactone. This is a steroid with a chemical structure similar to that of aldosterone. It binds competitively with cytoplasmic receptors for aldosterone[27] and thereby interferes with aldosterone-stimulated sodium reabsorption in the collecting duct (see Chapter 9, Figure 9.40). Inhibition of sodium reabsorption is associated with a marked decrease in urinary excretion of potassium and hydrogen ions. The inhibitory effect of spironolactone can be reversed, however, by raising the blood level of aldosterone. Therefore, spironolactone can be used as an effective diuretic in normal subjects or in patients on a low-sodium diet, but not in adrenalectomized patients or in individuals on a high-sodium diet.[58]

Triamterene and Amiloride. These two drugs are structurally dissimilar, but have similar effects on renal sodium transport.[29] Triamterene is effective from the peritubular side and irreversibly inhibits transbasolateral PD in the collecting duct. In contrast, amiloride is effective from the luminal side and reversibly inhibits transluminal PD in the "distal tubule" and collecting duct.

Of these two diuretics, the renal effect of amiloride has been most extensively studied. Evidence indicates that amiloride blocks sodium entry at the luminal membrane. The resulting decrease in luminal negativity reduces transepithelial PD, thereby decreasing potassium and hydrogen ions secretion into the lumen. The overall effect of amiloride is to reduce net sodium reabsorption and to increase net potassium retention. For this reason, amiloride is called a *potassium-sparing* diuretic. Available evidence suggests a similar mode of action for triamterene. Both amiloride and triamterene are orally effective antikaliuretics. They operate independently of aldosterone and hence can be used as effective diuretics in adrenalectomized patients.

The Xanthine Diuretics

This class of drugs is among the oldest of the modern diuretics. It include caffeine, theobromine, and theophylline. Of these, theophylline is the one most frequently used.

As pointed out in Chapter 9, theophylline inhibits phosphodiesterase. This leads to intracellular accumulation of cAMP. Thus, the diuretic effect of theophylline may be related to the inhibition of sodium reabsorption by cAMP, an idea confirmed by the recent observation that dibutyryl cAMP (an analog that readily enters the cell) inhibits sodium reabsorption in the proximal tubule.[1] The modes and sites of action of caffeine and theobromine are not known.

Since xanthine diuretics affect a variety of organ systems, including the cardiovascular system, the mechanism of their renal action remains obscure. At present, they are not used as standard diuretics in clinical medicine.

Summary

Diuresis may be induced either by interference with sodium chloride reabsorption (osmotic diuresis) or by interference with water reabsorption (water diuresis). Most of the clinically used diuretic drugs exert their effects by interfering with sodium chloride reabsorption. Hence, they all behave as osmotic diuretics. Of these, carbonic anhydrase inhibitors, such as acetazolamide and chlorothiazide, exert their effects on sodium bicarbonate reabsorption in the proximal and "distal" tubules. Furosemide and ethacrynic acid are among the most potent loop diuretics which exert their effects primarily in the thick ascending limb of Henle's loop. Antikaliuretic agents inhibit sodium-potassium exchange in the connecting tubule and collecting duct either by antagonizing the action of aldosterone (e.g., spironolactone) or by a mechanism unrelated to aldosterone (e.g., triamterene and amiloride).

Problems

12.1. There are three types of nephrons within the human kidney. (a) Name them and give their relative population ratio; (b) briefly outline the similarities and differences in their blood supply; and (c) briefly describe the functional contribution of each to the regulation of volume and composition of body fluids.

12.2. Describe the cross-sectional organization of the outer stripe, the inner stripe, and the inner zone of the renal medulla.

12.3. What are the specific features of the blood supply of the outer stripe of the outer medulla?

12.4. What is the functional relevance of the close association of the ascending vasa recta and the descending proximal tubules within the outer stripe?

12.5. What is the difference between a simple and a complex vascular bundle?

12.6. What is the reason for the peculiar shape of the inner medulla, namely, that it tapers from a broad base to a thin papilla?

12.7. Describe the distribution of thin limbs within the inner medulla.

12.8. Briefly outline the main features of the countercurrent multiplication principle as it applies to the kidney.

12.9. What are the major differences between the countercurrent multiplication and the countercurrent exchange processes?

12.10. Discuss the various types of experimental evidence that have been marshalled to support the operation of the countercurrent multiplication system within the kidney.

12.11. Explain the mechanism by which a countercurrent exchanger (a) may prevent a substance from entering a loop and (b) may trap a substance within a loop.

12.12. Describe the recycling routes for urea via short loops of Henle. Compare these routes in the simple and the complex types of the medulla.

12.13. Briefly describe the role of urea recirculation within the renal medulla in concentration and dilution of urine and indicate how this role is affected by ADH.

12.14. What is the role of the vasa recta in concentration and dilution of urine?

12.15. How does expansion of the extracellular fluid volume and hypertension affect the countercurrent multiplication process in Henle's loop? Do you expect the final urine to be more concentrated or dilute relative to normal?

12.16. Define osmolar and solute-free water clearances and briefly describe their potential clinical use in assessing the ability of the kidney to concentrate urine.

12.17. Define diuresis and briefly classify the major types of diuretic agents and the factors governing their use in research and clinical practice.

12.18. Of the various commonly used diuretics, which ones are the most potent and why?

12.19. Why is Na_2SO_4 solution a more effective osmotic diuretic than NaCl solution?

References

1. Agus ZS, Puschett JB, Senesky D, Goldberg M: Mode of action of parathyroid hormone nd cylcic adenosine 3',5'monophosphate on renal tubular phosphate reabsorption in the dog. *J Clin Invest* 1971; 50:617–626.
2. Atherton JC, Hai MA, Thomas S: Effects of water diuresis and osmotic (mannitol) diuresis on urinary solute excretion by the conscious rat. *J Physiol* 1968; 197:395–410.
3. Au WYW, Raisz LG: Studies on the renal concentrating mechanism. V. Effect of diuretic agents. *J Clin Invest* 1960; 39:1302–1311.
4. Aukland K: Renal blood flow, in Thurau K (ed): *Kidney and Urinary Tract Physiology (International Review of Physiology)*. Baltimore, University Park Press, 1976, vol 11, pp 23–79.
5. Baumgartl H, Leichtweiss HP, Lubbers DW, et al: The oxygen supply of the dog kidney: Measurements of intrarenal PO_2. *Microvasc Res* 1972; 4:247–257.
6. Bennett CM, Clapp JR, Berliner RW: Micropuncture study of the proximal and distal tubule of the dog. *Am J Physiol* 1967; 213:1254–1262.
7. Bennett CM, Brenner BM, Berliner RW: Micropuncture study of nephron function in the Rhesus monkey. *J Clin Invest* 1968; 47:203–216.
8. Berliner RW, Levinsky NG, Davidson DG, Eden M: Dilution and concentration of the urine and the action of antidiuretic hormone. *Am J Med* 1958; 24:730–743.
9. Beyer KH: The mechanism of action of chlorothiazide. *Ann NY Acad Sci* 1958; 71:363–379.
10. Bohmann SO: The ultrastructure of the renal medulla and the interstitial cells, in Mandal AK, Bohmann SO (eds): *The Renal Papilla and Hypertension*. New York, Plenum Medical Book Company, 1980, pp 7–33.
11. Bonventre JV, Lechene C: Renal medullary concentrating process: an integrative hypothesis. *Am J Physiol* 1980; 239:F578–588.
12. Bray GA, Preston AS: Effect of urea on urine concentration in the rat. *J Clin Invest* 1961; 40:1952–1960.
13. Brezis M, Rosen S, Silva P, Epstein FH: Renal ischemia: A new perspective. *Kidney Int* 1984; 26:375–383.
14. Clapp JR: Urea reabsorption by the proximal tubule of the dog. *Proc Soc Exp Biol Med* 1965; 120:521–523.
15. Clapp JR, Robinson RR: Distal sites of action of diuretic drugs in the nephron. *Am J Physiol* 1968; 215:225–235.
16. Cohen JJ, Kamm DE: Renal metabolism: Relation to renal function, in Brenner BM, Rector FC Jr (eds): *The Kidney*. Philadelphia, WB Saunders Co, 1981, pp 126-214.
17. Cohen HJ, Marsh DJ, Kyser B: Autoregulation in vasa recta of rat kidney. *Am J Physiol* 1983; 245:F32–F40.
18. De Rouffignac C: Physiological role of the loop of Henle in urinary concentration. *Kidney Int* 1972; 2:297–303.
19. De Rouffignac C, Morel F: Micropuncture study of water, electrolyte and urea movements along the loops of Henle in Psammomys. *J Clin Invest* 1969; 48:474–486.
20. De Wardener HE, Mills IH, Clapham WF, Hayter CJ: Studies on the efferent mechanism of the sodium diuresis which follows the administration of intravenous saline in the dog. *Clin Sci* 1961; 21:249–258.
21. Dirks JH, Cirksena WJ, Berliner RW: The effect of saline infusion on sodium reabsorption by the proximal tubule of the dog. *J Clin Invest* 1965; 44:1160–1170.
22. Dirks JH, Seely JF: Effect of saline infusion and furosemide on the dog distal nephron. *Am J Physiol* 1970; 219:114–121.
23. Duarte CG, Chomety F, Giebisch G: Effect of amiloride, ouabain, and furosemide on distal tubular function in the rat. *Am J Physiol* 1971; 221:632–639.
24. Earley LE, Kahn M, Orloff J: The effects of infusions of chlorothiazide on urinary dilution and concentration in the dog. *J Clin Invest* 1961; 40:857–866.
25. Ernst SA, Mills JE: Basolateral plasma membrane localization of oubain-sensitive sodium transport sites in the secretory epithelium of the avian salt gland. *J Cell Biol* 1977; 75:74–94.
26. Ernst SA, Schreiber JH: Ultrastructural localization of Na^+-K^+-ATPase in rat and rabbit kidney medulla. *J Cell Biol* 1981; 91:803–813.
27. Forte LR: Effect of mineralocorticoid agonists and antagonists on binding of ^{3}H-aldosterone to adernalectomized rat kidney plasma membrane. *Life Sci* 1972; 11:461–473.
28. Gamble JL, McKhann CF, Butler AM, Tuthill E: An economy of water in renal function referal to urea. *Am J Physiol* 1934; 109:139–154.
29. Gatzy J: The effect of K^+ sparing diuretics on ion

transport across the excised toad bladder. *J Pharmacol Exp Ther* 1971; 176:586–594.
30. Goldberg M: Real tubular sites of action of diuretics, in Fisher JW, Cafruny EJ (eds): *Renal Pharmacology*. New York, Appleton-Century-Crofts, 1971, pp 99–119.
31. Gottschalk CW: Micropuncture studies of tubular function in the mammalian kidney. *Physiologist* 1961; 4:33–55.
32. Gottschalk CW, Mylle M: Micropuncture study of the mammalian urinary concentrating mechanism: Evidence for the countercurrent hypothesis. *Am J Physiol* 1959; 196:927–936.
33. Hargitay B, Kuhn W: Das Multiplikations-prinzipals Grundlage der Harnkonzentrierung in der Niere. *Z Elektrochem* 1951; 55:539–558.
34. Holliger CH, Lemley KV, Schmitt SL, et al: Direct determination of vasa recta blood flow in the rat renal papilla. *Circ Res* 1983; 53:401–413.
35. Imai K, Araki M: Internephron heterogeneity and interspecies differences in the function of the descending limbs of Henle's loop (DLH). Abstract. *Proc Int Union Physiol Sci*. vol 15, Sydney, Australia, 1983.
36. Jamison RL: Urinary concentration and dilution, in Brenner BM, Rector FC Jr (eds): *The Kidney*. Philadelphia, WB Saunders Co, 1976.
37. Jamison RL, Bennett CM, Berliner RW: Countercurrent multiplication by the thin loops of Henle. *Am J Physiol* 1967; 212:357–366.
38. Jamison RL, Kriz W: *Urinary Concentrating Mechanism. Structure and Function*. New York, Oxford University Press, 1982.
39. Jamison RL, Lacy FB: Evidence for urinary dilution by the collecting tubule. *Am J Physiol* 1972; 223:898–902.
40. Johnston CI, Davis JO, Howards SS, Wright FS: Cross-circulation experiments on the mechanism of the natriuresis during saline loading in the dog. *Circ Res* 1967; 20:1–10.
41. Kawamura S, Kokko JP: Urea secretion by the straight segment of the proximal tubule. *J Clin Invest* 1976; 58:604–612.
42. Knepper MA: Urea transport in nephron segments from medullary ray of rabbits. *Am J Physiol* 1983; 244:F502–F508.
43. Knepper MA, Burg M: Organization of nephron function. *Am J Physiol* 1983; 244:F579–F589.
44. Kokko JP: Sodium chloride and water transport in the descending limb of Henle. *J Clin Invest* 1970; 49:1838–1846.
45. Kokko JP, Rector FC Jr: Countercurrent multiplication system without active transport in inner medulla. *Kidney Int* 1972; 2:214–223.
46. Koushanpour E, Tarica RR, Stevens WF: Mathematical simulation of normal nephron function in rat and man. *J Theor Biol* 1971; 31:177–214.
47. Kriz W: Structural organization of the renal medullary counterflow system. *Fed Proc* 1983; 42:2379–2385.
48. Kriz W: Structural organization of the renal medulla: Comparative and functional aspects. *Am J Physiol* 1981; 241:R3–R16.
49. Kriz W, Lever AF: Renal countercurrent mechanisms: Structure and function. *Am Heart J* 1969; 78:101–118.
50. Kriz W, Schnermann J, Koepsell H: The position of short and long loops of Henle in the rat kidney. *Z Anat Entwickl.-Gesch* 1972; 138:301–319.
51. Kuhn W, Ramel A: Aktiver Salz-transport als moglicher (und wahr-scheinlicher) Einzeleffekt bei der Harn-konzentrierung in der Niere. *Helv Chim Acta* 1959; 42:628–660.
52. Kuhn W, Ryffel K: Herstellung konzentrierter Losunger aus verdunnten durch blosse Membranwirkung. (Ein Modellversuch zur Funktion der Niere) Hoppe-Seylers. *Z Physiol Chem* 1942; 276:145–178.
53. Kunau RT Jr, Weller DR, Webb HL: Clarification of the site of action of chlorothiazide in the rat nephron. *J Clin Invest* 1975; 56:401–407.
54. Landwehr DM, Klose RM, Giebisch G: Renal tubular sodium and water in the isotonic sodium chloride loaded rat. *Am J Physiol* 1967; 212:1327–1333.
55. Lassiter WE, Gottschalk CW, Mylle M: Micropuncture study of net transtubular movement of water and urea in nondiuretic mammalian kidney. *Am J Physiol* 1961; 200:1139–1147.
56. Lassiter WE, Mylle M, Gottschalk CW: Micropuncture study of urea transport in rat renal medulla. *Am J Physiol* 1966; 210:965–970.
57. Lemley KV, Schmitt SL, Holliger CH, et al: Prostaglandin synthesis inhibitors and vasa recta erythrocyte velocities in the rat. *Am J Physiol* 1984; 247:F562–F567.
58. Liddle GW: Specific and non-specific inhibition of mineralocorticoid activity. *Metabolism* 1961; 10:1021–1030.
59. Macey RI: Transport of water and urea in red blood cells. *Am J Physiol* 1984; 246:C195–C203.
60. Maren TH: Carbonic anhydrase: chemistry, physiology and inhibition. *Physiol Rev* 1967; 47:597–781.
61. Miwa T, Imai M: Flow-dependent water permeability of the rabbit descending limb of Henle's loop. *Am J Physiol* 1983; 245:F743–F754.
62. Mollendorff Wv: Der Exkretionsapparat, in: *Handbuch der mikroskopischen Anatomie des Menschen*, vol 7. Berlin, Springer-Verlag, 1930, pp 1–327.
63. Moore LC, Marsh DJ: How descending limb of Henle's loop permeability affects hypertonic urine formation. *Am J Physiol* 1980; 239:F57–F71.
64. Morgan T, Berliner RW: Permeability of loop of Henle, vasa recta and collecting duct to water, urea and sodium. *Am J Physiol* 1968; 215:108–115.
65. Morgan T, Tadokoro M, Martin D, Berliner RW: Effect of furosemide on Na^+ and K^+ transport studied by microperfusion of the rat nephron. *Am J Physiol* 1970; 218:292–297.
66. Muller-Suur R, Ulfendahl HR, Persson AEG: Evi-

dence for tubuloglomerular feedback in juxtamedullary nephrons. *Am J Physiol* 1983; 244:F425–F431.
67. Pennell JP, Lacy FB, Jamison RL: An in vivo study of the concentrating process in descending limb of Henle's loop. *Kidney Int* 1974; 5:337–347.
68. Rocha AS, Kokko JP: Permeability of medullary nephron segments to urea and water: Effect of vasopressin. *Kidney Int* 1974; 6:379–387.
69. Roch-Ramel F, Peters G: Renal transport of urea, in Gregor R, Lang F, Silbernagl S (eds): *Renal transport of organic substances*. Berlin, Springer Verlag, 1981, pp 134–153.
70. Sanjana V, Johnston PA, Troy JL, et al: Hydraulic and oncotic pressure measurements in the inner medulla of the mammalian kidney. *Am J Physiol* 1975; 228:1921–1926.
71. Sanjana VM, Johnston PA, Robertson CR, Jamison RL: An examination of the transcapillary water flux in the renal inner medulla. *Am J Physiol* 1976; 231:313–318.
72. Sardet C, Pisam M, Maetz J: The surface epithelium of teleostean fish gills. Cellular and junctional adaptations of the chloride cell in relation to salt adaptation. *J Cell Biol* 1979; 80:96–117.
73. Scholander PF: The wonderful net. *Sci Am* 1957; 196:96–108.
74. Schmidt-Nielsen B: Urea excretion in mammals. *Physiol Rev* 1958; 38:139–168.
75. Schmidt-Nielsen B, O'Dell R: Structure and concentrating mechanism in the mammalian kidney. *Am J Physiol* 1961; 200:1119–1124.
76. Seely JF, Dirks JH: Micropuncture study of hypertonic mannitol diuresis in the proximal and distal tubule of the dog kidney. *J Clin Invest* 1969; 48:2330–2339.
77. Seldin DW, Eknoyan G, Suki WW, Rector FC Jr: Localization of diuretic action from the pattern of water and electrolyte excretion. *Ann NY Acad Sci* 1966; 139:328–343.
78. Sonnenberg H: Medullary collecting duct function in antidiuretic and in salt-or water-diuretic rats. *Am J Physiol* 1974; 226:501–506.
79. Stein JH, Mauk RC, Boonjaren S, Ferris TF: Differences in the effect of furosemide and chlorothiazide on the distribution of renal cortical blood flow in the dog. *J Lab Clin Med* 1972; 79:995–1003.
80. Stephenson JL: The renal concentrating mechanism: Fundamental theoretical concepts. *Fed Proc* 1983; 42:2386–2391.
81. Stephenson JL: Countercurrent transport in the kidney. *Ann Rev Biophys Bioeng* 1978; 7:315–339.
82. Stephenson JL: Central core model of the renal counterflow system. *Kidney Int* 1972; 2:85–94.
83. Stephenson JL, Tewarson RP, Mejia R: Quantitative analysis of mass and energy balance in non-ideal models of the renal counterflow system. *Proc Natl Acad Sci USA* 1974; 71:1618–1622.
84. Stokes JB: Integrated actions of renal medullary prostaglandins in the control of water excretion. *Am J Physiol* 1981; 240:F471–F480.
85. Tabei K, Imai M: Ion selectivity in the upper portion of descending limb of long-loop nephron (LDLu) of the hamster. Abstract. *Proc Int Union Physiol Sci.* Los Angeles, 1984.
86. Thurau K: Renal hemodynamics. *Am J Med* 1964; 36:698–719.
87. Ullrich KJ, Jarausch KH: Untersuchungen zum Problem der Harnkonzentrierung und Harnverdunnung: uber die Verteilung der Elektrolyte. Harnstoff, Aminosauren und exogenem Kreatinin in Rinde und Mark der Hundeniere bei verschiedenen Diuresezustanden. *Arch Ges Physiol* 1956; 262:537–550.
88. Ullrich KJ, Schmidt-Nielsen B, O'Dell R, et al: Micropuncture study of composition of the proximal and distal tubular fluid in rat kidney. *Am J Physiol* 1963; 204:527–531.
89. Wesson LG Jr: *Physiology of the Human Kidney*. New York, Grune & Stratton, 1969.
90. Wirz H, Hargitay B, Kuhn W: Lokalisation des Konzentrierungsprozesses in der Niere durch direkte Kryoskopie. *Helv Physiol Pharmacol Acta* 1951; 9:196–207.
91. Wirz H, Dirix R: Urinary concentration and dilution, in Orloff J, Berliner RW (eds): *Handbook of Physiology, Section 8, Renal Physiology*. Washington DC, American Physiological Society, 1973, pp 415–430.
92. Wright FS, Schnermann J: Interference with feedback control of glomerular filtration rate by furosemide, triflocin, and cyanide. *J Clin Invest* 1974; 53:1695–1708.
93. Zimmerhackl B, Robertson CR, Jamison RL: Effect of arginine vasopressin (AVP) on vasa recta blood flow. *Proc Am Soc Nephrol*. 204A, 1984.
94. Zimmerhackl B, Robertson CR, Jamison RL: The microcirculation of the renal medulla. *Circ Res* 1985; 57:657–667.

13

Renal Regulation of Extracellular Volume and Osmolality

Although each complex mammalian physiological system is developed to do a specific task, several of them are organized to perform one or more integrated functions. Of these functions, none is more essential to a normal life than that of *stabilizing* and maintaining a constant *internal environment* despite a wide variety of disturbances. Failure to achieve this stability or *homeostasis*, resulting from a breakdown in the various regulatory processes involved, poses a definite threat to life.

In a multicellular organism, such as man, several elaborate and specialized organ systems have the common task of minimizing or eliminating such a threat. In Chapter 1 we presented the salient, organizational features of these organ systems and collectively called them the *renal-body fluid regulating system*. As was depicted in Figure 1.1, this system is composed of metabolic, gastrointestinal, cardiovascular, pulmonary, cutaneous and renal subsystems. Although basic knowledge of the function of each of these subsystems is necessary, it was pointed out that only through an understanding of the interaction among them can one hope to gain an insight into the complex regulatory mechanisms involved in the body fluid homeostasis.

In the preceding chapters, we presented a systematic analysis of the components of body fluids, the various aspects of renal function, and their roles in the overall regulation of the constancy of the internal environment in the face of a wide variety of disturbances. Briefly, we learned that the distribution of water and solutes within the various body fluid compartments, in the steady-state of normality, depends on the adjustment of rates of fluid influx into and out of the body. We pointed out that of these, only the rate of renal efflux is subject to internal regulation in response to body needs in the face of external perturbations. The renal adjustment of fluid efflux was seen to be a function of the factors governing the formation of the glomerular filtrate and its subsequent sequential modification by tubular reabsorptive and secretory processes. Hence, in the final analysis, regulation of renal effluxes is ultimately determined by those factors—both renal and extrarenal in origin—that regulate glomerular filtration rate (GFR) and renal transport rate. Furthermore, since the volume and composition of the renal efflux reflect renal adjustment of these parameters in the blood compartment, which is in dynamic equilibrium with the other body fluid compartments, the regulation of renal efflux is the key to the ultimate regulation of the volume and composition of body fluids.

As mentioned in the preceding chapters, renal regulation of volume and osmolality of body fluids is influenced by the functional state of gastrointestinal, cardiovascular, pulmonary, and neuroendocrine systems. Of these, the gastrointestinal system influences primarily the rate of normal delivery of water and solutes into the body and hence indirectly influences kidney function. On the other hand, the cardiovascular, pulmonary, and neuroendocrine systems exert a direct influence on the intrarenal and extrarenal mechanisms responsible for the maintenance of body fluid homeostasis.

In this chapter we attempt to synthesize the materials presented in the preceding chapters by considering the functional characteristics and, more importantly, the pertinent input-output relationships of these subsystems, using the already familiar *functional* or *control diagram*.

Major Components of the Renal-Body Fluid Regulating System

Figure 13.1 presents schematically a simplified functional diagram of the renal-body fluid regulating sys-

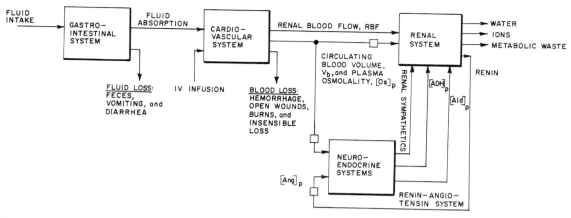

FIGURE 13.1. A simplified functional diagram of the major components of the renal-body fluid regulating system. (From Koushanpour.[88])

tem. To understand what this diagram represents, let us briefly examine each subsystem and identify its functional role in the overall operation of the renal-body fluid regulating system.

For the purpose of this presentation, we may divide the renal-body fluid regulating system into four separate but interacting subsystems:

1. The *gastrointestinal (GI) system* is the most important route of fluid entry into and a normal route of fluid loss from the body. Thus, fluid intake is the only input to this system, while fluid loss (normally as feces) and fluid absorption represent the two important outputs. Not shown is fluid lost through fistulas (e.g., gastric), which represents an abnormal but important output for the system. In pathological conditions such as vomiting and diarrhea, excessive fluid lost from the GI system may pose a serious threat to water and electrolyte balance. Because different kinds and amounts of solutes may be lost in each of these conditions, in adults reduction in the extracellular fluid volume (dehydration) usually leads to secondary and acute changes in acid-base balance (see Chapter 11). Therefore, alteration in the fluid intake and excessive fluid loss in disease are two of the ways by which the operation of the renal-body fluid regulating system can be disturbed.

2. The *cardiovascular system* is composed of a circulatory apparatus and a tissue exchanger, which are not shown in the functional diagram. The circulatory apparatus consists of a double heart, which mechanically pumps the blood containing fluid and nutrients absorbed by the GI system to the various tissues and organs of the body, including the neuroendocrine and renal systems. The tissue exchanger consists of the thin-walled capillaries, which supply all tissues. Water, electrolytes, and nutrients move in both directions across the capillary walls as a result of regional hydrostatic pressure gradients and electrochemical forces. In the functional diagram, two inputs are shown for the cardiovascular system: (a) the fluid absorbed from the GI system and (b) intravenous infusion. The latter is, of course, a mode of forcing fluid into the body in circumstances when normal fluid intake via the GI system is not possible. Because absorption of orally ingested fluid is associated with a considerable time delay, the renal response to such a volume expansion is also characterized by a time delay. In contrast, intravenous fluid infusion causes an immediate expansion of the extracellular fluid volume, with little or no time delay in renal response. Although the cardiovascular system has multiple outputs, for the purpose of the present discussion we have shown only three: (a) the renal blood flow (RBF); (b) the circulating blood volume, V_b, and plasma osmolality, $[Os]_p$; and (c) the blood loss, as in hemorrhage, open wounds, burns, and insensible loss. Hemorrhage, either arterial or venous, involves a loss of both plasma and blood cells. If not controlled, it leads to hypotension, circulatory shock, and death. As mentioned in Chapter 3, fluids discharged from open wounds and burns are lost from the interstitial space. Since the interstitial fluid is similar to plasma, except for the amount of proteins, its loss is reflected in the intravascular volume, a factor that must be recognized when fluid replacement is indicated. The insensible fluid loss occurs through the pulmonary and cutaneous sub-

systems. However, because it ultimately comes from the blood, it is included here for completeness. Note that the second output serves as an input to both the renal and the neuroendocrine systems. A *small box* on any *arrow* indicates that the signals being conveyed are *monitored* by appropriate receptors associated with the respective organ systems.

3. The *neuroendocrine systems* consist of three major components: the autonomic nervous system, the hypothalamus-pituitary system, and the adrenal gland. As depicted in the functional diagram, two inputs impinge on this system: one is the circulatory blood volume and plasma osmolality and the second is the output of the renin-angiotensin system, namely, angiotensin II, $[Ang]_p$. Once again, the small box on both of the input arrows indicates that the neuroendocrine system monitors the circulating blood volume, the plasma osmolality, and the blood level of angiotensin II. As we shall learn shortly, changes in V_b are monitored by peripheral *stretch receptors* in both high-pressure arterial and low-pressure venous sides of the circulation, which in turn modify the sympathetic outflow to the kidney and indirectly alter the rate of synthesis and release of antidiuretic hormone (ADH), and therefore also its plasma concentration, $[ADH]_p$. In contrast, changes in $[Os]_p$ are monitored by the hypothalamic *osmoreceptors*, thereby directly altering the rates of synthesis and release of ADH and hence its plasma concentration. Changes in the plasma level of angiotensin II produce a direct effect on the rate of aldosterone synthesis by the zona glomerulosa of the adrenal cortex and its release into the circulation, and consequently on its plasma concentration, $[Ald]_p$. Thus, as shown, the three output signals emerging from the neuroendocrine systems are the sympathetic outflow to the kidney, $[ADH]_p$, and $[Ald]_p$.

4. The *renal system* represents the final and most important component of the renal-body fluid regulating system. As we learned, the renal system indirectly maintains the constancy of the internal environment by directly monitoring the volume and osmolality of the circulating blood, thereby stabilizing the volume and osmolality of the extracellular fluid compartment. The kidney does this by carefully adjusting the rate of elimination of water, ions, and metabolic wastes. As shown, the output signals from the neuroendocrine systems modulate the renal function, and thereby the rate of excretion of water and ions.

As depicted in Figure 13.1, the renal system represents the primary regulatory organ whose function is directly modified by two other systems, namely, the cardiovascular and neuroendocrine systems. These two systems represent the *extrarenal* components of the renal-body fluid regulating system. Let us now examine in detail the functional organization and operational characteristics of the intrarenal and extrarenal components of the renal-body fluid regulating system, beginning with a review of the renal component.

The Renal Component

Figure 13.2 presents a functional diagram of the renal regulator developed on the basis of the materials presented in the preceding chapters. Note that in this and subsequent diagrams, a *plus sign* next to an input *arrow* impinging on a *box* means that the forcing increases the output signal, whereas a *minus sign* implies just the opposite.

The upper portion of Figure 13.2 depicts the four major components of the *renal regulator*: (1) the glomerular ultrafilter, (2) the proximal tubule, (3) Henle's loop, and (4) the distal tubule and collecting duct. To understand the operation of the renal regulator as it pertains to the present discussion, let us analyze it component by component, beginning with the glomerular ultrafilter component.

Consistent with what we have learned earlier (see Chapter 5), GFR is proportional to the effective ultrafiltration pressure, (ΔP_f), with the ultrafiltration coefficient k_g as the limiting proportionality constant. The magnitude of ΔP_f is determined by the algebraic sum (depicted by a *circle* with a *plus sign* inside) of the glomerular capillary blood pressure, P_g, and the glomerular capillary protein oncotic pressure, π_g. The magnitude of P_g is a function of the systemic arterial pressure, P_{AS} and the resistance of the renal vascular bed. The resistance of the renal vascular bed is a positive function of the renal sympathetic activity. The magnitude of π_g is a function of plasma protein oncotic pressure, π_p (not shown), which in turn is dependent on the rate of protein intake or loss. The osmolality of the filtrate, $[Os]_f$, emerging from the glomerular ultrafilter, is equal to the osmolality of the protein-free fraction of plasma, $[Os]_p$. The latter is determined by the intake of the osmotically active solute, Os, and the circulating blood volume, V_b. This relationship is depicted in the diagram by a *circle* with a *division sign* inside.

The functional diagram reveals that an important determinant of both $[Os]_p$ and π_p is the *volume of*

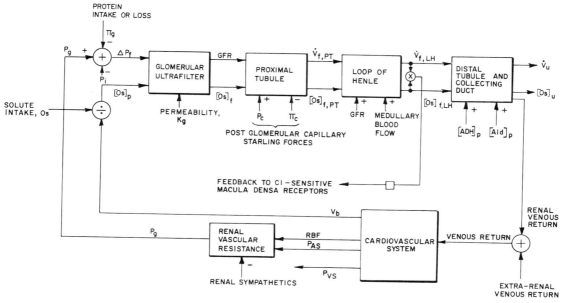

FIGURE 13.2. A functional diagram of the renal regulator. (From Koushanpour.[88])

circulating blood, whose magnitude is determined by the *cardiac output*, which in turn depends on the *venous return*. The latter is determined by the algebraic sum of the renal and extrarenal venous returns. The magnitude of the former depends on the algebraic sum of the tubular processing of the filtrate by all the functioning nephrons. The magnitude of the latter depends on three major factors: (1) the rate of oral intake of water, (2) the rate of extrarenal loss of water and solutes, such as insensible loss, respiration, sweat, and stool, and (3) fortuitous gain or loss of water and solutes, such as intravenous infusion, hemorrhage, vomiting, diarrhea, gastric fistula, and plasma loss through burned skin and open wounds.

In the *proximal tubule* (PT) approximately two thirds of the filtered water and solutes is reabsorbed iso-osmotically, yielding a filtrate tubular volume flow of $\dot{V}_{f,PT}$ and a filtrate osmolality of $[Os]_{f,PT}$ at the end of this segment. As shown, the reabsorption of the filtrate along this nephron segment depends mainly on the transepithelial transport of sodium, with reabsorption rate being influenced by changes in GFR and in postglomerular capillary pressure, P_c, and protein oncotic pressure, π_c, the so-called Starling forces. Thus, as described in Chapter 9, an increase in GFR, as by intravenous fluid infusion, increases the velocity of filtrate flow and reduces the time required to establish the sodium diffusion gradient across the luminal membrane, thereby decreasing the net transepithelial sodium and hence filtrate reabsorption. Likewise, an increase in P_g, besides causing an increase in ΔP_f and hence GFR, also induces further hemodynamic changes in the postglomerular capillaries (the peritubular capillary network), which leads to a reduction of filtrate reabsorption in the proximal tubule. The rise in P_g is transmitted to the peritubular capillary network, where it induces a proportional rise in the capillary hydrostatic pressure, P_c. Since the transcapillary movement of the filtrate from the peritubular interstitium to the blood depends largely on the hydrostatic pressure gradient (see Chapter 9), the rise in P_c opposes such a gradient. Hence, an increase in either GFR or P_g, or both, tends to reduce filtrate reabsorption in the proximal tubule, resulting in a net urinary excretion of solute (primarily sodium) and water. Conversely, an increase in π_c, induced by high protein intake, selective increase in efferent arteriolar resistance, or excessive loss of water (hemoconcentration), increases filtrate reabsorption in this nephron segment.

In the *loop of Henle* (LH) the filtrate volume and its osmolality undergo considerable modification, consistent with the operation of the countercurrent multiplication process described earlier (see Chapters 9 and 12). As shown, the volume, $\dot{V}_{f,LH}$, and osmolality, $[Os]_{f,LH}$, of the filtrate emerging from the ascending limb are influenced by changes in the GFR and the renal medullary blood flow.

An increase in GFR reduces the net reabsorption of sodium chloride from the thick ascending limb, thereby increasing the sodium chloride load, $\dot{V}_{f,LH} \cdot [NaCl]_{f,LH}$, emerging from this nephron segment. This is depicted by the *circle* with the *multiplication sign* inside. Likewise, an increase in the medullary blood flow causes solute washout from the medullary interstitium by the mechanism described in Chapter 12 and hence reduces the longitudinal gradient in the medullary interstitium. As shown, changes in the sodium chloride concentration or sodium chloride load (depicted by a *circle* with a *multiplication sign* inside) of the fluid emerging from the loop of Henle by way of the thick ascending limb are monitored (depicted by a small box on the arrow) by the chloride-sensitive tubular receptors at the macula densa. These receptors constitute the afferent limb of a negative feedback mechanism, which plays an important role in the regulation of GFR. The physiological role of the macula densa receptors in the renal regulation of the extracellular volume and osmolality, is discussed later in this chapter.

In the *distal tubule* and *collecting duct* the filtrate undergoes its final change in volume and osmolality. The final volume, $\dot{V}_u$, and osmolality, $[Os]_u$, of the excreted urine are the result of two simultaneous processes: (1) the exposure of the tubular fluid traversing along the collecting duct to the surrounding medullary interstitium, having a steep osmolar concentration gradient along its longitudinal axis, and (2) the exposure of the peritubular side of the collecting duct cell membrane to the circulating levels of aldosterone and ADH. As described in Chapter 12, the steeper the longitudinal osmolality gradient in the medullary interstitium, the smaller would be the volume and the greater would be the osmolality of the excreted urine. An increase in the blood level of either aldosterone or ADH would also yield the same result. A factor that limits the final volume and osmolality of the excreted urine is the limited adaptive reabsorptive capacity of the distal tubule and collecting duct. Thus, an increase in the volume of filtrate entering these nephron segments would overwhelm their reabsorptive capacity, leading to increased renal excretion of solute and water.

Note that the renal processing of the filtrate at each tubular segment returns to the blood nearly all the filtered solutes and water. This is depicted by the *arrow* emerging from the distal tubule and collecting duct and labeled "renal venous return." Thus, in this manner, the renal regulator stabilizes directly the blood volume and its osmolality and indirectly those of the extravascular and intracellular fluid compartments.

The Extrarenal Components

As depicted in Figure 13.1, the renal regulation of blood volume and osmolality is influenced by the normal functioning of three major organ systems: the gastrointestinal, cardiovascular, and neuroendocrine. The effects of alterations in the GI function in health and disease on body fluid homeostasis are mediated via the induced changes in the volume and osmolality of the circulating blood. The influence of the GI system on renal function is indirect and for the sake of simplicity we shall not discuss its function to any extent. Therefore, in this discussion and hereafter, we shall refer to the cardiovascular and the neuroendocrine systems as the extrarenal components of the renal-body fluid regulating system. Let us now examine the role of each separately.

The Cardiovascular System

As depicted in Figure 13.2, the cardiovascular system supplies a relatively constant fraction of cardiac output, as RBF, to both kidneys for processing. As discussed in Chapter 6, the constancy of RBF, for a given systemic arterial pressure (P_{AS}) is determined by the intrarenal vascular resistance, which in part is influenced by the activity of renal sympathetic nerves. The magnitude of RBF is in part determined by the cardiac output, which in turn depends on the venous return to the heart. The latter is in part determined by the renal venous return whose volume and osmolality are subject to renal regulation.

In terms of body fluid homeostasis, the cardiovascular system provides continuous information about two parameters of the circulating blood: *volume* and *osmolality*. Changes in blood volume are translated into changes in systemic arterial pressure (P_{AS}) and venous (P_{VS}) pressure. Changes in P_{AS} are continuously monitored by the *stretch receptors* situated in the high-pressure vascular bed (carotid sinus and aortic arch baroreceptors) and in the kidney (afferent arteriolar baroreceptors). Changes in P_{VS} are monitored by the *stretch receptors* situated in the low-pressure vascular bed (left and right atria). The afferent information arising from these receptors is then monitored by the appropriate components of the neuroendocrine system, leading to alterations in the blood levels of aldosterone and ADH. Changes in blood osmolality, $[Os]_p$, are likewise monitored by the neuroendocrine system, which also leads to alteration of blood levels of these hormones.

The Neuroendocrine Systems

As mentioned earlier, the neuroendocrine systems consist of three major components: the autonomic nervous system, the hypothalamus-pituitary system, and the adrenal gland. The input to these systems consists of changes in P_{AS}, P_{VS}, and $[Os]_p$, as well as changes in $[Ang]_p$. As depicted in Figure 13.1, the changes in these parameters will induce changes in the sympathetic outflow to the kidney and in the blood levels of aldosterone and ADH, leading to alterations in renal excretion of sodium and water and, hence, to alterations in the volume and osmolality of the blood.

Summary

The foregoing analysis reveals two important operational features of the renal-body fluid regulating system. (1) The *renal system* plays a central role in maintaining the constancy of the internal environment, a direct consequence of stabilizing the volume and composition of the blood. (2) The function of the renal regulator is partly modified by the *cardiovascular* and *neuroendocrine* systems.

As we proceed in this chapter, it is important to bear in mind these essential features, as outlined above, which are fully described below. Because they occupy the central theme of this chapter, we might restate them briefly. The kidneys stabilize the volume and composition of the blood and, hence, indirectly the extracellular fluid. Changes in volume and composition of the blood initiate compensatory responses from the cardiovascular and neuroendocrine systems, which in turn modify renal function.

With this brief outline of the functions of the major components of the renal-body fluid regulating system, we are now ready to consider in detail the various factors involved in the regulation of volume and osmolality of the blood and hence indirectly that of the extracellular fluids.

Regulation of Osmolality Versus Volume

Since the osmolality of the blood and hence of the extracellular fluid is in part determined by its volume, it is of special interest to inquire into the nature of interplay of the regulatory mechanisms involved in stabilizing the osmolality and volume of the extracellular fluid. In general, as long as the quantity of the osmotically active solutes (primarily sodium) in the extracellular fluid remains unaltered, any change in its volume due to fortuitous loss or gain of fluid is regulated by adjustment of renal efflux of osmotically active solutes. On the other hand, if there is a loss of osmotically active solutes, which is the case in most instances of fluid disturbance, restoration of the osmolality of the extracellular fluid volume depends on the readjustment of its volume. This is accomplished by adjustment of renal efflux of water.

The question of whether the regulation of volume precedes regulation of osmolality, or vice versa, has been the subject of intensive investigation. However, the issue, which is of considerable academic and clinical interest, continues to be controversial and has not been resolved. The bulk of the available experimental evidence (see reviews by Gauer et al,[49] Share and Claybaugh,[142] Goetz et al,[54] Menninger[101]) appears to favor the concept that within normal physiological range, regulation of plasma osmolality is more important than regulation of blood volume. Thus, an increase in plasma osmolality secondary to excessive loss of water more than sodium, such as may occur in heavy sweating in exercise, will stimulate ADH release and augment renal reabsorption of water. Concurrently, reduction in the "effective" circulating blood volume will stimulate the renin-angiotensin-aldosterone mechanism to augment renal reabsorption of sodium. The net result of the operation of these two mechanisms is to return the volume and osmolality of the extracellular fluid volume to normal. Therefore, it seems reasonable to state that the renal regulation of volume and osmolality of the blood, and hence of the extracellular fluid, are ultimately determined by the dynamic interplay of those renal and extrarenal factors that modify the normal rates of synthesis and release of ADH and aldosterone and their cellular action in the collecting duct epithelia.

Since ADH is primarily concerned with renal reabsorption of water, its regulation will affect primarily the plasma osmolality and secondarily the blood volume. On the other hand, because aldosterone is primarily concerned with renal reabsorption of sodium as isotonic solution, its regulation will affect primarily the blood volume and secondarily the plasma osmolality. Consequently, to facilitate presentation, we shall discuss the factors regulating plasma levels of ADH (and hence osmolality) and aldosterone (and therefore volume) separately.

Regulation of Plasma Osmolality: Control of Water Excretion

Regulation of plasma osmolality largely depends on the dynamic balance between the rates of influx and efflux of water into and out of the body. Of these fluxes, only the rate of renal efflux is subject to internal regulation.

Regulation of renal excretion of water is ultimately determined by the factors that influence the rates of synthesis and release of ADH into the blood and its cellular action in the collecting duct epithelia. There are two physiological stimuli for the synthesis and release of ADH: (1) the *plasma osmolality*, which acts via the osmoreceptors located in the anterior hypothalamus near the supraoptic nuclei; and (2) the *blood volume*, which, through induced changes in the systemic arterial (P_{AS}) and venous (P_{VS}) pressures, acts via the stretch or volume receptors located in the carotid sinus, aortic arch, and left atrium. Let us now consider the response characteristics of these receptors and their relative contribution to the rates of synthesis and release of ADH and hence the renal excretion of water.

Osmotic Regulation of ADH Release

Hypothalamic Osmoreceptors

As mentioned in Chapter 9, as early as 1947, Verney,[166] in a series of now classic experiments, demonstrated that a sustained five-minute infusion of a hypertonic solution of saline or sucrose (but not urea, which enters the cells) into the carotid artery of conscious dogs reduced a preestablished water diuresis. The changes in renal response were brought about by as little as 1% to 2% change in the osmolality of the perfusing arterial blood. On the basis of these findings, Verney proposed that there must be receptor cells in or near the hypothalamus sensitive to changes in the effective osmotic pressure of the extracellular fluid bathing them. He called these cells the *osmoreceptors*. Subsequent studies[76, 77] located these osmoreceptors in the vicinity of the supraoptic nucleus of the hypothalamus. Recent studies generally confirmed the osmoreceptor concept and further characterized the relationship between plasma osmolality and ADH release.

Using radioimmunoassay technique for measuring ADH, Robertson and co-workers[126] found a close correlation between plasma ADH concentration ($[ADH]_p$) and plasma osmolality ($[Os]_p$) in patients in various states of hydration. Using linear regression analysis, they defined an *osmotic threshold* of 280 mOsm/kg for ADH release (the intercept of the regression line with the horizontal axis) and a *sensitivity* for osmoreceptors (the slope of the linear regression). Subsequent studies in conscious sheep[169] have suggested that an exponential rather than a linear equation can best describe the relationship between $[ADH]_p$ and $[Os]_p$.

In the threshold model there is an abrupt increase in $[ADH]_p$ as $[Os]_p$ increases above 280 mOsm/kg, whereas in the exponential model $[ADH]_p$ changes gradually. Regardless of the model used to describe the plasma ADH-osmolality relationship, the data show that the omoreceptors are very sensitivie to changes in osmolality. Thus, a 1% increase in plasma osmolality results in an average change of 1 pg/mL in plasma ADH level, an amount sufficient to alter urinary concentration and flow rate. The extreme sensitivity of the osmoreceptors, as compared with other stimuli for ADH release, is the basis for the statement that the central osmoreceptors are the primary regulators of ADH synthesis and release. It is not known with certainty whether ADH is secreted continuously or episodically in response to osmotic stimulation.

According to present concepts,[130] the osmoreceptor neurons are stimulated by osmotically induced changes in their water contents. Thus, the effect of a given solute on stimulating the osmoreceptors depends on the rate the solute crosses the blood-brain barrier and enters the cells. Consequently, a solute that enters the cells slowly can create an effective osmotic gradient for water, thereby stimulating the osmoreceptors. However, a solute that enters the cell rapidly can not create an effective osmotic gradient, and hence can not stimulate the osmoreceptors. In short, stimulation of osmoreceptors is proportional to the level of their dehydration. Thus, the presence of an impermeable solute can create a greater osmotic gradient for water than a semipermeable or a completely permeable solute. Urea is known to cross the blood-brain barrier slowly but can readily enter the cell, thereby producing a relatively weak stimulus for osmotic release of ADH. This singular finding has led to the concept that most, if not all, of the osmoreceptors are located outside the blood-brain barrier.

The osmotically induced rise in [ADH]$_p$ will cause an increase in osmotic reabsorption of water from the collecting duct, thereby reducing C$_{H_2O}$. Although the details of how the increase in the osmolality of the extracellular fluid, to which the osmoreceptors are exposed, leads to an increase in neural discharge and eventual increase in synthesis and release of ADH remain entirely speculative, considerable evidence supports the osmoreceptor concept. Without being exhaustive, we shall cite two recent relevant studies as examples. For a more detailed review of evidence supporting the osmoreceptors theory, the interested reader is referred to a recent review by Schrier and co-workers.[130]

The most direct evidence supporting the osmoreceptor concept comes from studies of Johnson and associates.[79] They found that in conscious sheep, a decrease in the plasma osmolality of 1.2% produced a 2 to 1 μU/mL reduction in [ADH]$_p$, a change sufficient to bring about a detectable water diuresis. Further support for the osmoreceptor concept comes from electrophysiological studies of Durham and Novin.[40] They recorded slow potential changes from the region of the supraoptic nucleus in response to infusion of hypertonic solution into the rabbit carotid artery. From their observations they concluded that the supraoptic nucleus contains osmoreceptors that respond to rapid changes in osmolality of the circulating body fluids.

In summary, as depicted in Figure 13.3, changes in plasma osmolality, [Os]$_p$, are monitored by the osmoreceptors in the supraoptic nucleus of the hypothalamus, which in turn regulate the synthesis and release of ADH into the blood. The resulting change in the plasma concentration of ADH will in turn modulate the rate of osmotic reabsorption of water from the collecting duct. Thus, an increase in [Os]$_p$ leads to an increase in ADH release and hence in [ADH]$_p$, with eventual decrease in solute-free water clearance (C$_{H_2O}$) and hence antidiuresis. Conversely, a decrease in [Os]$_p$ will produce the opposite effects: decrease in [ADH]$_p$, increase in C$_{H_2O}$ and water diuresis. In the diagram the *negative sign* next to the *arrow* labeled [ADH]$_p$ indicates that ADH normally *decreases* C$_{H_2O}$ by increasing osmotic reabsorption of water in the distally located nephron segments.

Of clinical interest is the marked water diuresis (in excess of 2.5 L/d) seen in patients with central diabetes insipidus. This could result from lesion(s) in or near the supraoptic nucleus of the hypothalamus-pituitary system, leading to a marked decrease in synthesis and release of ADH and hence an increase in C$_{H_2O}$. The insatiable urge of these patients to drink water (polydipsia) has been considered secondary to a primary polyuria (marked excretion of hypotonic urine) and dehydration resulting from disruption of the synthesis and release of ADH mechanism.

Thirst Center

Numerous studies in conscious rats[152] and goats[3] have provided convincing evidence that in addition to osmoregulation, the hypothalamus plays an important role in the regulation of water drinking in thirst. The thirst sensitive center is believed to be located in the anterolateral region of the hypothalamus. Stimulation of this region in rats and goats elicits drinking behavior, while lesions in this area abolish the normal drinking response to water deprivation or administration of hypertonic saline.

Although there appears to be a direct relationship between plasma osmolality and thirst (and hence water intake), available evidence indicates that thirst is regulated by a separate group of osmoreceptor neurons that overlaps with those controlling ADH release.[124] The osmotic threshold for stimulating the thirst mechanism is always set some 10 to 15 mOsm/kg higher than that for ADH release. This implies that sensation of thirst is not experienced until [Os]$_p$ increases to a level sufficient to elevate [ADH]$_p$ to 5 pg/mL, a concentration that results in maximal antidiuresis.

Available evidence indicates that the role of the thirst center in water balance is equal in importance to that of the ADH-releasing supraoptic nucleus of the hypothalamus. Both centers appear to be activated by the same afferent stimuli—namely, changes in plasma osmolality and blood volume. However, the thirst center responds by controlling the oral influx of water, whereas the ADH-releasing center responds by regulating the renal efflux of water. In this manner, through the dynamic, coordinated, integrated dual actions of these centers, there is a continuous adjustment of total body water and hence of the osmolality of the extracellular fluid in the face of various disturbing stimuli. To preserve the simplicity of presentation, we have not included the thirst center and its input-output signals in Figure 13.3.

Hepatic Osmoreceptors

There is growing evidence that the liver plays an important role in osmoregulation, presumably by monitoring the osmotic pressure of the hepatic portal

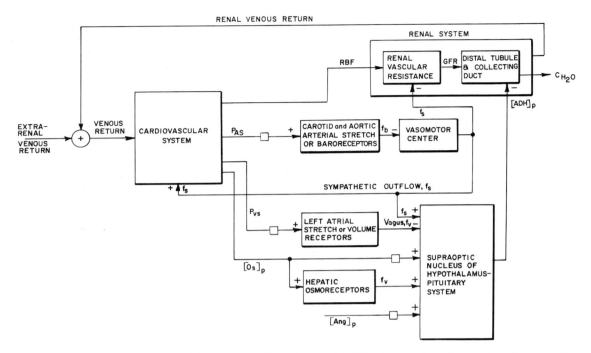

FIGURE 13.3. A functional diagram of the extrarenal regulation of ADH release and renal excretion of water. (From Koushanpour.[88])

venous blood. The existence of hepatic osmoreceptors was first postulated by Haberich[65] after observing that infusion of hypertonic saline solution into the hepatic portal vein of conscious rats resulted in antidiuresis, whereas infusion of water resulted in diuresis. As depicted in Figure 13.3, Haberich suggested that the renal response was mediated via a reflex modification of the ADH-releasing system by the hepatic branch of the vagus nerve (f_v). Further support for the existence of hepatic osmoreceptors has come from studies of Niijima,[113] who recorded the frequency of impulses from the hepatic branch of vagus nerve of isolated guinea pig livers perfused with hypertonic solutions of saline, glucose, sucrose, and mannose. He found that the frequency of afferent impulses increased in direct proportion to the increase in the osmolality of the perfusion fluid. However, in contrast to the hypothalamic osmoreceptors, a minimum of 6% increase in the osmolality of the perfusing fluid was required to stimulate the portal osmoreceptors.

Available evidence indicates that the hepatic receptors are more sensitive to changes in concentration of sodium chloride in the portal vein blood than in osmolality. Passo and associates[118] reported that infusion of hypertonic saline, but not hypertonic sucrose, into the hepatic portal vein of anesthetized cats for 90 minutes produced a significant natriuresis 30 minutes after the onset of infusion. However, infusion of both hypertonic saline and sucrose solutions into the femoral vein failed to alter the urinary excretion of sodium. Furthermore, they showed that bilateral vagotomy abolished the response to hepatic portal vein infusion. From these results, they concluded that, at least in the cat, the liver contains specific receptors that monitor sodium chloride concentration in the portal vein and that, via the vagus nerve, can reflexly alter the urinary excretion of sodium. Further support for the existence of hepatic sodium receptors comes from a study by Andrews and Orbach[5] who found an increase in the frequency of afferent impulses recorded from the hepatic branch of the vagus nerve of isolated rabbit livers perfused with hypertonic saline solution.

A recent study in rat[151] has further delineated the possible location of the hepatic osmoreceptors and the mechanism of their activation. They showed that superfusion of hepatic portal vein adventitia with hypertonic saline increased neural activity recorded from the hypothalamo-neurohypophysial system. Furthermore, superfusion with verapamil, which blocks calcium entry into the cell, reversibly inhibited whereas superfusion with acetylcholine activated the response to osmotic stimuli. These findings are consistent with the hypothesis that exposure to hypertonic solution induces water efflux from the osmoreceptor cells, which, coupled with calcium influx, results in an increase in the intracellular calcium. The latter somehow causes the release of acetylcholine, which excites the hepatic vagal afferent nerves that project to the hypothalamo-neurohypophysial region. It is further postulated that the release of ADH following activation of the hepatic osmoreceptors acts to inhibit sodium absorption from the GI tract, thereby reducing forthcoming changes in plasma sodium concentration.

Taken together, the above studies suggest that the hepatic osmoreceptors act to buffer large fluctuations in the osmolality of the systemic blood, and therefore the extracellular fluid, following food intake. Further, this mechanism may provide a protective mechanism against overhydration. At present, the question of whether the hepatic receptors are sensitive to sodium chloride concentration or osmolality of the hepatic portal vein blood is not fully resolved. Furthermore, the extent and relative contribution of the hepatic osmoreceptors to the overall regulation of water balance remains to be determined.

Nonosmotic Regulation of ADH Release

As depicted in Figure 13.3, there are two groups of stretch or volume receptors whose activities influence the rate of ADH synthesis and release. The first group is located in the *left atrium*, where they monitor the volume of blood returning to the left ventricle. Because of their anatomical location, the left atrial receptors have also been called the low-pressure venous vascular bed receptors. The second group is found in two locations. One set is located in the *carotid sinus*, which is a dilatation of the internal carotid artery at the point where the common carotid artery divides into the external and internal carotid arteries. Another set is located in the *aortic arch*, just proximal to the exit of the aorta from the left ventricle. Because of their anatomical locations, the carotid sinus and aortic arch receptors have also been called the high-pressure arterial vascular bed baroreceptors. That these two groups of receptors represent the afferent limb of a reflex mechanism for the control of ADH release is now well established. The afferent neurons arising from these receptors make primary synapse in the nuclei tractus solitari in the brainstem. From there, postsynaptic neurons that are partly noradrenergic project to regions of the paraventricular and supraoptic nuclei.[127] Under basal normovolemic and normotensive conditions, these neurons exert an inhibitory influence on ADH secretion since their elimination results in an acute increase in ADH secretion.[16, 156]

Left Atrial Stretch Receptors

Henry and associates[70] were the first to demonstrate the role of the left atrial stretch-sensitive volume receptors in the reflex regulation of ADH release. They showed that inflation of a small rubber balloon inserted into the left atrial appendage of an anesthetized dog produced an increase in urine flow after a latent period of 5 to 10 minutes. The reflex nature of the renal response was subsequently established by Henry and Pearce,[70] who showed that cooling the vagus nerve abolished the urinary response to left atrial balloon distention. Since the increase in urine flow in these studies was due largely to an increase in water excretion, they suggested that the renal response was a consequence of a reflex inhibition of ADH release by left atrial balloon distention. Although these conclusions were based largely on indirect evidence, subsequent studies generally have supported their validity and have further delineated the nature of the reflex pathway and its response characteristics. The following represents a summary of some of the more important studies that have provided the experimental evidence for the left atrial receptor component of the reflex regulation of ADH release.

Baisset and Montastruc[9] provided the first direct evidence that the increase in the left atrial pressure by balloon inflation produced a decrease in the antidiuretic activity of the plasma. This was subsequently confirmed by Shu'ayb and associates,[145] who observed an increase in urine flow and a decrease in plasma concentration of ADH in anesthetized dogs following an increase in left atrial pressure by balloon inflation. Moreover, they showed that both renal and hormonal effects could be abolished by vagotomy, thereby confirming the reflex nature of both responses. In a similar study Arndt and associates[6] observed hemodynamic changes, in addition to hormonal and renal responses, in anesthetized dogs following an increase in left atrial pressure by balloon inflation. They found that the increases in both osmolar and solute-free water clearances were accompanied by an increase in clearances of inulin and *p*-aminohippuric acid (PAH), despite a reduction in cardiac output and arterial blood pressure. That the hemodynamic changes were in part responsible for the renal and hormonal responses has further been verified in unanesthetized dogs by Lydtin and Hamilton.[97] These authors found that increasing the left atrial pressure by tightening a pursestring suture around the mitral valve caused an increase in arterial blood pressure, renal blood flow, urine flow, and urinary sodium excretion.

Although the studies cited above provide substantial support for the existence of a left atrial stretch receptor reflex regulating ADH release, they do not rule out the contributing effects of accompanying hemodynamic changes. The latter could have resulted from large increases in left atrial pressure, which were used in the above studies. Moreover, the methods used to determine plasma ADH activity were variable and unsatisfactory. To resolve the issue and to determine whether the left atrial stretch reflex is normally of physiological significance, Johnson and associates[80] reexamined the problem by determining the renal and hormonal response to much smaller increases in the left atrial pressure. They found that small increases in the left atrial transmural pressure (P_{LA}) up to 7 cm of water produced by balloon inflation in anesthetized dogs produced a linear decrease in the peripheral arterial plasma concentration of ADH. The latter was determined by bioassay in the ethanol-anesthetized rat after extraction and concentration of hormone. The decrease in $[ADH]_p$ was associated with a significant increase in urine flow and a decrease in urine-to-plasma (U/P) osmolality ratio and a negative solute-free water clearance ($T^c_{H_2O}$). The left atrial distention-induced renal and hormonal responses were not affected by concomitant changes in plasma osmolality or intravenous infusion of ADH. From these findings they concluded that the left atrial stretch-sensitive volume receptors play an important role in the regulation of plasma ADH level.

Considerable evidence[48,49] has generally established that the vagal afferents enter the *vasomotor center* in the medulla oblongata with secondary connection to the supraoptic region of the hypothalamus, which controls the rate of synthesis and release of ADH. The impulses in the vagal afferents have an inhibitory effect on the hypothalamus and hence on ADH release. Since the vasomotor center (which, as described below, is the site of neural control of the cardiovascular system) also receives afferent inhibitory impulses from the carotid sinus and aortic arch baroreceptors, it has been suggested[142] that the vagal afferents may in part determine the extent of information transmitted by the arterial baroreceptors. Thus, the influence of the baroreceptor afferents on the vasomotor center becomes dominant only when the impulse activity in vagal afferents is reduced.

There are a number of other stimuli besides direct changes in blood volume that bring about reflex changes in $[ADH]_p$ and C_{H_2O}. The most important of these are changes in the ambient temperature and body posture. Studies in human subjects[133] have shown that these stimuli bring about a redistribution of blood in the low-pressure intrathoracic vessels, which is monitored by the left atrial volume receptors, thereby causing reflex variations in $[ADH]_p$ and C_{H_2O}. Thus, exposure to cold or heat caused a marked decrease or increase in $[ADH]_p$, respectively, without any accompanying change in $[Na^+]_p$, $[Cl^-]_p$, or total solute concentrations. Since there was no change in plasma osmolality, changes in $[ADH]_p$ could not be mediated via the hypothalamic osmoreceptors. These changes, however, may be explained on the basis of redistribution of blood volume. It is well known that cooling causes peripheral vasoconstriction, which leads to an increase in the central blood volume. The resulting increase in the venous return will eventually cause an increase in the left atrial pressure and the intrathoracic "volume receptor" activity, which in turn causes inhibition of the hypothalamic centers. However, during exposure to heat, marked peripheral vasodilation results in a decrease in the central blood volume, leading to a decrease in the activity of volume receptors and an eventual increase in ADH release. In short, exposure to cold causes diuresis, whereas exposure to heat results in antidiuresis.

Postural changes lead to a redistribution of central blood volume and hence alter the activity of volume receptors. Thus, in *recumbent* position, the increase in central blood volume leads to an increase in left atrial pressure and inhibition of ADH release. In a sitting position, the left atrial pressure decreases somewhat, leading to eventual increase in ADH release. On *standing*, the left atrial pressure falls markedly, resulting in a lesser inhibition of the hypothalamus and an eventual increase in ADH release. Thus, assuming an upright position results in antidiuresis, whereas a recumbent position leads to diuresis. During sleep, the production of a concentrated urine is for the most part due to a reduction in blood pressure, which more than offsets the effect of reduced ADH release. Other agents that stimulate ADH release are emotion, exercise, anesthesia, and nicotine. The agents inhibiting ADH release include alcohol, caffeine, and CO_2 inhalation.

Finally, Moore[105] found that the diluting and con-

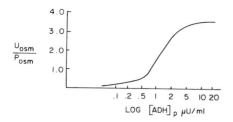

FIGURE 13.4. Relationship between U_{Osm}/P_{Osm} and the logarithm of plasma ADH concentration in man. The curve approximates the data of Moore,[105] which were fitted by the equation: $Y = 0.11 + 1.35 X - 0.12 X^2$.

centrating ability of the intact human kidney, as indicated by the U/P osmolality ratio, was a *sigmoidal* function of plasma ADH concentration. As shown in Figure 13.4, when $[ADH]_p = 0.7$ μU/mL, the kidney neither dilutes nor concentrates urine (U/P = 1.0), allowing the kidney to excrete about 18 L of water per day. When $[ADH]_p$ increases from 0.7 to 4.0 μU/mL, the U/P osmolality ratio increases linearly. However, an increase in $[ADH]_p$ above this level has no effect on the concentrating ability of the kidney. The significance of this sigmoidal relationship is that the maximal renal water conservation could be achieved by the release of moderate amounts of ADH. This allows for a normal kidney function in spite of moderate daily bodily activity, fluctuations in body and ambient temperatures, fortuitous fluid intake or loss, and postural changes.

To summarize, as depicted in Figure 13.3, an increase in the blood volume leads to an increase in the systemic venous pressure, P_{VS}, and eventually the left atrial pressure, P_{LA}, which is monitored (indicated by a small *box* on the *arrow*) by the left atrial stretch receptors. Stimulation of these receptors leads to an increase in the frequency of impulses in the vagal afferents, f_v, to the hypothalamic centers. Apparently, the vagal impulses arriving in these regions inhibit the synthesis and release of ADH by the supraoptic nucleus of the hypothalamus, thereby reducing the plasma concentration of ADH. The decrease in $[ADH]_p$ leads to a decreased osmotic reabsorption of water from the collecting duct and hence to an increase in solute-free water clearance (C_{H_2O}) and diuresis.

Carotid Sinus and Aortic Arch Baroreceptors

As depicted in Figure 13.3, a number of studies[72,85] have established that changes in the systemic arterial pressure (P_{AS}) are monitored (indicated by small *box* on the *arrow*) by the *stretch-sensitive* receptors located in the carotid sinus and aortic arch. Thus, an increase in P_{AS} stimulates these receptors, resulting in an increased frequency of impulses traveling in the baroreceptor component of the Hering nerve (for the carotid sinus receptors) and the depressor nerve component of the vagosympathetic trunk (for the aortic arch receptors). It has been found that the mean and pulse pressure components of the blood pressure interact nonadditively, so that at higher mean pressures the superimposed pulse pressure is a less effective stimulus in eliciting response from these receptors.[89] The afferent impulses arising from these baroreceptor nerves (f_b) will then enter the vasomotor center, where they inhibit the sympathetic nuclei (both the cardioaccelerator and vasoconstrictor centers) and by reciprocal inhibition stimulate the parasympathetic nuclei (the cardioinhibitor center). The result would be a net decrease in the frequency of sympathetic outflow (f_s) to (a) the kidneys, (b) the heart and blood vessels (indicated by the *box* labeled "cardiovascular system"), and (c) the hypothalamus-pituitary system. The decrease in sympathetic outflow to the cardiovascular system will lead to bradycardia and reduced vasoconstriction, the combined effects of which are to restore P_{AS} toward normal and hence close the negative feedback loop. Note that, as depicted here, the vasomotor center constitutes the site where the negative feedback occurs. Conversely, a decrease in P_{AS} results in a decrease in f_b and hence less inhibition of the vasomotor center. This would lead to a net increase in sympathetic outflow, which by the same mechanism leads to restoration of blood pressure toward normal. In short, changes in P_{AS}, brought about by, for example, changes in *blood volume*, will induce appropriate compensatory responses from the baroreceptor reflex system.

That the baroreceptors do indeed play a significant role in the regulation of blood volume is now well established. Perlmutt[121] observed that bilateral occlusion of common carotid arteries (a maneuver that reduces the blood pressure at the carotid sinus) for five minutes in vagotomized (left atrial and aortic arch receptors eliminated), hydrated dogs reduced solute-free water clearance after a latent period of 5 to 15 minutes. Share and Levy[139] showed that bilateral occlusion of common carotid arteries in vagotomized dogs resulted in an increase in plasma ADH concentration. This effect was blocked by carotid sinus denervation. In a subsequent study, Share[137] showed that if the systemic arterial blood pressure was maintained at a constant level, bilateral occlusion of common carotid arteries increased $[ADH]_p$ even in dogs with intact vagi. Subsequently, Share[138] showed that the rise in $[ADH]_p$, which normally occurs after bilateral common carotid occlusion, could be prevented by simultaneous inflation of a balloon in the left atrium. Furthermore, when the carotid sinuses of an anesthetized, vagotomized dog were perfused at normal mean and pulse pressure, no rise in $[ADH]_p$ occurred following hemorrhages of up to 50% of the blood volume. In another study, Share and Levy[140] showed that variations in arterial pulse pressure at the carotid sinus influence ADH release. Thus, they observed an increase in $[ADH]_p$ when flow through the carotid sinuses of the dog was changed from pulsatile to nonpulsatile. The effectiveness of nonpulsatile flow as a stimulus to baroreceptors is now well established.[89]

These studies indicate that a decrease in P_{AS}, acting through the carotid sinus baroreceptor mechanisms, brings about a reflex increase in $[ADH]_p$. The latter effect will eventually lead to a reduction in C_{H_2O} and antidiuresis. The opposite result, namely, water diuresis, follows an increase in P_{AS}. Essentially similar results have been obtained for the aortic arch baroreceptors.[142]

At present, the information about the neural pathways mediating the sympathetic inhibition of ADH release is meager. For simplicity of presentation we have depicted it as a single pathway in the block diagram. Finally, as shown, sympathetic outflow to the kidneys modifies the renal vascular resistance and hence GFR. Thus, an increase in sympathetic outflow should lead to a decrease in C_{H_2O}. This is in part substantiated by the studies of Karim and associates,[82] who found a decrease in renal sympathetic activity with left atrial balloon inflation.

Thus, it appears that changes in P_{AS} and P_{VS} resulting from a deliberate or sudden change or shift in blood volume induce a dual effect upon the renal system: (a) alteration of the sympathetic outflow to the kidneys and (b) modification of the rates of synthesis and release of ADH by the hypothalamus-pituitary system. Their integrative effects at the hypothalamus-pituitary and renal systems would induce a coordinated corrective response in the renal system.

A number of studies have attempted to determine the relative potency and contribution of the low-pressure and high-pressure vascular bed stretch receptors in the reflex regulation of ADH release. Furthermore, using radioimmunoassay technique for measuring plasma ADH levels in rats, dogs, monkeys, and humans, they have characterized the acute relationship between changes in blood pressure and ADH release (see review by Menninger[101]). The magnitude of ADH response to hypotension was found to be species dependent. The ADH response was found to be more responsive to a decrement of intracardiac pressure in dogs than it was in primates, including humans. Despite quantitative difference in the magnitude of ADH release, the increase in $[ADH]_p$ is found to be an exponential function of the degree of hypotension. Thus, a 5% to 10% decrease in blood pressure had little effect on $[ADH]_p$, whereas a 20% to 30% decrease in blood pressure produced a significant increase in ADH release and hence $[ADH]_p$. The ADH release in response to changes in blood volume appeared to be quantitatively and qualitatively similar to that of changes in blood pressure. Thus, acute nonhypotensive (less than 10%) reduction of blood volume had little or no effect on ADH secretion. However, an acute hypotensive (10% to 15%) reduction of blood volume doubled $[ADH]_p$, and a 20% reduction of blood volume caused marked elevation of $[ADH]_p$. Conversely, an acute increase in blood volume or pressure caused suppression of ADH release. However, the effect of increase in blood volume and pressure on ADH release has not been fully characterized. Finally, some studies have examined the chronic effect of changes in blood volume and pressure on ADH release in humans.[130] It has been found that patients with chronic hypovolemia and hypervolemia, due to deficient or excess secretion of aldosterone, show changes in $[ADH]_p$ that is similar to those produced by acute changes in blood volume. In contrast, patients with uncomplicated essential hypertension have normal $[ADH]_p$, indicating that the high-pressure stretch receptors can adapt (*reset*) in time to the elevated pressure.[87, 125] At present the details of the mechanisms mediating the chronic effect of changes in blood volume and pressure on ADH secretion remains incompletely defined.

Taken together, these findings suggest that the left atrial receptors are more sensitive to small changes in blood volume than the carotid sinus and aortic arch receptors. That is, the left atrial receptors constitute the first line of defense against nonhypotensive changes in blood volume. Thus, as long as blood loss does not exceed 10% of the blood volume, the hemorrhage is nonhypotensive and does not activate the high-pressure vascular bed receptors. The renal response is essentially mediated through the left atrial stretch receptors. However, when hemorrhage is hypotensive—that is, when it exceeds 10% of the blood volume—the reduced P_{AS} activates the arterial baroreceptor reflex, thereby initiating compensatory adjustment of renal response by the mechanism described above and depicted in Figure 13.3.

Finally, it should be noted that compared with the effect of osmotic stimuli on ADH secretion, the volume and pressure stimuli have relatively minimal effect on ADH secretion. This difference, coupled with the marked differences in the sensitiity of the osmoregulatory and nonosmoregulatory mechanisms, largely accounts for the relative contribution of the two regulatory systems to the overall regulation of ADH release under normal and pathologic conditions. In view of the available evidence, it is reasonable to state that the osmoregulatory mechanism is responsible for day-to-day fine adjustment of ADH secretion in response to normal fluctuations in fluid balance, whereas the nonosmoregulatory mechanism acts to buffer large fluctuations in fluid balance caused by abnormal disturbance in blood volume and pressure and hence ADH release. Additionally, the nonosmotic mechanism mediates the effects of a variety of pharmacological and pathological effects on ADH secretion. These include the effect of administration of diuretics, isoproterenol, nicotine, and nitroprusside, all of which are thought to stimulate ADH release by lowering blood volume or pressure,[11, 29] and norepinephrine, which suppresses ADH secretion by increasing the arterial blood pressure.[143] Moreover, upright posture, hemorrhage, sodium deficiency states, congestive heart failure, cirrhosis, and nephrosis probably stimulate ADH secretion secondary to a reduction in the total or effective blood volume, whereas orthostatic hypotension and vagovagal reactions stimulate ADH secretion secondary to a reduction in the arterial blood pressure.

Interactions Between Osmotic and Volume Stimuli

Having delineated the functional characteristics of the elements of the ADH-releasing system, we are now ready to inquire into the question of interaction between them and to assess their relative roles in the regulation of water balance and hence the plasma osmolality. This question has been studied by two groups of investigators in conscious sheep and by another group in man. Zehr and associates[173] found that water deprivation in conscious sheep reduced blood volume and lowered the left atrial pressure, thereby reflexly decreasing the vagal inhibition of ADH release from the hypothalamus-pituitary system. Concurrently, the increased plasma osmolality stimulated the osmoreceptors to increase ADH release. Thus, in water deprivation both osmotic and volume elements acted together to stimulate ADH release and hence to increase renal conservation of water. In contrast, iso-osmotic or hypo-osmotic expansion of the extracellular fluid volume in either the normally hydrated or dehydrated sheep resulted in a reduction in ADH release regardless of the direction of changes in plasma osmolality. From these findings they concluded that the left atrial stretch receptors play a major role in ADH release, a conclusion not warranted in view of the subsequent findings of Johnson and co-workers.[79] These investigators, using similar preparations, compared in conscious sheep the effects of iso-osmotic changes in blood volume and iso-volemic changes in plasma osmolality separately, and combinations of the volume and osmolality stimuli. When applied separately, iso-osmotic changes in blood volume or iso-volemic changes in plasma osmolality elicited appropriate, expected responses from the volume and osmotic receptor elements of the ADH-releasing system. When the volume and osmotic stimuli were combined, the effect on ADH release was approximately additive. Thus, plasma ADH concentration remained unchanged when a 1.2% reduction of plasma osmolality was combined with a hemorrhage of about 10% of blood volume. From these studies, they concluded that with small changes in blood volume and plasma osmolality neither receptor element appears to dominate the other in the control of ADH release. Similar conclusions were reached by Moses and Miller,[107] who studied the osmotic threshold for ADH release in human subjects whose plasma osmolality was elevated by either hypertonic saline infusion or water deprivation. They found that in hydrated subjects, the plasma osmolality at which the water diuresis was inhibited was higher when the plasma osmolality was increased by hypertonic saline infusion (which also increases the extracellular fluid volume) than when it was increased by water deprivation (which also decreases the extracellular fluid volume).

The interaction between the osmotic and nonosmotic stimuli for ADH release, and hence renal excretion of water, has recently been reexamined with a mathematical model of the renal-body fluid regulating system in dog.[86] The model, which incorporated the major nonlinearities of fluid assimilation, exchange, distribution, and excretion (see Figure 13.1), was used to simulate the renal response to small changes in blood volume and plasma osmolality. The model provided continuous simulation curves, which agreed reasonably well with the available transient and steady-state experimental data. The model predicted that stimulation of volume receptors via changes in left atrial pressure accounted for only 15% to 20% of changes in ADH secretion rate, whereas stimulation of the osmotic receptors via changes in plasma osmolality accounted for the remaining 80% to 85% of changes in ADH secretion rate. Thus, it appears that regulation of ADH secretion is largely dependent on plasma osmolality during perturbations which do not appreciably alter the cardiovascular blood volume.

In summary, available evidence indicates that both volume and osmotic elements of the ADH-releasing system act in concert to maintain the volume and osmolality of the extracellular fluid via the action of ADH on renal excretion of water. Moreover, within the normal physiological limits of changes in blood volume and osmolality, the effect of the two elements on ADH release is additive with the osmotic stimuli as the major determinant of ADH release.

The question of central interaction between the osmotic and nonosmotic stimuli for ADH release has been examined in two recent studies. Halter and coworkers,[66] studying patients with "essential hypernatremia," reported that these patients showed no osmotic-mediated ADH release during hypertonic saline infusion, while their nonosmotic ADH release (as tested by hypotension with trimethaphan) was normal. This finding suggests that the osmotic and nonosmotic pathways must be anatomically discrete and separate from each other. This conclusion is supported by electrophysiological studies in rat.[81] These investigators, recording from single supraoptic neurons, found an increase in activity in response to both the osmotic (intracarotid infusion of hypertonic saline) and nonosmotic (carotid occlusion) stimuli, but not to isotonic saline infusion. These results are consistent with the concept that the osmotic and nonosmotic stimuli have separate and anatomically discrete projections into the magnocellular neurosecretory cells located in the regions of the supraoptic and paraventricular nuclei of the hypothalamus. Further electrophysiological and histochemical studies are needed to fully resolve the existence of anatomically separate pathways for osmotic and nonosmotic stimulation of ADH release.

Role of Renin-Angiotensin System in ADH Release

In 1949 Gaunt and associates[50] reported that patients with adrenal insufficiency have an impaired ability to excrete a water load. Since then, there has been growing evidence that the renin-angiotensin system plays an important role in the regulation of water balance through its effect on ADH release.

It is now well established[73] that a reduction in blood volume by hemorrhage in the dog stimulates renin release and hence elevates the plasma concentration of angiotensin, $[Ang]_p$. Furthermore, similar alterations in blood volume have also been shown to stimulate ADH release and to increase its plasma level.[71] In man, variations in the redistribution of blood volume induced by postural changes have been shown to produce directional changes in both renin release[24] and plasma level of ADH.[133] Share and Travis[141] reported that plasma level of ADH increased in adrenal-insufficient dogs who had undergone bilateral adrenalectomy. Acute injection of glucocorticoids resulted in a decrease in ADH release. These authors attributed the rise in $[ADH]_p$ to a reduction in blood volume and blood pressure during the period of adrenal insufficiency. However, it is not known whether these responses of ADH secretion rate to adrenal function were due to an increase in the rate of ADH release or to a decrease in the rate of its removal.

Collectively, these studies indicate the possible existence of a functional interaction between the renin-angiotensin system and the control of ADH release. However, the nature of the interaction and a causal relationship, if any, remained to be determined.

Bonjour and Malvin[17] subsequently reexamined this question and they provided direct evidence for the control of ADH release by the renin-angiotensin system. They found that intravenous infusion of a small, nonpressor dose of angiotensin in conscious dog significantly increased $[ADH]_p$. Similar results were obtained by infusion of renin, suggesting that both exogenously administered and endogenously formed angiotensin were equally effective in stimulating ADH release. From these results, they concluded that there exists an intimate, causal relationship between the renin-angiotensin system and the control of ADH release. In a subsequent study from the same laboratory,[108] an attempt was made to define the locus of the receptor for angiotensin stimulation of ADH release. Intracarotid infusion of angiotensin in anesthetized dog was found to be a more potent stimulus of ADH release than its intravenous infusion. From this finding, they suggested that angiotensin can enter the brain tissue and that the angiotensin receptor must be located in the central nervous system. This concept was further supported by the observation that infusion of a small amount of angiotensin II directly into the ventriculocisternal system stimulated ADH release. Since the paraventricular nucleus of the hypothalamus—a nucleus associated with ADH release—lies near the ventricular space, they hypothesized that this nucleus is the site of the angiotensin receptor. At present, this hypothesis lacks direct neurophysiological confirmation.

A subsequent study by Tagawa and associates[153] has provided compelling evidence for a negative feedback relationship between the renin-angiotensin and plasma ADH concentration. They found that an increase in $[ADH]_p$ ranging from 0.4 to 4.4 μU/mL inhibited renin secretion in unanesthetized, sodium-deprived dogs. That this response may be of physiological importance was demonstrated by the fact that a 1.2 μU/mL increase in $[ADH]_p$ resulted in a 30% reduction in renin release.

Taken together, the above studies suggest that there exists a negative feedback between the renin-angiotensin system and ADH release, with the locus of interaction being in or near the hypothalamus. Thus, as depicted in Figure 13.3, an increase in $[Ang]_p$ stimulates ADH release, which in turn enhances the renal reabsorption of water, thereby closing the feedback loop. Although the studies cited above advance the hypothesis that the renin-angiotensin system plays an important role in water balance and hence osmoregulation, its importance in stimulating the volume and osmotic elements of the ADH-releasing system has been questioned in a subsequent review of the subject by Share and Claybaugh.[142] They suggest that the interrelationships between ADH, renin, and angiotensin may serve to minimize fluctuations in their plasma concentrations under resting conditions. Variations in the blood volume or plasma osmolality would then elicit appropriate responses from the volume receptors and osmoreceptors, thereby overriding this negative feedback relationship and bringing about appropriate changes in the secretion of ADH and renin.

In a subsequent study, Shade and Share[136] provided further evidence for the above point of view. Reporting their inability to repeat the results obtained by Bonjour and Malvin[17] and Mouw and co-workers,[108] they suggest that angiotensin II may potentiate the effect of a known stimulus for ADH release, rather than directly stimulating its release. Thus, they found that intravenous infusion of small amounts of angiotensin II potentiated the release of ADH in bilaterally nephrectomized dogs subjected to nonhypotensive hemorrhage. However, this potentiating effect in response to the volume stimulus was found to be much smaller compared with that reported by Shimizu and associates[144] for an osmotic stimulus. These latter investigators found that simultaneous intravenous infusion of angiotensin II and hypertonic solution caused a more significant increase in ADH release than infusion of hypertonic saline alone.

From these and other studies, Shade and Share[136] concluded that "volume and osmotic control of ADH are to some degree independent and ... angiotensin does not affect the final common pathway for ADH release. Thus, angiotensin may potentiate osmotically stimulated ADH release, but has little or no effect on ADH release resulting from decreases in blood volume".

In conclusion, although the role of the renin-angiotensin system on ADH release and its possible role in water and electrolyte balance represents an attractive idea, the available evidence is conflicting. The unequivocal existence of such a relationship remains to be demonstrated.

Regulation of Blood Volume: Control of Sodium Excretion

As described in Chapter 2, sodium is the major cation of the extracellular fluid (ECF) compartment. Accordingly, the availability of this ion, as determined by the dynamic balance between the rates of its influx into and efflux out of the body, will largely determine the volume of water in the ECF compartment.

Since the exchangeable sodium, which constitutes the major fraction of the total body sodium, is primarily confined in the ECF compartment, its regulation will ultimately determine the volume and hydrostatic pressure of the ECF compartment and its extravascular and intravascular components. Hence, in the final analysis, the regulation of blood volume and its associated hydrostatic pressure depends primarily on the careful adjustment of the dynamic balance between the rates of influx and efflux of sodium into and out of the body. As mentioned previously, of these fluxes, only the rate of renal efflux is subject to internal regulation. Thus, in the absence of vomiting, diarrhea, or excessive sweating, the kidney can precisely adjust the excretion rate of sodium to match the wide variations in its intake, thereby stabilizing the blood volume, and therefore the ECF volume, within normal limits.

The smooth operation of the mechanisms involved in the renal excretion of sodium depends on their continuous adjustment by two extrarenal mechanisms that respond to small changes in volume, hydrostatic pressure, and composition of the circulating blood. The first is a *humoral* mechanism consisting of the renin-angiotensin-aldosterone system, which plays the major role in the simultaneous regulation of all of the above three blood variables. The second is a *neural* mechanism, consisting of the right atrial stretch-sensitive volume receptors, which plays a minor but important role in regulating the above variables by reflexly modifying renin secretion by the kidney. It should be noted that the common pathway for the effects of both mechanisms is the fine adjustment of the rates of synthesis and release of aldosterone into the blood and its cellular action on the collecting duct epithelia of the nephron. It therefore follows that the renal regulation of blood volume ultimately depends on the careful adjustment of the renal excretion of sodium, which in turn depends on the factors that modify the rates of aldosterone synthesis and release.

Let us now consider the stimulus-response characteristics of these two extrarenal mechanisms, component by component, as well as their relative contributions to the renal excretion of sodium and hence the regulation of blood volume.

The Renin-Angiotensin-Aldosterone System

Historical Perspective

Our current knowledge of the role of the renin-angiotensin-aldosterone system in the renal regulation of blood volume is derived largely from those studies that were designed to delineate the causes of hypertension in general and of renal hypertension in particular. Thus, to better understand recent developments and their relevance to the regulation of blood volume (discussed later in this section), a brief summary of these early milestone studies is in order. For a more extensive treatment, the interested reader is referred to an excellent review by Laragh and Sealey.[91]

Although in 1827 Richard Bright[22] was first to recognize that hypertension may be a renal disease, the first "milestone" in the modern history of renal hypertension occurred about 70 years later. In 1898 Tigerstedt and Bergman[157] discovered that injection of a crude saline extract of rabbit kidney into anesthetized rabbits caused an increase in blood pressure. They attributed the hypertensive effect to a substance that they named *renin*. That the kidney is indeed involved in hypertension was subsequently reaffirmed by studies of Volhard and Fahr,[167] who defined an association between renal necrotizing arteriolitis and malignant hypertension.

The second milestone occurred in 1938, when Goldblatt[55] demonstrated that constriction of the renal artery in dogs and rabbits caused a chronic, sustained elevation of blood pressure. The third milestone occurred in 1939 when both Page[115] and Braun-Menendez and colleagues[19] independently discovered that renin is an enzyme that causes release of a pressor substance called *angiotensin* from its circulating plasma substrate, α-2-globulin. Subsequent studies by Goormaghtigh[58] showed that renin secretion in both human and animal renal hypertension is associated with changes in the granularity of the *granular cells* of the juxtaglomerular apparatus (see Chapter 6). This directly implicated the kidney in the development of hypertension. As mentioned in Chapter 6, granular cells are sites of synthesis, storage, and release of renin.

The next milestone came between 1956 and 1957, when Skeggs and his associates[147] and Elliot and Peart[42] independently determined the amino acid sequence of angiotensin, which was later synthesized by Bumpus and his group.[27] Then, in 1960 Laragh and his associates[90] showed that infusion of angiotensin II in man stimulates *aldosterone* release from the adrenal cortex. Thus, the link between renin, angiotensin, and aldosterone and their possible roles in the regulation of blood volume, composition, and hydrostatic pressure became firmly established.

Let us now consider some of the important physiological characteristics of the components of the renin-angiotensin-aldosterone system as revealed by subsequent studies.

Components of the System

Renin

Renin is a proteolytic enzyme with a molecular weight of about 40,000. It is produced primarily by the granular cells of the juxtaglomerular apparatus of the kidney in response to a variety of stimuli (see Figure 13.6). In addition, renin is also produced by two extrarenal organs, namely the submaxillary gland and the uterus, with the pregnant uterus exhibiting a greater renin-like activity than the nonpregnant uterus. However, the available evidence indicates that these extrarenal sources of renin normally contribute very little to the circulating plasma levels of renin.

Despite its molecular size, renin is found in urine, suggesting that it is partially filtered at the glomerulus. However, since its urinary excretion is normally less than 1% of the amount filtered, it is almost completely reabsorbed by the renal tubules.[23]

Plasma renin activity (PRA) may be quantitated either by radioimmunoassay of generated angiotensin I or bioassay of generated angiotensin II. The latter is usually expressed in *Goldblatt units*. A Goldblatt unit is defined as the amount of renin that when injected into a conscious, trained dog raises the blood pressure by 30 mm Hg.[56] Using either method, Sealey and associates[131] found a consistent dynamic relationship between the daily rate of renal sodium excretion ($\dot{E}_{Na}$) and midday PRA measured in 52 normal ambulatory subjects. In these subjects, PRA ranged from 0.5 to 2.8 ng/mL/h when $\dot{E}_{Na}$ was above 150 mEq/d, from 1.4 to 6.3 ng/mL/h when $\dot{E}_{Na}$ was between 50 and 150 mEq/d, and up to 21 ng/mL/h when $\dot{E}_{Na}$ was below 50 mEq/d.

Several studies, reviewed by Lee,[94] have shown that the circulatory half-life of renin ranges from 10 to 20 minutes in rat, with somewhat higher values in dogs (45 to 79 minutes) and man (42 to 120 minutes). The variations in the circulatory half-life indicate the rapidity with which renin is secreted in response to various stimuli, such as hemorrhage, exercise, and changes in posture.

Normally, the plasma renin level is determined by a dynamic balance between the rate of its production and the rate of its destruction or removal from the circulation. Direct measurements of arteriovenous differences for renin across several vascular beds in dogs after elevation of arterial plasma renin (induced by stimulation of endogenous renin secretion by acute salt depletion) or infusion of exogenous renin have clearly shown that the liver is the major site of renin inactivation.[68] In these and subsequent studies[128] the metabolic clearance of renin approached its hepatic clearance, suggesting that the rate of renin inactivation is a direct function of the hepatic blood flow. Thus, changes in plasma renin level in disease, such as congestive heart failure or liver cirrhosis, may reflect a reduction in normal hepatic renin inactivation.

Renin as such has no known physiological effect other than causing liberation of antiotensin I from its plasma substrate. Intravenous infusion of renin causes an increase in blood pressure after a latent period of 15 to 20 seconds. The elevation of blood pressure is gradual and may last 30 minutes or longer, depending on the amount of renin injected. In contrast, intravenous infusion of angiotensin II causes an immediate rise in blood pressure that is short-lived.[94] The differences in the pressor action of renin and angiotensin II are attributed to the enzymatic properties of renin and the kinetics of endogenous liberation of angiotensin I from plasma renin substrate and its subsequent conversion to angiotensin II.

It is well known that repeated injection of renin over a short period of time results in progressively less elevation of blood pressure, a phenomenon called *renin tachyphylaxis*. Although the underlying mechanisms are not presently known, angiotensin tachyphylaxis and saturation of the vascular receptor sites, as well as prostaglandin release, have been implicated as contributing factors in renin tachyphylaxis.

Several studies[94] have demonstrated that endogenous renin secretion is highly correlated with the degree of granulation of the granular cells of the juxtaglomerular apparatus. Thus, hypergranulation is linked to hypersecretion of renin, whereas hypogranulation or degranulation is linked to hyposecretion. Experimentally, several conditions are known to produce hypergranulation of granular cells and thus hypersecretion of renin. They include renal ischemia, prolonged anoxia, sodium deficiency, hyponatremia, experimental ascites, pregnancy, and adrenal insufficiency. In contrast, hypogranulation or degranulation, and hence hyposecretion of renin, occurs after sodium loading, increased arterial blood pressure, and expansion of body fluids by overtransfusion.

Renin Substrate

Since the classic studies of Page[115] and Braun-Menendez and associates,[19] who identified the α-2-globulin fraction of plasma as the renin substrate, a number of subsequent studies have attempted to identify the structure of the renin substrate molecule and the nature of the renin-renin substrate reaction. From degradation studies of horse renin substrate, Skeggs and co-workers[146] identified the renin substrate as a 14-amino acid residue polypeptide (Figure 13.5). Incubation of this tetradecapeptide with renin yielded angiotensin I, whose amino acid sequence was identical to the first ten amino acids from the *N*-terminal group of the renin substrate. Furthermore, they showed that renin acts on the leucyl-leucyl bond of the tetradecapeptide renin substrate to yield angiotensin I (a decapeptide) and a tetrapeptide residue.

In a subsequent study, Sealey and associates[131]

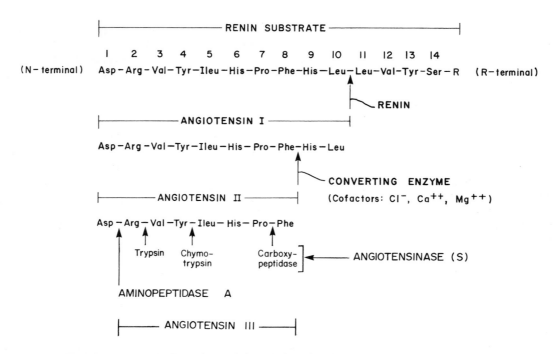

FIGURE 13.5. Steps involved in the formation and degradation of angiotensin II. (Modified from Koushanpour.[88])

reported a value of 1,500 ng/mL for the concentration of renin substrate in human plasma, an amount capable of yielding 6 mg of angiotensin on complete conversion by renin. Their findings clearly indicate that the concentration of renin substrate is normally far in excess of that needed to maintain the normal plasma level of angiotensin II. Furthermore, it shows that the large concentration of renin substrate excludes it from exerting a rate-limiting effect in the kinetics of the renin-renin substrate reactions.

Most of the available evidence implicate the liver as the main organ that manufactures the renin substrate. However, at present little is known about the mechanisms that maintain the plasma concentration of renin substrate at normal levels. In a review of the available evidence, Laragh and Sealey[91] suggest that a number of organ systems participate in the regulation of the plasma concentration of renin substrate. Of these, the kidney appears to be the most important. Thus, it has been found that bilateral nephrectomy in all species produces a rapid increase in the plasma concentration of the renin substrate. That this rise in substrate concentration may be a direct consequence of a reduction of renin synthesis following nephrectomy is not supported by recent experiments. For instance, it has been found that in salt-loaded or deoxycorticosterone-treated animals that exhibit a low plasma renin level, there is no accompanying rise in the plasma renin substrate concentration.

At present, little is known about the mechanisms responsible for the elevation of the plasma renin substrate concentration following nephrectomy. The most likely explanation offered to date is that of Laragh and Sealy,[91] who suggest that the kidney tissue may metabolize or utilize renin substrate, a factor accounting for the rise in its plasma concentration after nephrectomy. The precise nature of such a renal metabolism or utilization of renin substrate, however, remains to be determined.

In contrast to nephrectomy, the plasma concentration of the renin substrate is sharply reduced after adrenalectomy or hypophysectomy. In both cases the reduction in the substrate concentration is attributed to a marked increase in plasma renin level consequent to excessive loss of sodium during adrenal insufficiency. Renin substrate concentration is also lowered in patients with cirrhosis of the liver. Since in these patients plasma renin is usually elevated owing to a reduction in hepatic renin inactivation, it is not known whether the decrease in renin substrate level reflects a reduction in the capacity of the liver to synthesize the substrate or if it results from some other pathophysiological factors.

Angiotensins and Converting Enzymes

As mentioned above and depicted in Figure 13.5, renin, by acting on its plasma substrate, splits off four amino acids from the R-terminal group, yielding a physiologically inactive decapeptide called angiotensin *I*. On passage through the lung, the peptidase enzymes found chiefly in this tissue split off the terminal histidyl-leucine amino acids from angiotensin I, yielding a potent octapeptide pressor substance called angiotensin *II*. Hydrolysis of angiotensin II primarily by tissue aminopeptidase A yields the heptapeptide called angiotensin *III*. Similar conversion of angiotensin I to angiotensin II and angiotensin III can also take place by the action of peptidases found in the kidney tissue and blood. Collectively, these peptidases are called the *angiotensin-converting enzymes*.

The structure of the amino acid sequence of horse angiotensin I was the first to be elucidated (Figure 13.5). Subsequently, the amino acid compositions of angiotensin I isolated from bovine, hog, and human blood were determined. Of these, the amino acid sequence of human, hog, and horse angiotensin I was found to be identical. However, bovine angiotensin differed from these in that valine replaced isoleucine in the number 5 position.[91]

Several studies reviewed by Lee[94] showed that the biological activity of angiotensin II is highly dependent on the structural integrity of the C-terminal amino acid phenylalanine (Phe). Its modification or replacement results in the virtual loss of potency of angiotensin II. In contrast, the structural integrity of the N-terminal amino acid aspargine (Asp) does not appear to be as crucial to angiotensin activity, although its modification or replacement does somewhat reduce its potency. Such information has led to the synthesis of a variety of angiotensin II analogs for both research and clinical applications.

A number of studies have shown that the lung is the major organ that contains the angiotensin-converting enzymes. Ng and Vane[112] were the first to show that angiotensin I is rapidly converted into angiotensin II in a single passage through the lungs. These in-vivo studies have been further confirmed by in-vitro studies of Huggins and Thampi,[75] who obtained angiotensin II on incubating angiotensin I with lung tissues. Subsequent studies in intact anesthetized dogs by Oparil and associates[114] revealed that in addition to the lungs, renal tissue and blood can also convert tritium-labeled angiotensin I into angiotensin II. Furthermore, they found that both lungs and kidneys can convert angiotensin I into angiotensin II much more rapidly than does the plasma. They found no evidence for hepatic conversion of angiotensin I into angiotensin II.

Kinetic studies[74] reaffirmed the relative difference in potency and speed of reaction between the lung and plasma angiotensin-converting enzymes. Both enzymes had a broad, flattened pH curve, with a pH optimum of 7.25. However, there was a marked difference between the speed of conversion and the apparent K_m. Lung tissues required only eight minutes of incubation to convert angiotensin I into angiotensin II, whereas plasma enzyme required 45 minutes of incubation. Furthermore, the apparent K_m for plasma enzyme was 4.8×10^{-5} M, whereas that for the lung enzyme was 5.2×10^{-6} M. Also, both enzymes required chloride ion as cofactor.

As shown in Figure 13.5, in addition to the monovalent ion chloride, the converting enzyme requires the divalent cations calcium and magnesium for its activation. The enzyme is readily inhibited by chelating agents such as ethylenediaminetetraacetic acid (EDTA), suggesting that it is a metalloenzyme.

Plasma concentration of angiotensin II, $[Ang]_p$, in either arterial or venous blood varies considerably, depending on the techniques used to measure it and on the state of sodium balance. Gocke and associates[53] developed a sensitive and specific radioimmunoassay method capable of measuring angiotensin II level directly from 0.1 mL of human plasma. Using this method, they found that $[Ang]_p$ was directly related to the plasma renin activity and inversely related to the state of sodium balance. Depending on the degree of sodium depletion, values for $[Ang]_p$ ranged from 42 to 139 μg/mL in 29 normal human subjects studied. In these subjects, the greater the daily sodium excretion rate, and hence the greater the degree of sodium depletion, the smaller were the values for $[Ang]_p$. These results provide direct confirmation of the interrelationship between renin, angiotensin, and sodium balance.

Angiotensin II has a very short circulating half-life—about 1 to 3 minutes in super infusion and about 15 to 20 seconds in normal infusion. This suggests a mechanism for its rapid removal either by binding to receptor tissues or by degradation into inactive products by plasma angiotensinases, or both.

A number of experiments have shown that nearly 70% of the infused angiotensin II is removed in one passage through the liver.[12] Similarly, it has been found that angiotensin II is removed at the same rate by the kidney, but it is not destroyed as it passes through the lungs.[112] Studies with intact-cell and broken-cell preparations have revealed selective uptakes of angiotensin I and II by other organs.[57] Both kidney and uterus showed a greater capacity to bind angiotensin II than angiotensin I, whereas adrenal cortex showed a greater capacity to bind angiotensin I than angiotensin II.

As shown in Figure 13.5, two major *angiotensinases*, capable of inactivating the circulating angiotensin II, have been identified in the plasma. Khairallah and associates[83] identified an *angiotensinase A*, also called *aminopeptidase*, which has a pH optimum of 7.5. It requires calcium for its activation and is inhibited by chelating agents, such as EDTA. Pickens and his group[122] isolated a second plasma angiotensinase, which they called *endopeptidase*. It has a pH optimum of 5.5 and is inhibited by diisopropyl fluorophosphate (DFP).

Angiotensin II has three well-known physiological effects: (1) It causes vasoconstriction of the arterioles. As a vasoconstrictor, it is about ten times more potent than norepinephrine on a weight basis and about 50 times more potent on a molar basis. (2) It exerts a direct effect on the kidney; low doses cause sodium retention, whereas high doses cause natriuresis. The mechanisms mediating the renal response are not well understood. (3) It acts on the adrenal cortex, causing an increase in aldosterone secretion, and it stimulates the release of epinephrine and norepinephrine from the adrenal medulla.

The vasoconstrictor effect of angiotensin II is believed to be mediated via one or a combination of four pathways: (a) Direct action on the arteriolar smooth-muscle cells; (b) activation of the sympathetic branch of the autonomic nervous system; (c) activation of the central nervous system; and (d) stimulation of catecholamine release from the adrenal medulla. Currently it is not known how much of the initial or delayed cardiovascular effect of angiotensin II is due to its *direct* action on the vascular smooth-muscle receptors and how much is due to its *indirect* activation of adrenal medulla and the autonomic and central nervous systems. Angiotensin III has a much shorter biological half-life than does angiotensin II.[119] Angiotensin III accelerates catecholamine biosynthesis and is a potent steroidogenic agent. Both angiotensins II and III exert qualitatively and quantitatively similar effects on the renal vasulature and prostaglandin synthesis. However, both have less of a vasocontrcitive effect on the systemic vasculature, with angiotensin III having less of an effect as compared with angiotensin II.

Pressor responsiveness to angiotensin appears to be modified in a variety of conditions in which there are associated changes in body fluid volume.[91] It is reduced in situations in which there is volume depletion, such as hemorrhage, sodium depletion, normal pregnancy, and pathological states of cirrhosis, nephrosis, and congestive heart failure. The last three conditions are characterized by transudation of fluid out of the circulation, thereby reducing the effective volume of the circulating blood. In contrast, the pressor responsiveness is increased in situations where there is volume expansion, such as when sodium is given in excess or when sodium intake is combined with administration of aldosterone.

Angiotensin II exerts a dual effect on the kidney: It causes parallel dose-dependent changes in both renal hemodynamics and sodium and water excretion. Thus, intravenous infusion of a small amount of angiotensin II (0.5 ng/kg/min) causes a significant decrease in RPF and urine volume in man without any associated changes in the systemic arterial blood pressure.[39] Furthermore, intrarenal injection of a small amount of angiotensin that does not elevate the systemic arterial blood pressure causes renal vasoconstriction.[99]

Angiotensin II has been shown to be a potent stimulator of aldosterone release from the adrenal cortex in man,[2,90] sheep,[15] and dog.[30] The stimulatory effect of angiotensin II on sheep and dog adrenal glands was demonstrated by direct perfusion studies. This effect was found to be dependent on the electrolyte composition of the perfusate. Thus, raising the sodium concentration in the perfusate depressed the response to angiotensin, whereas raising the potassium concentration or decreasing the sodium concentration potentiated its stimulatory effect.

Like renin, repeated injection of large doses of angiotensin II results in progressively less elevation of blood pressure. This *angiotensin tachyphylaxis* has been attributed to a large extent to the saturation of the vascular receptor sites by the presence of excess hormone as well as to the release of prostaglandins.

Results of several experiments in both animal and man have shown that although infusion of large doses of angiotensin causes tachyphylaxis, continuous infusion of small doses results in progressive elevation of blood pressure. This is attributed to (a) angiotensin stimulation of aldosterone secretion, leading to eventual salt and water retention, (b) increased sympathetic activity, or (c) angiotensin-induced vascular damage. The mechanism mediating the latter effect remains obscure.

Aldosterone

In Chapter 9, while describing the possible mechanisms of renal action of aldosterone, we stated that this hormone is synthesized exclusively in the zona glomerulosa of the adrenal cortex. Furthermore, as was depicted in Figure 9.39, we indicated that the biosynthesis of aldosterone is influenced primarily by changes in sodium balance whose effects were mediated via induced changes in the plasma levels of angiotensin II and potassium, with ACTH exerting only a supportive role in the steroidogenesis. In this section, we shall briefly discuss the major points of evidence that have established the role of angiotensin II as a *trophic hormone* for the stimulation of aldosterone release from the adrenal cortex. For a comprehensive analysis of the earlier as well as the recent literature on the subject, the interested reader is referred to excellent reviews by Tobian,[159] Davis,[35] Page and McCubbin,[116] Laragh and Sealey,[91] Peach,[119] and Reid and co-workers.[123]

Beginning in the late 1950s, several studies led to the formulation of a negative feedback hypothesis for the control of renin release and its effects on aldosterone secretion. In 1958 Gross[62] found an *inverse* relationship between sodium balance and renin content in the rat kidney. This information, coupled with the knowledge that sodium metabolism is primarily a renal function and that it is influenced by aldosterone, led to the concept that aldosterone secretion is somehow influenced by renin release. Furthermore, since changes in sodium balance were known to be associated with changes in the extracellular fluid and plasma volumes, it was deduced further that both aldosterone secretion and renin release were inversely related to the changes in the blood volume. Thus, the concept of aldosterone as a *volume control regulator* was born.

In support of this concept Tobian[159] found that the renin content of the granular cells of the juxtaglomerular apparatus was correlated with the degree of granulation of these cells, which, in turn, was a function of the degree of the renal arterial perfusion pressure. Thus, a reduction in the perfusion pressure induced by partial occlusion of the renal artery caused hypergranulation, and therefore increased renin release, whereas an increase in the renal perfusion pressure induced by elevation of systemic arterial pressure resulted in degranulation, and hence reduced renin release. Since the renal arterial perfusion pressure is ultimately a function of the blood volume, which in turn depends on the renal sodium metabolism and sodium balance, it follows that renin release and consequently aldosterone secretion are inversely related to the renal arterial perfusion pressure, or some function thereof, induced by changes in the blood volume.

The direct role of angiotensin as the trophic hormone stimulating aldosterone secretion was subsequently defined by a number of investigators. Using the classic endocrine ablation techniques, in a series of experiments Davis and associates[38] found that acute blood loss, which normally stimulated aldosterone secretion in hypophysectomized dogs, failed to evoke the same response in nephrectomized-hypophysectomized animals. However, injection of saline extracts of both kidneys of each animal resulted in a striking increase in aldosterone secretion. This suggested that the kidney is the source of the aldosterone-stimulating agent.

That the stimulating agent is indeed angiotensin II was confirmed by Davis and his group[38] in dogs and by Laragh and associates[90] and Genest and co-workers[51] in man. They found that intravenous infusion of synthetic angiotensin II in both dogs and man caused an increase in aldosterone secretion as well as urinary excretion. Furthermore, in the dog experiments, both renin and angiotensin II stimulated steroidogenesis.

Subsequent studies by Ames and his associates[2] extended these earlier observations by studying the effect of prolonged infusion of angiotensin II on sodium balance, aldosterone secretion, and arterial blood pressure in normal subjects and in cirrhotic patients with ascites. They found that in contrast to studies in dogs,[38] infusion of angiotensin in doses required for stimulation of aldosterone secretion always produced a small but definite elevation of the arterial blood pressure. This suggested that the trophic action of angiotensin appeared to be associated with its pressor effect. However, since infusion of pressor-equivalent amount of norepinephrine had no trophic effect on aldosterone secretion, they concluded that the pressor agents in general do not stimulate aldosterone secretion. They further observed that both the pressor and the trophic effect of exogenous angiotensin were related to the state of sodium balance.

Thus, in normal subjects, prolonged infusion of angiotensin resulted in sustained sodium chloride retention. However, sodium depletion reduced the pressor responsiveness of angiotensin and enhanced the effectiveness of antipressor drugs. This suggested that the state of sodium balance is an important determinant of the vascular responsiveness to angiohtensin. In contrast, in cirrhotic patients with ascites, prolonged infusion of angiotensin produced natriuresis, with diuresis of edema fluid. Furthermore, these patients exhibited pressor unresponsiveness and tachyphylaxis to angiotensin.

From these studies, Ames and co-workers[2] proposed the now well-accepted negative feedback concept of the renin-angiotensin-aldosterone system, whose function is to prevent sodium depletion and hypotension. Accordingly, loss of body sodium or a reduction in arterial blood pressure would eventually lead to a reduction in renal perfusion pressure, thereby stimulating renin release and angiotensin II production. The latter causes sodium chloride retention both directly, by acting on the kidney (see Chapter 9 for the mechanism), and indirectly, by increasing aldosterone production. The resulting sodium chloride retention causes expansion of the blood volume and elevation of the arterial blood pressure. The net effect of this would be the restoration of the renal perfusion pressure toward normal and the return of the renin-angiotensin-aldosterone system back to the normal operating point.

Although potassium has been shown to serve as a strong stimulus for aldosterone secretion, potassium homeostasis appears not to be the sole mechanism. Thus, potassium metabolism has been found to be normal in patients with hyperaldosteronism (e.g., congestive heart failure, cirrhosis, and nephrosis), who also have renal sodium chloride retention and edema.[91]

Studies with radioactive-labeled aldosterone have shown that the hormone is distributed in both the intracellular and the extracellular fluid compartments.[154] These same studies have further shown that the circulatory half-life of aldosterone is about 30 minutes under normal conditions. However, it is found to be increased in patients with hepatic cirrhosis, a factor that may in part explain the formation of ascites and edema fluids in these patients.[34]

As stated in Chapter 9, the liver is the main site of inactivation of aldosterone. Furthermore, the metabolic clearance of aldosterone approaches its hepatic clearance, and it is influenced by postural changes. It is reduced in the upright position to about one half of that found in the supine position.

Aldosterone secretion rate has been found to range from 20 to 200 μg/d in human subjects with normal sodium intake.[92, 93] These studies have further shown that aldosterone secretion fell to between 15 and 80 μg/d when subjects were placed on a high sodium diet, but increased to as much as 1,000 μg/d after a period of sodium deprivation. Thus, it appears that the aldosterone secretion rate and hence its plasma concentration are highly sensitive to the state of sodium balance, a property resembling those described for both renin and angiotensin II.

Control of Renin Release

From the evidence presented in the preceding section, we learned that (1) renin is synthesized and stored in and released from the granular cells of the juxtaglomerular (JG) apparatus (Chapter 5, Figure 5.3); (2) there is a continuous release of renin from the kidney, which somehow is balanced by a continuous hepatic synthesis of the renin substrate; (3) the rate of formation of angiotensin follows first-order kinetics, that is, its synthesis is proportional to the concentrations of renin and its plasma substrate; (4) the circulatory half-life of renin is considerably longer than that of angiotensin, so that neither the quantity of renin nor that of its plasma substrate are the limiting factors in the synthesis of angiotensin; and (5) depending on the state of body sodium and potassium balance, the synthesis of angiotensin and its short circulatory half-life will ultimately determine the rates of synthesis and release of aldosterone.

Since the discovery in 1960 of the role of the renin-angiotensin system in the control of aldosterone secretion, there has been a growing interest in delineating the nature of the mechanisms that control renin release. Elucidation of these mechanisms is important in understanding the pathophysiology of disturbances in water and electrolyte balance and the etiology of edema formation in such diseases as congestive heart failure, liver cirrhosis, and nephrosis, as well as the etiology of initiation and maintenance of benign and malignant hypertension.

Numerous studies in both man and animals (see reviews by Vander,[162] Lee,[94] Davis,[36] Laragh and Sealey,[91] Peach,[119] Reid et al,[123] Gibbons et al,[52]) have clearly established that at least three groups of mechanisms are involved in the control of renin release. As depicted in Figure 13.6, these include (1) an intrinsic mechanism with two intrarenal receptors, (2) the renal sympathetic nerves, and (3) a number of humoral agents. The two intrarenal receptors are (a) a vascular receptor, located in the media of the wall of the afferent arteriole, sensitive to changes in stretch or wall tension, and (b) a tubular receptor, located in the macula densa region, sensitive to changes in chloride concentration and tubular flow rate. Available evidence suggests a primary role for the intrarenal vascular receptors in the control of renin release. The renal sympathetic nerves, although not essential to the control of renin release, exert a modulating effect via β-adrenergic receptors on the magnitude of renin release in response to a given stimulus. The major humoral agents that control renin release include catecholamines (epinephrine and norepinephrine), ADH, and angiotension II, as well as changes in body sodium and potassium balance. Thus, it is clear that renin release is controlled by a multiplicity of factors involving the interplay of all three of these mechanisms.

In this section, we shall consider in some detail each of these mechanisms and the major experimental evidence supporting them, as well as their strengths and weaknesses.

The Intrinsic Mechanisms Controlling Renin Release

The Intrarenal Vascular Receptor Theory. With the milestone discovery of Goldblatt[55] that partial constriction of the renal artery produces a chronic, sustained elevation of the arterial blood pressure came the suggestion that this hypertension resulted from an excessive increase in renin release induced by renal ischemia (reduced blood flow) consequent to renal artery constriction. Subsequent studies,[120] however, revealed that the renal ischemia per se is not essential for renin release, since significant changes in renin secretion do occur during the autoregulation of renal blood flow, where changes in flow are minimized in the face of marked changes in the renal perfusion pressure. This has raised the question that perhaps a reduction in the *renal perfusion pressure*, or some function thereof, rather than renal ischemia, was the stimulus for renin release and the subsequent development of hypertension following renal artery stenosis.

The original idea that a reduction in the renal perfusion pressure stimulates some intrarenal baroreceptors to release renin was proposed independently by Goormaghtigh[59] and Braun-Menendez and co-workers.[18] Direct experimental evidence, however, in support of this theory came some 14 years later. In 1959, Tobian and associates[160] observed that the degree of granularity of the granular cells, which closely correlates with renin release, was *inversely* related to the degree of renal perfusion pressure, and hence the degree of stretch in the wall of the afferent arteriole. On the basis of these findings, Tobian and his group formulated the currently accepted *intrarenal vascular receptor theory* of renin release.

According to this theory, the granular cells of the juxtaglomerular apparatus, located in the media of the afferent arteriole (Figure 5.3), are sensitive to changes in stretch in the wall of this vessel. Thus, as shown in Figure 13.6, any factor that decreases the renal perfusion pressure, such as hemorrhage, will tend to reduce the stretch in the wall of the afferent arteriole, which will tend to *increase* stimulation (stimulation is indicated by a *plus sign*) of the vascular receptors, which monitor (designated by small *box* on the *arrow*) changes in the wall tension. This will in turn increase stimulation of the granular cells, thereby *increasing* renin release. In contrast, any factor that tends to increase the stretch in the wall of the afferent arteriole, such as an increase in the renal perfusion pressure, will tend to *reduce* stimulation of the vascular receptors. This will in turn reduce stimulation of the granular cells, thereby *decreasing* renin release. In short, *the release of renin from the granular cells is inversely related to the degree of stretch in the wall of the afferent arteriole containing these cells.*

Although the underlying mechanisms that somehow translate changes in the wall stretch into renin secretion remain unknown, a considerable body of experimental evidence has been marshalled for the support of the baroreceptor theory of renin release. The following represents highlights of some of the more important of these studies, which are summarized in part in Figure 13.6 {see also Chapter 6).

As stated above, the intrarenal baroreceptor theory rests on the premise that renin release is inversely related to changes in the *renal perfusion pressure*. Since the latter is normally a direct function of the systemic arterial blood pressure, it follows that renin release should also be inversely related to changes in the *systemic arterial blood pressure*. Thus, one would expect that *acute hemorrhage* sufficient to reduce the systemic blood pressure should increase

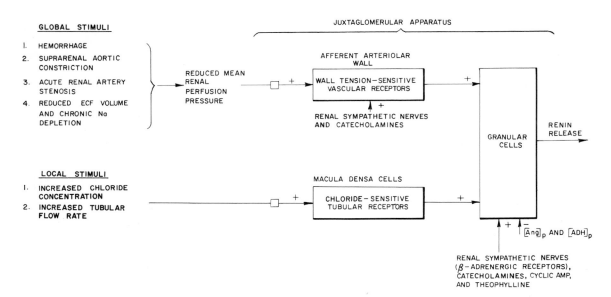

FIGURE 13.6. The multiple factors involved in the control of renin release from the granular cells of the juxtaglomerular apparatus. (From Koushanpour.[88])

renin release, whereas an increase in the systemic blood pressure, as by an expansion of the *ECF volume*, should decrease renin release.

In support of the above concept, numerous studies in rats, rabbits, and dogs[94] have shown that both hypotensive and nonhypotensive acute hemorrhage causes an increase in renin release. These findings, coupled with the fact that acute hemorrhage also causes a reflex stimulation of the sympathetic nervous system, suggest that at least two mechanisms may mediate renin release in response to hemorrhage. The degree of participation of these two mechanisms is dependent on the severity of hemorrhage and on the associated hypotension. Thus, in the case of mild, nonhypotensive hemorrhage, the increase in renin release is entirely due to activation of the renal sympathetics, whereas in the case of severe, hypotensive hemorrhage, both hypotension and renal sympathetic stimulation summate to produce a further increment in renin release.

Studies in the rabbit have been particularly revealing in elucidating the site as well as the time-course of renin release during acute hemorrhage. McKenzie and co-workers[100] found that within seconds after the onset of the hemorrhage, there was a large increase in plasma renin activity, which reached a value five times the control level 40 minutes later. This response was abolished 48 hours after bilateral nephrectomy, suggesting that the kidney may be the source of the released renin. The rapidity of this response led these investigators to suggest that renin release may have been due either to a direct stimulation of an intrarenal baroreceptor mechanism or to an indirect activation of the sympathetic nervous system, or both.

That *hypotension*, and not the hemorrhage per se, is the stimulus for renin release is supported by two different experiments. First, it has been found that an indirect reduction in the systemic arterial blood pressure produced by intravenous infusion of vasodilator drugs, such as histamine, sodium nitroprusside, and tetraethylammonium, causes an increase in renin release.[94] The effect of the last drug, which is a sympathetic ganglion blocker, also implicates the role of the sympathetic nervous system in renin release; but more about this will be presented later. Second, it has been found that a small, direct reduction in the renal perfusion pressure, produced by graded suprarenal aortic constriction, causes an increase in renin release.[149] In this study, renin release was found to be highly sensitive to changes in the mean blood pressure and not the pulse pressure. Thus, there was no change in renin release when the pulse pressure was reduced, but renin release was increased when the mean renal perfusion pressure was reduced by as little as 10 mm Hg. Moreover, when the pulse pressure was abolished by a combination of suprarenal aortic constriction and vagotomy, renin release was not affected as long as the mean blood pressure remained constant. These latter results provide compelling evidence that a decrease in *mean renal perfusion pressure* rather than in the pulse pressure is the primary stimulus for renin release. This is consistent with Tobian's stretch-sensitive baroreceptor hypothesis, since the juxtaglomerular granular cells, which are located in the media of the afferent arterioles, undergo the same changes in stretch that affect the walls of these vessels.

The above evidence strongly supports the concept that changes in the mean renal perfusion pressure, induced by either acute hemorrhage, vasodilator drug infusion, suprarenal aortic constriction, renal artery stenosis, or chronic sodium depletion, constitute the primary stimulus for renin release via the intrarenal vascular receptor mechanism (Figure 13.6). However, because of the nature of the experimental maneuvers used, these studies do not exclude the role of chloride-sensitive macula densa receptors in renin release. Thus, it is not clear whether the increase in renin release occurs as a consequence of a reduction in the renal perfusion pressure or of a reduction in GFR (and hence increased delivery of chloride to the macula densa cells). Or, alternatively, is the increase in renin release a consequence of a decreased stretch of the afferent arteriolar wall, or is it due to a change in some tubular function, such as a reduction in the sodium chloride load delivered to the "distal tubule" in the vicinity of the macula densa cells? To answer these questions and to dissociate the effect of the reduced renal perfusion pressure from that of the reduced sodium chloride load at the macula densa, Blaine and associates[13] developed a *nonfiltering* kidney model in the dog by constricting the left renal artery for two hours and ligating and cutting the left ureter. After two hours of renal ischemia, the renal artery constriction was released and the animal was maintained on peritoneal dialysis for three days. On the fourth day, the right kidney was removed, and the response of the nonfiltering kidney to acute hemorrhage and aortic constriction was tested. This procedure resulted in cessation of the glomerular filtration rate and caused excessive tubular damage, thereby eliminating sodium chloride delivery to the macula densa region. Hence, the model made it possible to study the response characteristics of the isolated intrarenal vascular receptors in the absence of functioning macula densa receptors. Blaine and associates found an increase in renin secretion (as determined from the measurements of the arterial-venous renin concentration) from the nonfiltering kidney in response to both an acute hemorrhage, of about 20 mL/kg of body weight, and suprarenal aortic constriction, which reduced renal perfusion pressure to a level of 40 to 80 mm Hg.

Although these findings strongly support the role of the intrarenal vascular receptors in renin release, they do not exclude the possibility that increased renin secretion, particularly in acute hemorrhage, may have been mediated by either stimulation of renal sympathetics[163] or increased catecholamine release from the adrenal medulla.[168] To eliminate these possibilities, Blaine and co-workers[14] studied the response of the denervated. nonfiltering kidney in dogs that were bilaterally adrenalectomized. They observed a striking increase in renin secretion in this preparation in response to the same degree of acute hemorrhage and graded suprarenal aortic constriction. These findings provide compelling evidence in support of the baroreceptor theory and indicate further that the renin release was prompted only by activation of the intrarenal vascular receptors in the absence of any changes in the glomerular filtration rate, sodium chloride delivery to the macula densa region, activation of the renal sympathetics, or catecholamine release. Moreover, this evidence strongly suggest that the intrarenal baroreceptor mechanism appears to be the primary receptor mechanism for the control of renin release and that it can override any signal from the macula densa receptors.

In a subsequent study from the same laboratory[172] an attempt was made to localize the site of the intrarenal vascular receptors and to elucidate the nature of the signal perceived by these receptors. As mentioned in Chapter 6, it is well known that the intrarenal injection of papaverine (a smooth-muscle relaxant) blocks renal autoregulation, which is an afferent arteriolar function (Figure 6.1). Thus, these investigators reasoned that papaverine-induced dilation of the afferent arteriole should minimize the capacity of these vessels to constrict and hence to undergo a decrease in stretch following an acute hemorrhage. To test this hypothesis, papaverine was infused into the renal artery of denervated, nonfiltering kidney of dogs subjected to acute hemorrhage. They found that infusion of 4 mg/min of papaverine, sufficient to produce a maximal rise in renal blood flow, completely blocked renin secretion in response to acute hemorrhage. The increase in renal blood flow occurred with little change in renal perfusion pressure, reflecting a reduction in the renal vascular resistance. Since papaverine is known to dilate the renal afferent arterioles, then these results are consistent with the baroreceptor theory of renin release and suggest further that the afferent arteriole is the locus of the receptors involved.

The evidence cited above raises an important question. What is the specific signal perceived by these intravascular receptors? As mentioned earlier, Tobian[159] suggested that the receptors are sensitive to the stretch transmitted through the wall of the afferent arterioles. Thus, in many situations, such as acute hemorrhage or sodium depletion, the increase in renin release is usually a consequence of a decrease in the stretch in the afferent arteriolar wall, a phenomenon that is also associated with an increase in the renal vascular resistance. However, there are exceptions to this inverse relationship between the diameter of the afferent arteriole and renin release. The most notable example is a decrease, not an increase, in renal vascular resistance, associated with an increase in renin release seen during the renal autoregulation of blood flow. Such findings, first reported by Skinner and his associates[148] in normal dogs subjected to graded suprarenal aortic or renal artery constriction, were subsequently confirmed by Ayers and co-workers[8] in conscious dogs with renal hypertension. The latter group found that partial constriction of the renal artery initially produced vasodilation, which was accompanied by a rise in plasma renin level. This was then followed by a gradual vasoconstriction, which returned the plasma renin to normal levels during the chronic phase of the benign hypertension. Additionally, they found that intravenous infusion of the same dosages of the vasodilator drugs dopamine, isoproterenol, nitroprusside, and acetylcholine caused a marked increase in renin release in hypertensive dogs, but only slightly so in normal dogs.

Subsequently, Eide and co-workers[41] reinvestigated the relationship between the diameter of the renal afferent arteriole and renin release. They measured renin secretion in response to stepwise reduction of the renal perfusion pressure in anesthetized dogs below the range of the autoregulation of renal blood flow (Figure 6.1). They found that renin release increased and reached its highest value when the renal perfusion pressure fell below the autoregulation range of 66 mm Hg. The renin release remained constant at this high level even when the perfusion pressure was further lowered below 66 mm Hg. This response was not significantly altered when sodium excretion was increased by intravenous infusion of the osmotic diuretic mannitol. This maneuver has been shown to blunt the increase in renin release following suprarenal aortic constriction, presumably by excluding the influence of the macula densa receptors on renin release.[164]

As discussed in Chapter 6, Fray[46], based on his previous studies of renin release in isolated rat kidney, presented a mathematical stretch receptor model to describe the stimulus-response relationship between the renal perfusion pressure and renin release. In this model he related renin release to the elastic modulus of the afferent arteriole, the transmural pressure across the wall of the afferent arteriole, and, more importantly, to the ratio of the internal and external radii of the afferent arteriole. Since the juxtagranular (JG) cells originate from the smooth-muscle cells of the afferent arteriole, he proposed that vasodilation (induced by papaverine) or an increase in renal perfusion may stretch the afferent arteriole and depolarize the JG cell membrane, thereby increasing intracellular (cytosolic) calcium and inhibiting renin release. Conversely, vasoconstriction or a decrease in renal perfusion pressure may decrease stretch of the afferent arteriole and hyperpolarize the JG cell membrane, thereby reducing intracellular (cytosolic) calcium and stimulating renin release. In support of this hypothesis, Fray[45] showed that verapamil, a calcium channel antagonist, blocks the inhibitory effect of an increase in renal perfusion on renin release. Other studies have shown that the increase in renin release induced by renal nerve stimulation and isoproterenol infusion (a β-adrenergic agonist) is associated with increase calcium efflux from JG cells,[67] whereas blockade of calcium efflux with lanthanum inhibited isoproterenol-induced renin release.[96] Finally, two recent studies[31, 47] showed that the cellular events controlled by intracellular calcium are mediated by calmodulin, a calcium-binding protein. Thus, they found that trifluoperazine, a calmodulin antagonist that inhibits hormone secretion in most systems, produced a dose-dependent increase in renin release. These investigators have, therefore, suggested that changes in calcium-calmodulin complex somehow regulate the effect of calcium on renin release.

Despite considerable evidence in support of the intrarenal baroreceptor theory of renin release, the exact nature of the signal perceived by these receptors remains unknown. As depicted in Figure 13.6, the intrarenal vascular receptors probably respond to changes in the wall tension in the afferent arteriole, and the magnitude of the receptor response (expressed as renin release) is influenced by at least three factors.[52] These include (1) variation in mean renal perfusion pressure, (2) dietary sodium intake, and (3) stimulation of renal sympathetics. Two recent studies in conscious dogs explored the separate and interactive effects of these variables on renin release. Farhi and co-workers[44] determined the stimulus-response curve relating renal perfusion pressure (RPP) to plasma renin activity (PRA) by varying RPP by adjusting an inflatable cuff around the renal artery and measuring renin release as PRA. They found that under normal sodium balance, the stimulus-response curve is characterized by two regions: one region in which renin secretion is relatively unresponsive to lowering of RPP, and another region in which renin release is a linear function of RPP. The RPP at which renin release begins to increase abruptly was only slightly below normal systemic blood pressure. Furthermore, they found that intravenous or intrarenal infusion of epinephrine caused the stimulus-response curve to be shifted to the right. Thus, epinephrine decreased the amount that the RPP must be lowered before PRA begins to rise. In a subsequent study,[43] they found that feeding the dogs a low-sodium diet also caused the stimulus-response curve to be shifted to the right. However, in contrast to epinephrine infusion, the slope of the linear segment was steeper and, therefore, not parallel to the control curve. Thus, low-sodium diet caused a change in the gain of the stimulus-response curve, although the percentage change in PRA was found to be the same for low-sodium diet and control curves. These findings clearly demonstrate that under normal physiological conditions, renin release is influenced by multiple factors. Of these, changes in renal perfusion pressure appear to be the primary stimulus for renin release, a finding consistent with the baroreceptor theory of renin release.

The Macula Densa Receptor Theory. The role of the macula densa in the tubuloglomerular feedback regulation of GFR and renin release has been fully discussed in Chapter 6. Briefly, this mechanism, which acts at the level of the single nephron, responds to variations in tubular chloride concentration and/or flow rate of the fluid emerging from the thick ascending limb. Thus, as depicted in Figure 13.6 (*lower panel*), an increase in tubular chloride concentration and/or flow rate, by an as yet undefined mechanism, stimulates the granular cells to release minute quantities of preformed renin into the interstitium of the parent afferent arteriole. The local conversion of this renin to angiotensin II promotes constriction of the afferent arteriole, thereby reducing the single-nephron GFR (SNGFR) and, therefore, the delivery of the filtrate to the thick ascending limb. In this manner, the macula densa receptors serve to adjust SNGFR to compensate for inadequate filtrate reabsorption in the "distal" nephron.

The Role of Renal Sympathetic Nerves in Renin Release

There is growing evidence that the sympathetic nervous system plays an important role in the control of renin release. Taquini and co-workers[155] found that denervation of one kidney in rats decreased renin content in that kidney as compared with the intact kidney. In the same year, Tobian[158] found that cutting the renal nerves reduced the granularity of the juxtaglomerular cells and hence their response to various stimuli. He suggested that renal sympathetics may be important in maintaining a basal level of renin production by the granular cells. Later, Hodge and associates[73] found that application of anesthetic agents to the renal nerves eliminated renin release in response to mild but not to severe hemorrhage, thus implicating a direct effect of these nerves on renin release. Subsequently, infusion of catecholamines in both man[60] and dogs[78] has also been shown to stimulate renin release. However, studies in conscious dogs have shown that under basal conditions, the renal sympathetic activity appears to be too low to influence renin release significantly.[63]

The above findings are further corroborated by several histological studies. Barajas and Latta,[10] using electron microscopy and histological techniques, showed that the juxtaglomerular apparatus receives a rich supply of nonmyelinated fibers believed to be sympathetics. More recently, using similar techniques, Muller and Barajas[109] reported that both adrenergic and cholinergic nerve endings are in contact with the basement membranes of both the proximal and "distal" tubules. However, the most direct evidence implicating the role of renal sympathetics in renin release is the finding that in anesthetized dogs direct electrical stimulation of renal nerves caused an increase in renal venous renin activity.[163] That the increase in renin release following renal nerve stimulation is indeed due to activation of the renal sympathetics has subsequently been confirmed by a variety of animal experiments.[33, 95]

Studies in man[60] have shown that both assumption of an upright posture and exercise increase renal sympathetic activity and produce renal arteriolar constriction, both maneuvers causing an increase in renin release. In addition, animal studies have clearly shown that activation of renal sympathetics exerts a modulating effect on the magnitude of renin release in response to a variety of stimuli. Thus, the increase in renin secretion following acute hemorrhage[28] and during mild sodium depletion[104] appears to be mediated in part by an increase in the renal sympathetic nerve activity.

The possible mechanisms mediating the effect of catecholamines and sympathetic stimulation on renin release have been investigated by a number of studies. Winer and associates[171] reported an increase in plasma renin activity in man after administration of ethacrynic acid and theophylline, and upon assumption of upright posture. The effect was blocked by propranolol (a β-adrenergic blocking drug) and phentolamine (an α-adrenergic blocker). Subsequently, Winer and co-workers[170] investigated the mechanism of action of sympathomimetic amines in dogs. They found that epinephrine, isoproterenol, and cAMP all increased renin secretion. These renin-stimulating effects were blocked by propranolol and phentolamine. From these studies, they suggested that intracellular accumulation of cAMP mediates renin secretion via both the α- and β-receptors and that the α- and β-blockers act at a step distal to cAMP production. In contrast, studies of Assaykeen and associates[7] and Passo and co-workers[17] favor the concept that the renin-stimulating effect of both renal sympathetic stimulation and catecholamines is mediated via β-adrenergic receptors. However, the specific mechanism mediating the effects remains to be elucidated.

Finally, recent studies have shown that in addition to catecholamines, dopamine may also be a neurotransmitter in the kidney.[84, 106] In addition, Mizoguchi and co-workers[103] showed that intrarenal infusion of dopamine causes a dose-dependent increase in renin secretion that is inhibited by specific dopamine blocking agents.

In summary, available evidence supports the concept that at least five different mechanisms may mediate the effects of renal sympathetics and catecholamines on renin release: (1) Both the renal sympathetics and catecholamines produce constriction of the renal arterioles, thereby causing renin release via stimulation of the vascular receptors. (2) As mentioned in Chapters 5 and 6, GFR is in part determined by the hydrostatic blood pressure in the afferent arteriole, whose constriction will reduce GFR. A decrease in GFR produces a decrease in the filtered load of sodium chloride, which has the potential of changing the amount of sodium chloride reaching the macula densa receptors. This, in turn, would alter the rate of renin release by the macula densa mechanism. (3) As depicted in Figure 13.6, both renal sympathetics and catecholamines directly act on the afferent arteriolar vascular receptors and the granular cells, thereby modulating the rate of renin release from the granular cells. (4) Stimulation by both catecholamines and sympathetics causes a redistribution of the renal blood flow, so that the blood is shifted from the renin-rich cortical and outer cortical areas to the relatively renin-poor inner cortical and medullary regions. (5) As depicted in Figure 13.6, the effect of renal sympathetics and catecholamines on renin release is believed to be mediated via β-adrenergic receptors. This finding, coupled with the evidence that catecholamines and cAMP can directly stimulate renin release,[102] suggests a possible link between β-adrenergic stimulation and intracellular cAMP accumulation. Thus, it is conceivable that activation of β-adrenergic receptors leads to stimulation of intracellular accumulation of cAMP by inhibition of phosphodiesterase. Inhibition of this enzyme by theophylline also potentiates cAMP stimulation of renin release.

Control of Renin Release by Some Selective Humoral Agents

As depicted in Figure 13.6, in addition to catecholamines, a number of other humoral agents, including some plasma electrolytes, influence the rate of renin release from the JG granular cells. In Chapter 9, in describing the factors influencing the biosynthetic pathway for aldosterone, we stated that changes in plasma concentrations of sodium and potassium influence aldosterone secretion and hence indirectly affect renin release (Figure 9.39).

A number of studies have demonstrated an inverse relationship between plasma sodium concentration and renin secretion in dogs.[110] Nash and associates[110] found that an increase in sodium concentration in the renal artery suppressed renin release, while a decrease stimulated renin secretion. Similar results have been reported for humans by Brown and co-workers,[25] who found an inverse relationship between plasma sodium concentration and plasma renin activity in 253 hypertensive patients. Newsome and Bartter[111] and Gordon and Pawsey[61] showed that changes in the extracellular fluid (ECF) volume produced the opposite effect on this inverse relationship. Thus, an expansion of ECF suppressed plasma renin activity, even though plasma sodium concentration was low, whereas contraction of ECF produced the opposite effect. This supports the concept that the effect of changes in ECF volume on renin release may be mediated via the vascular receptor mechanisms (Figure 13.6).

In contrast, results of several studies suggest that the effect of alterations in plasma sodium concentration on renin secretion may be mediated via the chloride-sensitive macula densa receptors. Thus, Nash and co-workers[110] found that stimulation of renin release by hyponatremic volume expansion, ureteral occlusion, or suprarenal aortic constriction was partially blocked by intrarenal infusion of hypertonic saline. Since in these experiments there were no detectable hemodynamic changes, they suggested that the macula densa receptor mediated the response. Essentially similar results were obtained by Shade and associates,[134] who found that intrarenal infusion of hypertonic saline decreases renin release in dogs with thoracic caval constriction (a procedure causing sodium retention) and one filtering kidney. In contrast, intrarenal infusion of hypertonic saline had no effect on renin secretion in dogs with thoracic caval constriction and one nonfiltering kidney. This latter finding strongly supports the concept that the tubular macula densa receptor may be the mechanism mediating the response to changes in plasma sodium concentration.

The effect of potassium metabolism on renin secretion has also been investigated by a number of studies. Brunner and associates[26] found that a high potassium intake resulted in a decrease in plasma renin activity in both normal and hypertensive subjects, whereas a low potassium intake produced the opposite effect. Moreover, they found that hyperkalemia, which inhibits renin secretion, stimulated aldosterone secretion, whereas hypokalemia produced the opposite effect. This relationship between chronic potassium loading and plasma renin activity has also been demonstrated in rats[132] and dogs.[1]

The possible mechanism whereby changes in the plasma potassium concentration influence renin secretion has been studied by a number of investigators. Vander[161] found that intrarenal infusion of potassium chloride into normal and salt-depleted dogs resulted in a decrease in renal vein renin activity. He suggested that the injected potassium inhibited sodium reabsorption in the proximal tubule, thereby increasing the sodium load reaching the macula densa region. However, micropuncture studies in dogs,[129] in which sufficient potassium chloride was given to depress renin release, showed that there was no increase in sodium delivery from the proximal tubules. These results do not support Vander's suggestion that potassium inhibits sodium reabsorption in the proximal tubule. They are, however, consistent with the concept that potassium inhibition of renin release may be mediated by a tubular mechanism, presumably the macula densa receptors. The specific nature of the mechanism mediating this response remains to be determined.

Finally, as depicted in Figure 13.6, ADH and angiotensin II directly inhibit (designated by a *negative sign*) renin release from the granular cells. This concept is supported by a number of studies in mammalian species, including man. Thus, Shade and his co-workers[135] demonstrated that intravenous infusion of ADH and angiotensin II inhibit renin release in sodium-depleted dogs with one nonfiltering kidney. These results suggest that both peptides act directly on the juxtaglomerular cells to inhibit renin release. Furthermore, they suggested that inhibition of renin release by angiotensin II involves a negative feedback loop, whereby the product of renin reaction with its substrate (angiotensin II) inhibits enzyme (renin) release.

Summary

The evidence presented in the preceding sections indicates that renin release is normally controlled by the interplay of intrinsic and extrinsic mechanisms. Of the intrinsic mechanisms, the intrarenal wall tension-sensitive vascular receptors, located in the afferent arterioles, play a prominent role in control of renin secretion. As depicted in Figure 13.6, the primary input to these receptors is a reduction in the mean renal perfusion pressure, induced by four classes of systemic perturbations. Moreover, the sensitivity of the vascular receptors to a given stimulus is modulated by the degree of activation of renal sympathetics as well as the circulating blood levels of catecholamines.

The chloride-sensitive tubular receptors, located in the macula densa region, serve as a secondary mechanism for control of renin release. Although tubular chloride concentration and/or flow rate at the level of the macula densa have been suggested as possible stimuli, the nature of the signal perceived by these receptors remains controversial. Nevertheless, the role of the macula densa receptors in an intrarenal negative feedback mechanism for the control of GFR in a single nephron appears to be well documented.

The major extrinsic factor influencing renin release is the renal sympathetic nerves. Although the activity of renal sympathetics appears not to be essential for renin release under basal conditions, available evidence indicates that these nerves exert a modulating influence on renin secretion. As depicted in Figure 13.6, the renal sympathetics exert their effects on both the vascular receptors and the renin-secreting granular cells. These effects are believed to be mediated via the beta-adrenergic receptors. Likewise, the catecholamines influence renin release by their effect on the renal arterioles and the juxtaglomerular cells.

Finally, dietary sodium and potassium ions and some humoral agents (ADH and angiotensin II) exert an influence on renin secretion. The available evidence indicates that the effect of sodium and potassium ions is mediated via the tubular macula densa receptors, whereas both ADH and angiotensin II appear to act directly on the renin-secreting granular cells via a negative feedback mechanism.

Right Atrial Stretch-Sensitive Receptors

As mentioned earlier, a second mechanism that influences urinary sodium excretion, and hence controls the blood volume, involves a reflex adjustment of renin release by the kidney. However, despite the evidence presented below in support of the existence of such a reflex, its role in the regulation of salt and water balance remains unresolved.

In 1957 Coleridge and co-workers[32] successfully isolated nerve fibers from the right atrium, whose frequency of discharge could be reduced or abolished by constriction of the superior or inferior vena cava. Subsequently, Anderson and associates[4] demonstrated that stretching the right atrium with sutures produced a decrease in the plasma concentration of aldosterone. These findings, coupled with the observation that constriction of the inferior vena cava caused an increase in plasma aldsoterone concentration,[37] suggested that this response may be mediated by a reflex mechanism, whose afferent limb is the stretch-sensitive volume receptors located in the right atrium. Subsequently, Gupta and co-workers[64] demonstrated that graded reductions in the blood volume were correlated with a decrease in the frequency of discharge recorded from the right atrial nerve fibers, thereby supporting the concept of the right atrial reflex.

Further evidence for this concept has been provided by Brennan and associates.[21] They showed that an increase in the right atrial pressure, produced by balloon inflation, resulted in a decrease in the plasma renin activity, presumably by a reflex suppression of renin secretion by the kidney. This finding provided a possible mechanism for the previously observed changes in the plasma aldosterone concentration. Subsequently, Stitzer and Malvin[150] examined the effect of alterations in the right atrial pressure on renal sodium excretion. They found that inflation of a balloon in the right atrium in dogs resulted in salt and water retention not attributable to any renal or systemic hemodynamic changes. They found no significant difference between the responses of the intact or denervated kidney to the right atrial balloon distention. From these results, they suggested that a binary mechanism may mediate the renal response to right atrial balloon inflation. One mechanism involves a reflex reduction in the renal excretion of salt and water, induced by a fall in the mean arterial blood pressure consequent to right atrial balloon inflation. Another mechanism involves a reflex stimulation of renin release and aldosterone production, which leads to renal salt and water retention. From the similarity of the responses of the intact and denervated kidney, they further suggested that both mechanisms are hormonally mediated. However, the nature of the hormone involved remains unresolved at present.

In summary, available evidence supports the concept of a right atrial receptor reflex system for the control of urinary salt and water excretion. However, the identification of the neural pathway involved, the hormonal mechanism mediating the effect, and its significance in the body fluid regulation remain to be determined.

Control of Sodium Excretion: A Synthesis

The evidence presented in the preceding sections indicates that two general types of stimuli cause renin release: The *first* involves changes in the arterial blood pressure and the extracellular fluid (ECF) volume; the latter is induced by changes in the total body sodium. Thus, a decrease in either the blood pressure or ECF volume stimulates renin release, whereas an increase would suppress renin release. The *second* involves changes in potassium metabolism; potassium depletion stimulates renin release, whereas potassium excess depresses renin release. In addition, the magnitude of renin release in response to these stimuli is modified by the degree of activation of renal sympathetics and the circulating blood levels of catecholamines. Moreover, changes in the various stimuli to which renin secretion responds are, in turn, initiated by changes in the plasma levels of angiotensin II, which produces two systemic effects: (1) It alters the vascular resistance, thereby modifying the arterial blood pressure, and (2) depending on its plasma concentration, it influences aldosterone secretion, thereby controlling sodium balance and hence ECF volume.

It is therefore clear that the effectiveness of angiotensin-mediated regulation of both the arterial blood pressure and the ECF volume will ultimately depend on the rate of renin release into the circulation and the factors affecting its blood levels. These, in turn, will determine the rate of production of angiotensin II and hence its effects on the blood vessels, adrenal cortex, and ultimately the kidney.

As depicted in Figure 13.7, once renin is released into the plasma compartment, the rate of formation of angiotensin II, and hence its plasma concentration ($[Ang]_p$), depends on several factors. Angiotensin synthesis and therefore $[Ang]_p$ are enhanced (designated by *plus signs*) in the presence of increased plasma levels of renin substrate concentration and increased pulmonary conversion rate (designated by *plus signs*). However, $[Ang]_p$ is reduced by an increase in plasma volume, hepatic blood flow, and plasma levels of angiotensinase (all designated by *negative signs*).

Available evidence indicates that depending on its plasma concentration, angiotensin II may exert three effects[20]: At very low plasma concentration it acts only locally in the renal vasculature to cause constriction of the renal afferent arteriole, thereby reducing the GFR (designated by a *negative sign*) in one or a few adjacent nephrons. At somewhat higher concentrations it acts on the smooth muscle cells lining the arterioles throughout the cardiovascular system, causing vasoconstriction and hence elevation of the systemic arterial blood pressure. At still higher concentrations, exceeding a threshold level (designated by a small *triangle* on the *arrow*), it stimulates the zona glomerulosa of the adrenal cortex to secrete aldosterone. The existence of a threshold concentration for aldosterone secretion is further supported by the findings of Ames and co-workers[2] that in patients in a state of positive sodium balance, infusion of a small amount of angiotensin II, insuffi-

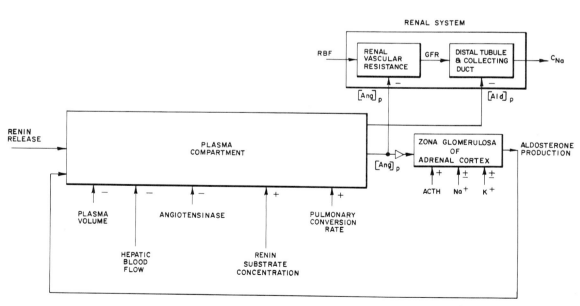

FIGURE 13.7. A functional diagram of the extrarenal regulation of aldosterone release and renal excretion of sodium. (From Koushanpour.[88])

cient to stimulate aldosterone production, was sufficient to maintain a hypertensive state.

The effectiveness of the trophic action of angiotensin II is enhanced by adrenocorticotropic hormone (ACTH), as well as through sodium and potassium metabolism. Thus, as discussed previously (Figure 9.39), aldosterone production is enhanced by potassium loading or sodium depletion, but is reduced by sodium loading or potassium depletion. These effects are indicated in Figure 13.7 by *plus* and *minus signs* next to the appropriate *arrows* impinging on the box labeled "zona glomerulosa of adrenal cortex." The aldosterone thus secreted enters the bloodstream, where for a given plasma volume it yields the circulating level of aldosterone ($[Ald]_p$). The latter, by the mechanism described in Chapter 9, acts on the principal cells of the collecting duct[98, 165] to enhance isotonic sodium chloride reabsorption, and hence reduce (designated by a *negative sign*) renal clearance of sodium (C_{Na}).

In summary, the plasma renin level depends to a large extent on sodium and potassium metabolism and on the state of body sodium and potassium balance. These, in turn, determine the ECF volume, plasma volume, arterial blood pressure, and renal perfusion pressure. Thus, for example, chronic low sodium intake leads to a reduction of the ECF volume and therefore of renal perfusion pressure as well. This leads to a reduction of the blood pressure in the afferent arteriole, which stimulates renin release by the intrarenal baroreceptor mechanism. The resulting increase in the plasma renin will eventually stimulate aldosterone secretion from the adrenal cortex. The elevation of the plasma aldosterone concentration will increase sodium and water reabsorption from the collecting duct as a compensatory mechanism, restoring the ECF volume to normal. The reduction of ECF volume will also reduce the hepatic blood flow, a factor that reduces the hepatic clearance of both renin and aldosterone. This would lead to a further increase in plasma concentration of aldosterone and hence to a further compensatory increase in renal reabsorption of sodium and water to restore ECF volume to normal.

Renal-Body Fluid Regulating System: A Synthesis

Having thus described the functional characteristics of the various components of the renal-body fluid regulating system, we are now ready to combine and synthesize them into a unifying functional diagram. As discussed previously and illustrated in Figure 13.8, the renal regulation of volume and osmolality of the body fluids is ultimately achieved by the interplay of three major systems: (1) The renal system with its ultrafilter and tubular components, shown in the upper portion of the figure; (2) the juxtaglomerular apparatus-adrenal cortex system, shown in the middle part of the figure, which controls the plasma volume and arterial blood pressure via the renin-angiotensin-aldosterone mechanism; and (3) the osmoreceptors-hypothalamus-pituitary system, shown in the lower part of the figure, which controls the plasma osmolality via the ADH mechanism.

As mentioned previously, the release of renin from the juxtaglomerular granular cells is controlled by two separate stimuli: one acting through the vascular receptors and the other through the tubular macula densa receptors. The vascular receptors respond to changes in the volume and hydrostatic pressure of the circulating blood, whereas the macula densa receptors are involved in an intrarenal negative feedback regulation of GFR. Accordingly, we may classify the various factors that influence renal regulation of volume and osmolality of body fluids under two general headings: First is the *intrarenal mechanism*, which operates via the chloride-sensitive macula densa receptors and the renin-secreting granular cells. This is depicted by the heavy broken pathways in the functional diagram. Second is the *extrarenal mechanisms*, consisting of the renin-angiotensin-aldosterone and ADH secreting systems. These are depicted by the other pathways in the functional diagram. Let us now consider the operation of each mechanism and its relative contribution to the renal regulation of volume and osmolality of body fluids, using the functional diagram in Figure 13.8 as our guide.

Intrarenal Mechanism

As depicted in this diagram, the intrarenal mechanism that regulates the volume and osmolality of excreted urine, and hence indirectly those of body fluids, involves monitoring the sodium chloride concentration and/or tubular flow rate emerging from the thick ascending limb of Henle's loop as it passes by the macula densa receptor cells. This process is depicted by a small *box* on the heavy broken *arrow* which impinges on the large *box* labeled "juxtaglomerular apparatus." The latter includes both the chloride-sensitive tubular macula densa receptors and the afferent arteriolar wall tension-sensitive vascular receptors. Thus, an increase in the chloride concentration in the tubular fluid stimulates (designated by a *plus sign*) the macula densa receptor component of the JG apparatus. This, in turn, will stimulate the

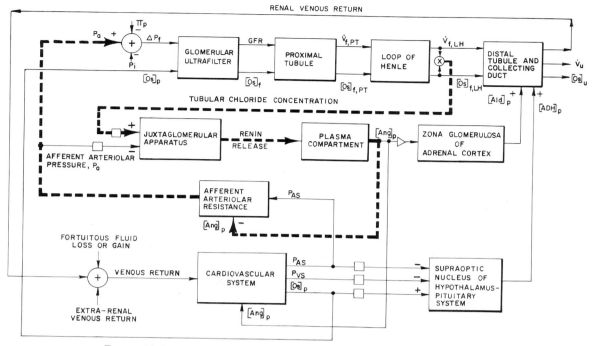

FIGURE 13.8. The renal-body fluid regulating system. (From Koushanpour.[88])

granular cells to release renin, by an unknown intermediate step, possibly via secondary stimulation of the interposed *mesangial cells*. The renin thus released into the plasma compartment combines with its plasma substrate angiotensinogen (found chiefly in the fraction of plasma containing α-2-globulin) to form the inactive decapeptide angiotensin I. This substance is then converted into the active, octapeptide vasopressor angiotensin II, by the action of plasma-converting enzyme present locally in the renal tissue. This local synthesis of angiotensin II raises its plasma concentration ($[Ang]_p$) in the afferent arterioles supplying the same as well as a few adjacent nephrons, thereby constricting these vessels. For a given systemic arterial blood pressure (P_{AS}), the resulting increase in the afferent arteriolar resistance reduces the afferent arteriolar pressure (P_a), lowering the effective filtration pressure (ΔP_f) and hence GFR. This effect is depicted by a *negative sign* on the *arrow* labeled "$[Ang]_p$", which impinges on the box labeled "afferent arteriolar resistance." The resulting momentary decrease in GFR slows the filtrate flow along the various nephron segments, and in particular along the thick ascending limb of Henle's loop. The increased tubular transit time increases the sodium chloride diffusion gradient at the luminal cell membrane, a factor tending to enhance the net transepithelial sodium chloride transport from the fluid in the thick ascending limb, thereby lowering its concentration in the fluid emerging from this segment. The reduced concentration of sodium chloride in the tubular fluid bathing the macula densa cells will tend to reduce the stimulation of these cells and their associated receptors, leading to reduced stimulation of the renin-secreting granular cells and subsequent reduction in angiotensin II synthesis. In this manner, small transient changes in the sodium chloride concentration in the fluid emerging from the thick ascending limb of Henle's loop, detected via the macula densa receptors-renin secreting granular cells system, by means of a negative feedback mechanism brings about a careful adjustment of the GFR in one or a few adjacent nephrons.

The importance of this mechanism in maintaining the constancy of the volume and osmolality of body fluids deserves special comment. As mentioned in Chapter 12, the fluid emerging from the thick ascending limb of Henle's loop and entering the early "distal tubule" is normally hypotonic. Small changes in the tonicity of this fluid will have a profound effect on the magnitude of the *transverse gradient* developed between the fluids in the ascending and descending limbs at the corticomedullary level, a factor that ultimately determines the magnitude of the generated longitudinal gradient within the medullary interstitium and hence the volume and osmolality of the excreted urine. Thus, operating at the level of single or few nephrons, the macula densa receptors, by monitoring the changes in the osmolality (due primarily to sodium chloride concentration) of the fluid emerging from the thick ascending limb, and via stimulation of the granular cells, can cause sufficient release of renin to adjust the GFR of individual nephrons. In this manner, the macula densa cells adjust filtrate flow through Henle's loop, a factor that affects sodium chloride cotransport in the thick ascending limb of Henle's loop, and therefore the osmolality of the tubular fluid leaving that segment of the nephron. The consequence of this *intrarenal regulation* of the tubuloglomerular function in individual nephrons is the fine adjustment of the all-important *transverse gradient* between the fluids in the ascending and descending limbs. The magnitude of this gradient, as mentioned in Chapter 12, for a given loop length and GFR, is the primary determinant of the final steady-state longitudinal gradient in the medullary interstitium, and hence the final volume and osmolality of the excreted urine.

In summary, the function of the *intrarenal mechanism*, as shown in Figure 13.8, is to adjust SNGFR in individual nephrons in response to changes in filtrate reabsorption in Henle's loop. Since the overall kidney function represents the algebraic sum of the glomerular-tubular balance of all the active nephrons, this intrarenal mechanism tends to coordinate and integrate the activity of the individual nephrons and consequently the overall kidney function under normal physiological conditions.

The operation of the intrarenal mechanism just described is mediated by small, transient changes in local renal concentration of angiotensin II ($[Ang]_p$), a fact supported by a number of studies cited earlier. This is depicted in the functional diagram by the first *branching point* (or tie point) on the *arrow* labeled "$[Ang]_p$," which connects the *box* labeled "plasma compartment" to that labeled "zona glomerulosa of adrenal cortex." Needless to say, the elevation of $[Ang]_p$ brought about by extrarenal mechanisms will *override* the intrarenal effect and, depending on the magnitude of $[Ang]_p$, will act on the cardiovascular system to cause systemic vasoconstriction and hypertension, stimulation of aldosterone secretion, or both. The extrarenal mechanisms mediating these effects are described next.

Extrarenal Mechanisms

As depicted in Figure 13.8, the extrarenal regulation of the final volume and osmolality of urine, and hence indirectly that of body fluids, involves careful adjustment of the rates of synthesis and release of ADH and aldosterone and their respective plasma concentrations. As discussed previously, the plasma concentration of the antidiuretic hormone, $[ADH]_p$, is controlled by the integrated effects of both *volume* stimuli (mediated via the systemic arterial and venous stretch-sensitive volume receptors) and *osmotic* stimuli (mediated via the hypothalamic and hepatic osmoreceptors) on the supraoptic nucleus of the hypothalamus-pituitary system. The volume and osmotic stimuli may be activated separately or together by fortuitous loss or gain of fluids having different proportions of water and electrolytes. Hence, these perturbations or forcings may be isotonic, hypotonic, or hypertonic. Furthermore, since the fortuitous fluid influx or efflux represents fluid entry or loss from the blood, they tend to affect the function of the cardiovascular system and its regulated variables, including the systemic arterial (P_{AS}) and venous (P_{VS}) pressures and plasma osmolality ($[Os]_p$). Induced changes in these cardiovascular variables are monitored by the appropriate components of the volume and osmotic receptors, which send afferent neural signals to the hypothalamic centers. The dynamic integration of these signals will eventually lead to appropriate directional changes in the rates of synthesis and release of ADH and its plasma concentration. The latter, acting on the principal cells of the collecting duct, will induce appropriate directional changes in the osmotic reabsorption of water and its renal clearance (C_{H_2O}). The net effect is the negative feedback regulation of the volume and osmolality of the circulating blood.

As shown in Figure 13.8, the other extrarenal mechanism influencing the renal regulation of volume and osmolality of body fluid is mediated via aldosterone-controlled isotonic sodium reabsorption in the collecting duct. As mentioned earlier, plasma concentration of aldosterone, $[Ald]_p$, is regulated primarily by plasma levels of angiotensin II and the state of sodium and potassium balance. Alterations in the latter two have profound effects on the volume and hydrostatic pressure of the circulating blood, which in turn influence the plasma levels of angiotensin II. Therefore, to facilitate presentation, we shall represent the effects of these three variables by a single process, namely, the renin-angiotensin-aldosterone mechanism.

As shown, any factor that tends to increase the afferent arteriolar pressure, P_a, will inhibit (*negative* sign) renin release from the juxtaglomerular apparatus and will reduce the synthesis of angiotensin II and hence its plasma concentration, $[Ang]_p$. Conversely, a decrease in P_a, induced by hemorrhage, sodium depletion, or renal artery stenosis, will stimulate the afferent arteriolar vascular receptors, which in turn will stimulate granular cells to release renin (Figure 12.6). Once released into the blood, the renin will initiate the synthesis of angiotensin I and II.

The extrarenal conversion of angiotensin I by the lung-converting enzyme, as the renal venous blood passes through the pulmonary tissues, produces the amount of angiotensin II required for its systemic effects. This extrarenal conversion of angiotensin I serves as a further check on the circulating plasma levels of angiotensin II and hence on its extrarenal effects. As depicted in Figure 13.7, besides pulmonary conversion rate, $[Ang]_p$ is further enhanced by the plasma concentration of renin substrate (angiotensinogen). However, $[Ang]_p$ is adversely affected by increases in plasma volume, hepatic blood flow, and plasma angiotensinases.

Depending on its plasma concentration, $[Ang]_p$, angiotensin II will have one or both systemic effects. At relatively low to moderate concentrations, it will act on the cardiovascular system, causing general systemic (including renal) vasoconstriction and systemic hypertension. This is depicted by the second branching point on the *arrow* connecting the *box* labeled "plasma compartment" to the *box* labeled "zona glomerulosa of adrenal cortex." At relatively higher concentrations, exceeding a *threshold* (*triangle* on the *arrow* labeled "$[Ang]_p$"), it stimulates the zona glomerulosa of the adrenal cortex to increase synthesis and release of aldosterone, raising its plasma concentration, $[Ald]_p$. The increase in $[Ald]_p$ will enhance isotonic sodium reabsorption from the collecting duct, thereby reducing the renal excretion of sodium and its clearance, C_{Na}, yielding a relatively hypertonic urine. The net result would be an increase in the plasma volume, renal perfusion pressure, and afferent arteriolar pressure, with ultimate reduction of the juxtaglomerular cell stimulation. Thus, the juxtaglomerular vascular receptors, acting as "volume" receptors, constitute an important element in a negative feedback system, which helps to restore the blood volume to normal.

In summary, as depicted in Figure 13.8, the final adjustment of urinary volume and osmolality, and hence that of body fluids, depends on the dynamic interplay of three regulatory processes: (1) The hemodynamic factors, which modify the input to the renal regulator; (2) the factors that influence the intrarenal regulation of glomerular-tubular balance, as well as urinary concentration and dilution mechanisms; and (3) the neuroendocrine systems, which through volume, pressure, and osmotic stimuli modify the plasma concentrations of ADH and aldosterone, thereby influencing the final volume and osmolality of the excreted urine.

Available evidence indicates that the first two mechanisms are involved in the minute-to-minute adjustment of the renal output and hence, indirectly, the volume and osmolality of body fluids. The neuroendocrine extrarenal mechanisms, on the other hand, are involved in the long-term adjustment of the extracellular fluid volume and osmolality, in response to wide variations in fluid influx or efflux under normal or abnormal conditions. Therefore, a thorough understanding of renal function under normal and disease conditions can only be achieved through an integrated analysis of the dynamic interactions of these three regulatory processes.

Problems

13.1. What are the stimuli and primary mechanisms for renin release (a) during nonhypotensive hemorrhage and (b) during hypotensive hemorrhage?

13.2. What extrarenal mechanisms are concerned in the regulation of (a) plasma osmolality and (b) plasma volume? Of these, which ones are most important in the normal regulation of plasma volume and osmolality?

13.3. What are the factors that regulate the blood levels of angiotensin II?

13.4. Define tachyphylaxis and explain the possible factors responsible for angiotensin II tachyphylaxis.

13.5. Describe briefly the factors controlling the formation of angiotensin II and its fate in the circulation.

13.6. What are the renal and systemic effects of angiotensin II? Are these effects concentration dependent?

13.7. What is the most important factor determining the pressor responsiveness of angiotensin II?

13.8. Describe briefly the renin-angiotensin-aldosterone system in the control of blood volume. What is believed to be the main function of this system under normal conditions?

13.9. Describe (a) the baroreceptor theory of renin release and (b) the macula densa theory of renin release. Cite briefly the strengths and weaknesses for each.

13.10. Describe briefly the theory of tubuloglomerular feedback control of GFR and its physiological significance.

13.11. What is the role of sympathetics in the control of renin release?

13.12. Name some of the humoral agents and their effects on renin release.

13.13. In each of the following circumstances, indicate the expected directional deviation from normal (+ for increase, − for decrease, and 0 for no change) in the rate of renin release.
   A. After intravenous infusion of 2 L of isotonic saline.
   B. Six hours after donating two pints of blood.
   C. Acute constriction of the renal artery.
   D. Chronic sodium deprivation from the diet.
   E. Assumption of upright posture.
   F. After intravenous infusion of a large dose of carbonic anhydrase inhibitor, acetazolamide.
   G. Chronic primary hyperaldosteronism.
   H. In congestive heart failure.

11.14. Of the two intrinsic intrarenal mechanisms for renin release, which is the primary mediator for the expected change in renin release in each of the circumstances in problem 13.13?

13.15. Indicate by a check mark whether the urine osmolality (and hence the urine concentration) increases or decreases in response to a sustained action of each of the following forcings:

Forcings	Urine Osmolality Increase	Decrease
A. An increase in GFR.		
B. A decrease in medullary blood flow.		
C. An increase in plasma ADH concentration.		
D. A decrease in plasma aldosterone concentration.		
E. Assumption of upright posture.		
F. In congestive heart failure.		
G. In severe sweating.		

13.16. Of the following conditions, indicate by a check mark the one in which the ADH release is the highest:
   A. Sitting position.
   B. Recumbent position.
   C. Upright position.

13.17. When a person stands up, there is increased venous pressure in the lower extremities. The plasma protein concentration rises and the volume of the legs increases, indicating a loss of water from the vascular compartment into the interstitial compartment. When the subject reclines, this fluid is returned to the vascular compartment. Describe what probably has happened in terms of capillary pressures.

13.18. In an unanesthetized dog, the arterial pressure may be raised reflexly to about 200 mm Hg for an hour by occluding both common carotid arteries. Despite the rise in arterial pressure and thus capillary hydrostatic pressure, there is no change in plasma volume. Why?

13.19. Muscular exercise has been shown to produce a marked change in water and electrolyte distribution. For example, in vigorous exercise there is an increase in extracellular fluid in the active tissue, which is reflected as an increase in plasma protein concentration, and in the percentage of red cells. A normal student running for 90 seconds at top speed may lose as much as 500 mL of fluid from the plasma during this run, but it will take 30 to 60 minutes for recovery. Explain, in terms of capillary function, why this time difference will occur.

13.20. A hospital ship is wrecked. In the excitement of getting ready to board the life raft, Mr. Jones, who has diabetes insipidus, received an extra injection of ADH (too much), while Mr. Smith, another patient with the same type of diabetes, failed to receive the required amount. Among the patients evacuated were a psychotic currently suffering from water intoxication, a cardiac patient on a low-salt diet, and a burn patient who has lost a lot of plasma. These five patients, along with their nurse, are set adrift with food but without drinking water. Describe the changes in water distribution (extracellular and intracellular) which would probably occur in each before drinking sea water. After three days, assuming all are still alive, they begin to drink sea water. How would this alter the water distribution in each of the six?

13.21. The following table gives findings in five patients as deviations from normal. From the list of disturbances given below, select the one that best describes the findings and then designate your choice by placing the appropriate letter in the column under diagnosis.

Patients	Total blood volume	Inulin clearance	Plasma renin level	Plasma ADH level	Sodium clearance	Diagnosis
1	−	−	+	+	+	
2	+	+	−	+	+	
3	−	−	+	+	−	
4	+	+	−	−	+	
5	+	−	+	−	−	

Disturbances:
A. Infusion of 2 L of 3 g/dL saline solution.
B. Congestive heart failure.
C. Two days following hemorrhage (about 20% of blood volume lost).
D. Excessive water ingestion.
E. Severe sweating.

13.22. The table given below gives the laboratory findings in four patients suspected of having renal disease. For comparison, the normal values for the same test are listed at the top of the table. (Assume permeability ratio for glucose is unity.)

Patients	Renal Function Tests		
	Inulin clearance (mL/min)	Plasma glucose concentration (mg/dL)	Maximum glucose reabsorption rate, $\dot{T}_{mG}$ (mg/min)
Normal	125	80	350
1. ___	75	80	60
2. ___	125	300	350
3. ___	125	80	90
4. ___	50	80	350

Match each patient's findings with the most likely disease from the list given below by placing the corresponding letter designating the disease in the space provided under "Patients" in the above table.
A. *Glomerulonephritis* is a disease characterized by a marked reduction in GFR, but no change in $\dot{T}_m$ for glucose.
B. Renal glycosuria is a disease characterized by a marked reduction in $\dot{T}_m$ for glucose but with normal GFR.
C. *Diabetes mellitus* is a disease characterized by an elevated plasma glucose concentration with normal GFR and $\dot{T}_m$ for glucose.
D. *Acute tubular necrosis* is a disease characterized by a decrease in GFR and extensive damage to tubules, leading to a marked reduction in $\dot{T}_m$ for glucose.

13.23. Using the data in the above table (Problem 13.22), answer the following questions:
A. The urine of which patient(s) gives a positive test for glucose? Enter patient(s)'s number(s) in the space provided. ___

B. If you had to diagnose the renal function in these patients by clearance of *p*-aminohippurate (C_{PAH}), indicate your expected findings for each patient in the following table as deviations from normal (+ for increase, − for decrease, 0 for no change) for C_{PAH} and tubular transport maximum for PAH ($\dot{T}_{mPAH}$).

Patients	C_{PAH}	$\dot{T}_{mPAH}$
1	___	___
2	___	___
3	___	___
4	___	___

C. In which of these patients, if any, do you expect endogenous creatinine clearance (C_{Cr}) to be least and most reduced?
(a) C_{Cr} is least reduced in ___.
(b) C_{Cr} is most reduced in ___.

13.24. In each of the following circumstances, if you expect urine flow to increase, mark (+); if you expect it to decrease, mark (−); and if you expect no change to occur, mark (0).
1. A rise in blood pressure at the glomerulus due to systemic hypertension.
2. A marked decrease in the plasma protein concentration.
3. An intravenous infusion of a large volume of an iso-oncotic solution (such as 6% dextran in normal saline).
4. An intravenous injection of a large dose of ethacrynic acid.
5. Consumption of three ounces of Old Kentucky Bourbon.
6. During a two-hour anesthesia for removal of an ovarian cyst.
7. In congestive heart failure.
8. An intravenous injection of isotonic saline suf-

Forcings	Responses											
	V_E	V_C	$\bar{P}_{AS}$	$\bar{P}_{VS}$	GFR	$[Na]_p$	π_p	$[ADH]_p$	$[Ald]_p$	$\dot{V}_u$	C_{Na}	C_{H_2O}
Isotonic influx (e.g., edema)												
Hypotonic influx (e.g., water loading)												
Hypertonic influx												
Isotonic efflux (e.g., burns, hemorrhage)												
Hypotonic efflux (e.g., sweating)												

V_E = extracellular fluid volume, V_C = intracellular fluid volume, $\bar{P}_{AS}$ = mean systemic arterial pressure, $\bar{P}_{VS}$ = mean systemic venous pressure, GFR = glomerular filtration rate, $[Na]_p$ = plasma sodium concentration, π_p = plasma protein oncotic pressure, $[ADH]_p$ = plasma concentration of antidiuretic hormone, $[Ald]_p$ = plasma concentration of aldosterone, $\dot{V}_u$ = urine flow rate, C_{Na} = sodium clearance, C_{H_2O} = solute-free water clearance.

ficient to increase the renal medullary blood flow.
9. A decrease in the active chloride transport in the thick ascending limb of Henle's loop.
10. In severe diarrhea.

13.25. For each of the five forcings listed, indicate deviation from normal (+ for increase, − for decrease, and 0 for no change) for the items listed in the table.

References

1. Abbrecht PH, Vander AJ: Effects of chronic potassium deficiency on plasma renin activity. *J Clin Invest* 1970; 49:1510–1516.
2. Ames RP, Borkowski AJ, Sicinski AM, Laragh JH: Prolonged infusions of angiotensin II and norepinephrine and blood pressure, electrolyte balance, and aldosterone and cortisol secretion in normal man and in cirrhosis with ascites. *J Clin Invest* 1965; 44:1171–1186.
3. Anderson B: Thirst and brain control of water balance. *Am Scientist* 1971; 59:408–415.
4. Anderson CH, McCally M, Farrell GL: The effects of atrial stretch on aldosterone secretion. *Endocrinology* 1959; 64:202–207.
5. Andrews WHH, Orbach J: Sodium receptors activating some nerves of perfused rabbit livers. *Am J Physiol* 1974; 227:1273–1275.
6. Arndt JO, Reineck H, Gauer OH: Ausscheidungsfunktion und Hamodynamik der Nieren bei Dehnung des linken Vorhofes am narkotisierten Hund. *Arch Ges Physiol* 1963; 277:1–15.
7. Assaykeen TA, Clayton PL, Goldfien A, Ganong WF: Effect of α- and β-adrenergic blocking agents on the renin response to hypoglycemia and epinephrine in dogs. *Endocrinology* 1970; 87:1318–1322.
8. Ayers CR, Harris RH Jr, Lefer LG: Control of renin release in experimental hypertension. *Circ Res* 1969; 24 (suppl I):103–112.
9. Baisset A, Montastruc P: Polyurie par distension auriculaire chez le chien; role de l'hormone antidiuretique. *J Physiol (Paris)* 1957; 49:33–36.
10. Barajas L, Latta H: Structure of the juxtaglomerular apparatus. *Circ Res* 1967; (suppl II) 20, 21:15–28.
11. Berl T, Cadnapaphornchai P, Harbottle JA, Schrier RW: Mechanism of stimulation of vasopressin release during β-adrenergic stimulation with isoproterenol. *J Clin Invest* 1974; 53:857–867.
12. Biron P, Meyer P, Panisset JC: Removal of angiotensins from the systemic circulation. *Can J Physiol Pharmacol* 1968; 46:175–178.
13. Blaine EH, Davis JO, Witty RT: Renin release after hemorrhage and after suprarenal aortic constriction in dogs without sodium delivery to the macula densa. *Circ Res* 1970; 27:1081–1089.
14. Blaine EH, Davis JO, Prewitt RL: Evidence for a renal vascular receptor in control of renin secretion. *Am J Physiol* 1971; 220:1593–1597.
15. Blair-West JR, Coghlan JP, Denton DA, et al: Hormonal stimulation of adrenal cortical secretion. *J Clin Invest* 1962; 41:1606–1627.
16. Blessing WW, Sved AF, Reis DJ: Destruction of noradrenergic neurons in rabbit brainstem. *Science* 1983; 217:661–663.
17. Bonjour JP, Malvin RL: Stimulation of ADH release by the renin-angiotensin system. *Am J Physiol* 1970; 218:1555–1559.
18. Braun-Menendez E, Fasciolo JC, Leloir LF, et al: *Renal Hypertension* (translated by Dexter L). Springfield IL, Charles C Thomas, 1946.
19. Braun-Menendez E, Fasciolo JC, Leloir LF, Munoz JM: La substancia hipertensora de la sangre del rinon isquemiado. *Rev Soc Arg Biol* 1939; 15:420–430.
20. Bravo EL, Khosla MC, Bumpus FM: Vascular and adrenocortical responses to a specific antagonist of angiotensin II. *Am J Physiol* 1975; 228:110-114.
21. Brennan LA Jr, Malvin RL, Jochim KE, Roberts DE: Influence of right and left atrial receptors on plasma concentrations of ADH and renin. *Am J Physiol* 1971; 221:273–278.
22. Bright R: Reports of medical cases selected with a view of illustrating the symptoms and cure of disease and by reference to morbid anatomy. Vol 1 Longmans Rees & Co, London, 1827. Reprinted in *Guy's Hospital Reports* 1836; 1:338–400.
23. Brown JJ, Davies DL, Lever AF, et al: A reninlike enzyme in normal human urine. *Lancet* 1964; 2:709–711.
24. Brown JJ, Davies DL, Lever AF, et al: Plasma renin concentration in relation to changes in posture. *Clin Sci* 1966; 30:278–284.
25. Brown JJ, Davies DL, Lever AF, Robertson JIS: Plasma renin concentration in human hypertension. I. Relationship between renin, sodium, and potassium. *Br Med J* 1965; 17:144–148.
26. Brunner HR, Baer L, Sealey JE, et al: Influence of potassium administration and of potassium deprivation on plasma renin in normal and hypertensive subjects. *J Clin Invest* 1970; 49:2128–2138.
27. Bumpus FM, Schwarz H, Page IH: Synthesis and pharmacology of the octapeptide angiotonin. *Science* 1957; 125:886–887.
28. Bunag RD, Page IH, McCubbin JW: Neural stimulation of release of renin. *Circ Res* 1966; 19:851–858.
29. Cadnapaphornchai P, Boykin L, Berl T, et al: Mechanism of effect of nicotine and renal water excretion. *Am J Physiol* 1974; 227:1216–1220.
30. Carpenter CCJ, Davis JO, Ayers CR: Relation of renin, angiotensin II and experimental renal hypertension to aldosterone secretion. *J Clin Invest* 1961; 40:2026–2042.
31. Churchill PC, Churchill MC: Effects of trifluopera-

zine on renin secretion of the kidney slices. *J Exp Pharm Ther* 1983; 224:68–72.
32. Coleridge JC, Hemingway A, Holmes R, Linden RJ: The location of atrial receptors in the dog: a physiological and histological study. *J Physiol (Lond)* 1957; 136:174–196.
33. Coote JH, Johns EJ, Macleod VH, Singer B: Effect of renal nerve stimulation, renal blood flow and adrenergic blockade on plasma renin activity in the cat. *J Physiol (Lond)* 1972; 226:15–36.
34. Coppage WS Jr, Island DP, Cooner AE, Liddle GW: The metabolism of aldosterone in normal subjects and in patients with hepatic cirrhosis. *J Clin Invest* 1962; 41:1672–1680.
35. Davis JO: The control of aldosterone secretion. *Physiologist* 1962; 5:65–86.
36. Davis JO: The control of renin release. *Am J Med* 1973; 55:333–350.
37. Davis JO, Carpenter CCJ, Ayers C, Bahn R: Relation of anterior pituitary function to aldosterone and corticosterone secretion in conscious dogs. *Am J Physiol* 1960; 199:212–216.
38. Davis JO, Hartroft PM, Titus EO, et al: The role of the renin-angiotensin system in the control of aldosterone secretion. *J Clin Invest* 1962; 41:378–379.
39. DeBono E, Lee G de J, Mottram FR, et al: The action of angiotensin in man. *Clin Sci* 1963; 25:123–157.
40. Durham RM, Novin D: Slow potential changes due to osmotic stimuli in supraoptic nucleus of the rabbit. *Am J Physiol* 1970; 219:293–298.
41. Eide I, Loyning E, Kiil F: Evidence for hemodynamic autoregulation of renin release. *Circ Res* 1973; 32:237–245.
42. Elliott DF, Peart WS: Amino acid sequence of hypertensin. *Biochem J* 1957; 65:246–254.
43. Farhi ER, Cant JR, Barger AC: Alteration of renal baroreceptor by salt intake in control of plasma renin activity in the conscious dog. *Am J Physiol* 1983; 245:F119–F122.
44. Farhi ER, Cant JR, Barger AC: Interactions between intrarenal epinephrine receptors and the renal baroreceptor in the control of PRA in conscious dogs. *Circ Res* 1982; 50:477–485.
45. Fray JCS: Mechanism by which renin secretion from perfused rat kidneys is stimulated by isoprenaline and inhibited by high perfusion pressure. *J Physiol (Lond)* 1980; 308:1–13.
46. Fray JCS: Stretch receptor model for renin release with evidence from perfused rat kidney. *Am J Physiol* 1976; 231:936–944.
47. Fray JCS, Lush DJ, Share DS, Valentine AND: Possible role of calmodulin in renin secretion from isolated rat kidneys and renal cells: Studies with trifluoperazine. *J Physiol (Lond)* 1983; 343:447–454.
48. Gauer OH, Henry JP: Circulatory basis of fluid volume control. *Physiol Rev* 1963; 43:423–481.
49. Gauer OH, Henry JP, Behn C: The regulation of extracellular fluid volume. *Ann Rev Physiol* 1970; 32:547–595.
50. Gaunt R, Birnie JH, Eversole WJ: Adrenal cortex and water metabolism. *Physiol Rev* 1949; 29:281–310.
51. Genest J, Biron P, Koiw E, et al: Adrenocortical hormones in human hypertension and their relation to angiotensin. *Circ Res* 1961; 9:775–791.
52. Gibbons GH, Dzau VJ, Farhi ER, Barger AC: Interaction of signals influencing renin release. *Ann Rev Physiol* 1984; 46:291–308.
53. Gocke DJ, Gerter J, Sherwood LM, Laragh JH: Physiological and pathological variations of plasma angiotensin II in man. Correlation with renin activity and sodium balance. *Circ Res* 1969; 24 (suppl I):131–146.
54. Goetz KL, Bond GC, Bloxham DD: Atrial receptors and renal function. *Physiol Rev* 1975; 55:157–205.
55. Goldblatt H: Studies in experimental hypertension: production of the malignant phase of hypertension. *J Exp Med* 1938; 67:809–826.
56. Goldblatt H, Katz YJ, Lewis HA, Richardson E: Studies on experimental hypertension. XX. The bioassay of renin. *J Exp Med* 1943; 77:309–314.
57. Goodfriend TL, Lin S-Y: Receptors for angiotensin I and II. *Circ Res* 1970; 27 (suppl I):163–174.
58. Goormaghtigh N: *La Fonction Endocrine des Arterioles Renales*. Louvain, Librarie Fonteyn, 1944.
59. Goormaghtigh N: Facts in favor of an endocrine function of renal arterioles. *J Pathol Bacteriol* 1945; 57:392–393.
60. Gordon RD, Kuchel O, Liddle GW, Island DP: Role of the sympathetic nervous system in regulating renin and aldosterone production in man. *J Clin Invest* 1967; 46:599–605.
61. Gordon RD, Pawsey CGK: Relative effects of serum sodium concentration and the state of body fluid balance on renin secretion. *J Clin Endocrin* 1971; 32:117–119.
62. Gross F: The renin-angiotensin system and hypertension. *Ann Intern Med* 1971; 75:777–787.
63. Gross R, Kirchheim H: Effects of bilateral carotid occlusion and auditory stimulation on renal blood flow and sympathetic nerve activity in the conscious dog. *Pfluegers Arch* 1980; 383:233–239.
64. Gupta PD, Henry JP, Sinclair R, Von Baumgarten R: Responses of atrial and aortic baroreceptors to nonhypotensive hemorrhage and to transfusion. *Am J Physiol* 1966; 211:1429–1437.
65. Haberich FJ: Osmoregulation in the portal circulation. *Fed Proc* 1968; 27:1137–1141.
66. Halter JB, Goldberg AP, Robertson GL, Porte D Jr: Selective osmoreceptor dysfunction in the syndrome of chronic hypernatremia. *J Clin Endocrinol Metab* 1977; 44:609–616.
67. Harada E, Rubin RP: Stimulation of renin secretion and calcium efflux from the isolated perfused cat

kidney by noradrenaline after prolonged calcium deprivation. *J Physiol (Lond)* 1978; 274:367–379.
68. Heacox R, Harvey AM, Vander AJ: Hepatic inactivation of renin. *Circ Res* 1967; 21:149–152.
69. Henry JP, Gauer OH, Reeves JL: Evidence of the atrial location of receptors influencing urine flow. *Circ Res* 1956; 4:85–90.
70. Henry JP, Pearce JW: The possible role of cardiac atrial stretch receptors in the induction of changes in urine flow. *J Physiol (Lond)* 1956; 131:572–585.
71. Henry JP, Gupta PD, Heehan JP, et al: The role of afferents from the low-pressure system in the release of antidiuretic hormone during non-hypotensive hemorrhage. *Can J Physiol Pharmacol* 1968; 46:287–295.
72. Heymans C, Neil E: *Reflexogenic Areas of the Cardiovascular System*. Boston, Little Brown & Co, 1958.
73. Hodge RL, Lowe RD, Vane JR: The effects of alteration of blood volume on the concentration of circulating angiotensin in anesthetized dogs. *J Physiol (Lond)* 1966; 185:613–626.
74. Huggins CG, Corcoran RJ, Gordon JS, et al: Kinetics of the plasma and lung angiotensin I converting enzymes. *Circ Res* 1970; 27 (suppl I):93–108.
75. Huggins CG, Thampi NS: A simple method for the determination of angiotensin I converting enzyme. *Life Sci* 1968; 7:633–639.
76. Jewell PA: The occurrence of vasiculated neurons in the hypothalamus of the dog. *J Physiol (Lond)* 1963; 121:167–181.
77. Jewell PA, Verney EB: An experimental attempt to determine the site of the neurohypophyseal osmoreceptors in the dog. *Trans R Soc Lond* 1957; 240:197–324.
78. Johnson JA, Davis JO, Witty RJ: Effects of catecholamines and renal nerve stimulation on renin secretion in the non-filtering kidney. *Circ Res* 1971; 29:646–653.
79. Johnson JA, Zehr JE, Moore WW: Effects of separate and concurrent osmotic and volume stimuli on plasma ADH in sheep. *Am J Physiol* 1970; 218:1273–1280.
80. Johnson JA, Moore WW, Segar WE: Small changes in left atrial pressure and plasma antidiuretic hormone titers in dogs. *Am J Physiol* 1969; 217:210–214.
81. Kannan H, Yagi R: Supraoptic neurosecretory neurons: Evidence for the existence of converging inputs both from carotid baroreceptors and osmoreceptors. *Brain Res* 1978; 145:385–390.
82. Karim E, Kidd C, Malpus CW, Penna PE: Effects of stimulation of the left atrial receptors on sympathetic efferent nerve fibres. *J Physiol (Lond)* 1971; 213:38P–39P.
83. Khairallah PA, Bumpus FM, Page IH, Smeby RR: Angiotensinase with a high degree of specificity in plasma and red cells. *Science* 1963; 140:672–674.
84. Kopp U, Bradley T, Hjemdahl P: Renal venous outflow and urinary excretion of norepinephrine, epinephrine, and dopamine during graded renal nerve stimulation. *Am J Physiol* 1983; 244:E52–E60.
85. Korner PI: Integrative neural cardiovascular control. *Physiol Rev* 1971; 51:312–367.
86. Koushanpour E, Stipp GK: Mathematical simulation of the body fluid regulating system in dog. *J Theor Biol* 1982; 99:203–235.
87. Koushanpour E, Kenfield KJ: Partition of carotid sinus baroreceptor response in dogs with chronic renal hypertension. *Circ Res* 1981; 48:267–273.
88. Koushanpour E: *Renal Physiology: Principles and Functions*, ed 1. Philadelphia, WB Saunders Co, 1976.
89. Koushanpour E, McGee JP: Effect of mean pressure on carotid sinus baroreceptor response to pulsatile pressure. *Am J Physiol* 1969; 216:559–603.
90. Laragh JH, Angers M, Kelly WG, Lieberman S: Hypotensive agents and pressor substances. The effect of epinephrine, norepinephrine, angiotensin II and others on the secretory rate of aldosterone in man. *JAMA* 1960; 174:234–240.
91. Laragh JH, Sealey JE: The renin-angiotensin-aldosterone hormonal system and regulation of sodium, potassium, and blood pressure homeostasis, in Orloff J, Berliner RW (eds): *Handbook of Physiology. Sec. 8, Renal Physiology*. Washington DC, American Physiological Society, 1973, pp 831–908.
92. Laragh JH, Sealey JE, Sommers SC: Patterns of adrenal secretion and urinary excretion of aldosterone and plasma renin activity in normal and hypertensive subjects. *Circ Res* 1966; 18 (suppl I):158–174.
93. Ledingham JGG, Bull MB, Laragh JH: The meaning of aldosteronism in hypertensive disease. *Circ. Res* 1967; 21 (suppl II):177–186.
94. Lee MR: *Renin and Hypertension; A Modern Synthesis*. London, Lloyd-Luke, Ltd, 1969.
95. Loeffler JR, Stockigt JR, Ganong WF: Effect of α and β-adrenergic blocking agents on the increase in renin secretion produced by stimulation of teh renal nerves. *Neuroendocrinology* 1972; 10:129–138.
96. Logan AG, Tenyi I, Peart WS, et al: The effect of lanthanum on renin secretion and renal vasoconstriction. *Proc R Soc Lond Ser B* 1977; 195:327–342.
97. Lydtin H, Hamilton WF: Effect of acute changes in left atrial pressure on urine flow in unanesthetized dogs. *Am J Physiol* 1964; 207:530–536.
98. Marver D, Schwartz MJ: Identification of mineralocorticoid target sites in the isolated rabbit cortical nephron. *Proc Natl Acad Sci USA* 1980; 77:3672–3676.
99. McGiff JC, Itskovitz HD: Loss of renal vasoconstrictor activity of angiotensin II during renal ischemia. *J Clin Invest* 1964; 43:2359–2367.
100. McKenzie JK, Lee MR, Cook WF: Effect of hemorrhage on arterial plasma renin activity in the rabbit. *Circ Res* 1966; 19:269–273.

101. Menninger RP: Current concepts of volume receptor regulation of vasopressin release. *Fed Proc* 1985; 44:55–58.
102. Michelakis AM, Caudle J, Liddle GW: In vitro stimulation of renin production by epinephrine, norepinephrine, and cyclic AMP. *Proc Soc Exp Biol Med* 1969; 130:748–753.
103. Mizoguchi H, Dzau VJ, Siwek LG, Barger AC: Effect of intrarenal administration of dopamine on renin release in conscious dogs. *Am J Physiol* 1983; 244:H39–H45.
104. Mogil RA, Itskovitz HD, Russell JH, Murphy JJ: Renal innervation and renin activity in salt metabolism and hypertension. *Am J Physiol* 1969; 216:693–696.
105. Moore WW: Antidiuretic hormone levels in normal subjects. *Fed Proc* 1971; 30:1387–1394.
106. Morgunov N, Baines AD: Renal nerves and catecholamine excretion. *Am J Physiol* 1981; 240:F75-F81.
107. Moses AM, Miller M: Osmotic threshold for vasopressin release as determined by saline infusion and by dehydration. *Neuroendocrinology* 1971; 7:219–226.
108. Mouw D, Bonjour JP, Malvin RL, Vander AJ: Central action of angiotensin in stimulating ADH release. *Am J Physiol* 1971; 220:239–242.
109. Muller J, Barajas L: Electron microscopic and histochemical evidence for a tubular innervation in the renal cortex of the monkey. *J Ultrastr Res* 1972; 41:533–549.
110. Nash FD, Rostorfer HH, Bailie MD, et al: Renin release: Relation to renal sodium load and dissociation from hemodynamic changes. *Circ Res* 1968; 22:473–487.
111. Newsome HH, Bartter FC: Plasma renin activity in relation to serum sodium concentration and body fluid balance. *J Clin Endocrin* 1968; 28:1704–1711.
112. Ng KKF, Vane JR: Fate of angiotensin I in the circulation. *Nature* 1968; 218:144–150.
113. Niijima A: Afferent discharges from osmoreceptors in the liver of the guinea pig. *Science* 1969; 166:1519–1520.
114. Oparil S, Sanders CA, Haber E: In vivo and in vitro conversion of angiotensin I to angiotensin II in dog blood. *Circ Res* 1970; 26:591–599.
115. Page IH: On the nature of the pressor action of renin. *J Exp Med* 1939; 70:521–542.
116. Page IH, McCubbin JW (eds): *Renal Hypertension.* Chicago, Year Book Publishers Inc, 1968.
117. Passo SS, Assaykeen TA, Otsuka K, et al: Effect of stimulation of the medulla oblongata on renin secretion in dogs. *Neuroendocrinology* 1971; 7:1–10.
118. Passo SS, Thornborough JR, Rothballer AB: Hepatic receptors in control of sodium excretion in anesthetized cats. *Am J Physiol* 1973; 224:373–375.
119. Peach MJ: Renin-angiotensin system: Biochemistry and mechanism of action. *Physiol Rev* 1977; 57:313–370.
120. Peart WS: Renin-angiotensin system. *N Eng J Med* 1975; 292:302–306.
121. Perlmutt JH: Relfex antidiuresis after occlusion of common carotid arteries in hydrated dogs. *Am J Physiol* 1963; 204:197–201.
122. Pickens PT, Bumpus FM, Lloyd AM, et al: Measurement of renin activity in human plasma. *Circ Res* 1965; 17:438–448.
123. Reid IA, Morris BJ, Ganong WF: The renin-angiotensin system. *Ann Rev Physiol* 1978; 40:377–410.
124. Robertson GL: The regulation of vasopressin function in health and disease, in: *Recent Progress in Hormone Research.* New York, Academic Press, 1977, vol 23, pp 333–385.
125. Robertson GL, Aycinena P: Neurogenic disorders of osmoregulation. *Am J Med* 1982; 72:339–353.
126. Robertson GL, Mahr EA, Athar S, Sinha T: Development and clinical application of a new method for the radioimmunoassay of arginine vasopressin in human plasma. *J Clin Invest* 1974; 52:2340–2352.
127. Sawchenko PE, Swanson LW: Central noradrenergic pathways for the integration of hypothalamic neuroendocrine and autonomic responses. *Science* 1981; 214:685–687.
128. Schneider EG, Davis JO, Baumber JS, Johnson JA: The hepatic metabolism of renin and aldosterone. *Circ Res* 1970; 27 (suppl I):175–183.
129. Schneider EG, Lynch RE, Willis LR, Knox FG: The effect of potassium infusion on proximal sodium reabsorption and renin release in the dog. *Kidney Int* 1972; 2:197–202.
130. Schrier RW, Berl T, Anderson RJ: Osmotic and nonosmotic control of vasopressin release. *Am J Physiol* 1979; 236:F321–F332.
131. Sealey JE, Gerten-Banes J, Laragh JH: The renin system: variations in man measured by radioimmunoassay or bioassay. *Kidney Int* 1972; 1:240–253.
132. Sealey JE, Clark I, Bull MB, Laragh JH: Potassium balance and the control of renin secretion. *J Clin Invest* 1970; 49:2119–2127.
133. Segar WF, Moore WW: The regulation of antidiuretic hormone release in man. I. Effect of changes in position and ambient temperature on blood ADH levels. *J Clin Invest* 1968; 47:2143–2151.
134. Shade RE, Davis JO, Johnson JA, Witty RT: Effects of renal arterial infusion of sodium and potassium on renin secretion in the dog. *Circ Res* 1972; 31:719–727.
135. Shade RE, Davis JO, Johnson JA, et al: Mechanism of action of angiotensin II and antidiuretic hormone on renin secretion. *Am J Physiol* 1973; 224:926–929.
136. Shade RE, Share L: Vasopressin release during nonhypotensive hemorrhage and angiotensin II infusion. *Am J Physiol* 1975; 228:149–154.

137. Share L: Effects of carotid occlusion and left atrial distention on plasma vasopressin titer. *Am J Physiol* 1965; 208:219–223.
138. Share L: Role of peripheral receptors in the increased release of vasopressin in response to hemorrhage. *Endocrinology* 1967; 81:1140–1146.
139. Share L, Levy MN: Cardiovascular receptors and blood titer of antidiuretic hormone. *Am J Physiol* 1962; 203:425–428.
140. Share L, Levy MN: Carotid sinus pulse, a determinant of plasma antidiuretic hormone concentration. *Am J Physiol* 1966; 211:721–724.
141. Share L, Travis RH: Interrelations between the adrenal cortex and the posterior pituitary. *Fed Proc* 1971; 30:1378–1382.
142. Share L, Claybaugh JR: Regulation of body fluids. *Ann Rev Physiol* 1972; 34:235–260.
143. Shimamoto K, Miyahara M: Effect of norepinephrine infusion on plasma vasopressin levels in normal human subjects. *J Clin Endocrinol Metab* 1976; 43:201–204.
144. Shimizu K, Share L, Claybaugh JR: Potentiation of angiotensin II of the vasopressin response to increasing plasma osmolality. *Endocrinology* 1973; 93:42–50.
145. Shu'ayb WZ, Moran WH, Zimmerman B: Studies of the mechanism of antidiuretic hormone secretion and post-commissurotomy dilutional syndrome. *Ann Surg* 1965; 162:690–701.
146. Skeggs LT Jr, Lentz KE, Hochstrasser H, Kahn JR: The chemistry of renin substrate. *Can Med Assoc J* 1964; 90:185–189.
147. Skeggs LT Jr, Lentz KE, Kahn JR, et al: Amino acid sequence of hypertensin II. *J Exp Med* 1956; 104:193–197.
148. Skinner SL, McCubbin JW, Page IH: Control of renin secretion. *Circ Res* 1964; 15:64–76.
149. Skinner SL, McCubbin JW, Page IH: Renal baroreceptor control of acute renin release in normotensive, nephrogenic, and neurogenic hypertensive dogs. *Circ Res* 1963; 15:522–531.
150. Stitzer SO, Malvin RL: Right atrium and renal sodium excretion. *Am J Physiol* 1975; 228:184–190.
151. Stoppini L, Baertschi AJ: Activation of portal-hepatic osmoreceptors in rats: Role of calcium, acetylcholine and cyclic AMP. *J Auton Nerv Syst* 1984; 11:297–308.
152. Stricker EM: Osmoregulation and volume regulation in rats: inhibition of hypovolemic thirst by water. *Am J Physiol* 1969; 217:98–105.
153. Tagawa H, Vander AJ, Bonjour JP, Malvin RL: Inhibition of renin secretion by vasopressin in unanesthetized sodium-deprived dogs. *Am J Physiol* 1971; 220:949–951.
154. Tait JF, Tait SAS, Little B, Laumas KR: The disappearance of 7-H3-d-aldosterone in the plasma of normal subjects. *J Clin Invest* 1961; 40:72–80.
155. Taquini AC, Blaquier P, Taquini AC Jr: On the production and role of renin. *Can Med Assoc J* 1964; 90:210–213.
156. Thames MD, Schmid PG: Cardiopulmonary receptors with vagal afferents tonically inhibit ADH release in the dog. *Am J Physiol* 1979; 237:H299–H304.
157. Tigerstedt R, Bergman PG: Niere und Krieslauf. *Skand Arch Physiol* 1898; 8:223–271.
158. Tobian L: Sodium, renal arterial distention and the juxtaglomerular apparatus. *Can Med Assoc J* 1964; 90:160–162.
159. Tobian L: Interrelationship of electrolytes, juxtaglomerular cells and hypertension. *Physiol Rev* 1960; 40: 280–312.
160. Tobian L, Tomboulian A, Janecek J: Effect of high perfusion pressures on the granulations of juxtaglomerular cells in an isolated kidney. *J Clin Invest* 1959; 38:605–610.
161. Vander AJ: Direct effects of potassium on renin secretion and renal function. *Am J Physiol* 1970; 219:455–459.
162. Vander AJ: Control of renin release. *Physiol Rev* 1967; 47:359–382.
163. Vander AJ: Effect of catecholamines and the renal nerves on renin secretion in anesthetized dogs. *Am J Physiol* 1965; 209:659–662.
164. Vander AJ, Miller R: Control of renin secretion in the anesthetized dog. *Am J Physiol* 1964; 207:537–546.
165. Vandewalle A, Farman N, Bencsath P, Bonvalet JP: Aldosterone binding along the rabbit nephron: An autoradiographic study on isolated tubules. *Am J Physiol* 1981; 240:F172–F179.
166. Verney EB: The antidiuretic hormone and the factors which determine its release. *Proc R Soc Lond Ser B* 1947; 135:25–106.
167. Volhard F, Fahr T: *Die Brightsche Nierenkrankheit: Klinik, Pathologie und Atlas*. Berlin, Springer, 1914.
168. Watts DT, Westfall V: Studies on peripheral blood catecholamine levels during hemorrhagic shock in dogs. *Proc Soc Exp Biol Med* 1964; 115:601–604.
169. Weitzman RE, Fisher DA: Log linear relationship between plasma arginine vasopressin and plasma osmolality. *Am J Physiol* 1977; 233:E37–E40.
170. Winer N, Chokshi DS, Walkenhorst WG: Effects of cyclic AMP, sympathomimetic amines and adrenergic receptor antagonists on renin secretion. *Circ Res* 1971; 29:239–248.
171. Winer N, Chokshi DS, Yoon MS, Freedman AD: Adrenergic receptor mediation of renin secretion. *J Clin Endocrin* 1969; 29:1168–1175.
172. Witty RT, Davis JO, Johnson JA, Prewitt RL: Effects of papaverine and hemorrhage on renin secretion in the nonfiltering kidney. *Am J Physiol* 1971; 221:1666–1671.
173. Zehr JE, Johnson JA, Moore WW: Left atrial pressure, plasma osmolality and ADH levels in the unanesthetized ewe. *Am J Physiol* 1969; 217:1672–1680.

Appendix A

Introduction to Quantitative Description of Biological Control Systems

The most apparent and yet fundamental characteristic of living organisms is their capacity to self-maintain, self-regulate, and self-reproduce. The ability to do each of these separate but related functions depends on the smooth, coordinated functioning of the different organ systems and their components. Because of the inherent complexity of these diverse regulatory functions and the multiplicity of their responses to a variety of external and internal stimuli, living organisms may be classified as *biological control systems*.

The ultimate goal in studying the biological control systems, such as the various organ systems of the human body, is to understand the mechanism of these diverse regulatry processes in health and to identify the causes of their failure in disease.

Because of the complex nature of the regulatory processes that compose the physiological systems, the most logical investigative approach is to examine systematically each component and to describe quantitatively its behavior. This investigative process is called *system analysis*. Having thus characterized the operation of each component, it should then be possible to describe the behavior of the overall system by combining the responses of the individual components. This process is called *system synthesis*.

The purpose of this Appendix is to introduce some basic principles and techniques of control system theory, aiming to provide the framework for a quantitative analysis and synthesis of the renal-body fluid regulating system.

General Description of a System

A regulatory biological control system consists of a group of interconnecting and interacting components for which there will be an identifiable *output* (response or dependent variable) that is related to a known *input* (stimulus or independent variable). The response of such a system to a normal or an abnormal stimulus depends on the *static* (time-invariant) and *dynamic* (time-dependent) properties of the components of the system. Understanding the static and the dynamic characteristics of such a response depends on a complete knowledge of the system as revealed by a description of (1) the input-output relationship of each of the components of the system, and (2) the arrangement and the nature of the paths connecting these components and the rules governing their relationships.

One way to begin such a system analysis and synthesis is to assemble the current knowledge about the overall function of the system and the arrangement of its components in the form of a block or flow diagram. In such a representation, each block serves as a simplified mathematical operator through which the input to the block is transformed into the output. Moreover, the block diagram representation is a powerful device for making the knowledge about the system explicit and rigorous. Figure A.1A shows a schematic block representation of an isolated system. Note that two arrows impinge on the end and the side of the block, representing two kinds of inputs, and one arrow leaves the block, as the output. As shown, the two inputs are called the *direct* and the *indirect*, respectively. Applying an input to a system is a way of forcing that system to respond. Hence the terms input, stimulus, and forcing may be used synonymously.

The direct input is the primary forcing, while the indirect input is a function of system properties. For example, the electrical activity recorded from the carotid sinus baroreceptor nerve (the output) is a

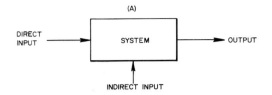

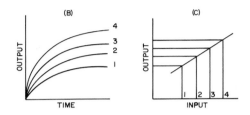

FIGURE A.1. (A) A schematic representation of an isolated system. (B) Time-course of the output of a hypothetical system to a series of four step inputs of increasing magnitude. (C) Steady-state input-output relationship for this system. (From Koushanpour.[2])

function of both the applied intrasinus pressure (the direct input) and the distensibility of the carotid sinus wall (the indirect input). Thus, when the effects of these two inputs are dissociated experimentally, it is found that for a given applied intrasinus pressure, the changes in the carotid sinus wall distensibility alter the amplitude and frequency of the action potentials recorded from the baroreceptor nerve.

Input-Output Relationships

The ultimate goal in the analysis of a complex biological system is to define, in mathematical terms, the input-output relationships of the overall system, as well as its components. Of particular interest is the time-behavior of the output as the input is varied. In general, the response of a system to a given input shows two distinct characteristics—an initial *transient* phase, followed by a *steady-state* phase. During the transient phase the response is changing with time, even if the stimulus is not, while during the steady-state the response remains time-invariant. The steady state response should not be confused with equilibrium, which characterizes the nonliving systems. In biological systems, equilibrium is achieved only at death.

Knowledge of how a system's response varies with time is provided by the time-constant (τ), or con-

stants, of each component of the system. Stated another way, the time-constant is the time required for the system's response to reach its final steady-state value. Therefore, calculation of the time-constant from the transient response allows us to predict the steady-state response of the system.

In general, the input-output relationship of a system may be described either mathematically in the form of an equation or visualized by a graph. Although the steady-state response of a system, or its components, may adequately be described by an algebraic expression, description of the transient response requires a *differential* equation, in which the dependent variable varies with time.

The response characteristics of a system may be determined by applying one or more of several types of forcings (Figure A.2). The most commonly used input is the *step forcing*, characterized by a sudden onset that is maintained until the steady-state response is reached. The presence of transient response during step forcing is due to the dynamics of the system and not the changing input. For example, a sudden bilateral occlusion of the common carotid arteries would result in a reflex rise of the systemic arterial blood pressure. The changes in the blood pressure, during the occlusion, usually show an initial transient phase of short duration followed by a steady-state response lasting as long as the occlusion persists.

The magnitude of both the transient and the steady state response to the step input is a function of the intensity of the forcing. Therefore, applying a series of step inputs of increasing intensities would result in a family of response curves, each with its own transient and steady-state characteristics. Figure A.1B shows the plot of the time-response of an hypothetical system to *four* step forcings of increasing magnitudes. The steady-state or static input-output relationship may be obtained by plotting the final steady-state value in each response curve against the magnitude of the step input. Such a plot is shown in Figure A.1B–C. Of course, this static input-output relationship may also be expressed algebraically by an equation, in this case by an equation for a straight line. The input-output relationships shown in Figure A.1C illustrates the response of a *linear* system in which the output is always a constant multiple of the input.

As depicted in Figure A.2, there are other types of forcings, with time-dependent patterns, which may be applied to determine the static as well as the dynamic response of a system. These include im-

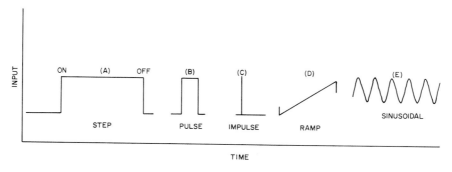

FIGURE A.2. Types of forcings used in analysis of physiological systems. (A) Step; (B) pulse; (C) impulse; (D) ramp; and (E) sinusoidal. (From Koushanpour.[2])

pulse, pulse, ramp, and sinusoidal forcings. The *impulse* is a forcing with essentially no duration at all. The *pulse* is similar in pattern to step forcing, except that it has a much shorter duration, hence the response has no chance to reach steay state. The intravenous injection of a drug or a dye soluton resembles qualitatively a pulse forcing. The *ramp* is a forcing characterized by a uniformly changing intensity. It is often used to determine the rate-sensitive properties of a system. The *sinusoidal* forcing is characterized by a cyclic variation of intensity (amplitude) with adjustable duration (period). The reciprocal of duration of each sinusoidal cycle is called the frequency, f, (equal to $\omega/2\pi$), where ω is the angular frequency in radians. The amplitude of a sinusoidal forcing may be described in four ways: the *peak* value (A_p), the *peak-to-peak* value ($2A_p$), the *average* value (A_{av}), and the *rms* (or root-mean-square) value (A_{rms}). These values may be calculated from the following expressions:

$$A_p = \frac{1}{2} \text{ (peak-to-peak amplitude)} \quad (A.1)$$

$$A_{av} = \frac{1}{\pi}\int_0^\pi A_p \sin \omega t = \frac{2A_p}{\pi} = 0.637 \, A_p \quad (A.2)$$

$$A_{rms} = \sqrt{\frac{A_p^2}{2}} = \frac{A_p}{2}\sqrt{2} = 0.707 \, A_p \quad (A.3)$$

The sinusoidal forcing is usually used to determine the dynamic resonse of a system.

Open-Loop and Closed-Loop Control Systems

There are two types of control systems: open-loop and closed-loop. In an open-loop system the input is not affected by the output, whereas in a closed-loop system a portion of the output signal is "fed back," thereby affecting the input. In general, every control system is a closed-loop system, but every closed-loop system is not necessarily a control system.

The first step in applying systems anaysis to the study of a closed-loop control system is to open the feedback loop and then determine the open-loop input-output relationship. Once every component of the system is thus characterized, it should be possible to synthesize the whole system and to determine its overall closed-loop static and dynamic response characteristics.

A self-regulating, physiological control system may be considered to consist of three components: *controlling* element, *controlled* element, and *feedback* element. Figure A.3 illusrates schematically the arrangement of these elements with their associated inputs and outputs. Grossly, a control system may be divided into two subsystems: a *controlling* system, consisting of the controlling and the feedback elements, and a *controlled* system, consisting of the controlled elements.

Operationally, the controlling element detects deviations of the stabilized variable or the output signal (S_o) monitored through the feedback element (S_f) from the input or the reference signal (S_i). The error signal, S_e (equal to $S_i - S_f$), thus produced causes the controlling element to issue appropriate corrective orders as the *controller* signal (S_c), to the controlled element. If the deviations in the stabilized variable are caused by a disturbing signal, adjustments through the feedback and the controlling elements should minimize the error and restore the output to the desired level.

The small *box* on the *arrow* representing the output

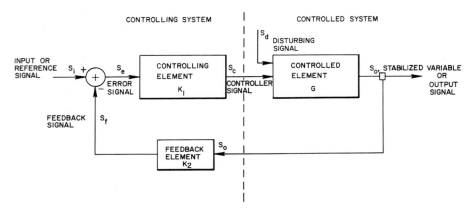

FIGURE A.3. A generalized closed-loop control system. (From Koushanpour.[2])

signal indicates that this signal is being monitored at that point by the feedback element. The letters K_1, K_2, and G inside the boxes represent mathematical operators and may be used to describe the input-output relationship of these elements. For example, $S_c = K_1 S_e$, $S_o = GS_c$, and $S_f = K_2 S_o$.

There are two types of closed-loop systems: regulator and servomechanism. A *regulator* is a closed-loop control system, such as that shown in Figure A.3, which stabilizes the output signal in the presence of a disturbing signal, by manipulting the feedback element. Thus, if a step-like disturbing signal is imposed on the system, the regulator responds by manipulating the feedback and the controlling elements so as to stabilize the output signal at its predisturbed level. Such a response characteristic is shown in Figure A.4. Note that imposing a sudden (step) disturbing signal causes a transient change in output, which levels off to a steady-state value, somewhat higher than its predisturbed level (*dashed lines*). This steady-state error in the output persists, owing to the imperfect operation of the regulator.

A *servomechanism* is a closed-loop control system that makes the output signal follow a changing input as closely as possible. Figure A.5 (*upper panel*) shows schematically the arrangement of such a system. Note that there is no feedback element in such a system; instead, a portion of the output is compared with the controller signal and the difference forces the controlled element. In this manner, a servomechanism control system can follow the changes in the input, but with a steady-state error due to imperfect operation. The response of a servomechanism to a step change in the input is shown in the lower panel of Figure A.5.

Both the regulator and the servomechanism control systems minimize the steady-state errors. The regulator minimizes the deviations of the output from a constant input, whereas the servomechanism minimizes deviations from a variable input.

Types of Controller Signals

The effectiveness of a regulator as a control system depends on the type and the sensitivity of the controller signal (S_c) used to force the controlled element. There are three types of controller signals that may be found singly or arranged in combination in a control system. They are *proportional*, *rate-sensitive*, and *integral* controller signals.

Proportional Controller

In a control system with a proportional controller there is a constant and linear relationship between the magnitude of the controller signal and the error signal. Using the symbols defined in Figure A.3, this relationship maybe expressed as

$$S_c = -K_p S_e \quad (A.4)$$

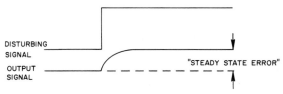

FIGURE A.4. Response of a regulator to a step disturbing signal. (From Koushanpour.[2])

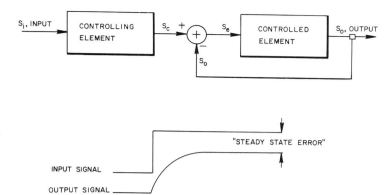

FIGURE A.5. *Upper panel*: A schematic diagram of a servomechanism closed-loop control system. *Lower panel*: Response of a servomechanism to a step input signal. (From Koushanpour.[2])

where K_p is the proportionality constant. Thus, deviations of the output signal from its stabilized level cause a continuous corrective response from the controlling element via the controller signal, the magnitude of which is linearly proportional to the magnitude of the error signal.

The steady-state response of a control system with a proportional controller to a step-disturbing signal will always show some error in the desired output. This *steady state error* is due to a *lag* in the response of the controlling element, which does not begin to respond until it detects an error that has already occurred. Then it responds by sending a controller signal whose magnitude is proportional to the magnitude of the error signal, but of the opposite sign. Hence, the time-course of the respone of such a system shows transient oscillations followed by a steady-state error, as shown in Figure A.15. To eliminate these undesired oscillations in the system's responce, a rate-sensitive controller may be added.

Rate-Sensitive Controller

In a control system with a rate-sensitive controller, the magnitude of the controller signal is proportional to the rate of change (time-derivative) of the error signal. This relationship may be expressed as

$$S_c = -K_r \left(\frac{dS_e}{dt}\right) \tag{A.5}$$

where K_r is the proportionality constant. Thus, the magnitude of the correction induced in the system's response is proportional to the rate of deviation of the output from the desired stabilized position, but with the opposite sign. The importance of a rate-sensitive controller in a control system is that it can stabilize the response during its dynamic rather than its steady-state phase. Hence, it adds stability to the system's operation by minimizing or eliminating the rapid oscillations in the system's response to a sudden, undesired disturbance. This is illustrated in Figure A.16.

Despite the rapid rate of correction in the resonse of a system with a rate-sensitive controller, the steady-state error persists. To eliminate this error an integral controller may be added.

Integral Controller

In a control system with an integral controller, the rate of the correction induced in the stabilized signal is proportional to the magnitude of the error signal. This may be expressed as

$$\frac{dS_c}{dt} = -K_I S_e \tag{A.6}$$

where K_I is the proportionality constant. On integrating this equation, we get

$$\int dS_c = -K_I S_e \int dt \tag{A.7}$$

$$S_c = -K_I S_e t + C_1 \tag{A.8}$$

where the symbol $\int$ indicates integration. The result of integration is given by Equation A.8, where C_1 is the constant of integration, and has a value of S_o, that is the value of S_c when $t = 0$. Substituting S_o for C_1 in Equation A.8, and rearranging terms, we get

$$S_c - S_o = -K_I S_e t \tag{A.9}$$

This equation states that the magnitude of the correction signal ($S_c - S_o$), produced by an integral controller, is proportional to the *summation* of the error signal over the interval of disturbance; hence, the idea of the *integral* controller. Consequently, the correction process, although very slow, continues until the steady-state error is abolished. This is illustrated in Figure A.17.

Mathematical Description of Response Characteristics of a Control System

A better understanding of a physiological control system will ultimately be linked to a mathematical description of its static and dynamic response characteristics. In addition, such a quantitative analysis allows us to (1) predict the future behavior of the system; (2) formulate a possible isomorphic model of the system; (3) synthesize rigorously the available information about the system; and (4) design future experiments that might be of great value in developing a complete understanding of the behavior of the system.

Mathematical characterization of a system's response must include a description of both the steady-state and the transient components. In general, although algebraic equations may be sufficient to characterize the static response, the transient or dynamic response can only be described by differential equations. In this section, without attempting to be complete, we discuss the basic steps involved in formulating the static as well as the dynamic input-output functions of a control system.

Static Characteristics

Analysis of the static characteristics involves the input-output description of a system in a steady-state condition. The simplest system for which this may be done is a single-component, open-loop system as shown in Figure A.1A. Physiologicaly, the carotid sinus baroreceptor, after its physical isolation from the circulation, may exemplify such an open-loop system. For this isolated system we could identify the intrasinus pressure (P) as the direct input, the carotid sinus wall distensibility (D) as the indirect input, and the action potentials in the Herring nerve (N) as the only output.

The first step in characterizing the static input-output function is to determine the open-loop *gain*, that is, the ratio of the output signal to the input signal. This is also called the system's *transfer function*. In the carotid sinus example, it has been shown that in response to static pressures the baroreceptors operate as "linear" pulse frequency modulated transducers. However, as shown in Figure A.6, when the isolated carotid sinus is forced with a pulsatile pressure, the amplitude and frequency of action potentials recorded from the Hering nerve increase (receptor recruitment) during the rising phase of pressure and decrease during its fallng phase. Since within the normal pressure range the changes in the recorded action potentials vary linearly with the forcing pressure, we may express the open-loop transfer function by

$$\frac{N(t)}{P(t)} = K(t) \tag{A.10}$$

where $K(t)$ is the transfer function; it is the value by which the input must be multiplied to generate the output. Note that in Equation A.10 both intrasinus pressure and the nerve action potentials are written as a function of time. Hence, the transfer function is also time-dependent. However, if we determine the steady-state response of the baroreceptors to a static (time-invariant) forcing, such as a step pressure increase or decrease, then a time-independent form of Equation A.10 yields the static transfer function:

$$\frac{N}{P} = K \tag{A.11}$$

If K is a direct proportion, the transfer function is linear and its value gives the "gain" of the system.

We can now extend the idea of open-loop transfer function to a closed-loop system. To illustrate the technique, we shall consider the carotid sinus baroreceptor as a component of a feedback system involved in the regulation of the systemic arterial pressure. The primary elements of this comlex feedback system are shown in their simplest form in Figure A.7. Note that in this scheme the controling element consists of the carotid sinus baroreceptor and the vasomotor center lumped as one component and the cardiovascular system, representing the controlled element, as the other component of the overall system. The input to the controlling element is the difference (ΔP) between the carotid sinus pressure (P_c) and the systemic arterial pressure (P_{AS}). An increase in the controller signal, determined by the magnitude of ΔP, causes a net inhibition of the vasomotor center and hence a decrease in the sympathetic outflow (N_S) to the cardiovascular system.

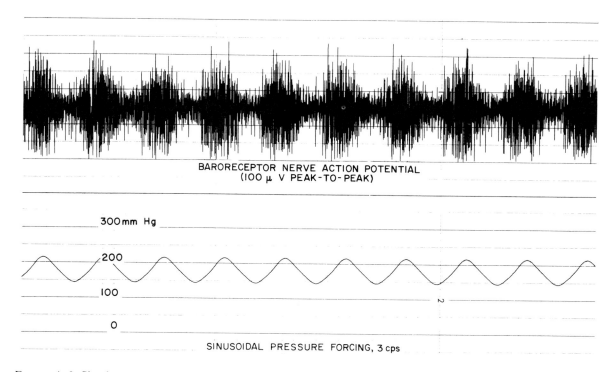

FIGURE A.6. Simultaneous records of the carotid sinus pressure and the action potentials in the Hering nerve. (From Koushanpour.[2])

The result is to decrease P_{AS} and to minimize ΔP, thereby restoring the blood pressure at the carotid sinus to the normal level.

The closed-loop gain of this system, defined as the ratio of the sympathetic outflow to the carotid sinus pressure (N_S/P_c), may be found as follows. First, we write a defining equation for the controller signal:

$$\Delta P = P_c - P_{AS} \qquad (A.12)$$

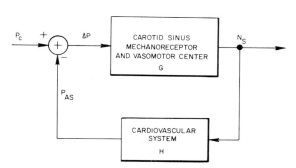

FIGURE A.7. A schematic diagram of the blood pressure control system. (From Koushanpour.[2])

Next, we write two equations expressing ΔP and P_{AS} as a function of static gains of the controling (G) and the controlled (H) elements, respectively:

$$\Delta P = \frac{N_S}{G} \qquad (A.13)$$

$$P_{AS} = H \cdot N_S \qquad (A.14)$$

Substituting Equations A.13 and A.14 into Equation A.12, and rearranging terms, we get

$$P_c = N_S \left(\frac{1 + GH}{G}\right) \qquad (A.15)$$

Rearranging Equation A.15 yields the closed-loop gain of the system:

$$\frac{N_S}{P_c} = \frac{G}{1 + GH} \qquad (A.16)$$

The product GH in Equation A.16 is a measure of the effectiveness of the regulatory control of the system. Hence, the larger the numerical value of GH, the closer the output (N_S) will follow changes in the input (P_c). Therefore, the value of GH deter-

FIGURE A.8. Lissajous' plots relating the averaged action potentials in the Hering nerve to pressure, as the frequency of the sinusoidal pressure input is varied from 0.5 to 10 c/s. (From Koushanpour.[2])

mines the closeness with which a physiological variable is controlled at a desired level.

Dynamic Characteristics

The time-dependent characteristics of an open-loop or a closed-loop control system can best be described if we resolve the dynamic transfer function, K(t), into its "gain" and "phase" components. As before, the gain is defined as the ratio of the amplitude of the output signal to that of the input signal. The phase is defined as the number of degrees (or the time unit per cycle of revolution) the output signal leads or lags the input.

There are several methods of characterizing the gain and phase components of the dynamic transfer function. In this section, we describe only two of these methods because their understanding and application requires minimal mathematical foundation. Both methods consist of determining the response of the system to a sinusoidal input at various frequencies. Since both the gain and the phase are time-dependent parameters, their numerical values vary as the frequency of the input sinusoids changes.

The first method for determining the dynamic transfer function and its components consists of plotting the output signal at a given frequency against the input signal. Since both the input and the output signals are functions of time, the resulting path traced out will not be a sinusoid but a loop. This is called a *dynamic* or a *Lissajous'* plot. The shape of the loop varies with the relative time phase of the input and output signals and with their relative frequency. Returning to the carotid sinus example, if we apply sinusoidal pressure of a given mean and peak-to-peak amplitude, but different frequencies, to the isolated (open-loop) system, we find that the action potentials recorded from the Hering nerve lag behind the input as the frequency of the input sinusoidal pressure increases. Figure A.8 illustrates the resulting Lissajous' plots of the computed electrical activity in the Hering nerve action potentials versus

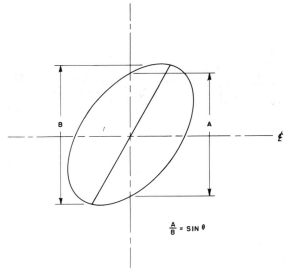

FIGURE A.9. Graphic determination of the gain and phase components of a transfer function from Lissajous' plot. (From Koushanpour.[2])

the intrasinus pressure as the frequency of the sinusoidal pressure is varied between 0.5 and 10 c/s.

To obtain the gain and phase values of the transfer function from these plots, we first draw a line bisecting each ellipsoid Lissajus' loop along its major axis (Figure A.9). Next, we draw two lines parallel to the horizontal axis through the two points intersecting the ellipsoid with the bisecting line. The vertical distance B between these two horizontal lines is a measure of the gain of the transfer function. The ratio of vertical distance A to that of B is equal to the sine of the phase difference between the input and the output. Figure A.9 shows schematically the graphic determination of the gain and phase components of the dynamic transfer function from Lissajous' plot of the input-output signals.

The second method for determining the dynamic transfer function involves the use of *Bode* analysis. Strictly speaking, this method is applicable ony for a linear system. A linear system, as defined earlier, is one whose output is always a constant multiple of the input. Under certain conditions, the Bode method could be extended to determine the transfer function of a *nonlinear system*, in which the output is not a constant multiple of the input.

In a linear system Bode analysis consists of plotting both gain and phase shifts against the logarithm of frequency of the input sinusoids. The advantage of Bode plot analysis, compared with the Lissajous' plot, is that it provides a direct method of deriving the differential equation describing the behavior of the system. To illustrate the technique, let us consider a linear system whose input-output relationship is described by the following differential equation:

$$y = x + a \frac{dx}{dt} \quad (A.17)$$

where y is the output, x is the input, and a is a proportionality constant. Both input and output signals are obviously functions of time. Equation A.17 states that the instantaneous value of the output is a function of the instantaneous value of the input, plus the rate of change of the input with respect to time. Suppose that we deliver a sinusoidal input of frequency $f = \omega/2\pi$ and peak amplitude A to this system. Then, according to Equation A.17, the input and output can be expressed by the following equations:

$$x(t) = A \sin \omega t \quad (A.18)$$

$$y(t) = A \sin \omega t + A\omega \cos \omega t \quad (A.19)$$

where ω is the angular frequency in radians, and ωt is the phase of the sinusoidal input. Equation A.19 was obtained by substituting $A \sin \omega t$ for x, and its derivative (rate of change with respect to time) for dx/dt in Equation A.17. Using trigonometric rules, Equation A.19 can be rewritten to show the sinusoidal form of the output, as follows:

$$y(t) = [A \sqrt{1 + (a\omega)^2}] \sin(\omega t + \tan^{-1} a\omega) \quad (A.20)$$

where the quantity in the bracket represents the amplitude of the output sinusoid, and $\tan^{-1}$ is the arc tangent. The gain of the system at each input frequency, $G(\omega)$, defined as the ratio of the output amplitude to that of the input at that frequency, is given by the following equation:

$$G(\omega) = \frac{A \sqrt{1 + (a\omega)^2}}{A} = \sqrt{1 + (a\omega)^2} \quad (A.21)$$

and the phase shift at each input frequency, $\phi(\omega)$, between the output and the input is given by

$$\phi(\omega) = (\omega t + \tan^{-1} a\omega) - (\omega t)$$
$$= \tan^{-1} a\omega \quad (A.22)$$

The gain and phase may be obtained by still another method, the *Laplace transformation*. This procedure consists of transforming the differential Equation A.17 from the time domain to the Laplace domain, that is, replacing (t) by (s). Applying this transformation, Equation A.17 becomes

$$y(s) = x(s) + asx(s) = (1 + as) \cdot x(s) \quad (A.23)$$

where $sx(s)$ is the Laplace transformation of dx/dt. Since Equation A.23 is the Laplace transformation of the system's differential equation, we can obtain the transfer function, in Laplace domain, as follows:

$$\frac{y(s)}{x(s)} = 1 + as \quad (A.24)$$

To obtain the gain and phase components of this transfer function, as a function of the input frequency, we transform Equation A.24 back to the complex frequency domain by replacing (s) with (jω):

$$\frac{y(j\omega)}{x(j\omega)} = (1 + aj\omega) \quad (A.25)$$

where $j = \sqrt{-1}$, $\omega = 2\pi f$, f is the real frequency in cycles per second, and $a = 1/\omega_c$ or the time-constant (τ) of the system. ω_c is the so-called *corner* or break frequency; it is the frequency at which the gain and phase curves change slope. The gain of the system, $G(\omega)$, is the magnitude of this equation, given by

$$G(\omega) = \sqrt{1 + (a\omega)^2} \quad (A.26)$$

and the phase shift, $\phi(\omega)$, is the phase angle of this equation, given by

$$\phi(\omega) = \tan^{-1}(a\omega) \quad (A.27)$$

Note that Equations A.26 and A.27 are exactly the same as Equations A.21 and A.22.

The Bode analysis consists of plotting Equations A.26 and A.27 against the logarithm of the input frequency. By convention, the gain values in such a plot are expressed in decibels (dB), and are defined by

$$G(dB) = 20 \log G \quad (A.28)$$

Converting the gain Equation A.26 to decibels yields:

$$G(dB) = 20 \log \sqrt{1 + (a\omega)^2}$$
$$= 10 \log [1 + (a\omega)^2] \quad (A.29)$$

At low frequencies, where $(a\omega)$ is very much less than 1, Equation A.29 becomes

$$G(dB) = 10 \log 1 = 0 \quad (A.30)$$

At high frequencies, where $(a\omega)$ is very much greater than 1, Equation A.29 becomes

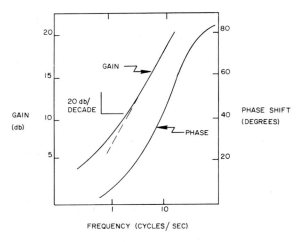

FIGURE A.10. Bode plot representation of Equation y = x + a dx/dt. (From Koushanpour.[2])

$$G(dB) = 10 \log (a\omega)^2 =$$
$$20 \log (a) + 20 \log (\omega) \quad (A.31)$$

Equation A.31, on a logarithmic frequency scale, plots as a straight line with a slope of 20 dB per log unit, or 20 dB per decade increase in frequency. At the corner frequency, where $a\omega = 1$, the gain obtained from Equation A.29 is

$$G(dB) = 10 \log 2 = 3 \text{ dB} \quad (A.32)$$

To understand the significance of these equations, we have plotted the gain and phase curves for the system described by Equation A.17 in Figure A.10. A portion of the gain curve covering a decade increase in frequency (between frequencies 1 and 10 c/s) is fitted to a straight line (*dashed line*) whose slope is 20 dB per decade. Note that the phase shift approaches zero degrees as the input frequency decreases, and it approaches 90 degrees as the input frequency increases. At the corner frequency, where $a\omega = 1$ and the gain is 3 dB, the phase shift is

$$\phi(\omega) = \tan^{-1} 1 = 45 \text{ degrees} \quad (A.33)$$

From the Bode plot of the gain and phase shift curves several observations may be made. (1) The gain curve can be approximated by a number of straight lines whose slopes will be a multiple of 20 dB/decade. (2) For each 20 dB/decade slope there is a phase shift of 45 degrees at the corner frequency. (3) The sign of the phase shift corresponds to the slope of the linearized gain curve. That is, if the slope is negative the phase shift will also be negative. (4) For a linear system, the gain and phase shifts are not independent. Thus, either gain or phase shift curve may be used to determine the shape of the other, as well as the transfer function.

As mentioned above, the advantage of the Bode analysis is that it allows us to determine the differential equation of the system. To illustrate, we use the gain and phase curves plotted in Figure A.10 as the starting point. In general, for each straight line with a multiple of 20 dB/decade slope there is associated a specific time-constant. In this example, since the gain curve is fitted to only one straight line, there is one time-constant ($\tau = 1/\omega_c = 1/2\pi f$). Since the gain has a slope of 20 dB/decade and the phase shift goes through 45 degrees at $f = 10$, $\tau = 0.016$ seconds. Also, a 20 dB/decade slope indicates that the gain at all frequencies must be multiplied by 20 dB. From these two pieces of information we can now write the system's transfer function:

$$\frac{y(s)}{x(s)} = 10 (1 + 0.016 \text{ s}) \quad (A.34)$$

where the factor of 10 is the gain constant that gives 20 dB gain at all frequencies without changing the phase shift. Transforming Equation A.34 into the time domain, we get

$$y = 10(x + 0.016 \frac{dx}{dt}) \quad (A.35)$$

This equation, except for a scale factor, is identical to Equation A.17.

Solution of System Differential Equations

The problem thus far has been to derive the system differential equation from empirical input-output data. The next phase in the analysis is to solve the differential equation. The solution consists of finding a functional relationship between the dependent variable, or the output y, and the independent variable time, t. There are two ways that such a solution may be obtained: by *analytical procedure* or by a *computer*. Before considering either method, let us examine the general form of a differential equation commonly encountered in the analysis of physiological systems:

$$A \frac{d^2y}{dt^2} + B \frac{dy}{dt} + C y = X(t) \quad (A.36)$$

This equation is called a *linear, ordinary, nonhomogeneous, second-order, first-degree differential equation with constant coefficients*. Let us examine the meaning of each term separately. It is a *differential* equation, because it has derivatives; that is, the value of the dependent variable in the equation is a function of time. It is *linear* because the dependent variable and its derivatives do not contain power or product terms. It is *ordinary* because it has only one independent variable, namely, time. If more than one independent variable were involved, it would be called a *partial* equation with respect to a specific independent variable. It is *nonhomogeneous* because it has a nonzero term on the right-hand side. This implies that the differential equation describes the response of a system to a forcing whose form is specified by the term on the right-hand side. It is *second-order* because the highest derivative is the second derivative of the dependent variable. It is *first-degree* because the algebraic power of its highest derivative is unity. Finally, it has *constant coefficients* because the coefficients A, B, and C do not vary with time.

In general, the solution of a differential equation, $y(t)$, is the sum of two components, $y_c(t)$ and $y_p(t)$:

$$y(t) = y_c(t) + y_p(t) \quad (A.37)$$

where $y_c(t)$ is called the *complementary* solution or transient (force-free) response and $y_p(t)$ is called the *particular* integral or forced response.

The coefficients in Equation A.36, when rearranged, describe some important physical characteristics of the system. Before we proceed with the solution, let us rearrange the coefficients and rewrite Equation A.36 in the so-called standard form,

$$\frac{1}{\omega_n^2} \frac{d^2y}{dt^2} + \frac{2\zeta}{\omega_n} \frac{dy}{dt} + y = \frac{1}{c} x(t) \quad (A.38)$$

where $\omega_n = \sqrt{C/A}$, which is the *undamped, natural frequency*, and zeta = $B/(2\sqrt{AC})$, which is the *damping ratio*. For example, if there is zero damping (i.e., $B = 0 = \zeta$), then, if the system is forced with a step input, it will oscillate sinusoidally indefinitely at the natural frequency, ω_n, and an amplitude equal to the magnitude of the step forcing. Now, if we introduce damping, and force the system with the same step input, it will oscillate sinusoidally at a frequency less than the natural frequency and at an amplitude that decays exponentially to zero.

Now, to obtain the force-free or the complementary solution, we replace $x(t)$ in Equation A.38 by zero:

$$\frac{1}{\omega_n^2} \frac{d^2y}{dt^2} + \frac{2\zeta}{\omega_n} \frac{dy}{dt} + y = 0 \quad (A.39)$$

TABLE A.1. Form of particular integral, $y_p(t)$, for different input functions, $x(t)$.

x(t)	Form of $y_p(t)$
Constant, k	K
Power, kt^n	$K_0 t^n + K_1 t^{n-1} + \ldots K_{n-1} t + k_0$
Real exponential, $ke^{\alpha t}$	$Ke^{\alpha t}$
Sine, $k \sin \omega t$	$K_0 \cos \omega t + K_1 \sin \omega t$
Cosine, $k \cos \omega t$	$K_0 \cos \omega t + K_1 \sin \omega t$

From Grodins, F.S.: Control Theory and Biological Systems. New York, Columbia University Press, 1963. Reprinted by permission.

The solution of this equation consists of the sum of two exponential terms (because it is a second-order equation):

$$y_c(t) = C_1 e^{\beta_1 t} + C_2 e^{\beta_2 t} \quad (A.40)$$

where β_1 and β_2 are the reciprocals of the time-constants and C_1 and C_2 are constants.

To obtain the forced response component of the solution, $y_p(t)$, we need to know the form of the input signal $x(t)$. For most cases, the input signal consists of one or more of the types listed in Table A.1. The final solution of the differential Equation A.38 is given by the following equation, after substituting an appropriate expression for $y_p(t)$ from Table A.1:

$$y(t) = C_1 e^{\beta_1 t} + C_2 e^{\beta_2 t} + y_p(t) \quad (A.41)$$

For a more detailed treatment of Equation A.36 and its solution the reader is referred to the selected refrences at the end of this Appendix.

Computer Solution of Response Characteristics of a Control System

The solution of a differential equation is not mathematically always as straightforward as that just described. Nevertheless, it is imperative to obtain some kind of a solution, even if incomplete, if we want to get an insight into the response characteristics of the system. To get around the mathematical difficulty of solving a differential equation, we can resort to electronic computers. There are two general types of computers: (1) an *analog* computer that performs addition, subtraction, multiplication, and division on *continuous* physical quantities in the form of electric voltages; and (2) a *digital* computer that performs the same mathematical operations on *discrete* physical data.

Most instruments used in biomedical research display their output in analog form. Very often a *transducer* of some sort is employed to convert the biomedical quantity being measured into electrical signals. Therefore, it is necessary to use an *analog-to-digital* converter before we can perform useful mathematical and statistical calculations on a digital computer. The fundamental difference between the two types of computers is one of speed. The advantage of the analog computer is that it allows the analysis and synthesis of a system with moderate speed and efficiency. However, there are certain types of problems that can be solved with greater speed and accuracy with a digital computer than with an analog computer. Therefore, the type of computer used depends on the type of problem to be solved. In this section, we provide a brief introduction to the use of the analog computer to simulate a control system. For the application of digital computers in the simulation and mathematical modeling of renal function the reader is referred to the work of Koushanpour and co-workers [4] and Koushanpour and Stipp.[3]

Our immediate goal now is to study the response characteristics of a system, regardless of the method we choose to solve the system's differential equation. In short, we do not care how we obtain the solution, but we are interested in what it tells us about the transient and steady state response of the system. The use of the analog computer not only allows us to solve a differential equation, it also enables us to observe the response of the system to any desired forcing without having to go through cumbersome mathematical manipulations required to solve one or more differential equations.

Let us now see how we can solve the differential equation A.36 by an analog computer. To simplify notation, it is customary to represent d/dt (the rate of change with respect to time or the *time-derivative*) of a variable by a *dot* placed above the letter designating it. Thus, we can represent dy/dt by ẏ (pronounced y-dot) and d²y/dt² by ÿ (pronunced y-double dots). Using this notation, we first consider the force-free form of Equation A.36, with x(t) = 0:

$$\ddot{y} = -\frac{B}{A}\dot{y} - \frac{C}{A}y \qquad (A.42)$$

Equation A.42 states that the second time-derivative of y is equal to the algebraic sum of y and its first time-derivative. The implication is that given ÿ, we could obtain ẏ and y by *inverse of differentiation*, which is *integration*. We can do this by an analogue computer using three of its basic electronic components: and *integrator*, a *summer*, and a *potentiometer*, which are shown symbolically in Figure

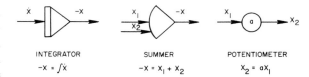

FIGURE A.11. Basic operating components of an analog computer. Note that both the integrator and the summer change the sign of the input signal. (From Koushanpour.[2])

A.11. The mathematical operation of each component is represented by the equation written below the symbol.

The first step in solving Equation A.42 is to represent it in terms of these analog computer circuit components. This is shown in Figure A.12. Since the analog computer operates on voltage, the solution, y(t), can be displayed on the face of a cathode ray oscilloscope. Therefore, it is possible to observe the behavior of ÿ, ẏ, and y as functions of time in response to any desired forcing.

We shall now investigate, by means of an analog computer, the static and dynamic response characteristics of the second-order system represented by Euation A.38, to a step input. First, we look at the response of the open-loop system, then close the loop and insert various types of controller signals and observe the static and dynamic responses. Figure A.13 shows the analog computer circuit and the different types of controller signals (S_c) used to stimulate the response of the second-order system represented by the equation

$$\ddot{y} + k_1\dot{y} + k_2 y = k_3 x(t) \qquad (A.43)$$

For the purpose of simulation, in the analog circuit diagram, we have assigned arbitrary values for the natural frequency and the damping ratio such that

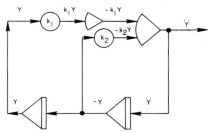

FIGURE A.12. Analog computer representation of Equation A.42, where k_1 = C/A and k_2 = B/A. (From Koushanpour.[2])

Computer Solution of Response Characteristics of a Control System

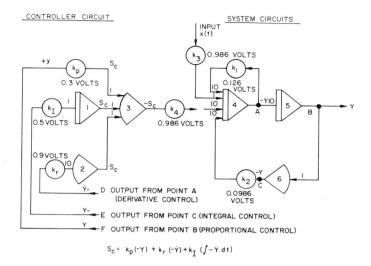

FIGURE A.13. Analog computer circuit representing a second-order control system with proportional, rate-sensitive, and integral controllers. (From Koushanpour.[2])

$$k_1 = 0.2\ \zeta\ \omega_n = 0.126 \text{ volt} \quad (A.44)$$
$$k_2 = 0.01\ \omega_n^2 = 0.0986 \text{ volt} \quad (A.45)$$
$$k_3 = 0.1\ \omega_n^2 = 0.986 \text{ volt} \quad (A.46)$$

These coefficient values are written below the potentiometers designated as k_1, k_2, k_3, on the right side of the circuit diagram. The numerical values 1 or 10 that appear at the input side of either an integrator or a summer are called the *gain factors*. They refer to the "gain" of the electronic amplifier which makes up the integrator or the summer.

The left side of Figure A.13 shows the analog computer circuits for the controller signals. The potentiometers labeled k_p, k_r, and k_I represent the proportionality coefficients for the proportional, rate-sensitive, and integral controller signals, respectively. The values of these coefficients are written below the appropriate potentiometers.

Figure A.14 shows the response of an open-loop system (the controller signal is disconnected) to a step input traced from photographs taken from the face of an oscilloscope. Both the input and output traces start at the same horzontal position. The output oscillates about the input and finally the two are superimposed during the steady state. Hence, in the absence of a controller signal, the output cannot be stabilized at the predisturbed position. When the loop is closed and a proportional controller is added to the controlling system, the steady-state error is reduced, as seen in Figure A.15. However, the output still shows oscillation before reaching steady state.

In terms of the analog computer simulation, shown in Figure A.13, the closing of the loop and the addition of a proportional controller is equivalent to connecting the output of the integrator (5) at B to the potentiometer k_p at F. The controller signal (S_c) is connected through potentiometer k_4 with a "gain" factor of 10 to the integrator (4). As mentioned pre-

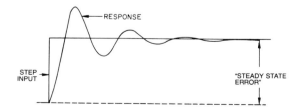

FIGURE A.14. Response of an open-loop control system to a step input. (From Koushanpour.[2])

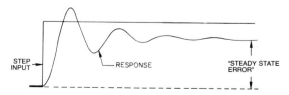

FIGURE A.15. Response of a closed-loop control system with a proportional controller to a step input. (From Koushanpour.[2])

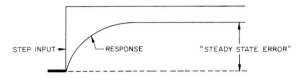

FIGURE A.16. Response of a closed-loop control system with proportional and rate-sensitive controllers to a step input. (From Koushanpour.[2])

viously in this Appendix, the response of a closed-loop control system, with a proportional controller, to a step input shows both transitory oscillations and steady-state errors. The transient oscillations in response can be eliminated by adding a rate-sensitive controller to the controlling system. In Figure A.13, this is equivalent to connecting the output of integrator (4) at A to the potentiometer k_r at D. When this is done, the response of the system to a step input is that shown in Figure A.16. Note that the only difference in response between a closed-loop control system with a proportional controller and one with both proportional and rate-sensitive controllers is that the transient oscillations are absent, but the steady-state error still persists.

The steady-state error can be eliminated by adding an integral controller to the controlling system. In Figure A.13, this is equivalent to connecting the output of the summer (6) at C to the potentiometer k_I at E. When this is done, the response of the system to a step input shows no oscillation and no steady-state error, as shown in Figure A.17.

For a more extensive treatment of the materials presented here, the reader may wish to consult the selected bibliography cited at the end of this Appendix.

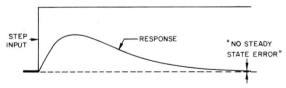

FIGURE A.17. Response of a closed-loop control system with proportional, rate-sensitive and integral controllers to a step input. (From Koushanpour.[2])

Problems

A.1. Discuss the similarities and differences between the flow and functional diagrams.

A.2. Define the various forcings commonly used to study the response characteristics of a biological system.

A.3. What is the difference between a transient and a steady-state response?

A.4. Define time-constant and briefly describe its physiological significance.

A.5. What distinguishes steady state from equilibrium?

A.6. What is the characteristic difference between a linear and a nonlinear system?

A.7. What is the difference between a control system and a closed-loop system?

A.8. What is the difference between a regulator and a servomechanism?

A.9. Define (a) controlling system, (b) controlled system, and (c) feedback.

A.10. What are the three types of controllers and their functions in a control system?

A.11. Define static gain of an open-loop system.

A.12. Define dynamic gain of an open-loop system and describe its components.

A.13. Describe three methods used to determine the dynamic transfer function of a linear system.

A.14. What are the main differences between an analog and a digital computer?

References

1. Grodins FS: *Control Theory and Biological Systems*. New York, Columbia University Press, 1963.
2. Koushanpour E: *Renal Physiology: Principles and Functions*, ed 1. Philadelphia, WB Saunders Co, 1976.
3. Koushanpour E, Stipp GK: Mathematical simulation of the body fluid regulating system in dog. *J Theor Biol* 1982; 99:203–235.
4. Koushanpour E, Tarica RR, Stevens WF: Mathematical simulation of normal nephron function in rat and man. *J Theor Biol* 1971; 31:177–214.

Bibliography

Cannon RH Jr: *Dynamics of Physical Systems*. New York, McGraw-Hill Book Co Inc, 1967.
Johnson CL: *Analog Computer Techniques*. New York, McGraw-Hill Book Co Inc, 1963.
Milhorn HT Jr: *The Application of Control Theory to Physiological Systems*. Philadelphia, WB Saunders Co, 1966.
Riggs DS: *The Mathematical Approach to Physiological Problems*. Baltimore, Williams & Wilkins Co, 1963.
Trimmer JD: *Response of Physical Systems*. New York, John Wiley & Sons Inc, 1950.

Appendix B

Mathematical Basis of Dilution Principle

As described in Chapter 2, the volume of a body fluid compartment, in practice, may be determined from a graphic analysis of a plot of the logarithm of the plasma concentration of the test substance against time. Experimental collection of the necessary data and the successful analysis of such a time-plot requires satisfctory resolution of two time-dependent characteristics of the test substance: (1) its *mixing* in the plasma and (2) its *distribution* into other compartment(s).

The initial nonlinearity seen in the semi-logarithmic lot of the plasma concentration against time (Figure 2.1) strongly suggests that the mixing of the test substance is not at all instantaneous, but actually follows an exponential decay process. Likewise, the distribution of the test substance into other compartment(s) is not instantaneous, but proceeds exponentially. Furthermore, while the test substance is being mixed in the plasma, it may also be penetrating and distributing into other compartment(s). Therefore, the time-dependent decrease in the plasma concentration of the test substance is in part determined by the rates at which the substance mixes in the plasma and distributes into other compartment(s). Hence, the overall shape of the semi-logarithmic plot of the plasma concentration against time depends largely on these two rates and their interaction.

To get a better insight into the effect of these processes on the application of the dilution principle, let us examine the mathematical basis of the often used *single-dose injection method* and the rationale for the graphical analysis of the resulting data.

Consider the hypothetical compartment illustrated in Figure B.1. As shown, $\dot{V}_i$ is the rate of volume flow into the compartment in milliliters per minute, $[x]_i$ is the concentration of the injected test substance in the inflow stream in milligrams per minute, $[\dot{x}]_o$ is the rate of change of concentration of the test substance in the compartment V in milligrams per milliliters per minute, and $\dot{V}_o$ is the rate of volume flow out of the compartment V in milliliters per minute.

Applying the principle of conservation of mass to this system, we see that the rate of change of mass of the test substance injected into the compartment is equal to the rate of the test substance entering the compartment minus the rate leaving the compartment, provided that the test substance is neither formed nor metabolized by the body. In mathematical notation, we can write

$$V \cdot [\dot{x}]_o = \dot{V}_i \cdot [x]_i \cdot \delta(t) - \dot{V}_o \cdot [x]_o \quad \text{(B.1)}$$

Rate of change of mass (mg/min) = rate of inflow of mass (mg/min) − rate of outflow of mass (mg/min)

where $\delta(t)$ is the *Dirac Delta function* and represents mathematically the fact that the test substance is administered as an impulse over a small time interval rather than continuously.

The first term on the right hand side of Equation B.1 is a measure of the amount of the administered test substance which when mixed completely in the volume V would yield the initial concentration, $[x(t=0)]_o$. This would be the concentration if the test substance had been delivered and mixed instantaneously in the compartment. In practice, for the period after the test substance has been injected, $\delta(t) = 0$—and hence the term containing this function—disappears, and Equation B.1 becomes

$$V \cdot [\dot{x}]_o + \dot{V}_o \cdot [x]_o = 0 \quad \text{(B.2)}$$

Dividing Equation B.2 through by $\dot{V}_o$ yields

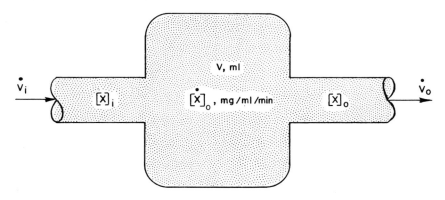

FIGURE B.1. An idealized representation of a body fluid compartment. (From Koushanpour.[1])

$$\frac{V}{\dot{V}_o}[\dot{x}]_o + [x]_o = 0 \quad (B.3)$$

Equation B.3 is a first-order linear differential equation with constant coefficients and with initial conditions as given in Figure B.1. The coefficient of the derivative term ($V/\dot{V}_o$) is called the time-constant (τ) or the *residence time*; it is the time during which the test substance remains in the compartment. In the steady state, when the rate of volume inflow is equal to the rate of volume outflow, $\dot{V}_i$ equals $\dot{V}_o$, and the above homogeneous differential equation would have the following force-free solution:

$$[x(t)]_o = [x(t=0)]_o e^{-(\dot{V}_o/V)t} \quad (B.4)$$

The coefficient of t in the exponent of the exponential term is the reciprocal of the time-constant ($\dot{V}_o/V = 1/\tau$) and hence makes the exponent dimensionless. Substituting this relationship, Equation B.4 may be written in a more conventional form:

$$[x(t)]_o = [x(t=0)]_o e^{-t/\tau} \quad (B.5)$$

Equation B.5 describes a process represented by an exponential decay. In studying the behavior of many biological systems, the physiologist is interested in the *half-time* ($t_{1/2}$) (also called *half-life*), which is the time required for half of the initially administered test substance to disappear from the compartment under study. To determine the half-time, Equation B.5 may be rearranged as follows:

$$[x(t_{1/2})]_o/[x(t=0)]_o = 1/2 = e^{-t_{1/2}/\tau} \quad (B.6)$$

Taking the natural logarithm of both sides of Equation B.6 and rearranging terms, we get

$$t_{1/2} = 0.693\,\tau \quad (B.7)$$

The time-constant τ, which appeared in Equations B.5 to B.7, is the time required for the concentration of the administered test substance to decrease to $1/e = 1/2.718 = 0.367$ or 36.7% of the initial value. The time-constant is also called the *turnover time* because it is the time required for the fluid flowing out at a rate of $\dot{V}_o$ to "turnover" and make one complete pass through the volume V.

To obtain the value of the time-constant ($\tau = t_{1/e}$), t is set equal to τ in Equation B.5. This makes the exponent of the exponential term equal to -1, and the exponential term e^{-1} may be written as $1/e$. Using these relationships, Equation B.5 may be rewritten to yield

$$[x(\tau)]_o/[x(t=0)]_o = 1/e \quad (B.8)$$

It should be noted that Equation B.4 in logarithmic form is similar to an equation for a straight line:

$$\ln[x(t)]_o = \ln[x(t=0)]_o - (\dot{V}_o/V)t \quad (B.9)$$

The semi-logarithmic plot of Equation B.9 is a straight line with a negative slope of magnitude $\dot{V}_o/V$, which is the reciprocal of the time-constant, and the value of $\ln[x(t=0)]_o$ as its ordinate intercept.

Equation B.9 is the basis for the semi-logarithmic plot of the plasma concentration of the test substance against time and the subsequent graphical analysis (linear extrapolation, etc.) of the results.

To get a better "feel" for the mathematical formulations just presented, try to answer the questions posed in the following problem.

Problem

B.1. The data listed below were obtained from an 80 kg patient after he had received 4.5 g of inulin in a single dose.

Time (min)	Plasma inulin concentration (mg/mL)
10	440
20	320
40	200
60	150
90	110
120	80
150	60
175	48
210	35
240	25

Calculate (a) the time-constant of the exponential decay curve, (b) the half-time for the disappearance of inulin, (c) volume of inulin space, and (d) the renal excretion rate of inulin.

Reference

1. Koushanpour E: *Renal Physiology: Principles and Functions*, ed 1. Philadelphia, WB Saunders Co, 1976.

Appendix C

Anatomical Nomenclatures

This table summarizes the nomenclature of the nephron and the collecting duct system as used by us in comparison with other frequently used names. A serpentine arrow means that the transition between

The Renal Tubule					
Main Portions	Segmentation		Abbr.	Cell types	Other frequently used denominations
PROXIMAL TUBULE	Convoluted Part	P1 segment (S1)	PCT	P1-cells	Proximal convoluted tubule ./. Pars convoluta
		P2 segment (S2)		P2-cells	
	Straight Part	P3 segment (S3)	PST	P3-cells	Proximal straight tubule ./. Pars recta ./. Descending thick limb
INTERMEDIATE TUBULE (THIN TUBULE)	Descending Thin Limb of a Short Loop		DTL	DTL-cells — Type I	Short descending thin limb
	Descending Thin Limb of a Long Loop	Upper Part		DTL-cells — Type II	Long descendig thin limb (upper and lower portions)
		Lower Part		DTL-cells — Type III	
		Prebend segment		ATL-cells — Type IV	
	Ascending Thin Limb (in long loops only)		ATL		Thin ascending limb
DISTAL TUBULE	Straight Part ./. Thick Ascending Limb	Medullary Straight Part	(MAL)	TAL-cells	Medullary thick ascending limb
		Cortical Straight Part	TAL		Cortical thick ascending limb
		Macula densa	(CAL)	MD-cells	
		Postmacula segment			DCT a* — Early distal tubule
	Convoluted Part ./. Distal Convoluted Tubule		DCT	DCT-cells + (IC-cells)	DCT b*
COLLECTING DUCT SYSTEM	Connecting Tubule		CNT	CNT-cells + IC-cells	DCT g* / CCT g* — Late distal tubule ./. Initial collecting tubule
	Cortical Collecting Duct		CCD	CD-cells (Principal cells) + IC-cells (Intercalated cells)	DCT 1* / CCT 1* — Cortical collecting tubule
	Outer Medullary Collecting Duct		OMCD		Outer medullary collecting tubule
	Inner Medullary Collecting Duct		IMCD	CD-cells + (IC-cells)	Papillary ducts ./. Ducts of Bellini

the two adjacent is gradual. A serpentine arrow combined with a cut surpetine arrow means that the transition in some species is gradual; in others, abrupt. Column three gives some frequently used abbreviations. Column four indicates the cell types. Note that the CNT, CCD, and OMCD are composed of two cell types. In addition, intercalated cells (IC-cells) may occur in the DCT (terminal portion) and the IMCD (beginning portion). Column five summarizes other frequently applied terms. The abbreviations marked by a star have been introduced by Morel and co-workers.[1] They mean: DCTa = distal convoluted tubule, initial portion; DCTb = distal convoluted tubule, bright portion; DCTg = distal convoluted tubule, granular portion; DCTl = distal convoluted tubule, light portion; CCTg = cortical collecting duct, granular portion; CCDl = cortical collecting duct, light portion.

Note that as denoted by the symbol (†) in the above table, the term "distal tubule" as has been used by physiologists comprises the distal convoluted tubule, the connecting tubule, and the cortical collecting duct. However, the term "distal nephron" comprises the distal convoluted tubule, the connecting tubule, and the entire collecting duct. In this book, whenever the general terms "distal tubule" and "distal nephron" are used, they are enclosed in quotation marks.

Reference

1. Morel F, Chabardes D, Imbert M: Functional segmentation of the rabbit distal tubule by microdetermination of hormone-dependent adenylate cyclase activity. *Kidney Int* 1976; 9:264–277.

Answers to Problems

Chapter 1

1.1. a. Osmolality = 308 mOsm/L; osmotic pressure = 5,243 mm Hg
b. Osmolality = 300 mOsm/L; osmotic pressure = 5,107 mm Hg
1.2. (a) 55.4 g; (b) 54 g
1.3. a. 3,000 mM/L; 3,000 mEq/L; 3,000 mOsm/L
b. 3,000 mM/L; 3,000 mEq/L; 6,000 mOsm/L
c. 500 mM/L; 500 mEq/L; 500 mOsm/L
1.4. 96 mOsm/L
1.5. 490 mOsm/L

Chapter 2

2.1. (a) 55.3%; (b) 24.5%; (c) 22.05 kg; (d) 1.05
2.2. 20.8%
2.3. 41.4 mEq/kg body weight
2.4. (a) 20.6 L; (b) 0.0425 mg/dL/min (natural logarithmic scale)
2.5. 25.4 L
2.6. a. Should not be toxic.
b. Should distribute uniformly within the compartment.
c. Should not be metabolized.
d. Should not alter the volume of the compartment.
2.7. (a) 14.4 kg; (b) 475 mOsm/L
2.8. (a) Sodium; (b) chloride; (c) potassium; (d) phosphate; (e) albumin; (f) large extracellular volume; (g) increase in the number of the cells; (h) increase in tissue with greater intracellular water; (i) increase in body fat

Chapter 3

3.1. (a) 2,000 mL; (b) 1,500 mL; (c) 857 mL
3.2. (a) 70 g in plasma; (b) 82 g in tissues
3.3. a. $[Os] = 334$ mOsm/L; $V_E = 19.9$ L; $V_C = 22.4$ L
b. $[Os] = 300$ mOsm/L; $V_E = 19$ L; $V_C = 25$ L
c. $[Os] = 298$ mOsm/L; $V_E = 20.8$ L; $V_C = 25.2$ L
3.4. a. $V_C = 29$ L; $V_E = 13$ L; $[Os]_p = 311$ mOsm/L; $[Na]_p = 155.5$ mEq/L
b. $V_C = 26.7$ L; $V_E = 21.3$ L; $[Os]_p = 337.5$ mOsm/L; $[Na]_p = 168.8$ mEq/L
c. $V_C = 31.3$ L; $V_E = 15.7$ L; $[Os]_p = 287$ mOsm/L; $[Na]_p = 143.5$ mEq/L
3.5. *Forcings*   *Condition or Disease*
1. Isotonic influx   Edema
2. Hypotonic efflux   Sweating
3. Hypertonic influx   Nephritis
4. Isotonic efflux   Hemorrhage
5. Hypotonic influx   Water intoxication

Chapter 5

5.1. a. + ; b. − ; c. + ; d. − ; e. + ; f. + ; g. +
5.2. a. 133.3 mEq/L
b. 0.93 for sodium and 1.10 for chloride
5.3. $(P_c - \pi_c)$ is lower in the lung, accounting for the relative dryness of this organ.
$(P_c - \pi_c)$ is higher in the kidney, accounting in part for the large quantity of fluid processed by this organ.
5.4. a. +; +; +; −
b. −; −; −; 0
c. +; +; +; 0
d. 0; 0; −; 0
e. 0; 0; +; 0

Chapter 7

7.1. a. 80.0 ; b. 0.2 ; c. 5.0 ; d. 0.75
7.2. a. 10; 0.2; b. 100; 0; c. 4; 1.0; d. 0; 4; e. 20; 5
7.3. 0
7.4. Arterial plasma concentration
7.5. a. 728; 129;   b. 480; 126;   c. 229; 275; d. 179; 224; e. 47.8; 129; f. 30.4; 140

Answers to Problems 373

7.6. 778 mL/min
7.7. 1,730 mL/min
7.8. 43%
7.9. 16.5%
7.10.
and
7.11. a. 565; 4,400; 0.165
 b. 380; 2,960; 0.162
 c. 229; 1,780; 0.155
 d. 179; 1,390; 0.162
 e. 85; 660; 0.167
 f. 50; 388; 0.180
7.12. a. 1; b. 3; c. 3; d. secretion
7.13. Only filtered and having a permeability ratio of unity.
7.14. Filtered, secreted, and having a high extraction ratio.
7.15. Filtration and secretion.
7.16. a. Glomerular filtration.
 b. Tubular reabsorption.
7.17. a. 1.0; b. 0.57; c. 0.257; d. 0.186; e. 0
7.18. a. 560; 280; 140
7.19. (a) 9 mg/dL; 45 mg/dL; 0.36/min; 0.57/min
 (b) 0.063 minutes; 0.02 minutes
7.20. a. Secreted at a rate of 40 mg/min. (Note negative sign in calculations.)
7.21. a. 130; 700; 0; 0; 130
 b. 650; 3,500; 55; 0.08; 375
7.22. a. 1, 3, 6; b. 2, 9; c. 1, 5, 8, 9; d. 1, 5, 8; e. 1, 5, 8, 9; f. 1, 10
7.23. a. 14 mEq/L; b. 13.5 mEq/L; c. 300 mg/min; d. 300 mg/dL; e. 20%
7.24. 35.5 mOsm/L
7.25. (a) 90 mg/dL; (b) 62%
7.26. (a) 300 mg/dL; (b) none
7.27. a. none; b. X; c. Z; d. Y

Chapter 8

8.1. a. For drug-receptor complex, $K_m = 25$.
 b. For inhibitor-receptor complex, $K_m = 47.5$.
 c. Since $V_{max} = 51.5$ for both, the inhibitor is a competitive one.
8.2. a. Ordinate intercept = $1/V_{max}$; a measure of the efficiency of the process.
 b. Abscissa-intercept = $-1/K_m$; a measure of the affinity of the enzyme and substrate.
8.3. $J_s = J_{CS} = -\dfrac{D}{X}[SC_1 - SC_2]$
8.4. When $S \gg K_{CS}$, then $J_s = -\dfrac{D}{X} \cdot C_T [1 - \dfrac{S_2}{S_2 + K_C}]$
8.5. a. $\dfrac{[Z_1]}{[C_1]} = \dfrac{[Z_2]}{[C_2]}$, that is, symmetrical metabolic reactions on both sides of the membrane.
 b. $K_{CS} = K_{ZS}$, that is, the two carriers have equal affinities for the substrate on both sides of the membrane.
8.6. Let $S_1 = 0.2$ M and $K_C = S_2 = 0.02$ M.
 Fraction of maximum flux = $[\dfrac{S_1}{S_1 + K_C} - \dfrac{S_2}{S_2 + K_C}] = \dfrac{9}{11}$

Chapter 9

9.7. 15.96 mEq/min or 95% of the filtered sodium is reabsorbed.
9.8. 1. $-$; 2. $+$; 3. $-$; 4. $+$; 5. $+$; 6. $-$; 7. $+$; 8. $-$; 9. $-$; 10. $+$; 11. $-$; 12. $+$; 13. $+$

Chapter 10

10.1. 650; 375
10.2. 5.7; 36.3
10.3. a. 65; 65; 350; 0; 0
 b. 130; 130; 700; 0; 0
 c. 375; 375; 2,020; 0; 0
 d. 650; 375; 3,500; 55; 0.0785
 e. 910; 375; 4,900; 76.5; 0.109
 f. 1,300; 375; 7,000; 92.5; 0.132
10.6. 288
10.7. 375
10.8. a. 260; 520; 780; 1,170
 b. 260; 375; 375; 375
 c. 0; 140; 400; 790
 d. 0; 27; 51; 68
10.9. It varies monotonically toward an asymptote.
10.10. a. Since $\dfrac{U_G V}{GFR \cdot P_G} = 1$, $C_G = C_{In} =$ GFR = 130
 b. $\dfrac{130}{700} = 0.186$ (see Figure 7.3)
10.11. a. 2.16; 11.84; 14; 700; 1
 b. 6.48; 35.52; 42; 700; 1
 c. 14.0; 77; 91; 700; 1
 d. 21.6; 77.9; 140; 497.5; 0.71
 e. 43.2; 77.3; 280; 300.0; 0.43
 f. 64.8; 77.7; 420; 238.0; 0.34
10.13. 13
10.14. 77.5
10.15. a. 4; 8; 16; 32; 54; 76
 b. 22; 46; 77.5; 77.5; 77.5; 77.5
 c. 28; 54; 94; 109; 130; 151
 d. 18.2%; 17.4%; 20.6%; 41.3%; 70%; 98%
10.16. Varies linearly and proportionally when plasma concentration increases above 13 mg/dL.
10.17. a. Since $U_{PAH} V = GFR \cdot P_{PAH} + \dot{T}_{mPAH}$, then, if $\dot{T}_{mPAH} \to 0$, $C_{PAH} = C_{In} =$ GFR (see Figure 7.3)
 b. 0.186
10.20. a.

10.21. a.
10.22. 13

Chapter 11

11.1. (a) 7.41 (b) 25.0 (c) 25.0
11.2. (a) 7.32 (b) 27.5 (c) 25.0 (d) acidemia (e) acidosis (f) respiratory (g) uncompensated (h) displaement
11.3. (a) 7.50 (b) 22.0 (c) 25.0 (d) alkalemia (e) alkalosis (f) respiratory (g) uncompensated (h) displacement
11.4. (a) Oppositely (b) together (c) of course
11.5. (a) 34.0 (b) 19.0 (c) metabolic (d) compensated for respiratory disturbance
11.6. (a) Down the $pCO_2 = 40$ line, until $pCO_2 = 40$, $[HCO_3^-]_p = 19$, and $pH < 7.32$.
(b) Low respiratory pathway until $pCO_2 = 35$, $[HCO_3^-]_p = 18$, and $pH = 7.32$.
(c) No, because respiratory regulator responds rapidly.
(d) Normal metabolic pathway.
11.7. (a) Up the $pCO_2 = 40$ line, until $pCO_2 = 40$, $[HCO_3^-]_p = 34.0$, and $pH > 7.50$.
(b) High gain respiratory pathway until $pCO_2 = 45.5$, $[HCO_3^-]_p = 35.0$, and $pH = 7.50$.
(c) No, because respiratory regulator responds rapidly.
(d) Normal metabolic pathway.
11.8. (a) Normal respiratory pathway.
(b) Uncompensated respiratory alkalosis.
11.9. (a) Yes, pO_2 is still low; (b) yes; (c) renal; (d) slow; (e) yes, integral control; (f) $[HCO_3^-]_{40}$; (g) high gain metabolic pathway; (h) 7.41, 16.5, 15, 24.5; (i) respiratory alkalosis, fully compensated; (j) more, because pH inhibition is removed; (k) better, because of still greater hyperventilation.
11.10. (a) Yes, pO_2 returns to normal; (b) still 19.0; (c) only moderate, because pH is now low; (d) low gain respiratory pathway; (e) 34, 7.32, 18; (f) metabolic acidosis, partially compensated (correcting the respiratory disturbance leaves the compensatory metabolic acidosis exposed).
11.11. (a) Will rise to normal by renal action; (b) normal metabolic pathway; (c) 7.41, 25, 40; (d) slow; (e) yes, back to normal, having followed a four-sided loop.
11.12. You will have to check yourself this time.
11.13. (a) Fully compensated respiratory acidosis and alkalosis.
(b) Respiratory alkalosis and metabolic acidosis.
(c) Metabolic acidosis, or compensated respiratory alkalosis.
(d) Metabolic acidosis or respiratory acidosis, but beware of *dissociated* disturbances.
(e) No, it takes more than one.
11.14. Do not forget combinations, like simultaneous respiratory and metabolic acidosis.

11.15. (a) High gain metabolic pathway.
(b) It will become abnormally high.
(c) Hinders–in fact, frustrates–renal compensation.
11.16. (a) The low gain respiratory pathway.
(b) It will become even lower.
(c) Hinders–in fact, frustrates–respiratory compensation.
11.17. (1) a. 6; b. 2; c. 1; d. 5; e. 7; f. 3; g. 4; h. 8
(2) 2, 3, 4, 5, and 1, 6, 7, 8
(3) 1 and 5
(4) 6, 7, 8 and 2, 3, 4
(5) 1, 2, 8 and 4, 5, 6
(6) 3, 4, 5 and 1, 7, 8
(7) 1, 2, 8
(8) 4, 5, 6
(9) 4, 5, 6, N
(10) 1, 7, 8
(11) 3, 4, 5
(12) 4

Chapter 13

13.13. (A) − ; (B) + ; (C) + ; (D) + ; (E) + ; (F) + ; (G) − ; (H) +
13.14. Vascular receptors for all except F, the response of which is mediated via the macula densa receptors.
13.15. (A) Decrease; (B) increase; (C) increase; (D) decrease; (E) increase; (F) increase; (G) increase
13.16. (C) Upright position
13.21. (1) E ; (2) A ; (3) C ; (4) D ; (5) B
13.22. (1) D ; (2) C ; (3) B ; (4) A
13.23. (A) 2 and 3
(B) Patient 1: − , − ; patient 2: 0 , 0 ; patient 3: − , − ; patient 4: − , 0
(C) a. Patient 2
b. Patient 1
13.24. (1) + ; (2) + ; (3) + ; (4) + ; (5) + ; (6) − ; (7) − ; (8) + ; (9) + ; (10) −
13.25. (1) + , 0 , + , + , + , 0 , − , − , − , + , + , +
(2) + , + , + , + , + , − , − , − , − , + , + , +
(3) + , − , + , + , + , + , − , + , − , − , + , −
(4) − , 0 , − , − , − , 0 , 0 , + , + , − , − , −
(5) − , − , − , − , − , + , + , + , + , − , − , −

Appendix B

B.1. (a) $\tau = 1.6$ minutes.
(b) $t_{1/2} = 1.1$ minutes.
(c) Volume of inulin space $= 16.0$ L.
(d) Excretion rate $= 0.63$ mg/mL/min.

Index

Absorption curve, carbon dioxide, in true plasma, 241–245
Acceleration, competitive, 125
Acetazolamide, 303–304
  and bicarbonate reabsorption in collecting duct, 254
Acetoacetic acid
  production in diabetes, 240
  tubular reabsorption of, 229
Acetylcholine
  and renal blood flow, 64
  renal vasodilation from, 91
  and renin release in hypertensive dogs, 336
Acid
  excretion of, 250
    and alkali accumulation, 248
  fortuitous fluxes of, 249, 252
  nonvolatile or fixed, 240
    increased influx of, 245
  organic
    in fluid compartments, 14
    in plasma and ultrafiltrate, 68
  titrable, controlled excretion of, 258–259
  volatile, 240
Acid-base balance, 240–269
  and bicarbonate reabsorption, 149
  and chloride reabsorption, 150
  compensation in, 246
    cross-compensation, 252–253
      analysis of, 262
      partial, 252
    within-regulator, 246, 252
      analysis of, 262
  disorders of, 245–246
    compensated, 246
    diagnosis of, 262–267
    excretion of acid or alkali in, 250
    uncompensated, 246
  elements of, 245–246
  metabolic pathway in, 264–266
    completely cross-compensated, 266
    high-gain, 265
    low-gain, 265
    partially cross-compensated, 266
  pulmonary regulator of, 245, 246–248
    cross-compensation with renal regulator, 252–253, 262
    forcings and responses of, 247–248
  renal regulator of, 248–252
    cross-compensation with pulmonary regulator, 252–253, 262
    forcings and responses of, 251–252, 263
    mechanisms in, 253–262
  respiratory pathway in, 263–264
    completely cross-compensated, 264
    completely uncompensated, 264
    high-gain, 264
    low-gain, 264
Acidemia, 245
Acidification of urine, in collecting duct, 258
Acidosis, 245
  metabolic, 246
    ammonia excretion in, 261–262
    ammonia production in, 260
    causes of, 251
    citrate excretion in, 228
    dissociated, 191
    extracellular, 250
      in Addison's disease, 258
      causes of, 251, 252
      in hyperkalemia, 257
      induced in respiratory alkalosis, 265
    intracellular, 250
      causes of, 251
    low-gain respiratory pathway in, 264
    potassium secretion in, 191, 258
    renal response to, 252
    respiratory response to, 247
    sodium chloride excretion in, 228
  respiratory, 246
    bicarbonate reabsorption in, 257
    induced in metabolic alkalosis, 248
    metabolic alkalosis induced in, 253, 257, 265
    potassium secretion in, 191, 258
ACTH, and aldosterone biosynthesis, 197, 199
Actinomycin D
  and aldosterone-induced sodium transport, 202
  and antiatruiretic effect of aldosterone, 200
  and gene transcription, 190
  and proximal sodium reabsorption, 197
Activation energy, in conversion of molecules in chemical reactions, 118
Active transport, 16, 127–129, 214
  of glucose, 221
  of potassium, 148
  of soduim, 146, 271, 281
Adenosine monophosphate, cyclic. *See* cAMP
Adenosine triphosphatase. *See* ATPase
Adenylate cyclase
  and action of parathyroid hormone, 160
  vasopressin-sensitive, localization of, 196
Adipose tissue, water in, 10
Adrenal insufficiency
  acidosis in, 258
  and antidiuretic hormone levels in plasma, 324
  and renin levels in plasma, 328
Adrenergic agonists
  and responses in connecting tubule cells, 182
  and sodium reabsorption in proximal tubule, 161–161
Adrenergic receptor activation, and renin release, 338
Age
  and creatinine clearance, 103
  and daily turnover of fluid, 22
  and electrolytes in fluid compartments, 18
  and glomerular filtration rate, 100
  and phosphate requirements, 222
  and urea clearance, 102
  and uric acid levels, 229
Alanine, transport and reabsorption of, 227
Albumin, 18
  filtrate-to-plasma concentration ratio for, 66

Albumin (cont.)
　ratio to globulin, 18
　tubular reabsorption of, 230
Aldosterone, 331–332
　and activity of Na-K-ATPase, 152
　affecting basal infoldings in
　　collecting duct, 115
　angiotensin affecting release of, 326,
　　330
　and bicarbonate reabsorption, 258
　biosynthesis of, 196–197
　　ACTH affecting, 197, 199
　　in anephric man, 199
　　factors regulating, 197–200
　　heparin affecting, 199
　　rate of secretion in, 332
　　renin-angiotensin system affecting,
　　　198–199, 331, 341–342
　　sodium-potassium balance
　　　affecting, 199
　cellular action of, 200–204
　　induction theory of, 201–202
　　metabolic theory of, 203–204
　　permease theory of, 203
　　sodium pump theory of, 202–203
　diurnal rhythm of, 197
　and intracellular sodium
　　concentration, 203
　plasma levels of
　　neuroendocrine system affecting,
　　　312, 315
　　and regulation of blood volume
　　　and plasma osmolality, 315
　　and urine volume and osmolality,
　　　314
　and potassium secretion in distal
　　tubule, 189–190, 191–192
　prolonged administration of, effects
　　of, 200, 203
　site of action, 197
　and sodium-potassium exchange in
　　principal cells, 255
　and sodium reabsorption in collecting
　　duct, 184
　spironolactone as antagonist of, 305
　and transport in distal tubule and
　　collecting duct, 196–204
　as volume control regulator, 331
Aldosterone-induced protein, 202, 203
Aldosterone-receptor complex,
　cytoplasmic, 201
Aldosteronism, alkalosis in, 258
Alkalemia, 245
Alkali
　excretion of, 250
　and acid accumulation, 248
　fixed, increased influx of, 245
　fortuitous flux of, 249
Alkaline reserve, 245
Alkalosis, 245
　metabolic, 246
　　causes of, 251
　　citrate excretion in, 228
　　excretion of organic anions in, 227
　　extracellular, 250
　　　in aldosteronism, 258
　　　causes of, 251
　　　in Cushing's syndrome, 258
　　　hypokalemic, 257

high-gain respiratory pathway in,
　264
induced in respiratory acidosis,
　253, 257, 265
intracellular, 250
　causes of, 251, 252
　and potassium secretion in distal
　　tubule, 191, 258
　renal response in, 252
　respiratory response in, 247–248
respiratory, 246
　bicarbonate reabsorption in, 257
　at high altitudes, 253
　induced in metabolic acidosis, 247
　metabolic acidosis induced in, 265
　and potassium secretion in distal
　　tubule, 191, 258
Altitude
　and hyperventilation, 265
　and respiratory alkalosis, 253
Amiloride, 305–306
　and intracellular sodium
　　concentration, 203
　and potassium secretion in distal
　　tubule, 190–191
　sites of action of, 303
　and sodium-hydrogen exchange, 149
　and transepithelial potential
　　difference in distal convoluted
　　tubule, 182
Amino acids
　clearance of, 232
　generation of lysosomal hydrolysis,
　　231
　and sulfate reabsorption, 225
　titration curves for, 226–227
　transport and reabsorption of, 225–
　　227
　　acidic acids, 227
　　basic acids, 227
　　cotransport with sodium, 130,
　　　142, 227
　　iminoglycine acids, 227
　　neutral acids, 227
p-Aminohippurate
　cotransport with sodium, 233
　renal clearance of, 101, 105
　secretion of, 233–234
　titration curve for, 234
　tubular concentration of, 132–133
Aminopeptidase, 330
Ammonia
　excretion of, 258, 259–262
　nonionic diffusion of, 261
　synthesis and secretion of
　　in collecting duct, 260
　　in proximal tubule, 260
Ammonium chloride
　excretion of, 250
　metabolic acidosis from, 252
Amobarbital, and sodium transport, 204
cAMP
　and cellular action of antidiuretic
　　hormone, 194–196
　glomerular synthesis of, 64
　and hormone-receptor interactions,
　　123–124
　prostaglandins affecting formation of,
　　89

and renal action of parathyroid
　hormone, 223
and renin secretion, 338
as second messenger, 194
and sodium-hydrogen
　countertransport, 160
Amphotericin B, and sodium transport,
　204
Amplification of membranes
　basolateral, 114–115, 140, 164
　luminal, 114, 140, 164
Anatomy of kidney, 41–51
　nomenclature for, 370–371
Angiotensin I, 198, 329, 343
　converting enzyme, 81, 84, 198,
　　329, 343
　　inhibition of, and renal
　　　hemodynamics, 91
　formation of, 81
　in juxtaglomerular apparatus, 84
　liberation from renin substrate, 327,
　　329
Angiotensin II, 198, 326, 329–331,
　343
　and afferent arteriole resistance, 81
　and aldosterone synthesis and
　　release, 312, 326, 330, 331,
　　341–342
　and antidiuretic hormone release,
　　324–325
　formation and degradation of, 328
　by short feedback loop, 91
　glomerular receptors for, 64
　glomerular synthesis of, 64
　pressor action of, 327, 330
　and prostaglandin formation and
　　release, 88, 89
　and renal blood flow, 64
　renal effects of, 330
　and renin release, 339
　and tachyphylaxis, 327, 330
　tubular reabsorption of, 230, 231
　vasoconstrictor effect of, 88, 330,
　　341
Angiotensin III, 329
Angiotensinases, 330
Angiotensinogen, 81, 84, 91, 198, 343
Anions
　in fluid compartments, 13, 14
　organic, transport and reabsorption
　　of, 227–229
　in plasma and ultrafiltrate, 68
Anoxia, hyperventilation in, 265
Antibody formation, 18
Antidiuretic hormone
　cAMP affecting, 194–196
　antagonists of, sites of action of, 303
　biosynthesis and secretion of, 193
　cellular action of, 193–196
　extracellular volume changes
　　affecting, 155
　glomerular receptors for, 64
　inappropriate secretion of, 5, 36
　and intracellular sodium
　　concentration, 203
　and medullary blood flow rate, 297
　metabolism of, 193
　nonosmotic regulation of release,
　　319–323

carotid sinus and aortic arch
  baroreceptors in, 321–323
 interaction with osmotic stimuli,
  323–324, 344
 left atrial stretch receptors in, 319–
  321
 posture affecting, 320
 temperature affecting, 320
osmotic regulation of release, 316–
 319
 hepatic osmoreceptors in, 317–319
 hypothalamic osmoreceptors in,
  316–317
 interaction with volume stimuli,
  323–324, 344
 thirst center in, 317
physiologic effects on thick
 ascending limb of Henle's
 loop, 172–173
plasma concentration of, 193
 angiotensin affecting, 324–325
 neuroendocrine system affecting,
  312, 315
 and osmolality of fluid in distal
  tubule, 180
 and osmolality of plasma, 315,
  316
 and urine-to-plasma osmolality
  ratio, 321
 and urine volume and osmolality,
  314
 and potassium secretion in distal
  tubule, 192
 and prostaglandin synthesis, 196
prostaglandins affecting, 89
renin-angiotensin system affecting
 release of, 324–325
and renin release, 339
and sodium transport and
 reabsorption in thick
 ascending limb of Henle's
 loop, 170, 289
and transport in distal tubule and
 collecting duct, 192–196
and water permeability of collecting
 ducts, 286, 287, 290
and water reabsorption
 in collecting ducts, 299
 facultative, 101
Antiport system, 16
 sodium-hydrogen, 144, 145, 149
 sodium-potassium. See ATPase,
  sodium-potassium
Antipyrine, in total body water
 measurements, 9–10, 11
Anuria, in hemorrhagic shock, 69–70
Aortic arch stretch receptors, 314, 321–
 323
Apical membrane. See Luminal
 membrane
Arcades, renal, 44, 50, 51, 175
Arcuate arteries, 44
Arcuate veins, 45
Arginine, transport and reabsorption of,
 227
Arrhenius equation, 122
Arteries, renal, 44
Arterioles, 24
 afferent, 44, 53
  diameter of, and renin release, 336

plasma protein in, and glomerular
 filtration rate, 60, 63–64
receptors in, and control of renin
 release, 333–337
renin affecting, 81
smooth muscle stretch of, 76
blood pressure in, and
 hemodynamics of glomerular
 circulation, 68–69
efferent, 46, 53
 constriction from angiotensin II,
  88
 cortical, 46, 47–48
 juxtamedullary, 46, 47–48
 norepinephrine affecting, 91
innervation of, 89, 161
resistance of
 and autoregulation of renal blood
  flow, 75, 76
 and blood flow in glomerular
  capillaries, 68–69
 and feedback regulation of
  glomerular filtration, 81, 83
 renal nerve stimulation affecting,
  90
 vasoactive agents affecting, 64
terminal, 24
Arteriovenous concentration difference
 of substances, ratio to
 excretion rate, 104
Aspartic acid, transport and
 reabsorption of, 227
Asthma, hypoventilation in, 265
ATP in cells, and drug-receptor
 interactions, 124
ATPase, sodium-potassium, 16, 127–
 128, 255
 aldosterone affecting, 191–192, 202–
  203
 aldosterone-induced protein affecting,
  202
 in connecting tubule, 182
 in distal convoluted tubule, 173, 182
 hormones affecting, 152–153
 and potassium electrodiffusion, 148
 and proximal tubule function, 129,
  140, 143, 146
 in thick ascending limb of Henle's
  loop, 166, 169
 in thin limbs of Henle's loop, 164
Atrial natriuretic factor, 154, 204
Atrial stretch receptors
 left atrial, 314, 319–321
 right atrial, 314, 340
Autoregulation, renal, 73–89
 and medullary blood flow, 297
 myogenic mechanism in, 75–76, 89
 prostaglandins in, 88–89
 renin-angiotensin system in, 87–88,
  89, 91
 tubuloglomerular feedback
  mechanism in, 76–87, 89

Back-diffusion, and amino acid
 transport, 225
Barium, and potassuim secretion in
 distal tubule, 191
Baroreceptors. See Stretch receptors
Base

deficit of, 245
excess of, 245
Basement membrane, glomerular, 56–
 58
 filtration of neutral compounds
  affected by molecular size,
  58, 66
Basolateral membranes
 amplification of, 114–115, 140, 164
 in connecting tubule, 175
 electrical measurements in single
  tubular cells, 137
 in intercalated cells, 176
 in principal cells of cortical
  collecting duct, 177
  aldosterone affecting, 203
 in proximal tubule, 151
  glucose transport across, 218, 221
  selectivity of ion permeability in,
   152
  sodium-potassium exchange in,
   152–153
Bence Jones proteins, tubular
 reabsorption of, 230
Bicarbonate
 as buffer, 241
 cellular pool of, 250
 exchange with chloride in proximal
  tubule, 149
 excretion of, 250
 in fluid compartments, 14, 17–18
 Henderson-Hasselbalch equation for,
  243
 infusions in phenobarbital poisoning,
  261
 net flux of, 249
 and pH diagram, 243
 plasma concentrations of, 4, 68
  and potassium secretion in distal
   tubule, 191
  and reabsorption of bicarbonate,
   255–256
  standard content of true plasma,
   245, 248–250
 reabsorption of
  adrenocortical hormones affecting,
   258
  apparent capacity for, 255–256
  and carbon dioxide partial pressure
   in arterial blood, 256–257
  chloride concentrations affecting,
   258
  controlled, 254–255
  in distal tubule and collecting
   duct, 254–255
  factors affecting, 255–258
  maximum capacity for, 256
  potassium concentrations affecting,
   257–258
  in proximal tubule, 143, 144,
   148–149
   obligatory, 253–254
   volume expansion affecting, 156
  titration curve for, 256
 in ultrafiltrate, 68
Binding site saturation, in solute
 transport, 16
Biological control system, 353
 closed-loop, 355–356
 gain in, 358–359

Biological control system (*cont.*)
  computer studies of, 363–366
  controlled element in, 355
  controller signals in, 356–358
    integral, 357–358
    proportional, 356–357
    rate-sensitive, 356
  controlling element in, 355
  description of, 353–354
  dynamic characteristics of, 353, 360–362
  feedback element in, 355
  input in, 353
    direct, 353
    indirect, 353
    relation to output, 354–355
  mathematical description of, 358–363
  open-loop, 355–356
    gain in, 358
  output in, 353
    relation to input, 354–355
  regulator in, 356
  servomechanism in, 356
  solution of differential equations for, 362–363
  static characteristics of, 353, 358–360
  steady-state response in, 354
    error in, 357
  transfer function in, 358, 360
  transient response in, 354
Blood
  buffers in, 240–241
  water in, 10
Blood flow, renal
  algebraic expression of, 159
  autoregulation of, 73–89. *See also* Autoregulation, renal
  and cardiac output, 314
  distribution in cortex and medulla, 106–108
    measured with inert diffusible gas, 106–107
      tail subtraction technique in, 106
    measurement with radioactive microspheres, 107–108
  extrinsic regulation of, 89–92
  in glomerular capillaries, 68–70
  measurement of, 107
  and glomerular filtration rate, 73–74
  humoral mechanisms in, 90–91
  medullary, 106–108
    and concentration of urine, 296–298
  neural mechanisms in, 89–90
Blood pressure
  arteriolar, and hemodynamics of glomerular circulation, 68–69
  glomerular capillary, 59, 312
    and filtrate reabsorption in proximal tubule, 313
  renal artery
    and autoregulation of renal blood flow, 75–76, 85–86, 88
    and glomerular filtration rate, 74
    and renin release, 88
    and sodium excretion rate, 158

systemic
  and glomerular filtration rate, 69
  monitoring by stretch receptors, 314, 321
  renal denervation affecting, 90
  renin-angiotensin system affecting, 88
  and renin release, 334
Blood supply of kidney, 44–48
Blood-uring mass balance, 104
Blood volume regulation, 1–7, 310–315, 325–342. *See also* Fluid regulation
Bode plot analysis, 360–362
Body fluid regulation, 1–7, 310–315, 325–342. *See also* Fluid regulation
Boltzman's constant, 118
Bone, water in, 10
Bowman's capsule, 53, 54
Bowman's space, 53
Bradykinin
  and renal blood flow, 64
  transport of, 231
Brain, water content of, 10
Brush borders in tubular luminal membrane, 114, 141
  carbonic anhydrase in, 253
  glucose binding with, 221
  proteases in hydrolysis of small peptides, 231
Buffering
  active, 241
  passive, 241
Buffering capacity, 241
Buffers in blood, 240–241
  chemical, 240
  titration curve of, 240
Burns
  acidosis in, 251
  fluid loss in, 311

Calcitonin
  action on thick ascending limb of Henle's loop, 172, 173
  and transport in distal tubule and collecting duct, 182, 204
Calcium
  action in feedback response in glomerular filtration regulation, 83
  and activation of angiotensin-converting enzyme, 329
  and aldosterone secretion, 199
  clearance of, 232
  in fluid compartments, 14
  and glucose binding with luminal cell membrane, 221
  plasma concentrations of, 4, 68
    and phosphate levels in plasma, 224
  reabsorption of, parathyroid hormone affecting, 182
  and renin release, 336
  as second intracellular messenger, 199
  and smooth muscle contraction in afferent arterioles, 70
  in ultrafiltrate, 68

Calmodulin antagonist, affecting renin release, 336
Capillaries of kidney
  continuous, 30
  discontinuous (sinusoid), 30
  fenestrated
    glomerular, 48, 54
    peritubular, 48
    transport in, 30–31
  filtration flow in, 25–26, 29
  fluid exchange in, 23–33
    hemodynamic factors in, 25–29
    lymphatic system in, 31–32
    paracellular route in, 31
    pressure-flow relationships in, 26–28
    sieving process in, 23
    temporal and spatial factors in, 29
    and tissue fluid formation, 25–26
    transcellular routes in, 30–31
      fenestrae in, 30–31
      transmembranous, 30
      vesicles in, 30
  glomerular, 30, 42, 53. *See also* Glomerulus
  peritubular, 30, 48
  permeability of, 29–31
    alterations in, 33
  tight junctions of endothelium in, 31
Capillary pressure
  hydrostatic, 25
  isogravimetric, 27
Carbon dioxide
  absorption curve of true plasma, 241–245
  and bicarbonate content of body, 17
  elimination by lungs, 240, 258
  inhalation of, respiratory acidosis from, 265–266
  partial pressure of
    in arterial blood, 246
    and bicarbonate reabsorption, 256–257
    in inspired air, 246
    isobar plotting of, 243
    relation to pH, 242
    and standard bicarbonate content, 245
Carbonic acid, 240
  formation of, 253
Carbonic anhydrase
  and bicarbonate reabsorption in proximal tubule, 148–149
  in brush border of luminal membrane, 253
  in epithelia of thin limbs of Henle's loop, 164
  inhibitors of, 303–304
    sites of action of, 303
  in intercalated cells, 254
  in principal cells of outer medulla, 254
Cardiovascular system, and body fluid homeostasis, 311–312, 314–315
Carotid sinus stretch receptors, 314, 321–323
Carrier-mediated transport, 15–16, 117–130
  active, 127–129

of amino acids, 225–226, 227
of *p*-aminohippurate, 233
bimoleclar reactions in, 119–120
of citrate, 228
cotransport, 129–130, 142
downhill system in, 129–130, 142
drug or hormone-receptor interactions in, 123–124
enzyme-substrate complex in, 120–123
facilitated, 124–127
of glucose, 219, 221
of phosphate, 222, 224
of sulfate, 225
unimolecular reactions in, 118–119
uphill systems in, 128–129, 142
Carrier-substrate complex, 125
Cartilage, fluid volume in, 12
Catecholamines
and activity of Na-K-ATPase, 153
and filtrate reabsorption in proximal tubule, 161–162
renal vasoconstriction from, 91
and renin release from granular cells, 91, 337, 338
Cations
in fluid compartments, 13, 14
in plasma and ultrafiltrate, 68
Cellular fluid exchange, 33
Cerebrospinal fluid, secretion and absorption of, 32
Charge, molecular, and filtration of compounds affected by size, 58, 66
Chemical potential of water, 3
Chemical reactions
bimolecular, 119–120
first-order, 118
reversible, 118–119
unimolecular, 118–119
zero-order, 118
Chloride
and activation of angiotensin-converting enzyme, 329
clearance of, 232
cotransport with potassium, in luminal membrane of distal tubule, 189
cotransport with sodium, in thick ascending limb of Henle's loop, 169–171
exchange with bicarbonate, in proximal tubule, 149
in fluid compartments, 14, 17
macula densa receptors for, and renin release, 335, 339
movement across membranes, 13–15
permeability for
in ascending thin limb of Henle's loop, 289
in cortical collecting dust, 186
in proximal tubule, 152
plasma concentrations of, 4, 68
and bicarbonate reabsorption, 258
role in tubuloglomerular feedback mechanism, 83, 85
transport and reabsorption of
in distal tubule, 186
in proximal tubule, 143, 144, 149–150

in various nephron segments, 168
in tubular fluid, and renin release, 337, 343
in ultrafiltrate, 68
Chloride shift, 258
Chlorothiazide, 304, 305
and potassium secretion in distal tubule, 192
Cholesterol, aldosterone biosynthesis from, 197
Cirrhosis
aldosterone levels in, 332
glomerular filtration rate in, 70
and plasma renin substrate concentrations, 328
and pressor responsiveness to angiotensin, 330
Citrate
excretion of, 228
titration curve for, 224
transport and reabsorption of, 228
Clearance, 96–108
extraction ratio in, 97–98
glomerular filtration rate in, 98–104
osmolar, 299–300
permeability ratio in, 96
plasma volume flow in, 98
of solute-free water, 300–301, 317
total partition of, 100–101
tubular component in, 100
Clearance ratio, 215
Closed-loop control system, 355–356
Collagen, in glomerular basement membrane, 57
Collecting ducts, 41, 42–43, 173–204, 274, 277
acidification of urine in, 258
aldosterone affecting, 202–203
ammonia synthesis in, 260
basal infoldings in, 115
bicarbonate reabsorption in, 254–255
chloride transport and reabsorption in, 179, 186
cortical
morphology of, 176–177
transport characteristics of, 168
diuretics affecting, 305–306
lateral intercellular spaces in, 114
medullary
inner, 177–178
outer, 177–178
nomenclature for, 370
potassium reabsorption from, 185
potassium secretion in, 185
mechanism of, 186–192
in regulation of blood volume and osmolality, 314
role in concentration of urine, 289–290, 299
sodium-hydrogen exchange in, 255
sodium transport and reabsorption in, 183–184
urea recycling through, 182, 294–295
Colligative properties of solutions, 3
Competitive acceleration, 125
Competitive inhibition
of *p*-aminohippurate, 233
of carrier-substrate complex, 127
of enzyme-substrate complexes, 122

and glucose reabsorption, 219
Compliance, interstitial, 29
Computer studies of control system responses, 363–366
Concentration gradient
of carrier-substrate complex in membrane, 125
and diffusion of nonpolar molecules, 117
Concentration ratio, tubular fluid-to-plasma, 134
Concentration of urine, 273. *See also* Urine, concentration of
Conn's syndrome, alkalosis in, 258
Connecting tubule, 42
morphology of, 175–176
potassium secretion in, 185
mechanism of, 186–192
sodium pump and glycolytic enzyme activity in, 182
Connective tissue, fluid volume in, 12
Control systems, biological, 353. *See also* Biological control system
Convection, 15
and bidirectional flow of fluid, 23
Corpuscle, renal, 41–42, 50, 53. *See also* Glomerulus
Cortex, renal, 41
architectural pattern of, 50–51
blood flow in, 73, 106–108
Cortico-medullary border, 44
Cortisone levels
and bicarbonate reabsorption, 258
and phosphate reabsorption, 224
Cotransport, 16, 129–130, 142–143
chloride-potassium, 189
phosphate and sodium, 224
sodium-amino acid, 130, 142, 227
sodium and *p*-aminohippurate, 233
sodium and chloride, 169–171
sodium and glucose, 129, 142, 218, 220–221
sodium and water, 150–152
Countercurrent exchange, 282–283
between ascending vasa recta and descending proximal tubules, 271
between descending and ascending vasa recta, 272, 276
and medullary blood flow, 296–297
and oxygen concentrations, 298
and recycling of urea, 295
Countercurrent multiplier system in Henle's loop, 280–281, 313
and concentration of urine, 291–292, 298
experimental evidence for, 283–285
model of, 285–287
Countertransport, 125
and bicarbonate reabsorption in proximal tubule, 143
Coupled transport. *See* Cotransport
Creatinine, renal clearance of, 102, 103–104, 232
Cushing's syndrome, alkalosis in, 258
Cycloheximide
and aldosterone-induced sodium transport, 202
and antinatriuretic effect of aldosterone, 200

Cysteine, transport and reabsorption of, 227

Dark cells, intercalated, 175
Darrow-Yannet diagram, 37–38
Dehydration, 36–37
   and acid-base balance, 311
   hypertonic, 6, 36–37
      and potassium secretion in distal tubule, 191
   hypotonic, 36
   isotonic, 6, 36
Denervated kidney, 90, 91
   renin levels in, 337
   and sodium excretion, 161
Deoxycorticosterone acetate
   and conductances of sodium and potassium, 203
   and potassium secretion in distal tubule, 190
Desmosomes
   in cortical collecting duct, 177
   in lateral intercellular spaces, tubular, 114
Deuterium oxide, in total body water measurements, 9, 10
Dexamethasone, and sodium-hydrogen exchange, 149
Diabetes insipidus
   central, 301
      water diuresis in, 317
   effects of osmotic diuretics in, 302
   nephrogenic, 301
Diabetes mellitus
   acidosis in, 251
   glucosuria in, 219
   production of nonvolatile acids in, 240
Diarrhea, acidosis in, 251
Dichlorophenamide, and potassium secretion in distal tubule, 190
Diet
   and potassium secretion in distal nephron, 188
   protein in. See Protein, dietary
   sodium in. See Sodium, dietary
Diffusion, 2, 15
   and bidirectional flux of water and electrolytes, 23
   exchange, 138
   facilitated, 16, 124–127
      carrier-substrate complexes in, 125–126
      carrier saturation affecting, 126–127
      competitive inhibition affecting, 127
      substrate concentration affecting, 126
   Fick's law of, 116
   nonionic, of ammonia, 261
   passive, in proximal tubule
      and potassium transport, 148
      and sodium transport, 146
   in tubular transport, 114, 116–117
      nonpolar molecules in, 117
      polar molecules in, 117
Diffusion coefficient, 116
Dilution principle
   for filtrate volume flow rate, 99
   in fluid volume measurements, 8–9
   mathematical basis of, 367–369
   in solute concentration measurements, 9
Dilution of urine, 270, 293
   measurement of, 298–301
Dissociation constant
   for carrier-substrate complex, 125
   for enzyme-substrate complex, 121
Distal tubules, 173–204
   bicarbonate reabsorption in, 254–255
   chloride transport and reabsorption in, 186
   convoluted part of, 42, 44, 51, 173
      morphology of, 173–175
      transport characteristics of, 168
   diuretics affecting, 305–306
   fluid composition affecting potassium secretion, 187–190
   nomenclature for, 370
   osmolality of filtrate in, 180
   phosphate reabsorption in, 222
   potassium transport in, 185–186
   protein reabsorption in, 230
   in regulation of blood volume and osmolality, 314
   sodium transport and reabsorption in, 180–185
   straight part of, 44, 51, 162, 173, 274
   transport and reabsorption in, 178–186
      aldosterone affecting, 196–204
      antidiuretic hormone affecting, 192–196
      calcitonin affecting, 204
      facultative, 185
      hormonal regulation of, 192–204
      parathyroid hormone affecting, 204
   urea recycling through, 294–295
   uric acid reabsorption in, 230
Diuresis, 301–302
   controlled, 5
   mannitol-induced, 155
   osmotic, 301–302
      kaliuresis in, 190
      sodium excretion in, 182–183
   pressure, 74–75, 158
   urea-induced, 155
   water, 301, 317
Diuretics, 302–306
   classification of, 302
   osmotic, 301, 302–306
   potassium-sparing, 306
   and potassium secretion in distal tubule, 192
   and reabsorption in Henle's loop, 173
   sites of action, 302–306
      in distal tubule and collecting duct, 305–306
      in Henle's loop, 304–305
      in proximal tubule, 303–304
   xanthine, 306
Dopamine
   glomerular receptors for, 64
   natriuresis from, 162
   renal vasodilation from, 91
   and renin release in hypertensive dogs, 336
   and renin secretion, 338
Drag
   solute, 15
   solvent
      and chloride reabsorption in proximal tubule, 143
      and potassium reabsorption in proximal tubule, 147
Drinking behavior, thirst center affecting, 317
Drug-receptor interactions, 123–124

Edema, 32–33
   exchangeable sodium pool in, 16–17
   extracellular, 33
   interstitial volume in, 24, 32
   intracellular, 33
Effluxes
   and dehydration. See Dehydration
   disturbances in, 33
   hypertonic, in inappropriate secretion of antidiuretic hormone, 5
   hypotonic, 6
   isotonic, 6
   and net flux across membrane, 116
   regulation of, 21–33
   venous and urinary, and arterial influx, 104
Electrical potential, transepithelial, 136–138
   in collecting ducts, 186
   in distal convoluted tubule, 182
   and transport in proximal tubule, 143, 145–146
   in various nephron segments, 168
Electrochemical gradient, in active transport of solutes, 16
Electrodiffusion
   chloride, in proximal tubule, 149
   potassium, in proximal tubule, 148
Electrogenic transport of sodium, 182, 200
Electrolytes
   distribution in fluid compartments, 13–18
   disturbances in balance of, 33–35
   plasma concentrations of, 4
Electroneutral transport
   chloride, in proximal tubule, 150
   sodium chloride
      peritubular Starling forces affecting, 157
      in proximal tubule, 143, 145
      in thick ascending limb of Henle's loop, 169
Electroneutrality in fluid compartments, 13
   maintenance of, 14–15
Electrostatic properties of glomerular capillary wall, and filtration of compounds affected by size, 58, 66
Emphysema, hypoventilation in, 265
Endocytosis, and protein reabsorption, 230–231
Endolysosomes, protein digestion in, 231

Endopeptidase, 330
Endothelium of glomerular capillaries, 54
Enzyme-substrate complex, 120–123
 dissociation constant of, 121
 inhibitors of, 122–123
  competitive, 122
  noncompetitive, 122–123
  uncompetitive, 122
 irreversible dissociation of, 120
 reaction velocity in, 120
  inhibitors affecting, 122–123
  Lineweaver-Burk plots of, 121, 122, 123
  pH affecting, 122
  temperature affecting, 121–122
Epinephrine
 and plasma renin activity, 337
 renal vasoconstriction from, 91
 and renin secretion, 338
Epithelia, tubular
 leaky, 113, 141, 152
 tight junctions in, 112–113, 141, 151
 permeability of, 113
Epithelial cells, glomerular, 53, 54–56
Equilibrium
 filtration, 59
 filtration pressure, 60
Equilibrium constant, 119
 and splay in titration curve, 226
Equivalent concentrations, 3
Ethacrynic acid, 304–305
 and plasma renin activity, 338
 and potassium secretion in distal tubule, 192
 and sodium chloride transport in Henle's loop, 173
Exchange diffusion, 138
Exchangeable solutes, measurement of, 9
Exercise
 lactic acid production in, 240, 246
 and metabolic acidosis, 251
 proteinuria after, 66
 and renal sympathetic activity, 337
 and water loss in sweating, 315
Extracellular fluid
 electrolyte content of, 13, 14
 volume of, 11–12
  in fast-equilibrating phase, 11–12
  increased, 5
   and bicarbonate reabsorption, 256
   and glucose reabsorption, 220
   and potassium secretion in distal tubule, 190
   and reabsorption in proximal tubule, 155–157
   and sodium excretion, 107
  reduction of. See also
   Dehydration; Hemorrhage and excretion of concentrated urine, 107
  regulation of. See Fluid regulation
  in slow-equilibrating phase, 12
Extraction ratio, 97–98, 104
 ratio to filtration fraction, 105

Facilitated transport, 16, 124–127
Facultative reabsorption, in distal tubule, 185
Fanconi's syndrome, 226
Faraday constant, 14, 129
Fat, excess, determination of, 10–11
Feedback mechanism
 short loop in angiotensin II formation, 91
 tubuloglomerular, in renal autoregulation, 76–87
Fenestrated capillaries
 glomerular, 48, 54
 peritubular, 48
  transport in, 30–31
Fever
 hyperventilation in, 265
 insensible loss of water in, 22
Fibrinogen, formation of, 18
Fibronectin
 in mesangial matrix, 54
 in glomerular basement membrane, 57
Fick equation, 104
Fick's law of diffusion, 116
Filtered load, 100
Filtrate, glomerular, composition of, 66–68
 and permeability of ultrafilter, 96
Filtrate-to-plasma concentration ratios, for various substances, 66
Filtrate-to-plasma osmolality ratio, in proximal tubule, 143
Filtrate-urine mass balance, 99
Filtration equilibrium, 59
Filtration flow of fluid, 115
 inward, 26
  and hemodilution, 29
  oncotic absorptive pressure in, 26
 outward, 25–26
  and hemoconcentration, 29
  hydrostatic pressure in, 26
Filtration fraction, 58, 159
 ratio to extraction ratio, 105
 single-nephron, 61–63, 76
Filtration pressure
 effective, and glomerular filtration rate, 312, 343
 equilbruim, 60
Filtration rate, glomerular 58–71. See also Glomerlar filtration
Filtration slit, glomerular, 56
First-order chemical reaction, 118
 reversible, 118–119
Flow diagram of renal-body fluid regulating system, 1–2
Fluid
 administration of
  and hydration, 35–36
  precautions with, 34–35
 electrolyte content of, 13–18
 formation in tissues, 25–26
 mosaic model of cell membrane, 117
Fluid exchange
 capillary, 23–33
 cellular, 33
 external, 21–23
 in gastrointestinal system, 21
 internal, 23–33

 in kidneys, 22–33
 in metabolic system, 21–22
 in skin and lungs, 22
Fluid-to-plasma osmolality ratio, 283–285
Fluid regulation, 1–7, 310–315
 baroreceptors in, 321–322
 blood volume in, 325–342
 cardiovascular system in, 311–312, 314–315
 distal tubule and collecting duct in, 314
 disturbances in, 33–35
 effluxes in, 5–6
 extrarenal factors in, 314–315, 344–345
 flow diagram of, 1–2
 gastrointestinal system in, 311, 314
 influxes in, 5
 intrarenal mechanism in, 312–314, 342–344
 loop of Henle in, 313–314
 neuroendocrine system in, 312, 315
 osmolality versus volume in, 315
 proximal tubule in, 313
 renin-angiotensin-aldosterone system in, 326–340
 right atrial stretch-sensitive receptors in, 340
Fluid volume, 8–13
 cartilage, 12
 in dense connective tissue, 12
 extracellular, 11–12
 interstitial, 12
 intracellular, 13
 measurements of, 8–9
  constant-infusion equilibrium method in, 9
  dilution principle in, 8–9
  instantaneous concentration in, 9
  priming dose in, 9
  single dose injection method in, 8
 plasma, 12
 shifts of, 24
  Darrow-Yannet diagram of, 37–38
 total body water, 9–11
 transcellular, 13
Flux
 of solutes across membrane, 116
 net, 116
 unidirectional, 116
 of substrate and carrier-substrate complex, 126
Foot processes of podocytes, 56
Forcings, 354–355
 of pulmonary pH regulator, 247
 pulse, 5, 355
 of renal pH regulator, 251, 263
 step, 5, 354
Fractional solute clearance, and glomerular filtration rate, 65–66
Fructose transport, 219
Fumarate, and citrate reabsorption, 228
Function of kidney, 51
Furosemide, 304–305
 and feedback response in glomerular filtration regulation, 83
 and potassium secretion in distal tubule, 192

Furosemide (*cont.*)
  sites of action of, 303
  and sodium chloride cotransport in thick ascending limb of Henle's loop, 169, 170, 173

Galactose transport, 219
Gap junctions, in juxtaglomerular apparatus, 79
Gas exchanger, pulmonary, 246
Gastrointestinal system, in fluid and electrolyte homeostasis, 21, 311, 314
Gene transcription, 161, 202
  actinomycin D affecting, 190
Gibbs-Donnan effect, 13, 25
Gibbs-Donnan principle, 13–14
Gibbs-Donnan ratio, 67
Globulin, plasma, 18
  ratio to albumin, 18
Glomerular filtration, 58–71
  afferent arteriole plasma protein affecting, 63
  and barrier function, 57–58
  and bicarbonate reabsorption, 256
  and creatinine clearance, 103
  determinants of, 58–61
    factors affecting, 61–65
  and effective filtration pressure, 312, 343
  and fractional solute clearance, 65–66
  and glomerular filtration coefficient, 61, 62–63
  glomerular plasma flow rate affecting, 62
  and glucose reabsorption, 219–220
  hemodynamics of, 58–61
  historical background of, 58
  hormones and vasoactive agents affecting, 64–65
  macromolecules in
    charge affecting, 58, 66
    disease affecting, 66
    size affecting, 58, 65–66
  measurement of, 98–104
  and protein reabsorption, 231
  and reabsorption of filtrate
    in Henle's loop, 171–172
    in proximal tubule, 313
  regulation by juxtaglomerular apparatus, 81–89
  renal blood flow affecting, 73–74
  single-nephron, 59, 68
  and sodium reabsorption in proximal tubule, 153, 157–158
  and total renal clearance, 100–101
  tubuloglomerular feedback affecting, 337
  and urea clearance, 102–103
Glomerular filtration coefficient
  and glomerular filtration rate, 61, 62–63
  renal nerve stimulation affecting, 90
Glomerulonephritis, experimental, and fractional clearance of dextran sulfate, 66
Glomerulosclerosis, intercapillary, 219
Glomerulotubular balance, 85, 153

extracellular volume affecting, 256
for glucose, 220
Glomerulus, 42, 53
  blood flow in, 68–70
    measurement of, 107
    renal nerve stimulation affecting, 90
  permselectivity of, 58, 65–66
  structure of, 53–58
Glucagon, tubular reabsorption of, 230
Glucose
  cotransport with sodium, 129, 142, 218, 220–221
  filtrate-to-plasma concentration ratio for, 66
  infusions of
    and amino acid transport, 226
    diuresis from, 302
    and phosphate reabsorption, 223
    and sulfate reabsorption, 225
  maximum tubular reabsorptive capacity, 217
    factors affecting, 219–220
  renal clearance of, 101
  renal threshold concentration for, 218
  titration curves for, 217–218
  transport and reabsorption of, 217–222
    cellular mechanism of, 220–222
    kinetics of, 218–219
    plasma levels of glucose affecting, 218
  tubular concentration of, 132–133
Glucosuria, 218, 219
Glutamic acid, transport and reabsorption of, 227
Glutamine, ammonia production from, 260
Glycine, titration curve for, 226
Glycolytic enzyme activity
  in connecting tubule, 182
  in distal convoluted tubule, 182
Goldblatt units, for plasma renin activity, 327
Golgi apparatus, in proximal tubule epithelium, 141
Gout, uric acid levels in, 229, 230
Granular cells, juxtaglomerular, 80, 198
  renin synthesis and release in, 81, 326, 343
Gravity, specific, 10–11
Growth hormone, tubular reabsorption of, 230

Heart, water content of, 10
Heart failure
  fluid balance in, 33
  and pressor responsiveness to angiotensin, 330
Hemoconcentration, 29
Hemodilution, 29
Hemoglobin
  analysis with co-oximeter, 245
  as buffer, 241
  filtrate-to-plasma concentration ratio for, 66
  tubular reabsorption of, 230
Hemorrhage

blood loss in, 311
and excretion of concentrated urine, 107
and pressor responsiveness to angiotensin, 330
renin release in, 334
Henderson-Hasselbalch equation, 240–241, 242
  for bicarbonate in plasma, 243
Henle's loop, 42, 271
  ammonia reabsorption in, 260
  cortical, 162
  countercurrent multiplier system in, 283–287
  diuretics affecting, 304–305
  function of, 171
  long, 162, 287
  morphology of, 162–166
  phosphate reabsorption in, 222
  in regulation of blood volume and osmolality, 313–314
  short, 162, 287
  thick ascending limb, 44, 77, 164–166, 274
    cortical part, 165, 166, 289
      transport and reabsorption in, 171
    as diluting segment, 299
    medullary part, 165, 166, 289
      transport and reabsorption in, 170–171
    role in concentration of urine, 289
    sodium chloride reabsorption from, 289, 314
    transport characteristics in, 168, 169–171
  thin limbs of, 162–164
    ascending, 44, 164, 277, 287
      role in concentration of urine, 289
      transport characteristics in, 168, 169
    descending, 44, 162–164, 277, 288
      lower segment of, 288
        role in concentration of urine, 288
      transport characteristics in, 168
      upper segment of, 288
      urea recycling through, 182, 294, 295–296
      potassium re-entry into, 185
  transport and reabsorption in
    diuretics affecting, 173
    general characteristics of, 166–171
    glomerular filtration rate affecting, 171–172
    hormones affecting, 172–173
    regulation of, 171–172
    in thick ascending limb, 169–171
      in cortical portion, 171
      in medullary portion, 170–171
    in thin segments, 168–169
Henry's law of gas solubility, 242
Heparin, hypoaldosteronism from, 199
Heparin sulfate, in glomerular basement membrane, 57
Hilus, renal, 41
Histamine
  glomerular receptors for, 64
  and renin release, 334

Histidine, transport and reabsorption of, 227
Homeostasis of fluids and electrolytes, 21–33, 310
Hormone(s)
  affecting glomerular filtration rate, 64–65
  affecting transport and reabsorption
    in distal tubule and collecting duct, 184–185, 192–204
    in Henle's loop, 172–173
    in proximal tubule, 160–162
Hormone-receptor complex
  interactions in, 123–124
  thyroid hormone in, 161
Hydration, 35–36
  hypertonic, 5, 36
  hypotonic, 5, 35–36
    and potassium secretion in distal tubule, 191
  isotonic, 5, 35
  and urea clearance, 102
Hydraulic pressure, transcapillary
  and glomerular filtration rate, 59–61, 63
  renal nerve stimulation affecting, 90
Hydrocephalus, 32
Hydrogen
  secretion of
    aldosterone affecting, 202
    in intercalated cells, 175, 184
  sodium-hydrogen exchange
    and bicarbonate reabsorption, 144, 259
    in collecting duct, 255
    hormones affecting, 160
    and potassium secretion in distal tubule, 191
    in proximal tubule, 145, 149, 253, 255
Hydrostatic pressure
  capillary, 25
  interstitial, 25
    and lymphatic drainage, 31–32
    negative, 25, 28–29
β-Hydroxybutyric acid
  production in diabetes, 240
  tubular reabsorption of, 229
Hydroxyproline, transport and reabsorption of, 227
Hypertension
  factors in development of, 326
  renal denervation affecting, 90
Hypertonic dehydration, 6, 36–37
Hypertonic hydration, 5, 36
Hypertonic urine, 300
Hyperventilation, 264, 265
  blood reactions in, 245
Hypophysectomy, and plasma renin substrate concentrations, 328
Hypotension, and renin release, 334
Hypothalamus
  osmoreceptors in, 193, 312, 316–317
  thirst center in, 317
Hypotonic dehydration, 36
Hypotonic hydration, 5, 35–36
Hypotonic urine, 300
Hypoventilation, 264, 265
  blood reactions in, 244
Hysteria, hyperventilation in, 265

Immune globulins, 18
Impulse, as forcing type, 355
Indomethacin, and effects of prostaglandins on antidiuretic hormone, 89
Influxes. *See also* Hydration
  arterial, and venous and urinary effluxes, 104
  disturbances in, 33
  hypertonic, 5
  hypotonic, 5
  isotonic, 5
  and net flux across membrane, 116
  regulation of, 21–33
Infoldings, basal
  in connecting tubule, 175
  in cortical collecting duct, 177
  in medullary collecting duct, 178
  in thin limbs of Henle's loop, 164
  in tubular epithelia, 114–115, 140
Inner stripe of renal medulla, 44, 272
Insulin
  and activity of Na-K-ATPase, 152
  glomerular receptors for, 64
  tubular reabsorption of, 230
Integral controller, in control system, 357–358
Intercalated cells
  carbonic anhydrase in, 254
  in connecting tubule, 175
  in cortical collecting duct, 175, 176
  hydrogen secretion in, 175, 184
  in outer medullary collecting duct, 175, 178
  potassium secretion associated with, 186
Interdigitation, cellular
  in distal convoluted tubule, 173
  in proximal tubule, 115, 140
  in thick ascending limb of Henle's loop, 166
  in thin limbs of Henle's loop, 164
Interlobar arteries, 44
Interlobular arteries, 44, 45, 50
Interlobular veins, 45, 48, 50
Interstitial cell, lipid-laden, 277–279
Interstitial fluid, 1
  balance with plasma, 23–31
  electrolyte content of, 13, 14
  turnover time of
    in convective flux, 24
    in diffusive flux, 23
  volume of, 12
    in edema, 24, 32
    shifts of, 24
Interstitium, renal, 49
  in outer stripe of outer medulla, 272
  periarterial, 49
  peritubular, 49
Intestines, water content of, 10
Intracellular fluid, 2
  electrolyte content of, 13, 14
  volume of, 13
Intracranial pressure, saline infusions affecting, 32
Inulin
  clearance of, 65, 99–100, 103, 105–106, 215
  in extracellular volume measurements, 11

filtrate-to-plasma concentration ratio for, 66
tubular fluid-to-plasma concentration ratio for, 134
Iodopyracet, maximum tubular reabsorption capacity for, 234
Ion exchange, 2
Isogravimetric capillary pressure, 27
Isohydric principle, and buffering capacity, 241
Isoleucine, transport and reabsorption of, 227
Iso-osmotic solutions, 4
Iso-osmotic reabsorption in proximal tubule, 143, 146
Isoproterenol
  renal vasodilation from, 91
  and renin release in hypertensive dogs, 336
  and renin secretion, 338
Isotonic dehydradion, 6, 36
Isotonic hydration, 5, 35
Isotonic solutions, 4

Juxtaglomerular apparatus
  and feedback regulation of glomerular filtration rate, 81–85
  granular cells of, 80, 198
  renin synthesis and release in, 81, 326, 343
  morphology of, 77–81
Juxtamedullary nephrons, 44, 271

α-Ketoglutarate, transport and reabsorption of, 229
Krypton (^{85}Kr), in measurements of renal blood flow distribution, 106

Labyrinth, cortical, 44, 50
  convoluted tubules in, 44, 51
Lactic acid
  production in exercise, 240, 246
  and metabolic acidosis, 251
  tubular reabsorption of, 229
Laminin, in glomerular basement membrane, 57
Lanthanum, and renin release, 336
Laplace transformation, 361
Lateral cell processes
  in distal convoluted tubule, 173
  in thick ascending limb of Henle's loop, 164
Lateral intercellular spaces, in cortical collecting duct, 176–177
Lean body mass, 10
Leucine, transport and reabsorption of, 227
Lineweaver-Burk plots of reaction velocity, in enzyme-substrate complex, 121, 122, 123
Lipid bilayer, membrane, 117
Lipid-laden interstitial cells, in inner medulla, 277–279
Lissajous' plot, 360
Liver
  cirrhosis of. *See* Cirrhosis

Liver (cont.)
  in osmoregulation, 317–319
  water content of, 10
Luminal membrane
  amplification of, 114, 140, 164
  brush borders in, 114, 141
    carbonic anhydrase in, 253
    glucose binding with, 221
    proteases in hydrolysis of small peptides, 231
  in connecting tubule, 175
  in cortical collecting duct, 177
    aldosterone affecting, 203
  in distal convoluted tubule, 173
  electrical measurements in, 137
  in proximal tubule, 151
    in control of filtrate reabsorption, 153–155
    glucose interaction with, 221
    selectivity of ion permeability in, 152
Lungs
  angiotensin-converting enzyme in, 329
  carbon dioxide elimination in, 240, 258
  respiratory system in acid-base balance, 245, 246–248
  water content of, 10
Lymphatic system, 31–32, 49
  obstruction of, 34
Lysine
  titration curve for, 226
  transport and reabsorption of, 227
Lysosomes, fusion with endosomes, 231

Macromolecules, glomerular filtration of, 58, 65–66
Macula densa, 77–79
  chloride-sensitive receptors in, 314, 342–343
    affecting renin release, 335, 339
  and diluting function of thick ascending limb of Henle's loop, 289
  receptor theory of renin release, 337
  in tubularglomerular feedback mechanism, 81, 82–83
  physiological role of, 85
Magnesium
  and activation of angiotensin-converting enzyme, 329
  clearance of, 232
  in fluid compartments, 14
  and glucose binding with luminal cell membrane, 221
  plasma concentrations of, 4, 68
  in ultrafiltrate, 68
Malate, and citrate reabsorption, 228
Malonate, and citrate excretion, 228–229
Mannitol, 301–302
  and chloride concentration in proximal tubule, 150
  and distal tubular flow rate and potassium secretion, 190
  in phenobarbital poisoning, 261

and sodium reabsorption in proximal tubule, 154–155
Mass balance
  blood-urine, 104
  filtrate-urine, 99
  plasma-urine, 104
Medulla, renal, 41
  blood flow in, 73, 106–108
    affecting concentration of urine, 296–298
    functional coupling or separating of routes for, 297–298
  cortico-medullary border, 44
  inner, 276–280
    urine concentration in, 290, 291
  outer
    inner stripe of, 44, 272
    outer stripe of, 44, 271–272
    urine concentration in, 290, 291
  potassium recycling in, 185
  structural organization of, 270–280
  urea recycling in, 294–296
Medullary rays, 44, 50–51, 271
  straight tubules in, 44, 51
Membranes. See also Basolateral membranes; Luminal membrane
  amplification of, 114–115, 140
  lipid bilayer of, 117
  permeability of. See Permeability
  semipermeable, 117
  thickness affecting fluxes
    of solutes, 116
    of substrate and carrier, 126
Mesangial cells, 53–54, 343
  extraglomerular, 53, 77, 79–80
  role in golmerular filtration, 64, 65
Metabolism, and fluid balance, 21–22
Methionine, transport and reabsorption of, 227
Michaelis constant, 120–121
Microcirculation, anatomy of, 24
Microfolds, tubular
  in basolateral membrane, 114
  in lateral intercellular spaces, 114, 140
  in luminal membrane, 114
Microperfusion studies
  flow-through, 136
  stop-flow, in situ, 135–136
Micropuncture collection of tubular fluid, 133–135
Microspheres, radioactive, in regional blood flow measurements, 107
Microvilli
  in basolateral membrane, 114
  in brush borders, 114
  in cortical collecting duct, 177
  in distal convoluted tubule, 173
  in lateral intercellular spaces, 114
  in luminal membrane, 114
  in thick ascending limb of Henle's loop, 166
Mitochondria, in interdigitating cell processes, 115, 140
Molar concentrations, 3
Molecular size
  and concentrations in ultrafiltrate and plasma, 96

and diffusion of solutes, 117
and flux of solutes across membrane, 116
and glomerular filtration of macromolecules, 58, 65–66
Muscle, water in, 10
Myogenic mechanism, in renal autoregulation, 75–76, 89
Myoglobin, filtrate-to-plasma concentration ratio for, 66

Natriuresis
  diuretic-induced, 302
  dopamine-induced, 162
  in expansion of extracellular fluid volume, 107
  mannitol-induced, 155
  renal perfusion pressure affecting, 158
  saline-induced, 154, 156, 182–183, 301
  urea-induced, 155
Natriuretic factor, atrial, 154, 204
Natriuretic hormone, 154, 301
Nephrectomy, and plasma renin substrate concentrations, 328
Nephrons, 41–42
  cortical loops in, 44
  function of, 51
  juxtamedullary, 44
  long loops in, 44
  short loops in, 44
  types of, 44
Nephrosis, and pressor responsiveness to angiotensin, 330
Nephrotic syndrome, glomerular filtration rate in, 70
Nernst equation, 137, 138
Nerves, renal, 49–50, 161
  and renal circulation regulation, 89–90
Neuroendocrine system, and body fluid homeostasis, 312, 315
Neurophysin, 193
Nicotinamide adenine dinucleotide dehydrogenase, and sodium transport, 204
Nitroprusside, and renin release, 334, 336
Nomenclatures, anatomical, 370–371
Nonfiltering kidney model in dogs, 335
Norepinephrine
  and fluid reabsorption in proximal tubule, 162
  glomerular receptors for, 64
  and renal blood flow, 64
  renal vasoconstriction from, 91

Oncotic pressure, protein, 15, 25, 26, 312
  affecting glomerular filtration rate, 59–61, 69, 70
  glomerular filtration rate affecting, 158
  and reabsorption in proximal tubule, 157, 158–160, 313
  reduction of, 34
  and water extraction from descending vasa recta, 297, 298

Open-loop control system, 355–356
Ornithine, transport and reabsorption of, 227
Osmolality, 3
　of glomerular filtrate, 312
　of plasma, 312
　　regulation of, 310–315, 316–325
　of tubular fluid in Henle's loop, 167
　of urine, 22–23, 270, 314
　of vasa recta plasma, 296–297
Osmolality ratio, fluid-to-plasma, 283–285
　in distal convoluted tubule, 180
　in proximal tubule, 143
Osmolar clearance, 299–300
Osmolar concentration of urine, 299
Osmometer, 22
Osmoreceptors
　hepatic, 317–319
　hypothalamic, 193, 312, 316–317
Osmosis, 2, 3
　and water extraction, 273
Osmotic diuresis, 301–302
　kaliuresis in, 190
　sodium excretion in, 182–183
Osmotic pressure, 4
Osmotic threshold, for antidiuretic hormone release, 316
Ouabain
　and intracellular sodium concentration, 203
　and sodium transport and reabsorption
　　in collecting ducts, 184
　　in distal convoluted tubule, 182
　　in thick ascending limb of Henle's loop, 169, 170
Outer stripe of renal medulla, 44, 271–272
Oxidation of food, 21–22
Oxygen
　countercurrent exchange of, 298
　partial pressure of
　　in arterial blood, 246
　　in inspired air, 246
　renal consumption of, and sodium reabsorption in proximal tubule, 146–147
Oxytocin, 192

Papaverine
　and arteriolar resistance, 76
　and renin secretion in hemorrhage, 335
Papilla, renal, 41, 279
Papillary ducts, 42, 43
Paracellular transport route, 112–114, 141, 142, 152
　and capillary fluid exchange, 31
　permeability of, Starling forces affecting, 156–157
Parathyroid hormone
　action on thick ascending limb of Henle's loop, 172
　and filtrate reabsorption
　　in distal tubule and collecting duct, 204
　　in proximal tubule, 160–161
　glomerular receptors for, 64

　and phosphate reabsorption, 223
　and responses in connecting tubule cells, 182
　tubular reabsorption of, 230
Partition coefficient, and diffusion of nonpolar molecules, 117
Passive transport, 112, 117, 214
Pelvis, renal, 41
Perfusion pressure, renal
　and control of renin release, 333–337
　and plasma renin activity, 337
　and sodium excretion rate, 158
Peritubular cell membrane. *See* Basolateral membranes
Permeability
　of capillary wall, 29–31
　of cell membranes, 4
　of glomerular filtration barrier, 57
　of paracellular pathway, Starling forces affecting, 156–157
　of proximal tubule cell membranes, 152
　to substrate-carrier complex, 126
　of thick ascending limb of Henle's loop, 167
　of tight junctions, 113
　of various nephron segments, 168
Permeability coefficient, 116
Permeability ratio, 96
Permselectivity
　and diffusion of solutes, 117
　glomerular, and filtration of macromolecules, 58, 65–66
　and paracellular transport, 112
　Starling forces affecting, 156–157
Peroxisomes, in proximal tubule, 141
pH
　bicarbonate-pH diagram, 243
　of blood, 240
　　and carbon dioxide tension, 242
　　measurement of, 245
　　and potassium secretion in distal tubule, 191
　　pulmonary regulator of, 246–248
　　renal regulator of, 248–252
　error development in, 246, 252
　　correction of, rate of, 253
　intracellular, 250
　　and excretion of acid or alkali, 250, 262
　in proximal tubular fluid, 143, 258
　and chloride reabsorption, 150
　and reaction velocity in enzyme-substrate complex, 122
　relation to pK, 241, 258
　urinary, and ammonia excretion, 258, 261
Phagosomes, amino acid digestion in, 231
Phenobarbital poisoning, treatment of, 261
Phenol red, maximum tubular reabsorptive capacity for, 234
Phentolamine, blocking sympathetic stimulation of renin release, 338
Phenylalanine
　and angiotensin II activity, 329
　transport and reabsorption of, 227
Phlorizin

　and amino acid transport, 226
　and glucose reabsorption, 219
　and phosphate reabsorption, 223
Phosphate
　clearance of, 232
　cotransport with sodium, 224
　daily requirement of, 222
　excretion of, 250, 258
　　circadian rhythm for, 224
　in fluid compartments, 14
　inorganic, as buffer, 241
　maximum tubular reabsorption capacity for, 222
　　factors affecting, 222–224
　plasma concentrations of, 4, 68
　　and reabsorption of phosphate, 222
　and sulfate reabsorption, 225
　titration curves for, 222
　tubular reabsorption of, 222–224
　in ultrafiltrate, 68
Phosphaturia, 222
Phosphoric acid, neutralization of, 258
pK, relation to pH, 241, 258
Plasma, 1
　electrolyte content of, 4, 13, 14
　fluid balance with interstitial compartments, 23–31
　separated, 241
　true, 241
　turnover time for
　　in convective flux, 24
　　in diffusive flux, 23
　volume of, 12
　　shifts of, 24
Plasma flow, renal, 58
　algebraic expression of, 159
　compared to urine flow rate, 104
　and glomerular filtration rate, 60, 62, 64
　measurement of, 98
Plasma-urine mass balance, 104
Podocytes, glomerular, 53, 54–56
　foot processes of, 56
Portal vein, hepatic, osmoreceptors in, 318–319
Posture
　and aldosterone clearance, 332
　and blood volume changes, 320
　and plasma renin activity, 338
　and renal sympathetic activity, 337
Potassium
　and aldosterone secretion, 199, 332
　basolateral membrane permeability for, 152
　and cellular and plasma pH levels, 250
　concentrations in nephron segments
　　in human, 295
　　in rat, 294
　cotransport with chloride, in distal tubule, 189
　depletion of
　　and chloride concentration in proximal tubule, 150
　　in diarrhea, 251
　　in vomiting, 251
　dietary, and secretion of potassium in distal nephron, 188
　in fluid compartments, 14, 17
　kaliuresis in osmotic diuresis, 190

Potassium (cont.)
  Na-K-ATPase pump, 127–128. See also ATPase, sodium-potassium
  plasma concentrations of, 4, 68
    and bicarbonate reabsorption, 257–258
    and secretion of potassium in distal tubule, 191
  recycled in renal medulla, 185
  and renin secretion and release, 339, 341
  secretion in connecting tubule and collecting duct, 185
    in acidosis, 258
    aldosterone affecting, 202
    in alkalosis, 258
    composition of tubular fluid affecting, 187–190
    luminal factors affecting, 187–191
    mechanism of, 186–192
    peritubular factors affecting, 191–192
    sodium and potassium transport inhibitors affecting, 190–191
    and sodium reabsorption, 187
    tubular flow rate affecting, 190
  sodium-potassium exchange
    in collecting duct, 255
    in proximal tubule, 152–153
  transport and reabsorption of
    in cortical collecting duct, 168
    in distal tubule, 185–186
    inhibitors affecting potassium secretion in distal tubule, 190–191
    in proximal tubule, 147–148
  in ultrafiltrate, 68
Potential gradient, electrical, transepithelial, 136–138. See also Electrical potential, transepithelial
Pregnancy, and pressor responsiveness to angiotensin, 330
Pressure, oncotic, 15
Pressure-diuresis, 74–75, 158
Principal cells
  in cortical collecting duct, 176–177
    aldosterone action in, 197, 203
    potassium secretion associated with, 185–186
    sodium-potassium exchange in, 255
  in medullary collecting duct, 177, 178
  carbonic anhydrase in, 254
Proline, transport and reabsorption of, 227
Proportional controller, in control system, 356–357
Propranolol, blocking sympathetic stimulation of renin release, 338
Prostaglandins
  action on thick ascending limb of Henle's loop, 173
  affecting renal function, 88–89
  glomerular receptors for, 64
  glomerular synthesis of, 64
  and pressure diuresis, 74–75

in regulation of renal hemodynamics, 88–89
and renal blood flow, 64
  medullary, 297
and renin release from granular cells, 91
stimulation by antidiuretic hormone, 196
synthesis in lipid-laden interstitial cells, 279
Proteases, brush-border, in hydrolysis of small peptides, 231
Protein
  aldosterone-induced, 202, 203
  Bence Jones, tubular reabsorption of, 230
  dietary
    and filtrate reabsorption in proximal tubule, 313
    and glomerular filtration rate, 222
    and production of nonvolatile acids, 240
    and urea clearance, 102
  in fluid compartments, 14
  in interstitial space, 31
  in lymph, 31
  membrane impermeability to, 13–14
  oncotic pressure of. See Oncotic pressure, protein
  plasma, 4, 18, 68
    in afferent arterioles, and glomerular filtration rate, 60, 63–64
    as buffers, 241
  Tamm-Horsfall, 79
  transport in membranes, 114
  tubular reabsorption of, 230–231
  in ultrafiltrate, 68
Protein kinase, activation by cAMP, 196
Proteinuria, postexercise, 66
Prothrombin, formation of, 18
Proximal tubules, 138–162
  aldosterone action in, 197
  amino acid reabsorption in, 225
  p-aminohippurate secretion in, 233
  ammonia synthesis in, 260
  basolateral membrane of, 151
  bicarbonate reabsorption in, 143, 144, 148–149, 253–254
  chloride reabsorption in, 143, 144, 149–150
  convoluted part of, 42, 44, 51, 139
    transport characteristics in, 168
  cytoplasm constituents in, 141
  diuretics affecting, 303–304
  glucose reabsorption in, 217, 218–219
  lateral intercellular spaces in, 114, 141, 151
  luminal membrane of, 151
  morphology of, 138–142
  nomenclature for, 370
  phosphate reabsorption in, 222
  potassium reabsorption in, 147–148
  protein reabsorption in, 230
  in regulation of blood volume and osmolality, 313
  sodium-hydrogen exchange in, 253, 255

sodium reabsorption in, 143, 144–147
straight part of, 44, 50–51, 139, 162, 271
subsegments of, 139, 141
  transport functions in, 143
sulfate reabsorption in, 225
transport and reabsorption in, 142–144
  extracellular volume expansion affecting, 155–157
  general characteristics of, 142–144
  glomerular filtration rate affecting, 153, 157–158
  hormonal regulation of, 160–162
  intrarenal hemodynamics affecting, 158–160
  luminal control of, 153–155
  mannitol affecting, 154–155
  mechanisms in, 144–153
  methods of study, 132–138
  peritubular control of, 155–160
  plasma sodium concentrations affecting, 153–154
  regulation of, 153–162
  renal perfusion pressure affecting, 158
  urea affecting, 155
urea recycling through, 295–296
uric acid reabsorption in, 229–230
water reabsorption in, 143, 150–152
Pulse forcings, 5, 355
Pump, sodium-potassium. See ATPase, sodium-potassium
Puromycin, and antinatriuretic effect of aldosterone, 200
Pyramids, medullary, 41
Pyrazinamide suppression test, in hyperuricemia, 230

Radioactive microspheres, in regional blood flow measurements, 107
Ramp, as forcing type, 355
Rate constant, in chemical reactions, 118
  in backward reaction, 119
  in forward reaction, 119
Rate-sensitive controller, in control system, 357
Rays, medullary, 44, 50–51, 271
Reabsorption, tubular, 217–232
  distal, 178–186
  proximal, 142–144
Receptors
  chloride-sensitive, in macula densa, 314
  drug-receptor interactions, 123–124
  hormone-receptor interactions, 123–124
  osmoreceptors
    hepatic, 317–319
    hypothalamic, 193, 312, 316–317
  renal vascular, and control of renin release, 333–337
  stretch, 312, 314. See also Stretch receptors
Rejection ratio, tubular fluid, 134
Renal artery, 44

pressure in. *See* Blood pressure, renal artery
Renin, 326–327
  activation and release of, stimuli affecting, 84
  hypersecretion of, 327
  hyposecretion of, 327
  plasma renin activity, 327
    and renal perfusion pressure, 337
  pressor action of, 327
  release from granular cells, 81, 326, 343
    agents affecting, 91
    and aldosterone secretion, 331
    baroreceptor theory of, 333–337
    control of, 332–339
    in hemorrhage, 334
    humoral agents affecting, 338–339
    in hypotension, 334
    intrinsic mechanisms in, 333–337
    macula densa receptor theory of, 337
    mechanisms in, 340–342
    and renal perfusion pressure, 76
    sympathetic nervous system in, 334, 335, 337–338
  substrate concentrations, 327–328
  synthesis in granular cells, 80, 326, 343
  and tachyphylaxis, 327
Renin-angiotensin-aldosterone system, 326–340
  aldosterone in, 198–199, 331–332
  angiotensins and converting enzymes in, 329–331
  and antidiuretic hormone release, 324–325
  and arterial blood pressure regulation, 88
  components of, 326–332
  historical aspects of, 326
  in regulation of renal hemodynamics, 87–88, 89, 91
  renin in, 326–327
  renin substrate in, 327–328
  and tubuloglomerular feedback response, 81, 83–84
Respiratory centers, 246
  depression of, hypoventilation in, 265
Respiratory system, in acid-base balance, 245, 246–248
Rete mirabile, 272
Rheogenic process, in glucose transport, 221
RNA, messenger, 161, 202
Rotenone, and sodium transport, 204

Saline infusions
  in carotid artery, effects of, 316
  and cerebrospinal fluid pressure, 32
  and distal tubular flow rate and potassium secretion, 190
  and glucose reabsorption, 220
  in hepatic portal vein, effects of, 318
  natriuresis from, 154, 156, 182–183, 301
  and phosphate reabsorption, 224
  and renin release, 339

  and sulfate reabsorption, 225
  and tubular transit time, 156
Saralasin
  and effects of prostaglandins on antidiuretic hormone, 89
  and nephron plasma flow rate, 87
Saturation of binding sites, in solute transport, 16
Second messengers, intracellular
  cAMP, 194
  calcium, 199
Secretion, tubular, 232–234
Semipermeable membranes, 117
Serine, transport and reabsorption of, 227
Serotonin, glomerular receptors for, 64
Servomechanism control system, 356
Sex
  and total body water, 10
  and uric acid levels, 229
Shock, hemorrhagic, anuria in, 69–70
Sieving process, in capillary fluid exchange, 23
Sinus, renal, 41
Sinusoidal forcings, 355
Skin
  fluid efflux from, 22
  water in, 10
Slit membrane, glomerular, 56, 58
Sodium
  active transport of, 146, 271, 281
  and amino acid transport, 226
  balance of
    and aldosterone secretion, 331–332
    and angiotensin II levels in plasma, 329
  and chloride reabsorption in proximal tubule, 150
  concentrations in nephron segments
    in human, 295
    in rat, 294
  cotransport with amino acids, 130, 142, 227
  cotransport with p-aminohippurate, 233
  cotransport with chloride in thick ascending limb of Henle's loop, 169–170
    in cortical portion, 171
    in medullry portion, 170–171
  cotransport with glucose, 129, 142, 218, 220–221
  cotransport with phosphate, 224
  cotransport with water, 150–152
  dietary
    and aldosterone biosynthesis, 196, 199
    and plasma renin activity, 337
    and potassium secretion in distal nephron, 188, 190
  exchange with hydrogen. *See* Hydrogen, sodium-hydrogen exchange
  exchange with potassium
    in collecting duct, 255
    in proximal tubule, 152–153
  excretion of
    abnormal. *See* Natriuresis
    and blood volume regulation, 325–342

    control of, 340–342
    redistribution of renal blood flow affecting, 106
    renal denervation affecting, 161
  and extracellular fluid volume, 34
  in fluid compartments, 14, 16–17
  glomerular-tubular balance for, 153
  hepatic receptors for, 318
  intracellular pool of, 146
  movement across membranes, 13–15
  NA-K-ATPase pump, 127. *See also* ATPase, sodium-potassium
  passive diffusion in proximal tubule, 146
  permeability to
    in basolateral membrane, 152
    in thin limbs of Henle's loop, 168, 289
  plasma concentrations of, 4, 68
    and proximal reabsorption of sodium and water, 153–154
    and renin secretion, 338–339
  role in tubuloglomerular feedback mechanism, 81, 82–83
  transport and reabsorption of
    aldosterone affecting, 197, 200–204
    in distal tubule, 180–185
      aldosterone affecting, 197
      and potassium secretion, 187
      and glucose reabsorption rate, 220
      inhibitors affecting potassium secretion in distal tubule, 190–191
    prostaglandins affecting, 89
    in proximal tubule, 143, 144–147
      catecholamines affecting, 161–162
      extracellular volume expansion affecting, 155–157
      glomerular filtration rate affecting, 153, 157–158
      intrarenal hemodynamics affecting, 158–160
      plasma sodium levels affecting, 153–154
      renal perfusion pressure affecting, 158
    in thick ascending limb of Henle's loop, 289, 314
      glomerular filtration rate affecting, 171–172
      tubular flow rate affecting, 215
    in various nephron segments, 168
    tubular concentrationn of, 132–133
  in ultrafiltrate, 68
Sodium chloride mixing with urea, in inner medulla, 291–292, 297
Sodium pump, 16. *See also* ATPase, sodium-potassium
Solute drag, 15
Solute-free water, 299
  clearance of, 300–301, 317
  prostaglandins affecting, 89
Solute transport, tubular. *See* Tubules, transport and reabsorption in
Solute washout in diuresis
  in expansion of extracellular fluid volume, 107
  mannitol-unduced, 155

Solvent drag
  and chloride reabsorption in proximal tubule, 143
  and potassium reabsorption in proximal tubule, 147
Specific activity of solutes in plasma, 9
Specific gravity, 10–11
  of urine, 22–23
Sphincters, precapillary, 24
  opening and closing of, 29
Spironolactone, 305
  as aldosterone antagonist, 202
  sites of action of, 303
Splay of titration curves, 218
Spleen, water content of, 10
Starling forces, 313
  peritubular, and reabsorption in proximal tubule, 156–157
Steady-state response of systems, 354
  error in, 357
Stellate veins, 45
Step forcings, 5, 354
Stretch receptors
  and arterial blood pressure, 314, 321
  carotid sinus and aortic arch, 314, 321–323
  and control of renin release, 333–337
  left atrial, 314, 319–321
  peripheral, 312, 314
  right atrial, 314, 340
Structure of kidney, 41–51
Succinate, and citrate reabsorption, 228
Sucrose, filtrate-to-plasma concentration ratio for, 66
Sulfate
  and chloride concentration in proximal tubule, 150
  clearance of, 232
  diuresis from, 302
  excretion of, 258
  in fluid compartments, 14
  maximum tubular reabsorptive capacity for, 225
  plasma concentrations of, 4, 68
  and potassium secretion in distal nephron, 188–189, 190
  tubular reabsorption of, 224–225
  in ultrafiltrate, 68
Sulfuric acid, neutralization of, 258
Sweating, water loss in, 22, 315
  and hypertonic dehydration, 36–37
Sympathetic nervous system, and renin release, 334, 335, 337–338
Symport system, 16
System analysis, 353
System synthesis, 353

Tachyphylaxis
  angiotensin, 327, 330
  renin, 327
Tamm-Horsfall protein, 79
Temperature
  and blood volume changes, 320
  and chemical reaction rate, 118
  and flux of solutes across membrane, 116
  and reaction velocity in enzyme-substrate complex, 121–122
Tetraethylammonium, and renin release, 334

Theophylline
  diuretic effect of, 306
  and plasma renin activity, 338
Thiazide diuretics, 304, 305
  sites of action of, 303
Thiocyanate, in extracellular volume measurements, 11, 12
Thirst center, hypothalamic, 317
Threonine, transport and reabsorption of, 227
Thromboxanes, glomerular receptors for, 64
Thyrocalcitonin, and transport of calcium and phosphate, 224
Thyroid hormone
  and activity of Na-K-ATPase, 152
  and filtrate reabsorption in proximal tubule, 161
Tight junctions
  in capillary endothelium, 31
  in distal convoluted tubule, 173
  in macula densa, 79
  in renal tubular epithelium, 112–113, 141, 151
  and paracellular transport, 112–113, 142
  in thin limbs of Henle's loop, 164
Time
  delay in bicarbonate accumulation, 249
  and transfer function of control system, 358, 360
Time constant
  in chemical reactions, 118
  and steady-state response, 354
Titration curves, renal, 215–217
  for amino acid, 226–227
  for p-aminohippurate, 234
  for bicarbonate, 256
  for buffers, 240
  for citrate, 224
  for glucose, 217–218
  for α-ketoglutarate, 229
  for phosphate, 222
  splay of, 218
    for p-aminohippurate, 234
    for citrate, 228
    equilibrium constant affecting, 226
    for glucose, 218
    for phosphate, 222
    for uric acid, 230
  for sulfate, 224
  for uric acid, 230
Transcellular fluid, 11
  volume of, 13
Transcellular transport route, 112, 114–115, 142
  and capillary fluid exchange, 30–31
Transcription, gene, 161, 202
  actinomycin D affecting, 190
Transcytosis, 30
Transient response of system, 354
Translocation, genetic, 161, 202
Transport mechanisms, 15–16, 217–232. See also Tubules, transport and reabsorption in
  active, 16, 127–129, 214. See also Active transport
  primary, 16
  secondary, 16

carrier-mediated, 15–16, 117–130. See also Carrier-mediated transport
convection, 15
cotransport, 16, 129–130, 142–143
diffusion, 15, 114, 116–117
passive, 112, 117, 214
Triamterene, 305
  sites of action of, 303
Tricarboxylic acid cycle
  citrate in, 228
  α-ketoglutarate in, 229
Trifluoperazine, affecting renin release, 336
Tritiated water, in total body water measurements, 9, 10
Tryptophan, transport and reabsorption of, 227
Tubular fluid
  distal, composition affecting potassium secretion, 187–190
  glucose in, and reabsorption rate, 220
  hypotonicity in thick ascending limb of Henle's loop, 289, 299
  isotonicity in collecting ducts, 290
Tubular fluid-to-plasma concentration ratio
  for aldosterone, 197
  for amino acids, 226
  for chloride, 149–150, 186
  for inulin, 150
  for potassium, 147, 185
  potassium-to-inulin, 185
  for sodium, 143, 180
  sodium-to-inulin, 181–182
  for urea, 294
Tubules, renal, 41, 42–44
  connecting, 42. See also Connecting tubule
  distal, 173–204. See also Distal Tubules
  function of, 51
  maximum reabsorptive capacity, 214
  maximum secretory capacity, 214
  proximal, 138–162. See also Proximal tubules
  secretion in, 232–234
  transit time in
    in distal convoluted tubule, 183
    glomerular filtration rate affecting, 157–158
    and glucose reabsorption, 220
    and rate of tubular flow, 286
    saline infusion affecting, 156
    and sodium chloride transport, 343
  transport and reabsorption in, 217–232
    of acetoacetate, 229
    active, 16, 127–129, 214
    of amino acids, 225–227
    biophysical basis of, 115–130
    carrier-mediated, 15–16, 117–130
    characteristics in nephron segments, 168
    of citrate, 228–229
    classification of, 214–217
    cotransport, 16, 129–130, 142–143

diffusion in, 15, 114, 116–117
  nonpolar molecules in, 117
  polar molecules in, 117, 124
in distal tubule and collecting
    duct, 173–204
  general characteristics of, 178–186
downhill carrier system in, 129–130, 142
facilitated, 124–127
of glucose, 217–222
gradient-time-limited, 215
in Henle's loop, 162–173
of β-hydroxybutyrate, 229
and interstitial composition and
    osmolality, 107
of α-ketoglutarate, 229
of lactate, 229
maximum capacity for, 101
methods for study of, 132–138
  electrical potential gradients in, 136–138
  flow-through microperfusion in, 136
  free-flow micropuncture
      technique in, 133–135
  in-vitro perfusion of isolated
      tubules in, 136
  short-circuit current in, 136
  stop-flow analysis of whole
      kidney in, 132–133
  stop-flow microperfusion in, in-
      situ, 135–136
paracellular route in, 112–114, 141, 142, 152
  lateral intercellular spaces in, 113–114
  Starling forces affecting
      permeability in, 156–157
  tight junctions in, 112–113, 142
passive, 214
of phosphate, 222–224
of potassium, 147–148, 168, 185–186
of proteins, 230–231
in proximal tubule, 138–162
of sodium. See Sodium, transport
    and reabsorption of
structural basis of, 112–115
of sulfate, 224–225
and total renal clearance, 100–101
transcellular route in , 112, 114–115, 142
uphill carrier system in, 128–129, 142
of uric acid, 229–230
of water, 117
Tubuloglomerular feedback mechanism
    in renal autoregulation, 76–87, 89
  afferent arteriole resistance in, 83
  global stimuli in, 84, 87
  and glomerular filtration rate, 337
  local stimuli in, 84, 87
  renin-angiotensin system in, 83–84
  sodium chloride role in, 82–83, 85
  at whole-kidney and single-nephron
      level, 84
Tyrosine, transport and reabsorption of, 227

Ultrafiltration, glomerular, 58–71. *See
    also* Glomerular filtration
Urea
  clearance of, 102–103
    correction formula for, 102
    standard, 102
  and concentration of urine, 287, 293–296
  concentrations in nephron segments
    in human, 295
    in rat, 294
  and countercurrent multiplication
      process, 286
  diuretic effect of, 301, 302
  in proximal tubule, 155
  excretion of, 22
  filtrate-to-plasma concentration ratio
      for, 66
  fluxes of, 2
  mixing with sodium chloride, in
      inner medulla, 291–292, 297
  permeability to
    in collecting ducts, 178, 184, 290
      antidiuretic hormone affecting, 194
    in thin limbs of Henle's loop, 168, 169, 288
  production in metabolic system, 22
  recycling in renal medulla, 182, 294–296
  transport in various nephron
      segments, 168
  tubular reabsorption of, 102
Ureteral obstruction
  glomerular filtration rate in, 80
  for stop-flow analysis of tubular
      urine, 132–133
Uric acid
  clearance of, 232
  excretion of, 230
  secretion of, 230
  titration curve for, 230
  transport and reabsorption of, 229–230
Urine
  acidification in collecting duct, 258
  concentration of, 270
    antidiuretic hormone levels
        affecting, 321
    ascending thin limb of Henle's
        loop in, 289
    collecting ducts in, 289–290
    concepts of, 287–293
    descending thin limb of Henle's
        loop in, 288
    measurement of, 298–301
    medullary blood flow affecting, 296–298
    passive mechanism in, 288, 291–292, 293, 294
    thick ascending limb of Henle's
        loop in, 289
    urea role in, 287, 293–296
  dilution of, 270, 293
    measurement of, 298–301
  flow rate of
    compared to plasma flow rate, 104
    and urea clearance values, 102
  formation of, 58
  hypertonic, 300

hypotonic, 300
osmolality of, 22–23, 299, 314
specific gravity of, 22–23
volume of, 314
Urine-to-plasma osmolality ratio,
    antidiuretic hormone levels
    affecting, 321
Ussing's flux equation, 137, 138

Vacuolar apparatus, in proximal tubule
    epithelium, 141
Valine, transport and reabsorption of, 227
Vanadate, and activity of Na-K-
    ATPase, 153
Vasa recta
  ascending, 46, 48, 271, 287
    osmolality of blood in, 296
    urea recycling through, 295–296
  and countercurrent multiplication
      process, 286
  descending, 46, 272, 287
    osmolality of blood in, 296
    water extraction from, 297, 298
  innervation of, 161
Vascular bundles
  in inner stripe of outer medulla, 272, 275
    complex type of, 272
    simple type of, 272
    urea recycling in, 295
  in outer stripe of outer medulla, 271
Vasculature of kidney, 44–48
  receptors in, and control of renin
      release, 333–337
  wall structure of, 46–48
Vasoactive agents, and glomerular
    filtration rate, 64–65
Vasoconstrictor agents, affecting renal
    circulation, 91
Vasodilation, renal, prostaglandin-
    induced, 88
Vasodilator agents, affecting renal
    circulation, 91
Vasomotion, 24
Vasomotor center in medulla oblongata, 320, 321
Vasopressin, 192–193. *See also*
    Antidiuretic hormone
  antidiuretic receptor for, 196
  vascular receptor for, 196
Veins, reneal, 44–45
Venous return of fluid, obstruction of, 34
Ventilation, 246
  hyperventilation, 264, 265
  Hypoventilation, 264, 265
Ventilation equivalent, 246
Venules, 24
Veapamil
  and feedback response in regulation
      of glomerular filtration, 83
  and renin release, 336
Vesicular transport of solutes and
    water, 30
Vitamin D
  and citrate excretion, 229
  and phosphate reabsorption, 224
Volume of body fluid, 2–3

Volume of body fluid (*cont.*)
  regulation of. *See* Fluid regulation
Vomiting, alkalosis in, 251

Washout studies, in measurements of renal blood flow distribution, 106
Water
  concentrations in nephron segments
    in human, 295
    in rat, 294
  cotransport with sodium, 150–152
  distribution in tissues, 10
  diuresis from, 297, 301, 317
  excretion of
    affected by redistribution of renal blood flow, 106
    regulation of, 315–325
  filtrate-to-plasma concentration ratio for, 66
  fluxes of, 2
  insensible loss of, 22, 311–312
  intake of
    and potassium secretion in distal tubule, 191
    thirst affecting, 317
  intoxication, 5, 22
  loss in sweating, 22, 315
  permeability to
    in distal tubule and collecting duct, 180, 182, 290
      antidiuretic hormone affecting, 194, 286, 287
    in thin limbs of Henle's loop, 168, 288
  solute-free, 299
    clearance of, 300–301, 317
  total body water, 9–11
  transport and reabsorption of, 117, 273, 297, 298
    in collecting ducts, 290, 292
    diuretics affecting, 302
    facultative, 101
    and glucose reabsorption rate, 220
    in Henle's loop, 292
      glomerular filtration rate affecting, 171–172
    obligatory, 101, 150
    in proximal tubule, 143, 150–152
      catecholamines affecting, 161–162
    in various nephron segments, 168
Weight/volume unit, 3
Work required
  for chloride movement, 14
  for sodium movement, 14

Xanthine diuretics, 306
Xenon-133, in measurements of renal blood flow distribution, 106
Xylose transport, 219

Zero-order chemical reaction, 118

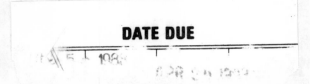